Advances in Experimental Medicine and Biology

Volume 842

More information about this series at http://www.springer.com/series/5584

Abhijit Chakrabarti • Avadhesha Surolia
Editors

Biochemical Roles of Eukaryotic Cell Surface Macromolecules

Springer

Preface

Understanding of cell–cell and cell–matrix interactions is critically important in embryonic development, tissue morphogenesis, growth, and differentiation and maintenance of cellular homeostasis. Dysregulation of these interactions contributes to the development and progression of several pathological conditions. Cell surface molecules particularly glycoconjugates have been identified as one of the key players involved in these cellular processes. Biochemical, immunochemical, cell biology, and molecular biology techniques and computational tools have been employed to establish structure–function relationship of these glycoconjugates. Recently, the focus has been on the Siglecs, G-protein-coupled receptors, and analysis of glycome, the entire complement of the saccharides of an organism. The complexity of sugars in terms of their structures, association with other molecules, such as proteins and lipids, their complex biosynthetic pathways along with distribution pattern and their dynamic nature, makes the study of glycome a challenging task.

The international symposium on the "Biochemical Role of Eukaryotic Cell surface Macromolecules" held in Kolkata, India in Jan 2014 was an opportunity to update comprehensively the major advances in these areas. Contributions from this meeting are presented in the book entitled "Biochemical Role of Cell surface Macromolecules" comprising 24 chapters that provide in-depth analysis of data on cell surface macromolecules in cellular function and their alteration associated with pathological conditions. All contributions are either comprehensive critical reviews or original research papers and cover the most relevant and recent topics related to functional role of cell surface molecules. These include contributions on glycome, biophysical, biochemical, and cell biological approaches to study cell membrane molecules, metabolism of glycoconjugates particularly of proteoglycans and glycoproteins and their implications to cell function.

It would not have been possible to complete this book but for the timely response of the contributors. We would like to acknowledge the efforts of all the contributors, and referees who critically reviewed the manuscripts. We are also grateful to Springer and Diana Ventimiglia for publishing this as a special volume of the Advances in Experimental Medicine and Biology series.

Kolkata, India Abhijit Chakrabarti
Bengaluru, India Avadhesha Surolia

Contents

1 **Human-Specific Evolutionary Changes in the Biology of Siglecs** 1
Flavio Schwarz, Jerry J. Fong, and Ajit Varki

2 **Structural Changes of GPI Anchor After Its Attachment to Proteins: Functional Significance** 17
Taroh Kinoshita

3 **Novel Insights in Membrane Biology Utilizing Fluorescence Recovery After Photobleaching** 27
Amitabha Chattopadhyay and Md. Jafurulla

4 **Defects in Erythrocyte Membrane Skeletal Architecture** 41
Avik Basu and Abhijit Chakrabarti

5 **Membrane Rafts in the Erythrocyte Membrane: A Novel Role of MPP1p55** 61
Aleksander F. Sikorski, Joanna Podkalicka, Walis Jones, and Agnieszka Biernatowska

6 **Immuno-Modulatory Role of Porins: Host Immune Responses, Signaling Mechanisms and Vaccine Potential** 79
Sanica C. Sakharwade, G.V.R. Krishna Prasad, and Arunika Mukhopadhaya

7 ***Vibrio cholerae* Cytolysin: Structure–Function Mechanism of an Atypical β-Barrel Pore-Forming Toxin** 109
Anand Kumar Rai and Kausik Chattopadhyay

8 **New Vis-Tas in Lactosylceramide Research** 127
Subroto Chatterjee, Sumita Mishra, and Sara Kimiko Suzuki

9 **Plasma Membrane-Associated Sialidase Confers Cancer Initiation, Promotion and Progression** 139
Taeko Miyagi, Kohta Takahashi, Kazuhiro Shiozaki, Kazunori Yamaguchi, and Masahiro Hosono

10 **A Signal with a Difference: The Role of GPI Anchor Signal Sequence in Dictating Conformation and Function of the Als5 Adhesin in *Candida albicans*** 147
Mohammad Faiz Ahmad, Pareeta Gajraj Mann, and Sneha Sudha Komath

11 **Novel Chondroitin Sulfate Oligosaccharide Motifs as Biomarkers: Insights into Their Involvement in Brain Development** 165
Kazuyuki Sugahara

12 **Role of Hyaluronidases in the Catabolism of Chondroitin Sulfate** 185
Shuhei Yamada

13 **Pattern Recognition in Legume Lectins to Extrapolate Amino Acid Variability to Sugar Specificity** 199
Nisha Jayaprakash Grandhi, Ashalatha Sreshty Mamidi, and Avadhesha Surolia

14 **Conformational Dynamics of Oligosaccharides Characterized by Paramagnetism-Assisted NMR Spectroscopy in Conjunction with Molecular Dynamics Simulation** 217
Ying Zhang, Takumi Yamaguchi, Tadashi Satoh, Maho Yagi-Utsumi, Yukiko Kamiya, Yoshitake Sakae, Yuko Okamoto, and Koichi Kato

15 **Characterization of Cholesterol Crystalline Domains in Model and Biological Membranes Using X-Ray Diffraction** 231
R. Preston Mason and Robert F. Jacob

16 **Role of Lipid-Mediated Effects in β_2-Adrenergic Receptor Dimerization** 247
Xavier Prasanna, Amitabha Chattopadhyay, and Durba Sengupta

17 **Effect of Temperature on the Phase Behaviour of Fully Saturated DAPC Lipid Bilayer: A Comparative Molecular Dynamics Simulation Study** 263
Ipsita Basu and Chaitali Mukhopadhyay

18 **Biophysical Characterization of the Interaction of *O*-acylcholines with the Major Bovine Seminal Plasma Protein, PDC-109** 279
Rajani S. Damai, Pradip K. Tarafdar, Bhanu Pratap Singh, S. Thirupathi Reddy, and Musti J. Swamy

19 Crystal Structure of Apo and Ligand Bound *Vibrio cholerae* Ribokinase (Vc-RK): Role of Monovalent Cation Induced Activation and Structural Flexibility in Sugar Phosphorylation 293
Rakhi Paul, Madhumita Dandopath Patra, and Udayaditya Sen

20 Synthetic Glycolipids and (p)ppGpp Analogs: Development of Inhibitors for Mycobacterial Growth, Biofilm and Stringent Response 309
Kirtimaan Syal, Krishnagopal Maiti, Kottari Naresh, Dipankar Chatterji, and N. Jayaraman

21 Regulations of Glycolipid: XI. Glycosyltransferase (GSL: GLTs) Genes Involved in SA-LeX and Related GSLs Biosynthesis in Carcinoma Cells by Biosimilar Apoptotic Agents: Potential Anticancer Drugs 329
Subhash Basu, Rui Ma, Joseph R. Moskal, and Manju Basu

22 *N*-Acetylglucosaminyl 1-Phosphate Transferase: An Excellent Target for Developing New Generation Breast Cancer Therapeutic 355
Aditi Banerjee, Juan A. Martinez, Maria O. Longas, Zhenbo Zhang, Jesus Santiago, Krishna Baksi, and Dipak K. Banerjee

23 Involvement of Vascular Endothelial Growth Factor in Serotonin 1A Receptor-Mediated Neuroproliferation in Neonatal Mouse Hippocampus 375
S. Samaddar, B. Ranasinghe, S.J. Tantry, P.R. Debata, and P. Banerjee

24 Structural Heterogeneity of Glycoform of Alpha-1 Acid Glycoprotein in Alcoholic Cirrhosis Patients 389
Goutam Mandal, Hirokazu Yagi, Koichi Kato, and Bishnu Pada Chatterjee

Correction to: *N*-Acetylglucosaminyl 1-Phosphate Transferase: An Excellent Target for Developing New Generation Breast Cancer Therapeutic C1

Index 403

Chapter 1
Human-Specific Evolutionary Changes in the Biology of Siglecs

Flavio Schwarz *, Jerry J. Fong *, and Ajit Varki

Introduction

Sialic acid-recognizing immunoglobulin-like lectins (Siglecs) are cell surface receptors that bind sialic acids, a class of monosaccharides found on the outermost end of glycans on a variety of glycoconjugates (Varki and Angata 2006; Crocker et al. 2007; Pillai et al. 2012). Sequence similarities and evolutionary conservation place Siglecs in two categories (Fig. 1.1). Sialoadhesin (Siglec-1), CD22 (Siglec-2), myelin-associated glycoprotein (Siglec-4), and Siglec-15 have orthologs in all mammalian species and relatively low (20–25 %) sequence similarity. In contrast, the CD33-related Siglecs (CD33rSiglecs, including Siglecs-3, -5 to -14, -16, and -17 in primates) form a large, rapidly evolving subfamily of genes that expanded in mammals by duplications involving a primordial cluster of *SIGLEC* genes (Fig. 1.2). Several mechanisms were involved in the rapid evolution of the CD33rSiglec subfamily: exon shuffling, gene duplication, gene conversion, deletion leading to pseudogenization, altered expression, and adaptive amino-acid substitutions in sialic acid recognition domains (Angata et al. 2004; Altheide et al. 2006).

CD33-related Siglecs are primarily expressed on immune cells (Lock et al. 2004), but specific members of the family are also found on other cell types in humans. For instance, Siglec-XII is present on epithelial cells (Mitra et al. 2011); Siglec-6 is expressed in the placental trophoblast (Brinkman-Van der Linden et al. 2007); and, Siglec-5 and -14 are found on amniotic epithelium (Ali et al. 2014). Siglec-3 and -11

*Author contributed equally with all other contributors.

F. Schwarz • J.J. Fong • A. Varki (✉)
Departments of Medicine, and Cellular and Molecular Medicine, Glycobiology Research and Training Center, University of California San Diego, La Jolla, CA, USA
e-mail: fschwarz@ucsd.edu; jjf002@ucsd.edu; a1varki@ucsd.edu

A. Chakrabarti, A. Surolia (eds.), *Biochemical Roles of Eukaryotic Cell Surface Macromolecules*, Advances in Experimental Medicine and Biology 842,
DOI 10.1007/978-3-319-11280-0_1

Editors
Abhijit Chakrabarti
Crystallography and Molecular Biology Division
Saha Institute of Nuclear Physics
Kolkata, India

Avadhesha Surolia
Molecular Biophysics Unit
Indian Institute of Science
Bengaluru, India

ISSN 0065-2598 ISSN 2214-8019 (electronic)
ISBN 978-3-319-11279-4 ISBN 978-3-319-11280-0 (eBook)
DOI 10.1007/978-3-319-11280-0
Springer Cham Heidelberg New York Dordrecht London

Library of Congress Control Number: 2014955053

Printed on acid-free paper

Springer is part of Springer Science+Business Media (www.springer.com)

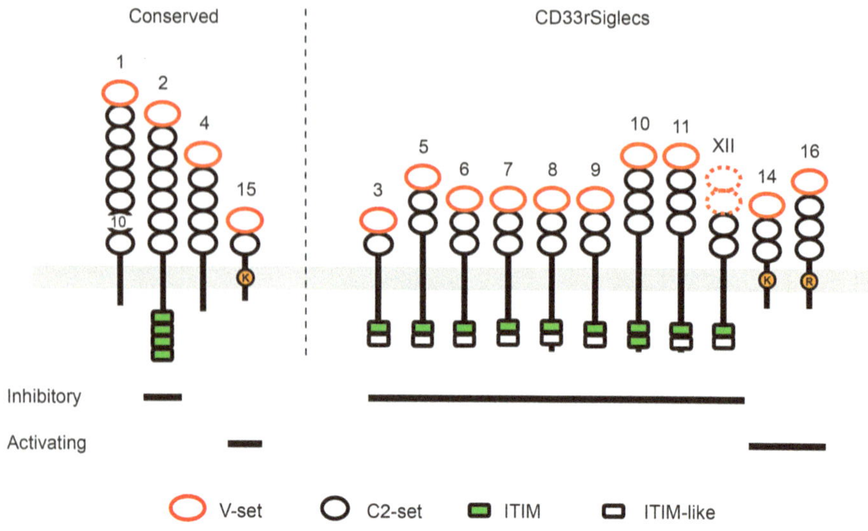

Fig. 1.1 The family of human Siglec receptors. Conserved (*left*) and CD33-related (*right*) Siglecs are cell surface receptors with a variable number of extracellular immunoglobulin-like domains. The outermost domain (V-set, in *red*) binds to sialylated molecules through a critical arginine residue. The V-set domains of Siglec-XII cannot bind sialic acid due to a mutation in a critical arginine residue, and are indicated with *dotted lines*. The transmembrane segment of Siglec-14, -15, and -16 contain a basic amino acid (lysine or arginine) that can interact with negatively charged amino acids of protein adapters. Siglecs may contain intracellular signaling motifs such as ITIM or ITIM-like. Structural elements for each protein were derived from the Uniprot database

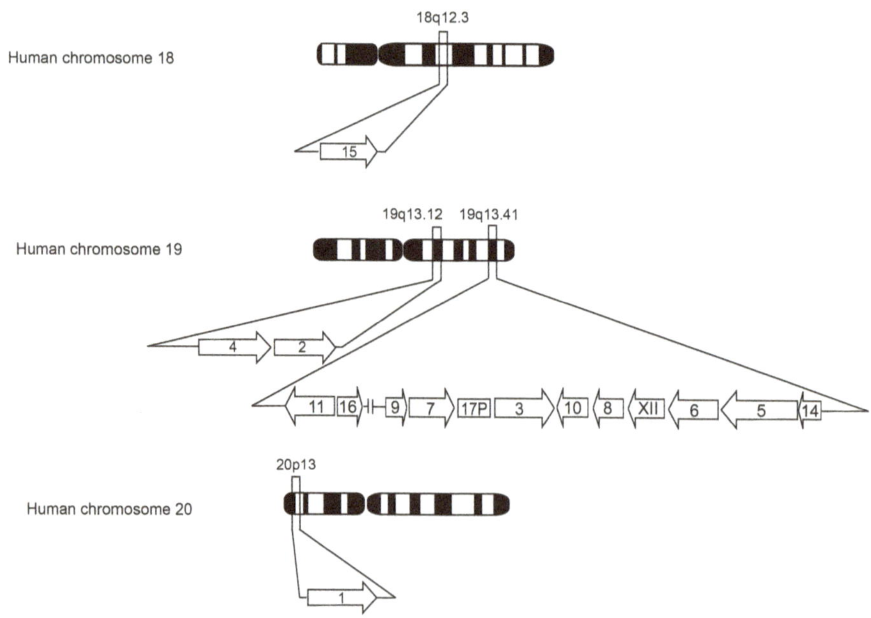

Fig. 1.2 Genomic localization of human *SIGLEC* genes. While genes encoding conserved Siglecs are found on different chromosomes, the CD33-related *SIGLEC* genes are clustered on the chromosome 19, along with multiple *SIGLEC* pseudogenes (only 17P is shown). Information on the localization, length and orientation of the genes was derived from the hg38 dataset of the UCSC Genome Browser

are also expressed in microglia, resident immune cells of the central nervous system, and influence their activity (Linnartz-Gerlach et al. 2014; Hayakawa et al. 2005; Malik et al. 2013; Griciuc et al. 2013).

Structurally, Siglecs are type-I membrane proteins with an extracellular N-terminus, a single transmembrane span, and an intracellular C-terminus (Varki and Angata 2006). The extracellular portion is composed of a V-set immunoglobulin-like domain, which binds to sialic acid-containing ligands, and one or more underlying C2-set immunoglobulin-like domains. The intracellular segment of many Siglecs contains immunoreceptor tyrosine-based inhibitory motifs (ITIMs), which can be phosphorylated by Src family kinases upon external ligand binding and recruit SHP-1 or SHP-2 tyrosine phosphatases (Pillai et al. 2012). These events lead to blockade of MAP kinase phosphorylation and eventually attenuate the cellular inflammatory response. Inhibitory Siglecs may also contain ITIM-like motifs (Crocker et al. 2007; Crocker and Varki 2001). Phosphorylation of the tyrosine residues of ITIM-like and ITIM motifs may occur sequentially and be required for efficient recruitment of SHP-1 or SHP-2 (Tourdot et al. 2013). Some of the CD33rSiglec ITIM-like motifs also contain a consensus sequence similar to those found in the signaling lymphocytic activation molecule (SLAM) receptors (Cannons et al. 2011). However, the contribution of these SLAM-like motifs of Siglec to the modulation of signaling has not been characterized, and it is not clear whether Siglecs can interact with SAP or EAT2 proteins.

More recently, a subset of Siglecs were found to lack ITIM motifs and instead engage DNAX-activation protein of 12 kDa (DAP12) through a positively charged residue in their transmembrane domains (Angata et al. 2006; Cao et al. 2008; Kameda et al. 2013; Takamiya et al. 2013; Ishida-Kitagawa et al. 2012). Upon engaging their ligands, these immuno-activating Siglecs augment inflammation by phosphorylation of the immunoreceptor tyrosine-based activating motifs (ITAMs) of DAP12 and enhancement of the MAP kinase signaling cascade (Lanier 2009). Thus, primate Siglecs may alternatively be categorized into three groups based on the features of the transmembrane and cytoplasmic tails: Siglecs (-1 and -4) that lack standard signaling motifs and are likely involved in adhesion and/or phagocytosis; Siglecs (-2, -3, -5, -6, -7, -8, -9, 10, -11, -XII) with the general ITIM consensus I/V/L/SxYxxL/V (Ravetch and Lanier 2000); and, Siglecs (-13, -14, -15, -16) that contain a positively charged amino acid (lysine or arginine) in the transmembrane span, which supports recruitment of a homodimer of DAP12.

Interestingly, some inhibitory and activating CD33rSiglec exist as pairs in primates. The sialic acid-binding properties of Siglecs -11/-16 and of -5/-14 are kept virtually identical by gene conversion (Angata et al. 2006; Wang et al. 2012a), but each member of these pairs mediates opposing signaling events (Ali et al. 2014). Moreover, some CD33rSiglecs have inactive alleles that segregate at intermediate frequency in some human populations, but the functional significance of these alleles is not yet known (Angata 2014; Cao et al. 2008; Mitra et al. 2011; Wang et al. 2012a). The general hypothesis is that CD33rSiglec variants adjust the inflammatory responsiveness of human immune cells, which need to limit reactions against "self," but also face pathogens expressing sialic acid-containing surface polysaccharides that can subvert the inhibitory Siglecs by mimicking self signals (Cao and Crocker

2011; Varki 2010; Barclay and Hatherley 2008; Barreiro and Quintana-Murci 2010). For instance, group B *Streptococcus* serotype III (GBS-III) produces a capsular polysaccharide containing Neu5Acα2-3Galβ1-4GlcNAc, a structure that precisely mimics the terminal sequences of many human glycoproteins. Binding of GBS-III to Siglec-9 leads to a reduction of reactive oxygen species and inflammatory cytokines (Carlin et al. 2009b). Other GBS serotypes have even evolved sialic acid-independent binding to Siglecs (Carlin et al. 2009a). Activating Siglecs may have evolved to prevent pathogen hijacking of inhibitory Siglecs: they are similar targets with opposite cellular effects. However, in the absence of a pathogen, the binding of self-molecules to activating Siglecs could generate unwanted inflammation. The benefit of activating Siglecs may therefore be context dependent: advantageous in the presence of a pathogen, but costly in its absence. This could explain the polymorphic loss of function alleles observed in human activating Siglecs-14 and -16.

SIGLEC genes are rapidly evolving in all taxa where they exist (Angata et al. 2004; Padler-Karavani et al. 2014). However, the abundance of Siglec changes in humans seems unusually high compared with other species (Table 1.1). For example, mouse and rat Siglecs appear nearly identical, and differences among nonhuman hominids (NHH) and other old world primates seem limited so far. Some changes may be a consequence of uniquely human sialic acid biology, the most

Table 1.1 Human-specific changes in *SIGLEC* genes and their encoded Siglec proteins

GENE	Type of change	Fixed	Phenotypic effects	Disease relevance
SIGLEC1	Expression change	Yes?	Altered expression in spleen lymphoid follicles	Altered response to sialylated pathogens?
SIGLEC3	Alternative splicing	No	Change in relative expression of two isoforms	Alters development of late onset Alzheimer's disease
SIGLEC5	Expression change	Yes?	T-cell over-reactivity due to low expression	Susceptibility to T-cell mediated disease?
	Gene conversion		Uniquely human expression in amniotic epithelium	Fetal susceptibility to sialylated bacteria
SIGLEC6	New expression	Yes	Uniquely human expression in placental trophoblast	Up-regulation in preeclampsia
SIGLEC9	Expression and ligand preference	Yes	Simultaneous balance between recognizing "self" sialic acids and escape from sialylated pathogens	Altered response to sialylated pathogens
SIGLEC11	Gene conversion	Yes	Uniquely human expression in microglia	Modulation of neurotoxicity in neurodegenerative disease. Impact on ovarian physiology?
	Change in ligand preference	Yes	Reduced binding to sialylated glycans. Novel ligands in brain and ovary	

(continued)

Table 1.1 (continued)

GENE	Type of change	Fixed	Phenotypic effects	Disease relevance
SIGLEC12	Mutation in arginine	Yes	Loss of sialic acid binding properties	Increased expression in carcinomas
	1-bp insertion in ORF	No	Expression loss in some individuals	
SIGLEC13	*Alu*-mediated deletion	Yes	Escape from sialylated bacteria?	Not applicable—all humans null
SIGLEC14	Gene deletion by fusion	No	Change in baseline state of innate immune response	Alters response to lung inflammation
SIGLEC16	4-bp deletion in ORF	No	Change in baseline state of innate immune response?	Modulation of neurotoxicity in neurodegenerative disease?
SIGLEC17	1-bp deletion in ORF	Yes	Escape from sialylated bacteria?	Not applicable—all humans null

prominent being the loss of the sialic acid Neu5Gc. Humans have a non-functional *CMAH* gene, and thus cannot synthesize the common mammalian sialic acid Neu5Gc via hydroxylation of the precursor sialic acid *N*-acetylneuraminic acid (Neu5Ac) (Chou et al. 2002). If the primary function of CD33rSiglecs is to recognize host sialic acid as "Self-Associated Molecular Pattern" (SAMPs) and send inhibitory signals to the immune cells via cytosolic ITIMs, loss of the ability to synthesize Neu5Gc in human ancestors may have resulted in loss of the ability to dampen the immune response and thus excessive immune activation (Varki 2010, 2011). Therefore, it is reasonable to suppose that loss of Neu5Gc in humans was followed by compensatory changes in Siglecs to adapt their binding preference to Neu5Ac, and to reduce or alter the scope for immune activation in response to self-molecules. Human pathogens that interact with Siglecs would also be expected to compensate following the loss of Neu5Gc. In the sections that follow, we consider human-specific changes in several of the Siglecs.

Human-Specific Changes in Siglecs

Siglec-1/Sialoadhesin

Siglec-1, also called sialoadhesin, is a macrophage receptor with 17 extracellular domains, a single transmembrane span, and no cytosolic signaling motif. The V-set domain seems to recognize only Neu5Ac and not Neu5Gc, and prefers α2-3 and α2-8 linkages (Collins et al. 1997; Hartnell et al. 2001). This overall pattern fits what is often found on bacteria. Considering all these features, it was suggested that one likely conserved function is to eliminate sialylated pathogens (Crocker et al. 1997). In keeping with this, Siglec-1 is found in mice and in primates at sites that would first encounter bacteria invading extracellular fluids, such as the sinuses of lymph nodes, spleen, and bone marrow (Crocker et al. 1991; Hartnell et al. 2001). However, Siglec-1 appears to be upregulated in the human spleen compared with the chimpanzee (Brinkman-Van der Linden et al. 2000). In fact, in chimpanzees, as in rodents, only a subset of splenic macrophages is Siglec-1 positive, whereas in humans the distribution is more widespread. One possible explanation is that humans are under increased selection pressure from bacteria that express sialic acids.

It has been shown that Siglec-1 on circulating monocytes binds to sialylated gp120 of HIV and facilitates entry of the virus into cells (Rempel et al. 2008; Zou et al. 2011). In this regard, it is interesting that HIV infection more often progresses to AIDS in humans.

Siglec-3/CD33

CD33 was first detected on human myeloid cells by a panel of monoclonal antibodies that recognize a 67 kDa glycoprotein uniquely expressed on cells of the hematopoietic system (Andrews et al. 1983). Later, human CD33 was found to bind to sialylated glycans, with a preference for Neu5Acα2-3Gal (Freeman et al. 1995), and catalogued as Siglec-3. This receptor is found on circulating monocytes, on subsets of B and activated T and NK cells, and on microglia (Hernandez-Caselles et al. 2006; Perez-Oliva et al. 2011). Notably, two forms of Siglec-3 are expressed (Hernandez-Caselles et al. 2006; Perez-Oliva et al. 2011). The full length human Siglec-3 (also named CD33M) is the 67 kDa protein that includes a V-set domain, a C2-set domain, and a transmembrane span followed by an ITIM domain. Alternative splicing of exon 2 generates an isoform CD33m that lacks the V-set domain, and is therefore unable to bind sialylated ligands.

Independent studies have shown that Siglec-3 levels are altered in patients with late-onset Alzheimer's disease (LOAD) (Bradshaw et al. 2013; Raj et al. 2014; Malik et al. 2013; Griciuc et al. 2013). A single nucleotide polymorphism (SNP) rs3865444 in the promoter region of human *SIGLEC3* associated with LOAD was shown to alter the ratio between the two Siglec-3 isoforms produced (Malik et al. 2013; Raj et al. 2014). Whereas the protective allele rs365444A results in a ratio CD33M:CD33m of 70:30, a higher expression of the full length protein CD33M was detected in LOAD brains (ratio 90:10). Interestingly, the rs365444 SNP was reported to be in high linkage disequilibrium with the rs12459419 SNP, which is physically found in the exon 2 and alters splicing efficiency of exon 2. Increased levels of CD33M (the form capable of binding sialic acid) in microglia are thought to suppress phagocytosis of Aβ42 peptide, possibly by blocking TREM2/DAP12-mediated activation, resulting in amyloid accumulation (Bradshaw et al. 2013; Griciuc et al. 2013; Malik et al. 2013). It was suggested that altered CD33 function could be involved in the presymptomatic phase of AD, in the middle or younger age (Bradshaw et al. 2013).

It is interesting to note that the complete pathology of LOAD is very rare in primates other than humans (Gearing et al. 1994; Perez et al. 2013; Varki et al. 2011). Chimpanzee Siglec-3 shares similar specificities for sialylated glycoconjugates as the human counterpart, but is detected on monocytes at lower levels than humans (Padler-Karavani et al. 2014). Currently, it is unknown whether Siglec-3 is expressed in microglia in chimpanzee, or whether the Siglec-3 mRNA transcript undergoes the same type of regulation.

So far, mice deficient in CD33 have not shown major morphological or histological abnormalities and very minor differences in biochemical and erythrocyte parameters (Brinkman-Van der Linden et al. 2003). However, mouse Siglec-3 has striking differences from the human counterpart. First, it is primarily expressed on granulocytes. Secondly, it does not bind to α2-3 sialylated ligands, but shows distinctive sialic acid-dependent binding only to the short *O*-linked glycans of certain mucins and weak binding to the sialyl-Tn epitope. Furthermore, mouse CD33 includes a positively charged amino acid in the transmembrane domain similar to activating Siglecs. Lastly, alternative splicing may generate two forms with different cytosolic tails, of which only one contains a canonical ITIM motif. The signaling properties of these two potential isoforms have not been elucidated. Overall, it seems likely that murine and human Siglec-3 receptors are functionally different.

Siglec-5 and -14

Because of ongoing gene conversion, Siglec-5 and -14 are 100 % identical in their V-set sialic acid binding domain, and differ in only one amino acid in the first underlying C2 domain. However, they transmit opposite intracellular signals upon ligand engagement: while Siglec-5 suppresses the immune response, Siglec-14 augments it (Angata et al. 2006; Ali et al. 2014).

Expression studies are complicated by the fact that nearly all known high affinity monoclonal antibodies against Siglec-5 cross-react with Siglec-14 (Angata et al. 2006). On the other hand, a monoclonal antibody that specifically recognizes only Siglec-14 with no cross reactivity to Siglec-5 has been reported (Yamanaka et al. 2009). Primates express Siglec-5 and -14 on myelomonocytic cells: neutrophils display both Siglec-5 and -14, but monocytes display only Siglec-14 under normal conditions (Yamanaka et al. 2009). However, humans, but not other primates, acquired the ability to express Siglec-5 and -14 on amnion (Ali et al. 2014): one of the few known examples of Siglecs with uniquely human expression on non-hematopoietic stem cell derived lineages. Also, chimpanzee lymphocytes such as $CD19^+$ B-cells and $CD4^+$ T-cells display relatively high levels of Siglec-5, but human T-cells display low or undetectable levels of any CD33rSiglec (Nguyen et al. 2006). Although the antibody used in this study cross-reacts with both Siglec-5 and -14, the subsequent studies demonstrating the immunosuppressive nature of the Siglec receptor on T-cell receptor activation highly suggested Siglec-5 rather than Siglec-14 (Nguyen et al. 2006). This cell-intrinsic Siglec expression difference between humans and other primates may be a contributing factor that explains why chimpanzee T-cells survive better than human T-cells after HIV-1 infection despite being equally susceptible to the virus (Soto et al. 2012). Differences in expression of Siglec-3, -7 and -9 on T-cells (high in chimpanzee, low in humans) might also affect the outcome of HIV-1 infection (Nguyen et al. 2006).

Group B *Streptococcus* (GBS), a leading cause of neonatal sepsis and death worldwide, is particularly noteworthy for its ability to engage Siglec-5 through its cell wall anchored β-protein to dampen the pro-inflammatory response (Carlin et al. 2007, 2009a, b). Such Siglec-pathogen interactions are believed to be a major driving force in the evolution of the *SIGLEC* gene cluster. *SIGLEC14* possibly emerged from a *SIGLEC5* gene duplication event, and was converted into a DAP12-binding immunoactivating receptor as an evolutionary response to combat the pathogens that subvert Siglec-5 (Angata et al. 2006; Ali et al. 2014).

Humans also have a unique *SIGLEC14* deletion polymorphism that has not so far been observed in other primates. The polymorphism occurs at varying frequencies based on the geographic origins of the population (Yamanaka et al. 2009). The deletion apparently resulted from an in-frame gene fusion that occurred between the *SIGLEC5* and *SIGLEC14* ORFs. The new gene product encodes a Siglec-5-like immunosuppressive receptor (designated as Siglec-14/5) regulated under the fully functional *SIGLEC14* gene promoter. Individuals homozygous for the wild-type ancestral allele encoding for both Siglec-5 and -14 are represented as *SIGLEC14* +/+, and those homozygous for the deletion polymorphism are represented as *SIGLEC14* –/–. Although *SIGLEC14* +/+ neutrophils display both Siglec-5 and -14, *SIGLEC14* –/– neutrophils display only the Siglec-14/5. In comparison, while *SIGLEC14* +/+ monocytes express only Siglec-14 but not -5, *SIGLEC14* –/– monocytes lose Siglec-14 but gain Siglec-14/5. *Ex vivo* experiments confirmed that *SIGLEC14* –/– myelomonocytic cells have a dampened inflammatory response in comparison to *SIGLEC14* +/+ cells when challenged with bacteria or endotoxin (Ali et al. 2014).

The selective forces responsible for establishing the *SIGLEC14* deletion polymorphism in the human genome are as yet unknown. However, *SIGLEC14* +/+ individuals are more susceptible to developing acute exacerbation of chronic obstructive pulmonary disease (COPD) compared to *SIGLEC14* –/– individuals (Angata et al. 2013). Although this complication of pulmonary inflammation has been largely linked to cigarette smoking in modern times, acute exacerbation of lung inflammation may develop for anyone with long term exposure to air pollution such as indoor smoke. The *SIGLEC14* polymorphism allele exists in all human populations worldwide, although at different frequencies. We speculate that the invention of indoor cooking, and consequently indoor smoke, could have been a selective force favoring the fusion allele. However, the ancestral *SIGLEC14* allele is not entirely evolutionary disadvantageous for modern humans. *SIGLEC14* –/– fetuses are more susceptible to preterm labor after GBS infection when compared to *SIGLEC14* +/+ infants, independent of the mother's *SIGLEC14* genotype (Ali et al. 2014). The *SIGLEC14* polymorphism might even have been maintained by heterozygote advantage since *SIGLEC14* +/– individuals would have an intermediate propensity for inflammation.

Siglec-6

Siglec-6 is a CD33rSiglec found on primate B-cells but also uniquely on human placental trophoblast (Brinkman-Van der Linden et al. 2007). Several differences found between human and primate sequences of *SIGLEC6* promoter regions may explain the changed expression pattern. There are conflicting reports regarding the changes in expression pattern in relation to the onset of labor (Brinkman-Van der Linden et al. 2007; Rumer et al. 2013).

Although Siglec-6 normally binds sialyl-Tn (Neu5Acα2-6GalNAcα1-), it can also interact with the non-glycosylated adipose-derived hormone leptin (Patel et al. 1999). In this regard, immunohistochemical analysis revealed that both wild type and arginine-mutated Siglec-6 recognize ligands in and adjacent to the placenta, including uterine endometrium, suggesting the presence of both sialic acid dependent and independent ligands (Brinkman-Van der Linden et al. 2007). Since leptin is also secreted from the placenta, it is likely to be sialic acid independent ligand recognized by the arginine-mutated Siglec-6. A further increase in Siglec-6 trophoblast expression is also associated with the uniquely human condition preeclampsia (Winn et al. 2009; Rumer et al. 2013). Both mRNA and protein expression of Siglec-6 are increased in placentas obtained from women who had preeclampsia as compared to the control group of pre-term labor. Mechanistically, Siglec-6 ligation with Glycodelin-A suppresses ERK signaling and subsequently trophoblast invasiveness. Taken together, these data suggest that Siglec-6 contributes to the human-specific aspects of reproductive biology. It may also be worthwhile to investigate whether Siglec-6 expression and function directly correlates with other immunoregulatory complications in pregnancy such as spontaneous abortion and recurrent miscarriage (Chatterjee et al. 2014).

Siglec-9

Siglec-9 is an ITIM-containing CD33rSiglec found primarily on human neutrophils and monocytes (Zhang et al. 2000). It is also expressed weakly on $CD4^+$ and $CD8^+$ T-cells, and but found at more modest levels on B-cells. Recent literature also revealed that Siglec-9 expression defines a subset of cytotoxic NK cell population (Jandus et al. 2014). Differences in Siglec-9 expression level were also found on circulating monocytes between gorilla, humans, and chimpanzees in descending order, although still abundant in all three primates (Padler-Karavani et al. 2014). Furthermore, immunohistochemical analysis showed that only human splenic macrophages display Siglec-9, but not those from chimpanzee or gorilla.

Siglec-9 ligand binding preferences also changed somewhat during evolution. As previously discussed, since the *CMAH* gene was functionally inactivated during human evolution, humans are the only known primates incapable of naturally synthesizing Neu5Gc. Interestingly, while gorilla and chimpanzee Siglec-9 somewhat preferred Neu5Gc sialylated ligands over Neu5Ac, human Siglec-9 slightly preferred Neu5Ac over Neu5Gc (Sonnenburg et al. 2004; Padler-Karavani et al. 2014). Thus, it is possible that Siglec-9 evolved to engage Neu5Ac with higher affinity after the human specific loss of synthesizing Neu5Gc.

In addition to Siglec-5, Group B *Streptococcus* (GBS) also hijacks Siglec-9's inhibitory properties through molecular mimicry of sialic acids (Carlin et al. 2007, 2009b). All three tested serotypes of GBS (Type Ia, Ib, and III) bound to human Siglec-9 at a stronger affinity over chimpanzee Siglec-9, but GBS was incapable of binding to baboon Siglec-9. Taken together, these properties may be a contributing factor for why GBS is a human-specific pathogen, as it can take advantage of human inhibitory Siglecs much easier than other primate Siglecs.

Siglec-9 may be a prime example of co-evolution between mammalian sialic acids, Siglecs, and the pathogens that exploit this receptor. Pathogens evolve to synthesize or acquire sialic acids identical or similar to ones displayed by the host to hide from the immune system by engaging inhibitory Siglecs. Meanwhile, the host evolves and alters their sialic acid repertoire to keep away from the pathogens. In order to recognize the "newly evolved" sialic acid SAMPs, Siglecs are also constantly adapting to keep up with this ongoing evolutionary arms race as well.

Siglec-11 and -16

SIGLEC11 and *SIGLEC16* genes are found head-to-head about 1 MB away from the *CD33rSIGLEC* gene cluster on human chromosome 19 (Angata et al. 2002; Hayakawa et al. 2005; Wang et al. 2012a; Cao et al. 2008). It has been suggested that *SIGLEC16* arose by an inverse duplication of inhibitory *SIGLEC11* and underwent subsequent pseudogenization. Then, two tandem and likely simultaneous gene conversions occurred from *SIGLEC16P* to the adjacent gene *SIGLEC11* with an intervening short segment being excluded, and ultimately resulting in the creation of an open reading frame. Both of the gene conversions have been dated to about 1–1.2 million years, after the emergence of the genus *Homo*, but prior to the emergence of the common ancestor of Denisovans and modern humans about 600,000 years ago (Wang et al. 2012a).

These extensive changes in the sequence of human Siglec-11 may explain the different affinity for sialylated glycans compared to the chimpanzee counterpart (Hayakawa et al. 2005), and the emergence of novel binding properties. For instance, Siglec-11 is detected on fibroblasts in ovaries of both human and chimpanzee (Wang et al. 2011). However, probing for Siglec-11 ligands revealed distinct and strong mast cell expression in human ovaries, and diffuse stromal ligands in chimpanzee ovaries. This dramatic difference in ligand specificity may have an impact on ovarian physiology.

One of the conversion events also changed the 5′ untranslated sequence, altering predicted transcription factor binding sites of *SIGLEC11* and, perhaps consequently, its expression pattern. Indeed, while Siglec-11 is found in both human and chimpanzee tissue macrophages, it is expressed in brain microglia only in humans (Hayakawa et al. 2005). *In vitro* studies have indicated that Siglec-11 suppresses the levels of proinflammatory cytokines, reduces phagocytosis of apoptotic neurons and alleviates microglia neurotoxicity (Wang and Neumann 2010). Interestingly, the neuroprotective effects were dependent on polysialic acid, a polymer important for maintenance of brain plasticity (Rutishauser 2008).

Like Siglec-5 and -14, Siglec-11 and -16 are also paired receptors (Cao et al. 2008). Their V-set and the first C2-set domains are 99 % identical in the amino acid sequence, and the other two C2-set domains share about 80 % sequence identity. However, the intracellular carboxyl-terminal region of Siglec-11 contains one ITIM, whereas Siglec-16 has a transmembrane domain and a short cytosolic tail that may associate with DAP12 (Cao et al. 2008). Therefore, whereas the two proteins are likely to recognize similar ligands, this binding will allegedly lead to opposite intracellular signaling cascades. To date, Siglec-11 and -16 mediated immunomodulation has not been well characterized.

It is interesting to note that whereas *SIGLEC11* is fixed in the human population, *SIGLEC16* is often pseudogenized due to a deletion of four nucleotides in the second exon that results in frameshift and premature termination of translation (Cao et al. 2008; Wang et al. 2012a). The frequency of the functional *SIGLEC16* allele varies among populations, adding another potential layer of complexity to the modulation of the immune responses.

Siglec-XII

Human *SIGLEC12* encodes for a Siglec receptor with two V-set domains (Yu et al. 2001). However, the arginine residues critical for glycan binding are substituted in both domains (Angata et al. 2001). Therefore, this receptor is unable to bind sialic acid, and it is referred by convention as Siglec-XII. The single nucleotide substitution (C to T) at the first position of the arginine codon in the outer most V-set domain is universal in humans but not found in other primates (Angata et al. 2001). It occurred prior to the common ancestor of all modern humans and sometime after the split of the hominin lineage from that of the common ancestor of chimpanzee and bonobos. In chimpanzee, Siglec-12 is strongly expressed in macrophages, and epithelial cells of prostate, pancreas, kidney, and stomach. Human Siglec-XII is generally less expressed in the same tissues (Mitra et al. 2011).

A SNP in human *SIGLEC12* produces a single nucleotide insertion in the exon that encodes the first V-set, which changes the open reading frame and results in a polypeptide of 115 amino acids (Mitra et al. 2011). The global frequency of this mutation was found to be 0.58, with allele frequencies ranging from 0.38 in Sub-Saharan Africa to 0.86 in Native American populations. About 40 % of human individuals are homozygous for the *SIGLEC12P*; heterozygosity reaches about 30 %. Siglec-XII was also observed to be overexpressed in prostate cancer (Mitra et al. 2011). However, no association was found with the inactivating SNP and incidence of prostate cancer. Interestingly, a study on polymorphic nonsense SNPs in the human genome found *SIGLEC12* to be one of the outliers, among 167 cases (Yngvadottir et al. 2009). In particular, it was noted that the inactivating SNP was not itself deleterious, suggesting a balancing selection or a selective sweep.

Siglec-13

Analysis of genomic BAC clones indicated that the primate *SIGLEC13* gene was missing from the human genome, but was present in chimpanzee and baboon (Angata et al. 2004). Comparative analysis of the genomic regions of human, chimpanzee, baboon and rhesus genomes identified five repetitive *Alu* elements in a 10 kb genomic region containing the *SIGLEC13* locus in the chimpanzee, baboon and rhesus genome. By contrast, the human genome includes a single composite *Alu* element occupying the region of 7 kb in the same genomic region. The composite element likely derives from recombination of *Alu* elements that resulted in the excision of *SIGLEC13* in humans. Indeed, analysis of monocytes from peripheral blood confirmed expression of Siglec-13 in chimpanzee, and absence in human monocytes (Wang et al. 2012b). Universal absence of Siglec-13 in humans was also confirmed in 28 HapMap human samples and in the common ancestral population of Neanderthals and Denisovan. It is interesting that despite recruiting DAP12, Siglec-13 reduces inflammation in response to pathogenic bacteria that can specifically interact with Siglec-13 (Wang et al. 2012b). This indicates that the ability of a Siglec receptor to alter inflammatory responses cannot be deduced only based on the primary sequence, and that the outcome of Siglec engagement might be more complex (Barrow and Trowsdale 2006; Hamerman and Lanier 2006).

Siglec-17

The primate *SIGLEC17* pseudogene (*SIGLEC17P*) exhibits high sequence similarity with *SIGLEC3* and was originally annotated as *SIGLEC3P* (Angata et al. 2004). This locus has a human-specific 1 bp deletion that alters the predicted open reading frame (ORF) and results in a truncated protein (Wang et al. 2012b). The rest of the ORF remains intact in all humans tested and the corresponding mRNA is strongly expressed in NK cells. Moreover, the human *SIGLEC17P* contains a human-unique missense mutation of the codon encoding an arginine residue that is required for sialic acid binding. The human *SIGLEC17P* allele was already present in the common ancestral populations of Neanderthals and Denisovans. Some other primate *SIGLEC17* genes underwent independent events of inactivation. *SIGLEC17* seems to be completely deleted in the rhesus and baboon genomes (Wang et al. 2012b). In contrast, New World monkeys carry a functional *SIGLEC17* ORF.

Conclusions and Perspectives

Comparative studies in mammals have revealed an expansion in the number of *SIGLEC* genes in primates, and an accumulation of multiple variations, particularly in the human lineage. Work of the last decade has revealed that pathogens exploit Siglec function by expressing sialylated and non-sialylated ligands, and thus constitute a major selective force for the evolution of *SIGLEC* genes. Therefore, pathogens likely shaped the *SIGLEC* gene family: expansion, deletion and polymorphic inactivation of activating receptors are signatures of past and ongoing selection. At the same time, current data suggest that some variations in Siglecs might have resulted in advantageous functions that were retained and contributed to human evolution. The human-specific *SIGLEC11* and *SIGLEC16* conversion events and subsequent recruitment of novel Siglec-11 dependent functions to the brain are paradigmatic and deserve further studies. The same is true of the recruitment of *SIGLEC6* expression to the placental trophoblast. Novel regulation of Siglec-3/CD33 at the V-set domain and its impact in the development of late onset alzheimer's disease calls for studies to define whether this phenomenon occurs in other primates, or if it is specific to humans. Future research should also address whether observed reduction of affinity for sialylated molecules in human Siglecs are accompanied by the emergence of alternative binding properties. Also, as the innate immune system impacts many human conditions such as cancer and obesity, it will be interesting to study the influence of Siglec polymorphic variations in the incidence and progression of various diseases. It is even possible that *SIGLEC* gene changes played key roles in population bottlenecks involved in human origins.

Acknowledgments We thank the members of the Varki lab group for fruitful discussions, and Yuko Naito-Matsui, Anel Lizcano, Shoib Siddiqui, Corinna Landig, Stevan Springer and John T. Ngo for comments on the manuscript. Research in the Varki laboratory is supported by grants from the NIH and by the NHLBI Program of Excellence in Glycosciences. F.S. is supported by a fellowship from the Novartis Foundation for medical-biological research.

References

Ali SR et al (2014) Siglec-5 and Siglec-14 are polymorphic paired receptors that modulate neutrophil and amnion signaling responses to group B Streptococcus. J Exp Med 211:1231–1242

Altheide TK et al (2006) System-wide genomic and biochemical comparisons of sialic acid biology among primates and rodents: evidence for two modes of rapid evolution. J Biol Chem 281:25689–25702

Andrews RG, Torok-Storb B, Bernstein ID (1983) Myeloid-associated differentiation antigens on stem cells and their progeny identified by monoclonal antibodies. Blood 62:124–132

Angata T (2014) Associations of genetic polymorphisms of Siglecs with human diseases. Glycobiology 24(9):785–793

Angata T, Varki NM, Varki A (2001) A second uniquely human mutation affecting sialic acid biology. J Biol Chem 276:40282–40287

Angata T et al (2002) Cloning and characterization of human Siglec-11. A recently evolved signaling molecule that can interact with SHP-1 and SHP-2 and is expressed by tissue macrophages, including brain microglia. J Biol Chem 277:24466–24474
Angata T et al (2004) Large-scale sequencing of the CD33-related Siglec gene cluster in five mammalian species reveals rapid evolution by multiple mechanisms. Proc Natl Acad Sci U S A 101:13251–13256
Angata T et al (2006) Discovery of Siglec-14, a novel sialic acid receptor undergoing concerted evolution with Siglec-5 in primates. FASEB J 20:1964–1973
Angata T et al (2013) Loss of Siglec-14 reduces the risk of chronic obstructive pulmonary disease exacerbation. Cell Mol Life Sci 70:3199–3210
Barclay AN, Hatherley D (2008) The counterbalance theory for evolution and function of paired receptors. Immunity 29:675–678
Barreiro LB, Quintana-Murci L (2010) From evolutionary genetics to human immunology: how selection shapes host defence genes. Nat Rev Genet 11:17–30
Barrow AD, Trowsdale J (2006) You say ITAM and I say ITIM, let's call the whole thing off: the ambiguity of immunoreceptor signalling. Eur J Immunol 36:1646–1653
Bradshaw EM et al (2013) CD33 Alzheimer's disease locus: altered monocyte function and amyloid biology. Nat Neurosci 16:848–850
Brinkman-Van der Linden ECM et al (2000) Loss of N-glycolylneuraminic acid in human evolution—implications for sialic acid recognition by siglecs. J Biol Chem 275:8633–8640
Brinkman-Van der Linden EC et al (2003) CD33/Siglec-3 binding specificity, expression pattern, and consequences of gene deletion in mice. Mol Cell Biol 23:4199–4206
Brinkman-Van der Linden EC et al (2007) Human-specific expression of Siglec-6 in the placenta. Glycobiology 17:922–931
Cannons JL, Tangye SG, Schwartzberg PL (2011) SLAM family receptors and SAP adaptors in immunity. Annu Rev Immunol 29:665–705
Cao H, Crocker PR (2011) Evolution of CD33-related siglecs: regulating host immune functions and escaping pathogen exploitation? Immunology 132:18–26
Cao H et al (2008) SIGLEC16 encodes a DAP12-associated receptor expressed in macrophages that evolved from its inhibitory counterpart SIGLEC11 and has functional and non-functional alleles in humans. Eur J Immunol 38:2303–2315
Carlin AF et al (2007) Group B streptococcal capsular sialic acids interact with siglecs (immunoglobulin-like lectins) on human leukocytes. J Bacteriol 89:1231–1237
Carlin AF et al (2009a) Group B Streptococcus suppression of phagocyte functions by protein-mediated engagement of human Siglec-5. J Exp Med 206:1691–1699
Carlin AF et al (2009b) Molecular mimicry of host sialylated glycans allows a bacterial pathogen to engage neutrophil Siglec-9 and dampen the innate immune response. Blood 113:3333–3336
Chatterjee P et al (2014) Regulation of the anti-inflammatory cytokines interleukin-4 and interleukin-10 during pregnancy. Front Immunol 5:253
Chou HH et al (2002) Inactivation of CMP-N-acetylneuraminic acid hydroxylase occurred prior to brain expansion during human evolution. Proc Natl Acad Sci U S A 99:11736–11741
Collins BE et al (1997) Binding specificities of the sialoadhesin family of I-type lectins. Sialic acid linkage and substructure requirements for binding of myelin-associated glycoprotein, Schwann cell myelin protein, and sialoadhesin. J Biol Chem 272:16889–16895
Crocker PR, Varki A (2001) Siglecs, sialic acids and innate immunity. Trends Immunol 22:337–342
Crocker PR et al (1991) Purification and properties of sialoadhesin, a sialic acid-binding receptor of murine tissue macrophages. EMBO J 10:1661–1669
Crocker PR et al (1997) The potential role of sialoadhesin as a macrophage recognition molecule in health and disease. Glycoconj J 14:601–609
Crocker PR, Paulson JC, Varki A (2007) Siglecs and their roles in the immune system. Nat Rev Immunol 7:255–266
Freeman SD et al (1995) Characterization of CD33 as a new member of the sialoadhesin family of cellular interaction molecules. Blood 85:2005–2012

Gearing M et al (1994) Neuropathology and apolipoprotein E profile of aged chimpanzees: implications for Alzheimer disease. Proc Natl Acad Sci U S A 91:9382–9386
Griciuc A et al (2013) Alzheimer's disease risk gene CD33 inhibits microglial uptake of amyloid beta. Neuron 78:631–643
Hamerman JA, Lanier LL (2006) Inhibition of immune responses by ITAM-bearing receptors. Sci STKE 2006:re1
Hartnell A et al (2001) Characterization of human sialoadhesin, a sialic acid binding receptor expressed by resident and inflammatory macrophage populations. Blood 97:288–296
Hayakawa T et al (2005) A human-specific gene in microglia. Science 309:1693
Hernandez-Caselles T et al (2006) A study of CD33 (SIGLEC-3) antigen expression and function on activated human T and NK cells: two isoforms of CD33 are generated by alternative splicing. J Leukoc Biol 79:46–58
Ishida-Kitagawa N et al (2012) Siglec-15 protein regulates formation of functional osteoclasts in concert with DNAX-activating protein of 12 kDa (DAP12). J Biol Chem 287:17493–17502
Jandus C et al (2014) Interactions between Siglec-7/9 receptors and ligands influence NK cell-dependent tumor immunosurveillance. J Clin Invest 124(4):1810–1820
Kameda Y et al (2013) Siglec-15 regulates osteoclast differentiation by modulating RANKL-induced phosphatidylinositol 3-kinase/Akt and Erk pathways in association with signaling adaptor DAP12. J Bone Miner Res 28(12):2463–2475
Lanier LL (2009) DAP10- and DAP12-associated receptors in innate immunity. Immunol Rev 227:150–160
Linnartz-Gerlach B, Kopatz J, Neumann H (2014) Siglec functions of microglia. Glycobiology 24(9):794–799
Lock K et al (2004) Expression of CD33-related siglecs on human mononuclear phagocytes, monocyte-derived dendritic cells and plasmacytoid dendritic cells. Immunobiology 209:199–207
Malik M et al (2013) CD33 Alzheimer's risk-altering polymorphism, CD33 expression, and exon 2 splicing. J Neurosci 33:13320–13325
Mitra N et al (2011) SIGLEC12, a human-specific segregating (pseudo)gene, encodes a signaling molecule expressed in prostate carcinomas. J Biol Chem 286:23003–23011
Nguyen DH et al (2006) Loss of Siglec expression on T lymphocytes during human evolution. Proc Natl Acad Sci U S A 103:7765–7770
Padler-Karavani V et al (2014) Rapid evolution of binding specificities and expression patterns of inhibitory CD33-related Siglecs in primates. FASEB J 28:1280–1293
Patel N et al (1999) OB-BP1/Siglec-6—a leptin- and sialic acid-binding protein of the immunoglobulin superfamily. J Biol Chem 274:22729–22738
Perez SE et al (2013) Alzheimer's disease pathology in the neocortex and hippocampus of the western lowland gorilla (Gorilla gorilla gorilla). J Comp Neurol 521:4318–4338
Perez-Oliva AB et al (2011) Epitope mapping, expression and post-translational modifications of two isoforms of CD33 (CD33M and CD33m) on lymphoid and myeloid human cells. Glycobiology 21:757–770
Pillai S et al (2012) Siglecs and immune regulation. Annu Rev Immunol 30:357–392
Raj T et al (2014) CD33: increased inclusion of exon 2 implicates the Ig V-set domain in Alzheimer's disease susceptibility. Hum Mol Genet 23:2729–2736
Ravetch JV, Lanier LL (2000) Immune inhibitory receptors. Science 290:84–89
Rempel H et al (2008) Sialoadhesin expressed on IFN-induced monocytes binds HIV-1 and enhances infectivity. PLoS One 3:e1967
Rumer KK et al (2013) Siglec-6 expression is increased in placentas from pregnancies complicated by preterm preeclampsia. Reprod Sci 20:646–653
Rutishauser U (2008) Polysialic acid in the plasticity of the developing and adult vertebrate nervous system. Nat Rev Neurosci 9:26–35
Sonnenburg JL, Altheide TK, Varki A (2004) A uniquely human consequence of domain-specific functional adaptation in a sialic acid-binding receptor. Glycobiology 14:339–346
Soto PC et al (2012) Cell-intrinsic mechanism involving Siglec-5 associated with divergent outcomes of HIV-1 infection in human and chimpanzee CD4 T cells. J Mol Med (Berl) 91:261–270

Takamiya R et al (2013) The interaction between Siglec-15 and tumor-associated sialyl-Tn antigen enhances TGF-beta secretion from monocytes/macrophages through the DAP12-Syk pathway. Glycobiology 23:178–187

Tourdot BE et al (2013) Immunoreceptor tyrosine-based inhibitory motif (ITIM)-mediated inhibitory signaling is regulated by sequential phosphorylation mediated by distinct nonreceptor tyrosine kinases: a case study involving PECAM-1. Biochemistry 52:2597–2608

Varki A (2010) Colloquium paper: uniquely human evolution of sialic acid genetics and biology. Proc Natl Acad Sci U S A 107(Suppl 2):8939–8946

Varki A (2011) Since there are PAMPs and DAMPs, there must be SAMPs? Glycan "self-associated molecular patterns" dampen innate immunity, but pathogens can mimic them. Glycobiology 21:1121–1124

Varki A, Angata T (2006) Siglecs—the major subfamily of I-type lectins. Glycobiology 16: 1R–27R

Varki NM et al (2011) Biomedical differences between human and nonhuman hominids: potential roles for uniquely human aspects of sialic acid biology. Annu Rev Pathol 6:365–393

Wang Y, Neumann H (2010) Alleviation of neurotoxicity by microglial human Siglec-11. J Neurosci 30:3482–3488

Wang X et al (2011) Expression of Siglec-11 by human and chimpanzee ovarian stromal cells, with uniquely human ligands: implications for human ovarian physiology and pathology. Glycobiology 21:1038–1048

Wang X et al (2012a) Evolution of Siglec-11 and Siglec-16 genes in hominins. Mol Biol Evol 29:2073–2086

Wang X et al (2012b) Specific inactivation of two immunomodulatory SIGLEC genes during human evolution. Proc Natl Acad Sci U S A 109:9935–9940

Winn VD et al (2009) Severe preeclampsia-related changes in gene expression at the maternal-fetal interface include sialic acid-binding immunoglobulin-like lectin-6 and pappalysin-2. Endocrinology 150:452–462

Yamanaka M et al (2009) Deletion polymorphism of SIGLEC14 and its functional implications. Glycobiology 19:841–846

Yngvadottir B et al (2009) A genome-wide survey of the prevalence and evolutionary forces acting on human nonsense SNPs. Am J Hum Genet 84:224–234

Yu Z et al (2001) Identification and characterization of S2V, a novel putative siglec that contains two V set Ig-like domains and recruits protein-tyrosine phosphatases SHPs. J Biol Chem 276: 23816–23824

Zhang JQ et al (2000) Siglec-9, a novel sialic acid binding member of the immunoglobulin superfamily expressed broadly on human blood leukocytes. J Biol Chem 275:22121–22126

Zou Z et al (2011) Siglecs facilitate HIV-1 infection of macrophages through adhesion with viral sialic acids. PLoS One 6:e24559

Chapter 2
Structural Changes of GPI Anchor After Its Attachment to Proteins: Functional Significance

Taroh Kinoshita

Introduction

One hundred and fifty or more of human proteins are post-translationally modified by a glycolipid, termed glycosylphosphatidylinositol (GPI) that anchors proteins to the outer leaflet of plasma membrane (Orlean and Menon 2007; Kinoshita et al. 2008). While GPI-anchored proteins (GPI-APs) have wide range of functions, such as hydrolytic enzymes, adhesion molecules, receptors and protease inhibitors, they share common membrane-anchors. GPI is synthesized in the endoplasmic reticulum (ER) from phosphatidylinositol (PI) via at least 11 steps (Fig. 2.1). Preassembled GPI is transferred en bloc by GPI transamidase to the C-terminus of proteins having a C-terminal GPI-attachment signal peptide. A unique feature of GPI-anchor is that GPI structure is dynamically modified during transport of GPI-APs to the cell surface. Structural remodeling of lipid and glycan moieties in the ER is critical for efficient recruitment of GPI-APs into ER-exit site and association with their cargo receptors for transportation to the Golgi apparatus (Kinoshita et al. 2013). Fatty acid remodeling of GPI in the Golgi is important for homodimerization and raft association. In this chapter, these structure-biology relationships are described.

T. Kinoshita (✉)
WPI Immunology Frontier Research Center and Research Institute for Microbial Diseases, Osaka University, 3-1 Yamada-oka, Suita, Osaka 565-0871, Japan
e-mail: tkinoshi@biken.osaka-u.ac.jp

A. Chakrabarti, A. Surolia (eds.), *Biochemical Roles of Eukaryotic Cell Surface Macromolecules*, Advances in Experimental Medicine and Biology 842,
DOI 10.1007/978-3-319-11280-0_2

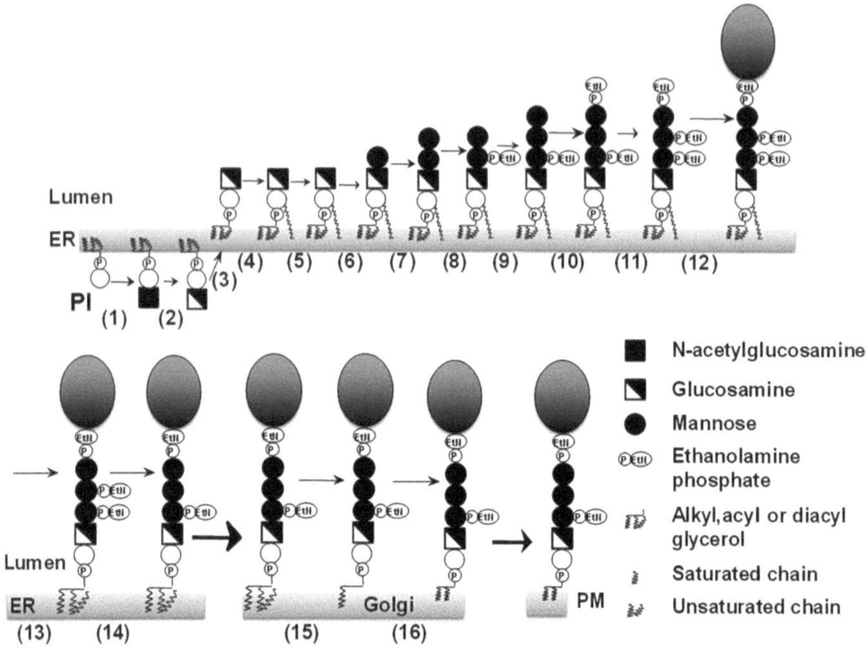

Fig. 2.1 Schematic representation of biosynthesis of GPI-APs in mammalian cells. GPI is synthesized from phosphatidylinositol (PI) in the ER (steps 1–11). Preassembled GPI is transferred to the protein's carboxyl-terminus (step 12). After two remodeling reactions in the ER (steps 13 and 14), GPI-AP is transported to the Golgi apparatus where GPI's sn2-linked fatty acid is remodeled (steps 15 and 16) before transport to the plasma membrane (PM)

Biosynthesis of GPI and Its Attachment to Proteins

GPI biosynthesis is initiated using cellular PI in the ER (Fig. 2.1). The first reaction is transfer of N-acetylglucosamine (GlcNAc) to PI from UDP-GlcNAc to generate first intermediate GlcNAc-PI (step 1). Step 1 is mediated by GPI-GlcNAc transferase (GPI-GnT) complex consisting of PIG-A, PIG-C, PIG-H, PIG-P, PIG-Q, PIG-Y and DPM2 proteins (Watanabe et al. 2000; Murakami et al. 2005). GlcNAc-PI is de-N-acetylated by deacetylase PIG-L to generate second intermediate glucosamine (GlcN)-PI (step 2) (Nakamura et al. 1997; Watanabe et al. 1999). These two reactions occur on the cytoplasmic side of the ER. GlcN-PI then flips into the luminal side with currently unknown mechanism (step 3). The inositol in GlcN-PI is acylated (mainly palmitoylated or myristoylated) to generate third intermediate GlcN-(acyl)PI (step 4). PIG-W, acyl-CoA-dependent acyltransferase, mediates step 4 (Murakami et al. 2003). The lipid moiety in GlcN-(acyl)PI is remodeled by currently unknown mechanism (step 5). See part 3 in the next section for lipid remodeling of GlcN-(acyl)PI. Two mannoses are transferred from dolichol-phosphate-mannose to generate Manα1-6Manα1-4GlcNα1-6(acyl)PI by PIG-M/PIG-X complex (Maeda et al. 2001; Ashida et al. 2005) and PIG-V (Kang et al. 2005), respectively (steps 6 and 7). Ethanolamine phosphate (EtNP) side branch is transferred by PIG-N from phosphatidylethanolamine to the first mannose generating Manα1-6(EtNP2) Manα1-4GlcNα1-6(acyl)PI (step 8) (Hong et al. 1999). Third mannose is then transferred by PIG-B from dolichol-phosphate-mannose to generate Manα1-2Manα1-6(EtNP2)Manα1-4GlcNα1-6(acyl)PI (step 9) (Takahashi et al. 1996). Two EtNP are sequentially transferred from phosphatidylethanolamine to the third and second mannoses by PIGO/PIGF complex (Inoue et al. 1993; Hong et al. 2000) and PIGG/PIGF complex (Shishioh et al. 2005), respectively, to generate EtNP6Manα1-2(EtNP6)Manα1-6(EtNP2)Manα1-4GlcNα1-6(acyl)PI (steps 10 and 11). The EtNP linked to the third mannose acts to bridge GPI to proteins.

The preassembled GPI is then transferred to the carboxyl terminus of proteins by GPI transamidase consisting of PIG-K, GPAA1, PIG-S, PIG-T and PIG-U (step 12) (Ohishi et al. 2000, 2001; Hong et al. 2003). Proteins that are to be GPI anchored have a GPI attachment signal peptide at their carboxyl terminus. The signal peptide is cleaved and replaced with the preassembled GPI. The amino group of the "bridging EtNP" makes an amide bond with the newly generated carboxyl terminus by transamidation (Orlean and Menon 2007).

GPI Remodeling Reactions That Occur After Attachment to Proteins

Inositol-Deacylation by PGAP1 in the ER

Shortly after attachment to proteins, the inositol-linked acyl chain is removed by PGAP1, inositol-deacylase, in the ER (step 13) (Tanaka et al. 2004). This lipid remodeling is necessary for efficient ER-to-Golgi transport of GPI-APs. Specifically, recognition of GPI-APs by the cargo receptor for recruitment into COPII-coated transport vesicles requires elimination of the acyl chain (Fujita et al. 2011). In PGAP1 defective Chinese hamster ovary (CHO) cells, ER-to-Golgi transport of DAF/CD55, a GPI-AP, is threefold slower than in wild-type cells due to inefficient binding to the cargo receptor (Tanaka et al. 2004).

Removal of an EtNP Side Branch from the Second Mannose by PGAP5 in the ER

In addition to inositol-deacylation, elimination of the EtNP from the second mannose is required for binding to the cargo receptor (step 14) (Fujita et al. 2011). In PGAP5 defective CHO cells, ER-to-plasma membrane transport of GPI-APs is fourfold slower (Fujita et al. 2009).

Fatty Acid Remodeling of GPI-AP in the Golgi

Cellular PI is exclusively diacyl form and its predominant species is 1-stearoyl(C18:0)-2-arachidonoyl(C20:4) PI. The structures of the PI moiety in GPI intermediates that were accumulated in mutant cells defective in each of the early biosynthetic steps, were determined by mass-spectrometry and were found that GlcNAc-PI and GlcN-PI have PI structures similar to cellular PI in both CHO cells and human lymphoma cells. In contrast, GlcN-(acyl)PI was a mixture of 1-alkyl-2-acyl and diacyl forms with the former being the major form. The major species of the former were 1-stearyl-2-oleoyl, -2-arachidonoyl, and -2-docosatetraenoyl PI. It was, therefore, suggested that the lipid moiety changes in GlcN-(acyl)PI (Houjou et al. 2007). It was also found that the acyl chain compositions of GlcN-PI and the diacyl form of GlcN-(acyl)PI are different, suggesting that diacyl GlcN-(acyl)PI is generated from diacylGlcN-PI and then converted to a mixture of 1-alkyl-2-acyl and diacyl GlcN-(acyl)PI (step 5 in Fig. 2.1) (Kanzawa et al. 2009). A possible mechanism of the conversion may be that the diacyl glycerol part is exchanged with 1-alkyl-2-acyl or diacyl glycerol derived from a putative donor phospholipid. Alternatively, phosphatidic acid part is exchanged. The putative donor phospholipid may contain 1-alkyl-2-acyl and diacyl forms. It was pointed out that phosphatidylethanolamine has chain compositions similar to that of the remodeled GlcN-(acyl)PI and is a candidate of the donor lipid (Kanzawa et al. 2009). The gene(s) required for this lipid remodeling in the ER have not yet been clarified.

It was then found that generation of the 1-alkyl-2-acyl form of GlcN-(acyl)PI is dependent upon a pathway in the peroxisome that generates 1-alkyl-glycerone-phosphate from dihydroxyacetone phosphate. Mutant CHO cells defective in synthesis of 1-alkyl-glycerone-phosphate generated only the diacyl form of GlcN-(acyl) PI (Kanzawa et al. 2009). It is likely that 1-alkyl-glycerone-phosphate is converted to the putative donor lipid in the ER, which is then used in lipid remodeling reaction (step 5 in Fig. 2.1). Consistent with these findings, fibroblasts from patients with peroxisomal disorders, Zellweger syndrome and rhizomelic chondrodysplasia punctata, were defective in generation of 1-alkyl-2-acyl form of GPI anchors, suggesting that a lack of or a decrease in the 1-alkyl-2-acyl form of GPI anchors might be related to some of the symptoms of these patients (Kanzawa et al. 2012).

1-Alkyl-2-acyl PI is the major form of protein-bound GPI-anchors in mammalian cells and diacyl PI is a minor form. Therefore, the profile of lipid moiety of GlcN-(acyl)PI after lipid remodeling is similar to that of the GPI-anchors of cell-surface GPI-APs. However, there is a major difference in the sn2-linked fatty acids. Mammalian GPI-APs usually have two saturated fatty chains, with a small fraction containing one unsaturated bond in an sn1-linked chain. The sn2-linked fatty acid is usually stearic acid (C18:0), while GlcN-(acyl)PI contains various unsaturated chains, such as oleic, arachidonic, and docosatetraenoic acids.

We established a CHO mutant cell line, termed clone C84 that synthesized GPI normally but showed greatly reduced surface expression of GPI-APs. We clarified the mechanisms of the abnormality by demonstrating that GPI-APs in C84 mutant cells were converted to the lyso-GPI form by losing a fatty acid before exiting the trans-Golgi network and that, after transport to the cell surface, the lyso-GPI-APs were cleaved by a phospholipase D, resulting in secretion of soluble GPI-APs lacking a phosphatidic acid moiety and reduced cell surface levels of GPI-APs. Based on the study of C84 CHO cells, we proposed that GPI fatty acid remodeling occurs, in which the sn2-linked fatty acid is exchanged from an unsaturated chain to a saturated chain (stearic acid) and that the lyso-GPI-AP found in C84 cells is an intermediate in the fatty acid remodeling (steps 15 and 16) (Tashima et al. 2006).

Tashima et al. cloned the gene responsible for C84 defect, termed *PGAP2*, by sorting C84 cells that restored the normal levels of GPI-APs after transfection of a cDNA library. PGAP2 is a 254-amino-acid membrane protein mainly expressed in the Golgi. PGAP2 is involved in the second step of GPI-AP fatty acid remodeling, in that the lyso-GPI-AP intermediate is reacylated by stearic acid (step 16) (Tashima et al. 2006). There is no significant sequence homology between PGAP2 and known acyltransferases, and the issue of whether PGAP2 is the acyltransferase itself or a regulatory protein remains to be determined.

Maeda et al. then hypothesized that there must be a gene, termed *PGAP3*, involved in the elimination of the unsaturated fatty acid (step 15) and that, if *PGAP3* is mutated in the *PGAP2*-defective cells, the decreased GPI-AP expression might be restored because two fatty chains are maintained. Indeed, Maeda et al. established a double-mutant CHO cell line expressing almost normal levels of GPI-APs from the *PGAP2*-defective C84 cells and determined that a mammalian homolog of yeast *PER1*, which was reported to be involved in similar fatty acid remodeling in yeast, is *PGAP3*. PGAP3 is a 320-amino-acid Golgi-resident protein with seven transmembrane domains. PGAP3 belongs to a hydrolase superfamily and is most likely to be GPI-AP-specific phospholipase A2, although the enzyme activity has yet to be demonstrated in vitro. These findings together demonstrated that fatty acid remodeling of GPI-APs occurs in the Golgi of mammalian cells (Maeda et al. 2007).

Biological Significance of Fatty Acid Remodeling

Raft association and homodimerization of GPI-APs: Raft association is the prominent characteristic of GPI-APs (Schroeder et al. 1994). GPI-APs expressed on the surface of *PGAP3*- and *PGAP2*-double defective CHO cells were not efficiently recovered in the detergent resistant membrane fraction (Maeda et al. 2007). Similarly inefficient recovery of GPI-APs into the detergent resistant membrane fraction was seen in peritoneal macrophages, spleen T-lymphocytes and embryonic fibroblasts from *Pgap3*-knockout mice (Murakami et al. 2012a; Wang et al. 2013). This profile of GPI-APs lacking fatty acid remodeling is compatible with the idea that two saturated fatty chains are required for raft association of GPI-APs and hence the presence of unsaturated chain is inhibitory (Schroeder et al. 1994).

It was reported that GPI-APs form transient homodimers on the cell surface with a life-time of 200 ms and that the homodimers are the major state on unstimulated cells.

For homodimerization, both protein and GPI parts are important. Upon ligand binding, these minimal rafts of GPI-APs make clusters with signaling capability (Suzuki et al. 2012). We confirmed homodimerization of GPI-APs and demonstrated that the fatty acid remodeling is necessary for the dimerization (Seong et al. 2013).

PGAP2 and PGAP3 deficiencies: Hypomorphic mutations were found in *PGAP2* gene in nine individuals with hyperphosphatasia with mental retardation syndrome (HPMRS, also termed Mabry syndrome) and non-syndromic intellectual disability, mainly by whole-exome sequencing (Hansen et al. 2013; Krawitz et al. 2013). HPMRS is an autosomal recessive disorder characterized by intellectual disability and elevated levels of serum alkaline phosphatase (alkaline phosphatases are GPI-APs), often accompanied by seizures, facial dysmorphism, and various anomalies such as brachytelephalangy. In 2010, Krawitz and colleagues identified hypomorphic mutations in *PIGV* by whole-exome sequencing of DNA samples from three patients with HPMRS (Krawitz et al. 2010). Blood granulocytes from some of the patients with *PIGV* mutations had partially reduced surface expression of CD16, a GPI-AP. Using an assay in which mutant *PIGV* cDNAs were transfected into *PIGV*-defective CHO cells to determine the ability to restore the surface expression of GPI-APs by flow cytometry, the functional effects of the mutations on PIG-V function were found to cause a partial loss of functional activity. Murakami et al. proposed a mechanism for the hyperphosphatasia based on an in vitro study with *PIGV*-defective CHO cells (Murakami et al. 2012b). In the ER of the *PIGV*-defective cells, the C-terminal GPI attachment signal peptide of a nascent protein, such as alkaline phosphatase, is cleaved by GPI transamidase and the major part of the protein is either secreted without GPI-anchoring or degraded by ER-associated degradation. The secretion accounts for the high serum levels of alkaline phosphatase. Subsequently, hypomorphic mutations in *PIGO* (Krawitz et al. 2012) and *PIGW* (Chiyonobu et al. 2014) were found in individuals with HPMRS.

It is not known due to unavailability of cell samples from affected individuals whether the surface levels of GPI-APs on cells from individuals with *PGAP2* mutations are decreased. The hyperphosphatasia is an indication that GPI-APs are released from the cells. The mechanism of the hyperphosphatasia in *PGAP2-deficiency* must be different from that in *PIGV*-deficiency. Alkalinephosphatase released from *PIGV*-defective cells has never been modified by GPI whereas one released from *PGAP2*-defective cells is once GPI-anchored and after cell surface expression is released. As Tashima et al. showed for *PGAP2*-defective CHO cells, when only the removal of the sn2-linked fatty acid by PGAP3 occurred because of inefficient PGAP2-dependent reacylation during fatty acid remodeling in the Golgi, GPI-APs became lyso-GPI-APs, and were transported to the cell surface and secreted (Tashima et al. 2006).

Four mutations in *PGAP3* were identified in five individuals with HPMRS (Howard et al. 2014). All four mutations caused severe reduction in cellular PGAP3 function either by mislocalization of the mutant PGAP3 proteins to the ER, a loss of activity, or non-sense mediated mRNA decay. In one of the individuals, reduction in the surface level of CD16 was confirmed. The exact mechanisms of reduction in the cell surface levels of GPI-APs and release of GPI-APs (hyperphosphatasia) in *PGAP3* deficiency are unclear at the moment. But GPI-APs bearing unremodeled fatty acids are not well associated with lipid rafts and may be released under some unknown conditions. Therefore, the mechanisms of secretion or hyperphosphatasia in *PIGV*-defective cells, *PGAP3*-defective cells and *PGAP2*-defective cells are different.

These results from studies on *PGAP2*- and *PGAP3*-deficiencies indicate that proper fatty acid remodeling in the Golgi is critical for stable cell surface expression of GPI-APs. Impairment in the fatty acid remodeling causes abnormalities in neuronal functions, such as intellectual disability and seizures.

References

Ashida H, Hong Y, Murakami Y, Shishioh N, Sugimoto N, Kim YU, Maeda Y, Kinoshita T (2005) Mammalian PIG-X and yeast Pbn1p are the essential components of glycosylphosphatidylinositol-mannosyltransferase I. Mol Biol Cell 16(3):1439–1448. doi:10.1091/mbc.E04-09-0802

Chiyonobu T, Inoue N, Morimoto M, Kinoshita T, Murakami Y (2014) Glycosylphosphatidylinositol (GPI) anchor deficiency caused by mutations in PIGW is associated with West syndrome and hyperphosphatasia with mental retardation syndrome. J Med Genet 51(3):203–207. doi:10.1136/jmedgenet-2013-102156

Fujita M, Maeda Y, Ra M, Yamaguchi Y, Taguchi R, Kinoshita T (2009) GPI glycan remodeling by PGAP5 regulates transport of GPI-anchored proteins from the ER to the Golgi. Cell 139(2):352–365. doi:10.1016/j.cell.2009.08.040

Fujita M, Watanabe R, Jaensch N, Romanova-Michaelides M, Satoh T, Kato M, Riezman H, Yamaguchi Y, Maeda Y, Kinoshita T (2011) Sorting of GPI-anchored proteins into ER exit sites by p24 proteins is dependent on remodeled GPI. J Cell Biol 194(1):61–75. doi:10.1083/jcb.201012074

Hansen L, Tawamie H, Murakami Y, Mang Y, ur Rehman S, Buchert R, Schaffer S, Muhammad S, Bak M, Nothen MM, Bennett EP, Maeda Y, Aigner M, Reis A, Kinoshita T, Tommerup N, Baig SM, Abou Jamra R (2013) Hypomorphic mutations in PGAP2, encoding a GPI-anchor-remodeling protein, cause autosomal-recessive intellectual disability. Am J Hum Genet 92(4):575–583. doi:10.1016/j.ajhg.2013.03.008

Hong Y, Maeda Y, Watanabe R, Ohishi K, Mishkind M, Riezman H, Kinoshita T (1999) Pig-n, a mammalian homologue of yeast Mcd4p, is involved in transferring phosphoethanolamine to the first mannose of the glycosylphosphatidylinositol. J Biol Chem 274:35099–35106

Hong Y, Maeda Y, Watanabe R, Inoue N, Ohishi K, Kinoshita T (2000) Requirement of PIG-F and PIG-O for transferring phosphoethanolamine to the third mannose in glycosylphosphatidylinositol. J Biol Chem 275:20911–20919

Hong Y, Ohishi K, Kang JY, Tanaka S, Inoue N, Nishimura J, Maeda Y, Kinoshita T (2003) Human PIG-U and yeast Cdc91p are the fifth subunit of GPI transamidase that attaches GPI-anchors to proteins. Mol Biol Cell 14:1780–1789

Houjou T, Hayakawa J, Watanabe R, Tashima Y, Maeda Y, Kinoshita T, Taguchi R (2007) Changes in molecular species profiles of glycosylphosphatidylinositol-anchor precursors in early stages of biosynthesis. J Lipid Res 48:1599–1606

Howard MF, Murakami Y, Pagnamenta AT, Daumer-Haas C, Fischer B, Hecht J, Keays DA, Knight SJ, Kolsch U, Kruger U, Leiz S, Maeda Y, Mitchell D, Mundlos S, Phillips JA 3rd, Robinson PN, Kini U, Taylor JC, Horn D, Kinoshita T, Krawitz PM (2014) Mutations in PGAP3 impair GPI-anchor maturation, causing a subtype of hyperphosphatasia with mental retardation. Am J Hum Genet 94(2):278–287. doi:10.1016/j.ajhg.2013.12.012

Inoue N, Kinoshita T, Orii T, Takeda J (1993) Cloning of a human gene, PIG-F, a component of glycosylphosphatidylinositol anchor biosynthesis, by a novel expression cloning strategy. J Biol Chem 268:6882–6885

Kang JY, Hong Y, Ashida H, Shishioh N, Murakami Y, Morita YS, Maeda Y, Kinoshita T (2005) PIG-V involved in transferring the second mannose in glycosylphosphatidylinositol. J Biol Chem 280(10):9489–9497. doi:10.1074/jbc.M413867200

Kanzawa N, Maeda Y, Ogiso H, Murakami Y, Taguchi R, Kinoshita T (2009) Peroxisome dependency of alkyl-containing GPI-anchor biosynthesis in the endoplasmic reticulum. Proc Natl Acad Sci U S A 106(42):17711–17716. doi:10.1073/pnas.0904762106

Kanzawa N, Shimozawa N, Wanders RJ, Ikeda K, Murakami Y, Waterham HR, Mukai S, Fujita M, Maeda Y, Taguchi R, Fujiki Y, Kinoshita T (2012) Defective lipid remodeling of GPI anchors in peroxisomal disorders, Zellweger syndrome, and rhizomelic chondrodysplasia punctata. J Lipid Res 53(4):653–663. doi:10.1194/jlr.M021204

Kinoshita T, Fujita M, Maeda Y (2008) Biosynthesis, remodelling and functions of mammalian GPI-anchored proteins: recent progress. J Biochem 144(3):287–294. doi:10.1093/jb/mvn090

Kinoshita T, Maeda Y, Fujita M (2013) Transport of glycosylphosphatidylinositol-anchored proteins from the endoplasmic reticulum. Biochim Biophys Acta 1833(11):2473–2478. doi:10.1016/j.bbamcr.2013.01.027

Krawitz PM, Schweiger MR, Rodelsperger C, Marcelis C, Kolsch U, Meisel C, Stephani F, Kinoshita T, Murakami Y, Bauer S, Isau M, Fischer A, Dahl A, Kerick M, Hecht J, Kohler S, Jager M, Grunhagen J, de Condor BJ, Doelken S, Brunner HG, Meinecke P, Passarge E, Thompson MD, Cole DE, Horn D, Roscioli T, Mundlos S, Robinson PN (2010) Identity-by-descent filtering of exome sequence data identifies PIGV mutations in hyperphosphatasia mental retardation syndrome. Nat Genet 42(10):827–829. doi:10.1038/ng.653

Krawitz PM, Murakami Y, Hecht J, Kruger U, Holder SE, Mortier GR, Delle Chiaie B, De Baere E, Thompson MD, Roscioli T, Kielbasa S, Kinoshita T, Mundlos S, Robinson PN, Horn D (2012) Mutations in PIGO, a member of the GPI-anchor-synthesis pathway, cause hyperphosphatasia with mental retardation. Am J Hum Genet 91(1):146–151. doi:10.1016/j.ajhg.2012.05.004

Krawitz PM, Murakami Y, Riess A, Hietala M, Kruger U, Zhu N, Kinoshita T, Mundlos S, Hecht J, Robinson PN, Horn D (2013) PGAP2 mutations, affecting the GPI-anchor-synthesis pathway, cause hyperphosphatasia with mental retardation syndrome. Am J Hum Genet 92(4):584–589. doi:10.1016/j.ajhg.2013.03.011

Maeda Y, Watanabe R, Harris CL, Hong Y, Ohishi K, Kinoshita K, Kinoshita T (2001) PIG-M transfers the first mannose to glycosylphosphatidylinositol on the lumenal side of the ER. EMBO J 20:250–261

Maeda Y, Tashima Y, Houjou T, Fujita M, Yoko-o T, Jigami Y, Taguchi R, Kinoshita T (2007) Fatty acid remodeling of GPI-anchored proteins is required for their raft association. Mol Biol Cell 18(4):1497–1506

Murakami Y, Siripanyapinyo U, Hong Y, Kang JY, Ishihara S, Nakakuma H, Maeda Y, Kinoshita T (2003) PIG-W is critical for inositol acylation but not for flipping of glycosylphosphatidylinositol-anchor. Mol Biol Cell 14(10):4285–4295

Murakami Y, Siripanyaphinyo U, Hong Y, Tashima Y, Maeda Y, Kinoshita T (2005) The initial enzyme for glycosylphosphatidylinositol biosynthesis requires PIG-Y, a seventh component. Mol Biol Cell 16(11):5236–5246

Murakami H, Wang Y, Hasuwa H, Maeda Y, Kinoshita T, Murakami Y (2012a) Enhanced response of T lymphocytes from Pgap3 knockout mouse: insight into roles of fatty acid remodeling of GPI anchored proteins. Biochem Biophys Res Commun 417(4):1235–1241. doi:10.1016/j.bbrc.2011.12.116

Murakami Y, Kanzawa N, Saito K, Krawitz PM, Mundlos S, Robinson PN, Karadimitris A, Maeda Y, Kinoshita T (2012b) Mechanism for release of alkaline phosphatase caused by glycosylphosphatidylinositol deficiency in patients with hyperphosphatasia mental retardation syndrome. J Biol Chem 287(9):6318–6325. doi:10.1074/jbc.M111.331090

Nakamura N, Inoue N, Watanabe R, Takahashi M, Takeda J, Stevens VL, Kinoshita T (1997) Expression cloning of PIG-L, a candidate N-acetylglucosaminyl-phosphatidylinositol deacetylase. J Biol Chem 272:15834–15840

Ohishi K, Inoue N, Maeda Y, Takeda J, Riezman H, Kinoshita T (2000) Gaa1p and gpi8p are components of a glycosylphosphatidylinositol (GPI) transamidase that mediates attachment of GPI to proteins. Mol Biol Cell 11:1523–1533

Ohishi K, Inoue N, Kinoshita T (2001) PIG-S and PIG-T, essential for GPI anchor attachment to proteins, form a complex with GAA1 and GPI8. EMBO J 20:4088–4098

Orlean P, Menon AK (2007) Thematic review series: lipid posttranslational modifications. GPI anchoring of protein in yeast and mammalian cells, or: how we learned to stop worrying and love glycophospholipids. J Lipid Res 48(5):993–1011

Schroeder R, London E, Brown D (1994) Interactions between saturated acyl chains confer detergent resistance on lipids and glycosylphosphatidylinositol (GPI)-anchored proteins: GPI-anchored proteins in liposomes and cells show similar behavior. Proc Natl Acad Sci U S A 91(25):12130–12134

Seong J, Wang Y, Kinoshita T, Maeda Y (2013) Implications of lipid moiety in oligomerization and immunoreactivities of GPI-anchored proteins. J Lipid Res 54(4):1077–1091. doi:10.1194/jlr.M034421

Shishioh N, Hong Y, Ohishi K, Ashida H, Maeda Y, Kinoshita T (2005) GPI7 is the second partner of PIG-F and involved in modification of glycosylphosphatidylinositol. J Biol Chem 280(10):9728–9734

Suzuki KG, Kasai RS, Hirosawa KM, Nemoto YL, Ishibashi M, Miwa Y, Fujiwara TK, Kusumi A (2012) Transient GPI-anchored protein homodimers are units for raft organization and function. Nat Chem Biol 8(9):774–783. doi:10.1038/nchembio.1028

Takahashi M, Inoue N, Ohishi K, Maeda Y, Nakamura N, Endo Y, Fujita T, Takeda J, Kinoshita T (1996) PIG-B, a membrane protein of the endoplasmic reticulum with a large lumenal domain, is involved in transferring the third mannose of the GPI anchor. EMBO J 15:4254–4261

Tanaka S, Maeda Y, Tashima Y, Kinoshita T (2004) Inositol deacylation of glycosylphosphatidylinositol-anchored proteins is mediated by mammalian PGAP1 and yeast Bst1p. J Biol Chem 279:14256–14263

Tashima Y, Taguchi R, Murata C, Ashida H, Kinoshita T, Maeda Y (2006) PGAP2 is essential for correct processing and stable expression of GPI-anchored proteins. Mol Biol Cell 17(3): 1410–1420

Wang Y, Murakami Y, Yasui T, Wakana S, Kikutani H, Kinoshita T, Maeda Y (2013) Significance of glycosylphosphatidylinositol-anchored protein enrichment in lipid rafts for the control of autoimmunity. J Biol Chem 288(35):25490–25499. doi:10.1074/jbc.M113.492611

Watanabe R, Ohishi K, Maeda Y, Nakamura N, Kinoshita T (1999) Mammalian PIG-L and its yeast homologue Gpi12p are N-acetylglucosaminylphosphatidylinositol de-N-acetylases essential in glycosylphosphatidylinositol biosynthesis. Biochem J 339:185–192

Watanabe R, Murakami Y, Marmor MD, Inoue N, Maeda Y, Hino J, Kangawa K, Julius M, Kinoshita T (2000) Initial enzyme for glycosylphosphatidylinositol biosynthesis requires PIG-P and is regulated by DPM2. EMBO J 19:4402–4411

Chapter 3
Novel Insights in Membrane Biology Utilizing Fluorescence Recovery After Photobleaching

Amitabha Chattopadhyay and Md. Jafurulla

Abbreviations

25-NBD-cholesterol	25-[*N*-[(7-nitrobenz-2-oxa-1,3-diazol-4-yl)methyl]amino]-27-norcholesterol
5-HT_{1A} receptor	5-Hydroxytryptamine-1A receptor
5-HT_{1A}R-EYFP	5-Hydroxytryptamine-1A receptor tagged to enhanced yellow fluorescent protein
$DiIC_{18}(3)$	1,1′-Dioctadecyl-3,3,3′,3′,-tetramethylindocarbocyanine perchlorate
EYFP	Enhanced yellow fluorescent protein
FAST DiI	1,1′-Dilinoleyl-3,3,3′,3′,-tetramethylindocarbocyanine 4-chlorobenzenesulfonate
FRAP	Fluorescence recovery after photobleaching
GFP	Green fluorescent protein
GPCR	G protein-coupled receptor
NBD-PE	1,2-Dipalmitoyl-*sn*-glycero-3-phosphoethanolamine-*N*-(7-nitrobenz-2-oxa-1,3-diazol-4-yl)
p-MPPI	4-(2′-Methoxy)phenyl-1-[2′-(*N*-2″-pyridinyl)-*p*-iodobenzamido]ethylpiperazine

A. Chattopadhyay (✉) • Md. Jafurulla
CSIR—Centre for Cellular and Molecular Biology, Uppal Road, Hyderabad 500 007, India
e-mail: amit@ccmb.res.in

A. Chakrabarti, A. Surolia (eds.), *Biochemical Roles of Eukaryotic Cell Surface Macromolecules*, Advances in Experimental Medicine and Biology 842,
DOI 10.1007/978-3-319-11280-0_3

The Dynamic Membrane

Biological membranes are complex two-dimensional, non-covalent assemblies of a diverse variety of lipids and proteins. They impart an identity to the cell and its organelles and represent an ideal milieu for the proper function of a diverse set of membrane proteins. A unique feature of biological membranes is their characteristic dynamics that gets manifested as lateral and rotational dynamics of the constituent lipids and proteins (Marguet et al. 2006; Baker et al. 2007a). It is becoming increasingly clear that membrane dynamics holds the key to membrane function. For example, the conformational dynamics of membrane receptors (such as G protein-coupled receptors (GPCRs)) is beginning to be appreciated in relation to their function (Nygaard et al. 2013; Schmidt et al. 2014). Understanding cellular signaling by membrane receptors in terms of their lateral dynamics represents a challenging area in contemporary biology (Calvert et al. 2001; Ganguly et al. 2008).

Fluorescence Recovery After Photobleaching

Fluorescence recovery after photobleaching (FRAP) represents a convenient approach to measure lateral (translational) diffusion and is widely used for measuring lateral diffusion of lipids and proteins in membranes (Edidin 1994; Lippincott-Schwartz et al. 2001; Klonis et al. 2002; Hagen et al. 2005). FRAP involves generation of a concentration gradient of fluorescent molecules by irreversibly photobleaching a fraction of fluorophores in the observation region (region of interest). The dissipation of this gradient with time owing to diffusion of fluorophores into the bleached region from the unbleached regions of the membrane is an indicator of the mobility of the fluorophores in the membrane. The recovery of fluorescence into the bleached area in FRAP experiments is described by two parameters, an apparent diffusion coefficient (D) and mobile fraction (M_f). The rate of fluorescence recovery provides an estimate of the lateral diffusion coefficient of diffusing molecules, whereas the extent of fluorescence recovery provides an estimate of the mobile fraction (in FRAP time scale). Figure 3.1 illustrates the basic principles of FRAP measurements. In this review, we will provide an overview of the range of research problems that could be addressed in membrane and receptor biology using FRAP, taking representative examples mostly from work carried out in our laboratory. This review is by no means an exhaustive review of FRAP methodology and its application in membrane biology.

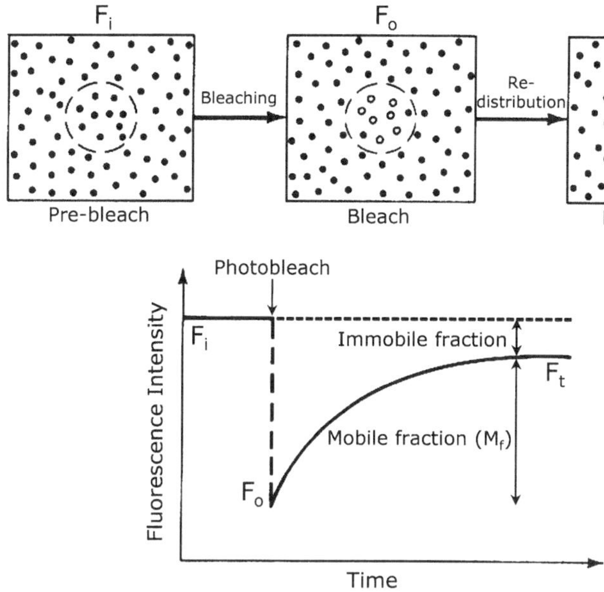

Fig. 3.1 Basic design of FRAP measurements. F_i represents the initial total fluorescence intensity in the region of interest (ROI) prior to photobleaching. A concentration gradient of fluorescent molecules is generated by photobleaching a population of fluorophores in the ROI (shown as a *dashed circle*) using a strong laser beam. F_o represents the total fluorescence intensity in the ROI immediately after photobleaching. The concentration gradient of fluorophores created this way gets dissipated with progress of time due to lateral diffusion of unbleached fluorophores (outside ROI) into the bleached region. F_t represents the total fluorescence intensity in the region at a given time (t) after photobleaching. Careful analysis of the rate of recovery of fluorescence (from F_o to F_t) yields lateral diffusion coefficient (D). Note that the diffusion coefficient obtained in case of membranes represents two-dimensional diffusion (since the membrane is considered to be two-dimensional). The extent of fluorescence recovery provides information on the fraction of molecules that are mobile in this time scale (termed mobile fraction, M_f)

Lipid Dynamics by FRAP

Fluorescently labeled lipid probes are widely used for measuring lipid dynamics in model and natural membranes. The DiI series of lipid analogues are commonly used probes for such measurements. The DiI analogues are composed of a polar indocarbocyanine headgroup and two hydrophobic alkyl chains (see Fig. 3.2) which impart an overall amphiphilic character. They have earlier been shown to preferentially partition into gel (ordered) or fluid (disordered) phases depending on the degree of matching between their acyl chain length and those of lipids that comprise the host membrane (Klausner and Wolf 1980; Spink et al. 1990; Kalipatnapu and Chattopadhyay 2004). DiIC_{18}(3) and *FAST* DiI (Fig. 3.2) represent two such probes

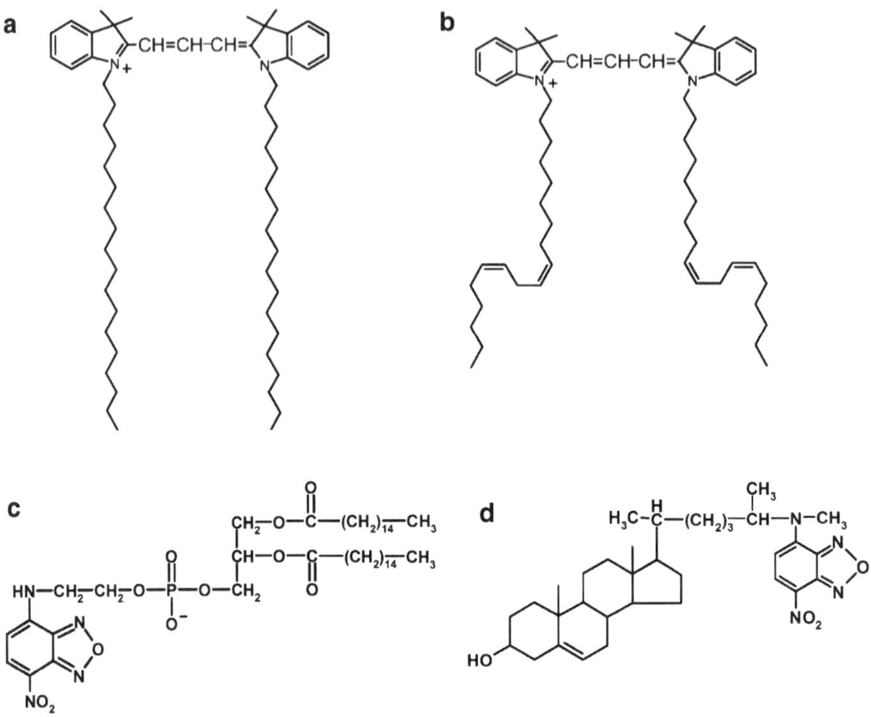

Fig. 3.2 Chemical structures of common fluorescent probes used for measuring lipid dynamics in membranes using FRAP: (**a**) $DiIC_{18}(3)$, (**b**) *FAST* DiI, (**c**) NBD-PE and (**d**) 25-NBD-cholesterol

that are similar in their intrinsic fluorescence properties but differ in their phase partitioning preference. Lateral diffusion characteristics of these probes in native hippocampal membranes have been analyzed in detail using FRAP (Pucadyil and Chattopadhyay 2006). The results show that mobility of these probes in hippocampal membranes varies with membrane cholesterol content. Lateral mobility was found to be higher in cholesterol-depleted membranes. These results could provide insight in the function of neuronal receptors present in these membranes. In another study, *FAST* DiI was used to monitor lateral diffusion in membranes of the wild type and *erg* mutants of the pathogenic yeast, *Candida albicans* (Mukhopadhyay et al. 2004). Interestingly, lipid diffusion in membranes of the wild type and *erg* mutants of *C. albicans* (mutants for ergosterol; *erg2* and *erg16*) correlate well with their drug resistance characteristics. These results represent the first report of analysis of lipid dynamics in *C. albicans* using FRAP. Another interesting application of FRAP to study lipid dynamics is the demonstration of the presence of cholesterol monomers and transbilayer dimers in membranes at low concentration (Pucadyil et al. 2007).

The fluorescent probes used in this case were NBD-PE and 25-NBD-cholesterol (see Fig. 3.2). The NBD group is a commonly used fluorescent lipid probe for studies with model and natural membranes (for a recent review, see Haldar and Chattopadhyay 2013).

GPCR Activation: Manifestations in Receptor Dynamics

The G protein-coupled receptor (GPCR) superfamily is the largest and most diverse protein family in mammals, involved in signal transduction across membranes (Pierce et al. 2002; Rosenbaum et al. 2009). GPCRs are seven transmembrane domain proteins and include >800 members which are encoded by ~5 % of human genes (Zhang et al. 2006). Since GPCRs regulate multiple cellular processes, they have emerged as major targets for the development of novel drug candidates in all clinical areas (Heilker et al. 2009). It is estimated that ~50 % of clinically prescribed drugs act as ligands of GPCRs (Schlyer and Horuk 2006). The serotonin$_{1A}$ (5-HT$_{1A}$) receptor is a representative member of the GPCR family and is implicated in the generation and modulation of various cognitive, behavioral and developmental functions (Pucadyil et al. 2005; Kalipatnapu and Chattopadhyay 2007; Müller et al. 2007). Ligands that bind to the serotonin$_{1A}$ receptor are reported to possess potential therapeutic effects in anxiety or stress-related disorders (Pucadyil et al. 2005). As a consequence, the serotonin$_{1A}$ receptor serves as an important target in the development of therapeutic agents for neuropsychiatric disorders such as anxiety and depression (Celada et al. 2013).

Signaling by GPCRs provides an efficient way for cells to communicate with each other and with their environment. This is achieved through the activation of GPCRs upon binding of ligands present in the extracellular environment that leads to transduction of signals to the interior of the cell through concerted changes in the transmembrane helices (Nygaard et al. 2013). Ligand stimulation of GPCRs generally leads to the recruitment and activation of the heterotrimeric G-proteins. The activation process stimulates the GDP-GTP exchange leading to the dissociation of the GTP-bound α-subunit and the $\beta\gamma$-dimer of the G-protein from the GPCR. This activation could lead to dissociation of G-proteins from the receptors, increasing receptor diffusion. This was validated by FRAP measurements of the serotonin$_{1A}$ receptor tagged to enhanced yellow fluorescent protein (5-HT$_{1A}$R-EYFP) upon activation of the receptor (Pucadyil et al. 2004; Pucadyil and Chattopadhyay 2007a). Figure 3.3 shows a representative FRAP experiment with 5-HT$_{1A}$R-EYFP in CHO cells. The results show that activation with the natural agonist serotonin resulted in a significant increase in the diffusion coefficient of the serotonin$_{1A}$ receptor, while treatment with the antagonist *p*-MPPI did not exhibit any significant difference (see Fig. 3.4). Interestingly, the increase in the diffusion

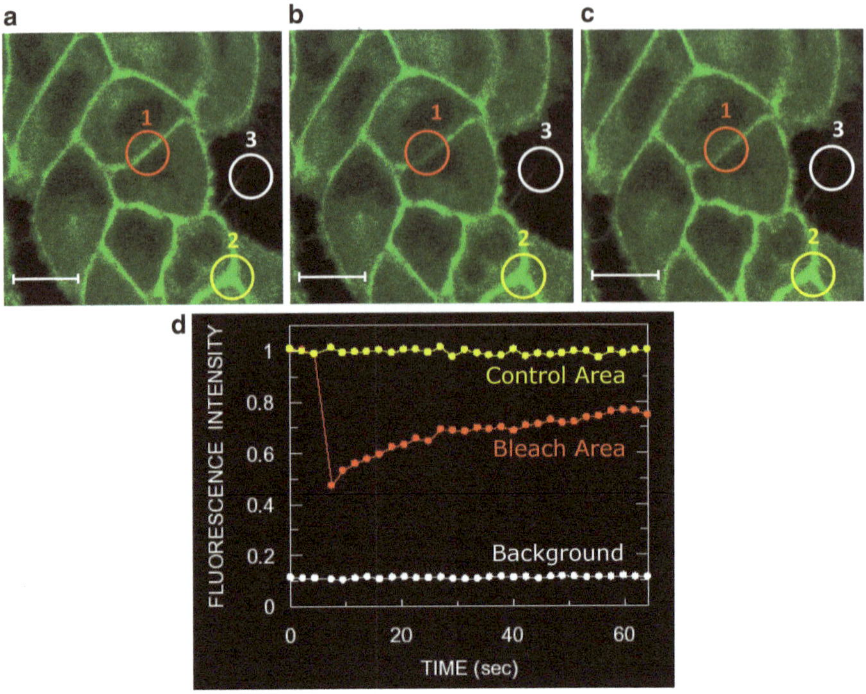

Fig. 3.3 Lateral dynamics of the serotonin$_{1A}$ receptor tagged to enhanced yellow fluorescent protein (5-HT$_{1A}$R-EYFP) stably expressed in CHO cells. The cellular periphery with distinct plasma membrane localization of 5-HT$_{1A}$R-EYFP was chosen for FRAP measurements. Typical images corresponding to (**a**) pre-bleach, (**b**) bleach and (**c**) post-bleach are shown. Regions 1, 2 and 3 represent bleach area, control area and background, respectively. The scale bar represents 10 μm. The plot in (**d**) shows a representative set of normalized fluorescence intensity of 5-HT$_{1A}$R-EYFP corresponding to regions 1 and 2, and normalized background intensity in region 3. The normalized fluorescence intensity in control area (2) was monitored for same duration of time, and shows no significant photobleaching

coefficient with serotonin could be reversed upon addition of *p*-MPPI. The observed increase in receptor diffusion coefficient upon stimulation with the agonist (but not with the antagonist) clearly suggested that activation of G-proteins resulted in an increase in mobility of the receptor. This was further supported by an increase in diffusion coefficient of the receptor in presence of mastoparan and AlF_4^- (see Fig. 3.4), both of which activate G-proteins in a receptor-independent manner. In addition, treatment of cells with pertussis toxin (PTX), that abolishes receptor and G-protein interaction, resulted in an increase in diffusion coefficient of the receptor. Taken together, these results show that receptor diffusion is dependent on its interaction with G-proteins.

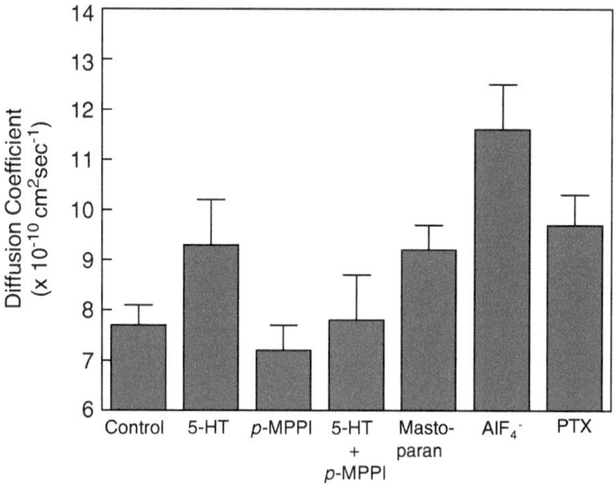

Fig. 3.4 Lateral diffusion coefficients of 5-HT_{1A}R-EYFP under various conditions. Serotonin (5-HT) and *p*-MPPI act as agonist and antagonist of the serotonin$_{1A}$ receptor. Both mastoparan and AlF_4^- activate G-proteins in a receptor-independent manner, whereas pertussis toxin (PTX) inactivates G-proteins of $G_{i/o}$ subtype, thereby abrogating interaction of G-proteins with the receptor. Note that the diffusion coefficient of the receptor exhibits an increase upon activation of G-proteins, irrespective of whether G-proteins are activated in receptor-dependent or independent manner. Adapted and modified from Pucadyil et al. (2004)

Dynamic Confinement of GPCRs Upon Cholesterol Depletion: Insight from Bleach Area-Dependent FRAP

An interesting source of cell membrane heterogeneity (domain) is the relative confinement of membrane components. From this perspective, cellular signaling could be viewed as a consequence of differential mobility of the various interacting partners (Peters 1988). The fluorescence recovery kinetics in FRAP measurements contains information on the area being monitored. This provides a handle to explore spatial organization of molecules in the cell membrane by systematically varying the area monitored in FRAP measurements (Edidin 1992). Differences in diffusion properties obtained from FRAP measurements performed with bleach areas of different sizes can be correlated to the presence of domains on the cell membrane, with dimensions that fall in the same range as the area monitored in these measurements (Yechiel and Edidin 1987; Edidin and Stroynowski 1991; Salomé et al. 1998; Cézanne et al. 2004; Baker et al. 2007b; Saulière-Nzeh Ndong et al. 2010). This interpretation is based on the following model (see below), and was earlier validated by simulations and FRAP experiments performed on physically domainized model membrane systems (Salomé et al. 1998).

The rate of fluorescence recovery provides an estimate of the apparent diffusion coefficient of molecules, while the extent to which fluorescence recovers provides an estimate of mobile fraction of molecules. In general, for molecules diffusing in

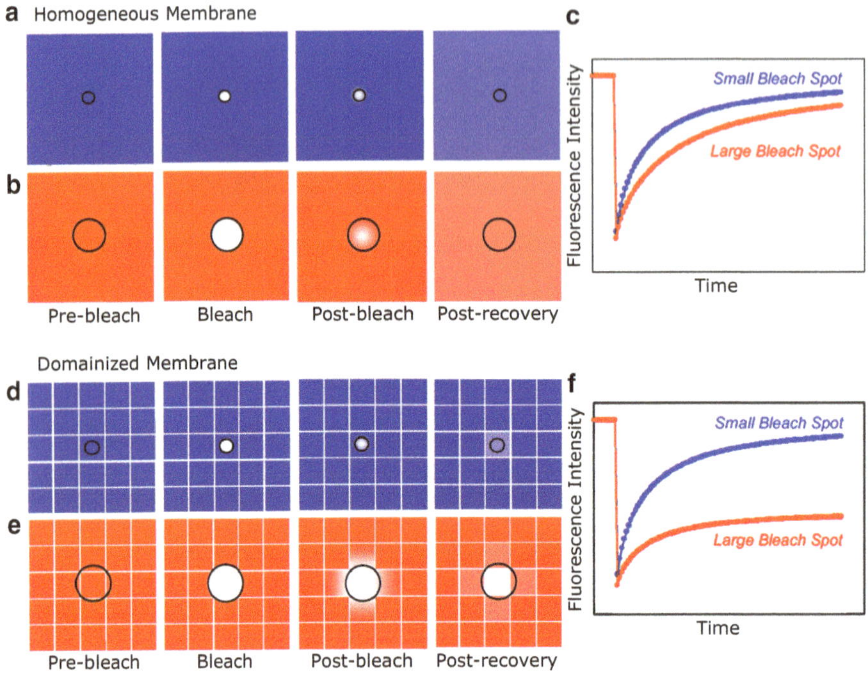

Fig. 3.5 Fluorescence recovery plots with a small or large bleach area performed on homogeneous or domainized membranes. The region of interest for FRAP is represented by a *circle*. The homogeneous membrane is characterized by random diffusion throughout the total area of the membrane in the experimental time scale. In contrast, diffusion on the domainized membrane is confined to closed areas (of comparable dimension as that of the bleached area) termed as 'domains'. The diffusion coefficient and mobile fraction in homogeneous membranes (panels (**a**) and (**b**)) would be independent of the size of the bleach area (see panel (**c**)). In contrast, these parameters would depend on the bleach area size in case of a domainized membrane (panel (**d**) and (**e**)). FRAP measurements on such a domainized membrane therefore would show an increase in diffusion coefficient and reduction in mobile fraction with increasing bleach area size (panel (**f**)). See text for details. Adapted and modified from Pucadyil and Chattopadhyay (2007b)

a homogeneous membrane, the diffusion coefficient is independent of the dimensions of the bleach area in FRAP measurements. A small bleach area (see Fig. 3.5a) would result in faster recovery of fluorescence while a large bleach area (Fig. 3.5b) would produce a slower fluorescence recovery. Yet, the rate of fluorescence recovery would be same in both cases, irrespective of the size of the bleach area. This means that the diffusion coefficient would remain same in both cases. In addition, if the bleached area is significantly smaller than the total area of the membrane, the extent of fluorescence recovery is the same in both cases resulting in a constant mobile fraction (Fig. 3.5c).

On the other hand, if diffusion was confined to closed domains of dimensions of the same scale as that of the bleach area, and static in FRAP time scale, diffusion coefficient would no longer be constant. A small bleach area (Fig. 3.5d) would tend

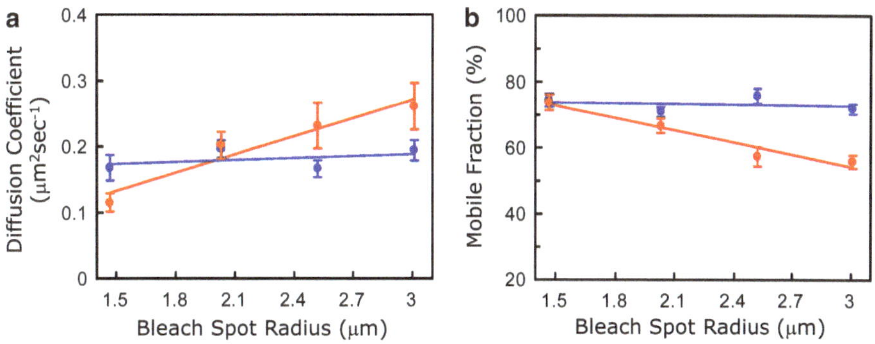

Fig. 3.6 Diffusion parameters from FRAP measurements with varying bleach area size. The (**a**) diffusion coefficient and (**b**) mobile fraction of 5-HT_{1A}R-EYFP obtained are shown for normal (*blue line*) and cholesterol-depleted (*red line*) cells. Adapted and modified from Pucadyil and Chattopadhyay (2007b)

to monitor diffusion properties of molecules within domains. Fluorescence recovery kinetics with a small bleach area on a domainized (heterogeneous) membrane therefore would be similar to that observed in a homogeneous membrane. On the other hand, a large bleach area (overlapping different domains to varying extents, shown in Fig. 3.5e) would result in non-uniform bleaching of domains since the bleached area would be partial for a few and complete for others. As a consequence, fluorescence recovery kinetics in the entire region of observation would not be proportional to the actual size of the bleach area. While kinetics of fluorescence recovery within domains would be proportional to the area bleached in these domains, the apparent diffusion coefficient would show an increase (since diffusion coefficient is calculated taking into account the actual size of the bleach area). Importantly, a large bleach area would reduce mobile fraction since it could bleach an entire domain resulting in total loss of fluorescence in such a domain (Fig. 3.5e, f).

Analysis of fluorescence recovery kinetics of 5-HT_{1A}R-EYFP in CHO cells with bleach areas of different sizes exhibited relatively constant diffusion coefficient and mobile fraction (Pucadyil and Chattopadhyay 2007b; see Fig. 3.6). This suggests that serotonin$_{1A}$ receptors experience a homogeneous membrane environment. Interestingly, FRAP experiments performed on cholesterol-depleted cells with an identical range of bleach area size showed a marked dependence of diffusion coefficient and mobile fraction of the receptor on the dimension of the bleach area (see Fig. 3.6). This characteristic dependence of diffusion coefficient and mobile fraction in cholesterol-depleted membranes is consistent with a model describing confined diffusion in a domainized membrane (see Fig. 3.5c, d) (Yechiel and Edidin 1987; Edidin and Stroynowski 1991; Salomé et al. 1998; Cézanne et al. 2004; Baker et al. 2007b; Saulière-Nzeh Ndong et al. 2010). The dependence of the lateral diffusion parameters on the bleach area size in cholesterol-depleted cells indicates that cholesterol depletion induces dynamic confinement of the receptor resulting in confined diffusion into domains.

Are Signaling and Dynamics Correlated?

Cellular signaling has been hypothesized to be a consequence of differential mobility of various interacting components. This forms the basis of the 'mobile receptor' hypothesis, which proposes that receptor-effector interactions at the plasma membrane are controlled by lateral mobility of the interacting components (Kahn 1976; Peters 1988). Although conceptually elegant, this hypothesis has been difficult to validate experimentally. This was addressed by monitoring lateral mobility of 5-HT_{1A}R-EYFP utilizing FRAP and measuring downstream signaling by the reduction in cellular cAMP level upon activation of the receptor under the same condition (Ganguly et al. 2008). Lateral diffusion of membrane lipids and proteins is known to be influenced by cytoskeletal proteins. Upon destabilization of the actin cytoskeleton by increasing concentrations of cytochalasin D, the mobile fraction of the receptor showed a significant increase, whereas diffusion coefficient remained constant (see Fig. 3.7a, b). This was accompanied by an increase in signaling by the receptor, as measured by reduction in cAMP (Fig. 3.7c). The fact that the change in

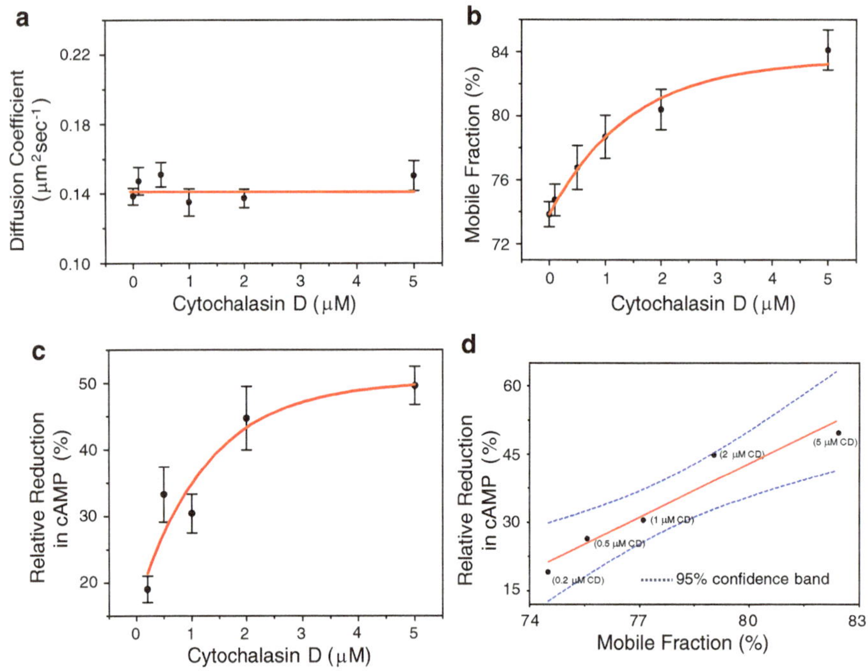

Fig. 3.7 A tight correlation between receptor dynamics and signaling. Effect of increasing cytoskeletal destabilization on (**a**) diffusion coefficient, (**b**) mobile fraction and (**c**) agonist-mediated signaling of the serotonin$_{1A}$ receptor. Cytoskeletal destabilization was achieved by treatment with cytochalasin D. Panel (**d**) shows that signaling of the receptor is strongly correlated with its dynamics (mobile fraction), with a correlation coefficient (r) ~ 0.95. Adapted and modified from Ganguly et al. (2008)

signaling was correlated with the change in receptor dynamics was supported by a positive correlation of ~0.95 obtained from a plot of these two parameters (see Fig. 3.7d). Such a tight correlation between the mobile fraction of the receptor and its signaling is supportive of the mobile receptor hypothesis.

Lateral Dynamics as Readout of Infection

The above example shows that lateral dynamics could be correlated with cellular signaling (Ganguly et al. 2008). Interestingly, a few studies have highlighted the correlation of lateral dynamics of host cell membrane proteins to infection by obligate intracellular parasites. For example, lateral dynamics has been related to the stage of infection of intracellular obligate parasites such as *Plasmodium falciparum*. In an elegant study, Parker et al. (2004) showed that the lateral diffusion coefficient and mobile fraction of host erythrocyte proteins (such as band 3 and glycophorin) depend on the stage of the infection. The diffusion coefficient and mobile fraction of these proteins were reported to be lower for mature stage-infected cells compared to ring stage-infected cells. The corresponding values of diffusion parameters were found to be the highest in case of uninfected cells. This observation points out the potential of lateral dynamics as an indicator of progress of infection. In another study, HIV-1 fusion and entry into target cells have been shown to be dependent on the lateral mobility of CD4 receptors (which serve as one of the receptors for viral entry) in host cell membranes (Rawat et al. 2008).

Conclusion and Future Perspectives

Although we have discussed only representative examples of the application of FRAP in membrane and receptor biology, it is clear that this approach is capable of providing a variety of information depending on experimental design and question asked. With the advent of confocal microscopy and our ability to optically section the cellular interior, FRAP is being increasingly used to explore dynamics of intracellular organelles (Lippincott-Schwartz et al. 2001; Aguila et al. 2011; Staras et al. 2013) using reporters such as GFP (Haldar and Chattopadhyay 2009). This is an exciting area of research and was not possible a few years back. A particularly exciting application is dynamics of nuclear proteins using FRAP (Dundr and Misteli 2003; Mariappan and Parnaik 2005). We envision that future applications of FRAP will involve generating a dynamic map of intracellular components and their modulation with differentiation and development, thereby enabling a novel dynamic view of cellular signaling and function in healthy and diseased states.

Acknowledgments We dedicate this paper to Prof. Michael Edidin (The Johns Hopkins University, Baltimore, MD) who pioneered the application of FRAP in biological membranes and in whose laboratory one of us (A.C.) learnt the nuts and bolts of FRAP measurements during a visit as a CSIR-Raman Fellow. Work in A.C.'s laboratory was supported by the Council of Scientific and Industrial Research, Govt. of India. A.C. is an Adjunct Professor at the Special Centre for Molecular Medicine of Jawaharlal Nehru University (New Delhi, India) and Indian Institute of Science Education and Research (Mohali, India), and Honorary Professor of the Jawaharlal Nehru Centre for Advanced Scientific Research (Bangalore, India). A.C. gratefully acknowledges J.C. Bose Fellowship (Dept. of Science and Technology, Govt. of India). Some of the work described in this article was carried out by former members of A.C.'s group whose contributions are gratefully acknowledged. We thank members of our laboratory for critically reading the manuscript.

References

Aguila B, Simaan M, Laporte SA (2011) Study of G protein-coupled receptor/β-arrestin interactions within endosomes using FRAP. Methods Mol Biol 756:371–380

Baker A, Saulière A, Dumas F, Millot C, Mazères S, Lopez A, Salomé L (2007a) Functional membrane diffusion of G-protein coupled receptors. Eur Biophys J 36:849–860

Baker A-M, Saulière A, Gaibelet G, Lagane B, Mazères S, Fourage M, Bachelerie F, Salomé L, Lopez A, Dumas F (2007b) CD4 interacts constitutively with multiple CCR5 at the plasma membrane of living cells. A fluorescence recovery after photobleaching at variable radii approach. J Biol Chem 282:35163–35168

Calvert PD, Govardovskii VI, Krasnoperova N, Anderson RE, Lem J, Makino CL (2001) Membrane protein diffusion sets the speed of rod phototransduction. Nature 411:90–94

Celada P, Bortolozzi A, Artigas F (2013) Serotonin 5-HT_{1A} receptors as targets for agents to treat psychiatric disorders: rationale and current status of research. CNS Drugs 27:703–716

Cézanne L, Lecat S, Lagane B, Millot C, Vollmer J-Y, Matthes H, Galzi J-L, Lopez A (2004) Dynamic confinement of NK2 receptors in the plasma membrane. Improved FRAP analysis and biological relevance. J Biol Chem 279:45057–45067

Dundr M, Misteli T (2003) Measuring dynamics of nuclear proteins by photobleaching. Curr Protoc Cell Biol 18:13.5.1–13.5.18

Edidin M (1992) Patches, posts and fences: proteins and plasma membrane domains. Trends Cell Biol 2:376–380

Edidin M (1994) Fluorescence photobleaching and recovery, FPR, in the analysis of membrane structure and dynamics. In: Damjanovich S, Edidin M, Szöllõsi J, Trón L (eds) Mobility and proximity in biological membranes. CRC, Boca Raton, FL pp 109–135

Edidin M, Stroynowski I (1991) Differences between the lateral organization of conventional and inositol phospholipid-anchored membrane proteins. A further definition of micrometer scale membrane domains. J Cell Biol 112:1143–1150

Ganguly S, Pucadyil TJ, Chattopadhyay A (2008) Actin cytoskeleton-dependent dynamics of the human $serotonin_{1A}$ receptor correlates with receptor signaling. Biophys J 95:451–463

Hagen GM, Roess DA, de León GC, Barisas BG (2005) High probe intensity photobleaching measurement of lateral diffusion in cell membranes. J Fluoresc 15:873–882

Haldar S, Chattopadhyay A (2009) Green fluorescent protein: a molecular lantern that illuminates the cellular interior. J Biosci 34:169–172

Haldar S, Chattopadhyay A (2013) Application of NBD-labeled lipids in membrane and cell biology. In: Mely Y, Duportail G (eds) Springer series on fluorescence, vol 13. Springer, Heidelberg, pp 37–50

Heilker R, Wolff M, Tautermann CS, Bieler M (2009) G-protein-coupled receptor-focused drug discovery using a target class platform approach. Drug Discov Today 14:231–240

Kahn CR (1976) Membrane receptors for hormones and neurotransmitters. J Cell Biol 70: 261–286

Kalipatnapu S, Chattopadhyay A (2004) A GFP fluorescence-based approach to determine detergent insolubility of the human serotonin$_{1A}$ receptor. FEBS Lett 576:455–460
Kalipatnapu S, Chattopadhyay A (2007) Membrane organization and function of the serotonin$_{1A}$ receptor. Cell Mol Neurobiol 27:1097–1116
Klausner RD, Wolf DE (1980) Selectivity of fluorescent lipid analogues for lipid domains. Biochemistry 19:6199–6203
Klonis N, Rug M, Harper I, Wickham M, Cowman A, Tilley L (2002) Fluorescence photobleaching analysis for the study of cellular dynamics. Eur Biophys J 31:36–51
Lippincott-Schwartz J, Snapp E, Kenworthy A (2001) Studying protein dynamics in living cells. Nat Rev Mol Cell Biol 2:444–456
Marguet D, Lenne P-F, Rigneault H, He H-T (2006) Dynamics in the plasma membrane: how to combine fluidity and order. EMBO J 25:3446–3457
Mariappan I, Parnaik VK (2005) Sequestration of pRb by cyclin D3 causes intranuclear reorganization of lamin A/C during muscle cell differentiation. Mol Biol Cell 16:1948–1960
Mukhopadhyay K, Prasad T, Saini P, Pucadyil TJ, Chattopadhyay A, Prasad R (2004) Membrane sphingolipid-ergosterol interactions are important determinants of multidrug resistance in *Candida albicans*. Antimicrob Agents Chemother 48:1778–1787
Müller CP, Carey RJ, Huston JP, De Souza Silva MA (2007) Serotonin and psychostimulant addiction: focus on 5-HT$_{1A}$-receptors. Prog Neurobiol 81:133–178
Nygaard R, Zou Y, Dror RO, Mildorf TJ, Arlow DH, Manglik A, Pan AC, Liu CW, Fung JJ, Bokoch MP, Thian FS, Kobilka TS, Shaw DE, Mueller L, Prosser RS, Kobilka BK (2013) The dynamic process of β_2-adrenergic receptor activation. Cell 152:532–542
Parker PD, Tilley L, Klonis N (2004) *Plasmodium falciparum* induces reorganization of host membrane proteins during intraerythrocytic growth. Blood 103:2404–2406
Peters R (1988) Lateral mobility of proteins and lipids in the red cell membrane and the activation of adenylate cyclase by β-adrenergic receptors. FEBS Lett 234:1–7
Pierce KL, Premont RT, Lefkowitz RJ (2002) Seven-transmembrane receptors. Nat Rev Mol Cell Biol 3:639–650
Pucadyil TJ, Chattopadhyay A (2006) Effect of cholesterol on lateral diffusion of fluorescent lipid probes in native hippocampal membranes. Chem Phys Lipids 143:11–21
Pucadyil TJ, Chattopadhyay A (2007a) The human serotonin$_{1A}$ receptor exhibits G-protein-dependent cell surface dynamics. Glycoconj J 24:25–31
Pucadyil TJ, Chattopadhyay A (2007b) Cholesterol depletion induces dynamic confinement of the G-protein coupled serotonin$_{1A}$ receptor in the plasma membrane of living cells. Biochim Biophys Acta 1768:655–668
Pucadyil TJ, Kalipatnapu S, Harikumar KG, Rangaraj N, Karnik SS, Chattopadhyay A (2004) G-protein-dependent cell surface dynamics of the human serotonin$_{1A}$ receptor tagged to yellow fluorescent protein. Biochemistry 43:15852–15862
Pucadyil TJ, Kalipatnapu S, Chattopadhyay A (2005) The serotonin$_{1A}$ receptor: a representative member of the serotonin receptor family. Cell Mol Neurobiol 25:553–580
Pucadyil TJ, Mukherjee S, Chattopadhyay A (2007) Organization and dynamics of NBD-labeled lipids in membranes analyzed by fluorescence recovery after photobleaching. J Phys Chem B 111:1975–1983
Rawat SS, Zimmerman C, Johnson BT, Cho E, Lockett SJ, Blumenthal R, Puri A (2008) Restricted lateral mobility of plasma membrane CD4 impairs HIV-1 envelope glycoprotein mediated fusion. Mol Membr Biol 25:83–94
Rosenbaum DM, Rasmussen SGF, Kobilka BK (2009) The structure and function of G-protein-coupled receptors. Nature 459:356–363
Salomé L, Cazeils JL, Lopez A, Tocanne JF (1998) Characterization of membrane domains by frap experiments at variable observation areas. Eur Biophys J 27:391–402
Saulière-Nzeh Ndong A, Millot C, Corbani M, Mazères S, Lopez A, Salomé L (2010) Agonist-selective dynamic compartmentalization of human mu opioid receptor as revealed by resolutive FRAP analysis. J Biol Chem 285:14514–14520
Schlyer S, Horuk R (2006) I want a new drug: G-protein-coupled receptors in drug development. Drug Discov Today 11:481–493

Schmidt P, Thomas L, Müller P, Scheidt HA, Huster D (2014) The G-protein-coupled neuropeptide Y receptor type 2 is highly dynamic in lipid membranes as revealed by solid-state NMR spectroscopy. Chemistry 20:4986–4992

Spink CH, Yeager MD, Feigenson GW (1990) Partitioning behavior of indocarbocyanine probes between coexisting gel and fluid phases in model membranes. Biochim Biophys Acta 1023:25–33

Staras K, Mikulincer D, Gitler D (2013) Monitoring and quantifying dynamic physiological processes in live neurons using fluorescence recovery after photobleaching. J Neurochem 126: 213–222

Yechiel E, Edidin M (1987) Micrometer-scale domains in fibroblast plasma membranes. J Cell Biol 105:755–760

Zhang Y, DeVries ME, Skolnick J (2006) Structure modeling of all identified G protein-coupled receptors in the human genome. PLoS Comput Biol 2:e13

Chapter 4
Defects in Erythrocyte Membrane Skeletal Architecture

Avik Basu and Abhijit Chakrabarti

Introduction

From thermodynamic point of view biological cell is an open system, where exchange of both energy and matter can occur through a permeable boundary, called cell membrane. Cell membrane consists of mainly lipids and proteins. According to fluid mosaic model it is a fluid lipid bilayer decorated with mobile protein molecules (Singer and Nicolson 1972). Perhaps this bilayer membrane description is an oversimplified view of the actual cellular boundary. In the truest sense it is not only a bilayer of lipids, but a tripartite complex of extra cellular matrix, plasma membrane and underlying membrane skeleton (MS). In this review we will restrict ourselves in description of MS part and our main focus will be the erythrocyte, as most of the literature available till date is on erythrocyte due to its relative simplicity. But MS and its components are present in almost every kind of cell and they are even present in sub cellular organelle like Golgi complex (Beck and Nelson 1998).

A. Basu
Biophysics and Structural Genomics Division, Saha Institute of Nuclear Physics,
1/AF Bidhannagar, Kolkata 700064, India

A. Chakrabarti, Ph.D. (✉)
Crystallography and Molecular Biology Division, Saha Institute of Nuclear Physics,
1/AF Bidhannagar, Kolkata 700064, India
e-mail: abhijit.chakrabarti@saha.ac.in

A. Chakrabarti, A. Surolia (eds.), *Biochemical Roles of Eukaryotic Cell Surface Macromolecules*, Advances in Experimental Medicine and Biology 842,
DOI 10.1007/978-3-319-11280-0_4

Membrane Skeleton

Underlying the plasma membranes of cells there is a self-assembled network of proteins termed the membrane skeleton. This network plays a major role in conferring a definite shape to the cells and organelles and maintaining mechanical properties of the membrane. Spectrin, ankyrin, actin, tropomyosin and few other proteins are the principal components of the MS. Lateral interactions among these proteins constitute the spectrin-based composite structure that is anchored to the bilayer through vertical interactions.

The MS of cells is likely to interact with a variety of integral membrane proteins and participate both in stable linkages as well as dynamic structures capable of rapid assembly and disassembly. The basis for diversity of roles for MS includes multiple, functionally distinct isoforms of spectrin, ankyrin and other associated proteins, regulation of protein interactions through phosphorylation and calcium/calmodulin, as well as differential expression of accessory proteins that determine the organization and localization of MS in cells. MS proteins are highly conserved from Drosophila to man and is likely to be involved in fundamental aspects of membrane structure requiring long range order and organization.

Membrane lipid bilayer–membrane skeleton interactions are thought to be responsible for the membrane integrity and its mechanical properties (e.g. very high linear elasticity while maintaining negligible extensibility). One of the most studied biological membranes is that of the erythrocyte. During its 120 day life time in the circulation this 8 μm diameter cell has to pass repeatedly through 2 μm capillaries and hence withstand and respond to very strong mechanical stresses. This makes the red cell an appealing model to study. Another reason is the simplicity of the cell and their membranes, i.e., mature mammalian erythrocytes are devoid of a nucleus and any organelles. The remarkable mechanical properties of the red cell membrane stem from the presence on the cytoplasmic surface of a dense, well organized protein network of MS.

The structure of the MS, the mutual interactions of its components and its interactions with membrane proteins in nonerythroid cells are known to a much lesser extent, largely because of their much higher structural complexity. However, since many animal cell membranes contain spectrin and spectrin-binding protein analogues (of erythrocyte membrane proteins), together with novel spectrin-binding proteins, the existence of a similar protein network, tightly associated with membrane proteins, is anticipated (Bennett 1985).

Erythroid Membrane Skeleton

Amongst all, the MS of mammalian erythrocytes was first visualized in electron micrographs of detergent extracted erythrocytes (Yu et al. 1973). Spectrin, actin, protein 4.1, protein 4.2, adducin, dematin, ankyrin, 55 Kd palmitoylated protein (p55), non-muscle tropomyosin and tropomodulin and few other proteins are the principal components of the MS (Mohandas and Gallagher 2008). Lateral interactions among these proteins constitute the spectrin-based composite structure that is anchored to the bilayer through vertical interactions.

The erythrocyte MS is organized as a polygonal network formed by five to seven extended spectrin molecules linked to short actin filaments (Liu et al. 1987; Shen et al. 1986). The spectrin-actin network of erythrocytes is coupled to the membrane bilayer primarily by association of spectrin with ankyrin, which in turn is bound to the cytoplasmic domain of the band 3 (anion exchanger-1) (Bennett and Stenbuck 1979a; Tyler et al. 1980). Band-3 dimers also are associated on their cytoplasmic surface with band 4.2 (Fig. 4.1 Ankyrin complex) (Yu and Steck 1975). Additional membrane connections are provided at the spectrin-actin junction by a complex between proteins 4.1 R, p55 and glycophorin C (Marfatia et al. 1994) (Fig. 4.1 4.1 complex). Several proteins responsible for capping actin and defining the length of actin filaments as well as stabilizing spectrin-actin complexes have been localized to spectrin actin junctions by electron microscopy (Derick et al. 1992). Protein 4.1 stabilizes spectrin-actin complexes (Tyler et al. 1980; Ungewickell et al. 1979). Dematin was initially identified as an endogenous kinase having actin bundling property (Husainchishti et al. 1989),anchors MS to membrane via Glucose transporter-1 with the help of adducin (Khan et al. 2008) and now shown to have roles in modulating spectrin actin interactions (Koshino et al. 2012). A nonmuscle isoform

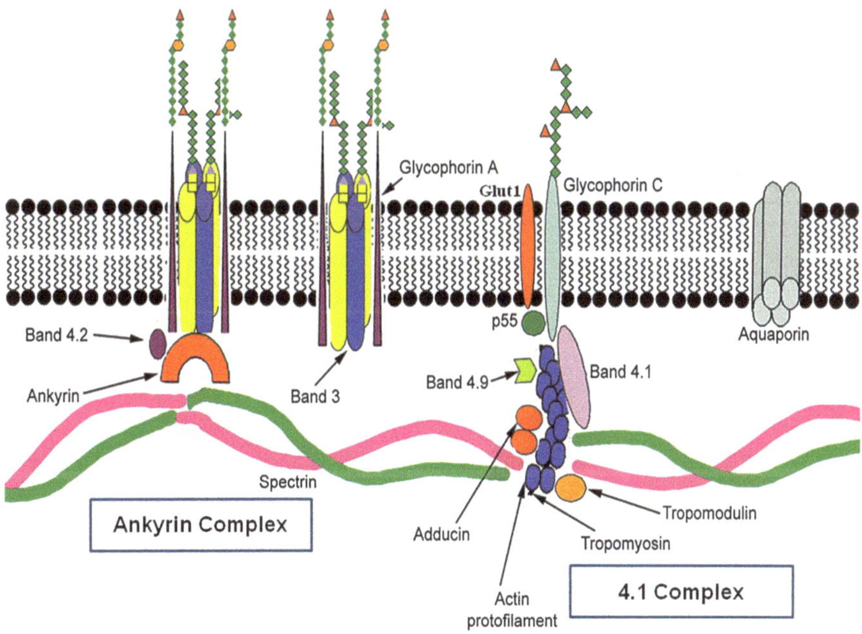

Fig. 4.1 Schematic illustration of our current understanding of erythrocyte membrane and membrane skeleton

of tropomyosin is associated with the sides of actin filaments (Fowler and Bennett 1984). Tropomyosin is of the same length as actin filaments visualized in electron micrographs and is a candidate to function as a morphometric ruler defining the length of actin filaments in erythrocyte membranes. Adducin associates with the fast-growing end of actin filaments in a complex that caps the filament and promotes assembly of spectrin (Gardner and Bennett 1987; Kuhlman et al. 1996). Tropomodulin caps the slow-growing end of actin filaments in a ternary complex involving tropomyosin (Fowler 1990; Weber et al. 1994). Figure 4.1 represents our current understanding of erythrocyte membrane and membrane skeleton organisation.

Idea of the organization of the spectrin-based MS of the human erythrocyte membrane has provided the biochemical equivalent of a high-resolution genetic pathway of interacting membrane structural proteins. The discovery that other tissues express isoforms of spectrin (Bennett et al. 1982) and ankyrin (Bennett 1979) suggested that the erythrocyte MS had a broad relevance for other cell types. However, although the basic structural principles established in erythrocytes are likely to apply in other tissues, the organization, protein interactions, and functions of spectrin based structures are considerably more diverse in other cells.

Lipid Bilayer: Membrane Skeleton Interactions

Though high affinity protein–protein interactions are mostly responsible for MS attachment to the membrane, however, there are many indications coming from various studies on cells, isolated membranes, and model systems that direct protein–lipid interactions contribute to the attachment of the MS to the membrane hydrophobic domain. Gratzer and coworkers showed, by intrinsic fluorescence quenching experiments, that spectrin, 4.1 and ankyrin reveal an affinity for hydrophobic compounds (Kahana et al. 1992). This supported the view that spectrin contains a number of hydrophobic sites. Numerous studies on the interaction of erythrocyte spectrin with membrane bilayer phospholipids from natural (erythrocyte) membranes, liposomes, or monolayer lipid films have been carried out e.g. the addition of liposome suspensions to spectrin solutions causes quenching of its intrinsic tryptophan fluorescence (Michalak et al. 1993; Sikorski et al. 1987), indicative of the lipid binding property of spectrin. The nature of lipid-binding by spectrin seems to be controlled by phosphorylation, since phosphorylation reduces the binding to lipid monolayer. Protein 4.1 was also found to contain hydrophobic region, indicating that it could bind phospholipids. Preferential binding of protein 4.1 to vesicles prepared from PS containing lipid mixtures was demonstrated and the interaction of protein 4.1 with PS-containing membranes was inhibited by Ca^{2+} (Sato and Ohnishi 1983). As outlined above, some of the facts concerning direct interactions of membrane skeletal proteins with membrane lipids have been known for many years. There is, however, increasing amount of data indicating the importance of these interactions as possible mechanisms of regulatory events within the cell.

Assembly of the Membrane-Skeleton of Erythroid Cells

One major principle to emerge from the analysis of the biochemical properties of the individual constituents of the MS is that of simple self assembly whereby all components carry sufficient information in their primary structure to allow them to reassemble spontaneously with each other under defined biochemical conditions in vitro. For example, purified spectrin subunits isolated by denaturation with chaotropic agents will spontaneously reassemble upon renaturation into heterodimers of the correct stoichiometry and antiparallel conformation (Ungewickell and Gratzer 1978; Yoshino and Marchesi 1984). Purified spectrin heterodimers bind with high affinity to ankyrin in vitro through the β-spectrin subunit and in turn this complex can bind to the cytoplasmic domain of the anion transporter band 3, in inverted ankyrin-depleted plasma membrane vesicles or onto purified molecules reconstituted in lipid vesicles (Bennett and Stenbuck 1979b, 1980; Ohanian et al. 1984; Ungewickell et al. 1979). Protein 4.1 will enhance the affinity of spectrin tetramers for actin in vitro in the form of a ternary complex, and it will bind to the cytoplasmic domain of glycophorin C in the presence of phosphoinositides (Marchesi 1985; Ohanian et al. 1984; Tyler et al. 1980).

One characteristic feature of the MS is that in the mature cell it is assembled exclusively on the cytoplasmic side of the plasma membrane. Thus despite the fact that all peripheral MS components can spontaneously self-assemble in vitro, in vivo nucleation of assembly must be spatially controlled during erythroid differentiation to occur only proximal to the plasma membrane without anomalous nucleation in the cytoplasm. This issue becomes more complicated if one considers that the erythroid-specific structure arises from a pre-existing structure inherent to undifferentiated progenitor cells. Therefore, temporal control must also play an important role in the emergence of the ultimate erythroid structural phenotype. Therefore substantially more information than that inherent for simple self-assembly is necessary to allow for formation of an erythroid cell's membrane structure.

The principle theme of this assembly is remodeling of preexisting membrane skeletal structure in early erythroid progenitor cells during differentiation. This is possibly achieved by posttranslational control of spectrin assembly mainly by changing its rate of degradation. The apparently wasteful overproduction of spectrin may be necessary to ensure that adequate amounts of heterodimers assemble onto the MS. Expression of the band 3 also plays a role in temporal control of peripheral membrane-skeleton assembly. Ankyrin and Protein 4.1 express at the later Stages of membrane-skeleton biosynthesis. Finally the expression of Actin and its assembly starts the process of nucleation of the MS.

Despite difficulties, certain important principles that govern this aspect of membrane biogenesis in differentiation have emerged. It is evident that the erythroid cell utilizes substantially more information than simple self-assembly in directing the assembly of its MS during differentiation. Not only does it remodel its preexisting structure present in its progenitors, but it uses both temporal and spatial information in directing assembly. Gradual stabilization of unstably assembled components appears to be one of the overriding mechanisms governing assembly. Inefficient though it may be, this mechanism of assembly may have evolved to maximize assembly of a multicomponent system that has to occur over a period of several days (Lazarides and Woods 1989).

Defects in Membrane Skeleton

Advancement in the characterization of the structure and function of erythrocyte MS proteins and their genes has led to substantial progress in the understanding of the molecular pathology of erythrocyte membrane skeletal disorders and also made possible the definition and characterization of mutations of MS proteins as a well-defined cause of hereditary hemolytic disease. Likewise, understanding of the molecular mechanisms underlying changes in erythrocyte deformability, structural integrity, and shape has advanced and often provide a clue to the pathobiology and diagnosis of the underlying disorder.

Palek and coworkers first suggested dividing all membrane interactions into two categories, vertical and horizontal interactions. Vertical interactions, which are perpendicular to the plane of the membrane, stabilize the lipid bilayer membrane. These interactions include spectrin-ankyrin-band 3 interactions, the protein 4.1-glycophorin C linkage, adducin,-dematin- glucose transporter-1 association and the weak interactions between the skeletal proteins and the lipids of the inner leaflet of the membrane lipid bilayer. On the other hand, horizontal interactions, which are parallel to the plane of the membrane, maintain the structural integrity of erythrocytes during exposure to shear stress. Horizontal interactions involve the site, where spectrin heterodimers assemble into tetramers and the association of spectrin heterodimers with actin and protein 4.1 within the junctional complex. Although the interactions are more complex than this simple horizontal and vertical interactions model, it serves as a useful starting place for understanding erythrocyte membrane protein interactions, particularly in reference to membrane-related disorders.

According to the vertical/horizontal model, Hereditary Spherocytosis (HS) is considered a disorder of vertical interactions. As one common feature of HS erythrocytes is a weakening of the vertical contacts between the skeleton and the overlying lipid bilayer membrane together with its integral proteins. On the other hand in most cases of Hereditary Elliptocytosis (HE), the principal lesion involves horizontal membrane protein associations, primarily spectrin dimer–dimer interactions. In this review we would revisit these two prevalent diseases along with perturbation of MS by malaria parasite.

Hereditary Spherocytosis

Hereditary spherocytosis (HS) is by far the most common congenital hemolytic anemia in northern European descendants, the hallmarks being anemia, intermittent jaundice, and splenomegaly. Although the eminent role of the spleen in the premature hemolysis of red cells in HS is unquestioned, the molecular events that cause splenic conditioning of spherocytes are unclear. Electron micrographs show that small membrane vesicles are shed during the formation of spherocytes.

Molecular Defects in HS

Meticulous work of several research groups has shown that HS is caused mainly by defects in the red cell membrane proteins ankyrin, spectrin, band 3, and protein 4.2. In dominant HS, nonsense and frame shift mutations of ankyrin, band 3, and β spectrin predominate whereas recessive HS is most often due to compound heterozygosity of defects in ankyrin, β spectrin, or protein 4.2. Common combinations include a defect in the promoter or 5′untranslated region of ankyrin paired with a missense mutation, a low expression allele of β spectrin plus a missense mutation, and various mutations in the gene for protein 4.2 (Eber and Lux 2004).

Defect in ankyrin is one of the most prevalent causes of HS, resulting in both dominant and recessive HS with prominent spherocytosis, hemolysis and anemia varying from mild to moderately severity (delGiudice et al. 1996; Randon et al. 1997). Due to its double linkage to β-spectrin and band 3, ankyrin plays a key role in the stabilization of the MS. Ankyrin has high affinity binding site for spectrin heterodimers, which are stable only when bound to the membrane (Woods and Lazarides 1986). Since it is present in limiting amounts, deficit of ankyrin leads to loss of both proteins and the deficiency of one protein is strictly correlated with that of the other and is proportional to clinical severity. Ankyrin mutations are located throughout the molecule, and nearly every family has its own mutation. Frame shift, splicing and nonsense mutations are frequent in dominant HS resulting in either unstable ankyrin transcripts or truncated peptides. In most cases the mutant mRNA is destroyed by nonsense-mediated mRNA decay and no abnormal protein is detectable. Promoter defects that disrupt transcription factor binding sites or insulator function and compound heterozygosity for ankyrin defects are common in recessive HS.

α-spectrin chains are produced in three- to fourfold excess compared with β-spectrin, that makes HS caused by α-spectrin mutations a recessive trait whereas that due to β-spectrin is dominant (Hanspal et al. 1991). A moderate reduction of α-spectrin production, as would be seen in a heterozygote, would not decrease formation of the spectrin α-β dimer. Patients with α-spectrin defects are often compound heterozygous for missense and low expression mutations with a marked reduction of spectrin dimer content. Clinically, only homozygous α-spectrin deficiency causes hemolytic anemia and patients with recessive HS and α-spectrin deficiency are rare (Agre et al. 1982, 1985, del Giudice et al. 2001; Wichterle et al. 1996). Monoallelic expression of β-spectrin occurs frequently in HS patients with spectrin deficiency (del Giudice et al. 1998, 2001; Hassoun et al. 1995), suggesting that null mutations including initiation codon disruption, frameshift and nonsense mutations, gene deletions and splicing defects of β-spectrin are common.

Many band 3 gene mutations, including missense and frame shift mutations, have been described in HS. Conserved arginine residues are frequent sites of mutations; examples include arginines R490C, R518C, R760Q, R808C, R808H, and R870W

(Dhermy et al. 1997). These extremely conserved sites are positioned at the internal boundaries of transmembrane segments, and substitution possibly interferes with co-translational insertion of band 3 into the membranes of the endoplasmic reticulum during synthesis of the protein. Mutations in the cytoplasmic domain of band 3 can interfere with its binding to other MS proteins, resulting in a functional defect. An amino acid substitution (G130A) in the cytoplasmic domain in band 3 Fukuoka possibly affects protein 4.2 binding (Kanzaki et al. 1997). HS due to band 3 is inherited dominantly and is generally milder than HS caused by ankyrin or spectrin mutations. Most band 3 deficient patients have a small number of mushroom-shaped cells in blood smears, sometimes called "pincered" cells, as if they had been pinched by a tweezers. These peculiar cells seem to occur only in band 3 mutants and characteristic of this type of defect.

HS related to 4.2 mutation is common in Japan but is rare in other populations. Due to various missence, nonsence, deletion or splicing mutations in band 3 binding N-terminus, 4.2 is either completely or almost completely absent in such cases of HS (Palek and Jarolim 1993).

Pathophysiology

The primary membrane lesions involve the "vertical interactions" loss between the skeleton and the bilayer, followed by vesiculation of the unsupported surface components. These processes, in turn, lead to progressive reduction in membrane surface area and to a "spherocyte," (Fig. 4.2). The observation that spectrin or spectrin-ankyrin deficiencies are frequent in HS has led to the hypothesis that interactions of spectrin with bilayer lipids or proteins are needed to stabilize the membrane. Budding off of spectrin-deficient areas would lead to HS. The alternate hypothesis proposes that the bilayer is stabilized by interactions between lipids and the band 3 molecules. In erythrocytes lacking band 3 the area between band 3 molecules would increase diminishing the stabilizing effect. Transient variations in the local density of band 3 could magnify this instability and allow unsupported lipids to be lost, resulting in HS. Spectrin- and ankyrin-deficient erythrocytes could become spherocytic by a analogous mechanism, as spectrin filaments hold band 3 molecules and limit their lateral movement (Corbett et al. 1994), a decrease in spectrin would let band 3s to diffuse and transiently cluster, promoting vesiculation. Both the mechanisms might operate to variable degrees in different diseases like first hypothesis dominating in spectrin and ankyrin defects and hypothesis 2 controlling in band 3 and protein 4.2 linked disorders.

In summary, the primary membrane defects involve deficiencies or defects of spectrin, ankyrin, protein 4.2, or band 3, but the etiologic relationship of these defects to surface loss is less obvious. Present assumption is that the MS (including band 3) may not adequately support all regions of the lipid bilayer in HS, leading to loss of small areas of untethered lipids and integral membrane proteins. Uncertain is whether the effect is directly due to deficiency of spectrin and ankyrin, or spectrin-ankyrin deficiency indirectly increases the lateral mobility of band 3

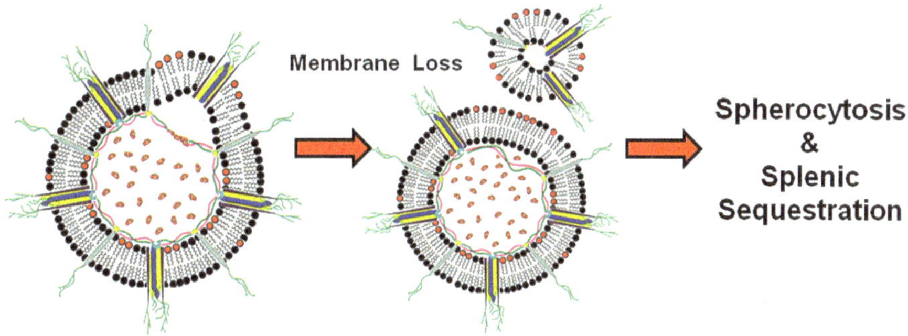

Fig. 4.2 Membrane loss by vesiculation in hereditary spherocytosis

molecules and decreases their stabilization of the lipid bilayer, or both. In addition, the loss of band 3 due to band 3 or protein 4.2 defects may directly diminish lipid anchoring in case of HS.

Laboratory Features and Treatment

For patients with typical HS, the diagnosis is mainly established by increased erythrocyte osmotic fragility and spherocytosis on the blood smear. Recently, high through put techniques like proteomics is also used to understand the disease (Demiralp et al. 2012; Peker et al. 2012; Polprasert et al. 2012; Saha et al. 2011). These studies not only reveals the involvement of major MS proteins in the disease but also tells about increased association of globin chains with the membrane and up regulation of redox regulators in the erythrocyte cytosol (Saha et al. 2011).

Although splenectomy cures almost all patients with HS, the indication for splenectomy must be carefully weighed, the spleen is important in controlling parasites and patients who are splenectomized have an increased risk of infections. Finally, the incidence of coronary heart disease and cerebral stroke is increased in older splenectomized HS patients. Recent detection and patient management of HS is elaborated in recent review (Da Costa et al. 2013) could be of potential interest.

Hereditary Elliptocytosis

Hereditary elliptocytosis (HE) is a collection of disorders characterized by the occurrence of elliptical-shaped erythrocytes on peripheral blood smear. HE and related disorders are associated with clinical, biochemical, and genetic heterogeneity where symptoms range from the asymptomatic carrier state to severe, transfusion-dependent hemolytic anemia. It is a prevalent in individuals of African and Mediterranean descent, presumably because elliptocytes give some protection to malaria.

The primary lesion in HE is mechanical weakness or fragility of the erythrocyte MS due to faults in α-spectrin, β-spectrin, or protein 4.1. Several mutations have been illustrated in the genes encoding these proteins, including point mutations, gene deletions and insertions, and mRNA processing defects (Da Costa et al. 2013). HE is inherited in an autosomal dominant fashion, with only odd occasions of de novo mutation (Palek and Lambert 1990). Usually, individuals heterozygous for an elliptocytocytic variant have asymptomatic elliptocytosis. Individuals homozygous or compound heterozygous for HE variants experience mild to severe hemolysis with moderate to marked anemia. Difference in severity amid families and in individuals of the same family has been attributed to different molecular lesions and/or modifier alleles or to other defects that alter disease manifestation (Coetzer et al. 1987, 1990, 1991; Discher et al. 1993).

Molecular Defects in HE

The major defect in HE is mechanical weakness or fragility of the erythrocyte MS, due to qualitative and quantitative defects in several MS proteins like α-spectrin, β-spectrin, protein 4.1, and even glycophorin C, identified in HE patients (Tse and Lux 1999). Spectrin tetramers and higher order oligomers are vital for erythrocyte membrane stability, as well as for erythrocyte shape and function. Local dissociation and reassociation of tetramers and dimers, respectively, may give the membrane the capability to pass through the microvessels. Defects that abate or interrupt the interactions in the self-association of spectrin or structural and functional defects of protein 4.1 that disrupt spectrin–actin interactions in the spectrin/protein 4.1//p55 junctional complex, perturb the integrity of the MS (Conboy et al. 1993; Delaunay and Dhermy 1993). Ultrastructural assessment of the elliptocyte MS reveals alteration of the normally uniform hexagonal lattice making erythrocytes mechanically unstable, leading to fragmentation and hemolysis. HE membranes, which are less tolerant to shear stress, are likely to suffer permanent deformation (Liu et al. 1993; Mohandas and Chasis 1993). Defective membrane skeleton facilitate shear stress-induced rearrangement of skeletal proteins after extended or recurring cellular deformation that prevents recovery of the normal biconcave shape. This is consistent with the fact that HE erythrocyte precursors are round, progressively becoming more elliptical with aging after release into the circulation (Rebuck and Van Slyck 1968).

The αβ-spectrin heterodimer self-association contact site between the opposed α- and β-spectrin chains is a combined "atypical" triple helical repeat in which two helices (helices A and B in the crystallographic structure of the repeat) are contributed by the C-terminus of β-spectrin, while the third helix is a portion of the N-terminus of α-spectrin (helix C) (Ipsaro et al. 2010). Missense mutations in the N-terminal region of α-spectrin are among the most common defects in HE (Delaunay and Dhermy 1993) and are commonly related with spectrin deficiency and the presence of elliptocytes on peripheral blood smear. Patients homozygous for these mutations have severe hemolytic anemia (Coetzer et al. 1987, 1990).

Mutations in the C-terminal region of β-spectrin associated with HE of variable clinical severity are truncations or point mutations that disrupt the spectrin self-association. Truncation mutations, which delete one or more of the β-spectrin phosphorylation sites, have included insertions, deletions, nonsense mutations, and exon skipping. In the homozygous state, they have been fatal or near fatal (Gallagher et al. 1997).

Influence of spectrin mutations outside the αβ-spectrin self-association contact site on membrane structure and function is rising. Current evidence suggests that spectrin self-association, spectrin-ankyrin binding, and ankyrin-band 3 binding are coupled in a positively cooperative manner (Giorgi et al. 2001). Mutations of linker sequences joining helices C and A, mutations in α-spectrin repeats 4–6 and truncating mutations that cause repeats 4–6 to fall out of register with the ankyrin binding region of β-spectrin have been proposed to perturb this cooperative coupling. A significant number of HE mutations associated with impaired spectrin self-association are located outside the self-association contact site in repeats 2, 3, 4, and 5, and a few are in repeats 8 and 9. The bulk of these mutations are situated in linker sequences joining helices C and A.

Although abnormalities of protein 4.1R are less common than spectrin mutations partial deficiency of protein 4.1R is associated with mild, dominant HE, while complete deficiency leads to severe hemolytic disease. Both quantitative and qualitative defects of protein 4.1R have been associated with HE. Partial protein 4.1R deficiency, known as protein 4.1R(−) trait occurs in heterozygotes that have mild HE with little or no hemolysis, prominent elliptocytosis, and minimal erythrocyte fragmentation, is a common cause of HE in some Arab and European populations (Mcguire et al. 1988). Complete protein 4.1R deficiency causes osmotically fragile erythrocytes with normal thermal stability and associated with significant hemolytic anemia, which may require transfusions or even splenectomy. Their membranes fragment much more quickly than normal at moderate sheer stresses but it can be completely restored by reconstituting the deficient erythrocytes with normal protein 4.1 or the protein 4.1/spectrin/actin binding site (Takakuwa et al. 1986). Protein 4.1-deficient membranes also lack protein p55 and are deficient in glycophorin C and D (Lambert et al. 1988). Protein 4.1R variants with abnormal molecular weights, primarily due to deletions or duplications of the exons encoding the spectrin-binding domain have also been described in case of HE. A truncated protein 4.1R was discovered in erythrocytes from a family with dominant, typical HE and mechanically unstable erythrocytes also an elongated protein 4.1R, protein $4.1^{\text{Hurdle-Mills}}$, was found in the erythrocytes of a Scottish-Irish family with dominant, typical HE (Mcguire et al. 1988). Mutations in the C-terminus of protein 4.1R associated with HE has also been identified, characterized by heterogeneity in clinical phenotype and degree of protein 4.1 deficiency (Dalla Venezia et al. 1998; Moriniere et al. 2000). In few cases, the mutant protein 4.1R mRNA was unstable. A large deletion in the protein 4.1R gene was identified in a family where a stable, truncated protein 4.1R mRNA was produced and unaltered tissue-specific alternative splicing was observed (Dalla Venezia et al. 1998).

Laboratory Features and Treatment

Cigar-shaped elliptocytes on peripheral blood smear are the diagnostic characteristic of HE. Some HE patients have thermally sensitive erythrocytes, fragmenting between 44 and 48 °C compared to normal erythrocytes fragment that spontaneously at 49 °C due to spectrin denaturation. Although therapy is rarely needed in HE, in severe cases transfusions or splenectomy may be required. Patients with significant hemolysis are recommended daily folic acid supplementation (Gallagher 2004).

Malaria and the Erythrocyte Membrane Skeleton Alterations

Alterations in MS is not always hereditary, environmental factors could also influence the structure and assembly of MS. Perhaps the most common example is the remodeling of the erythrocyte MS by malaria parasite *Plasmodium falciparum.* Malaria is the most severe and prevalent parasitic disease of humans, exerting a huge economic and social burden on society, particularly in the developing world. Malaria has exerted a powerful effect on human evolution and selection for resistance has led to the appearance and persistence of a number of inherited diseases like sickle cell disease, HE etc. Each year, up to 500 million people are infected with malaria parasites and half a million succumb as a consequence of the infection (Greenwood et al. 2008).

Intracellular growth of the parasite is accompanied by remarkable structural and functional changes in erythrocytes, some of which are strongly correlated with parasite-induced alterations to the erythrocyte MS (Cooke et al. 2004; Sherman 1985). Modification of MS protein organization and the adhesive, mechanical attributes of erythrocytes are of special relevance since these traits are directly linked to enhanced destruction of erythrocytes and to sequestration of parasitized erythrocytes (Macpherson et al. 1985) and the development of often fatal severe anemia and cerebral malaria (Miller et al. 2002).

Maturation of the malaria parasite causes striking structural and morphological changes in the infected red cell, including loss of the normal discoid shape, perturbations in the mechanical and adhesive properties of the cell, and alterations in the state of phosphorylation of red cell membrane skeletal proteins. The red cell becomes more spherical and its surface punctuated by up to 10,000 distinct electron-dense elevations called knobs that are associated with altered cellular adhesive properties of infected red cells. Maturing parasites produce proteins for export to the erythrocyte MS. Generally, exported parasite proteins are of large molecular mass, highly charged, and contain defined blocks of low complexity sequence, often in the form of tandemly repeated oligonucleotides (Cooke et al. 2001). The repeats are characteristically present in distinct regions, each composed of repeats of a particular sequence. The repeats often contain charged residues, either positive or negative. A common motif is a dipeptide of glutamic acid, and the repeat regions are typically modeled to be either alpha-helical, random coil, or coiled coil.

Likely the parasite proteins within the erythrocyte will be arranged in some form of multi-protein complexes. Most of the membrane-associated proteins, including PfEMP-3, RESA, MESA, FEST, and FIRA, appear to be distributed evenly around the skeleton, while others, such KAHRP and PfEMP1, tend to cluster at higher density beneath membrane knobs. In early maturing parasites, a number of these proteins and others such as PfSBP1 and MAHRP associate with discrete membrane-bound structures, known as Maurers' clefts, scattered throughout the erythrocyte cytoplasm (Cooke et al. 2006). Some, if not all, are recruited into these structures where they are either assembled into the cytoadherence complex and/or participate in assembly before insertion of the complex into the erythrocyte membrane or skeleton (Wickham et al. 2001).

Now it is known that MESA binds to protein 4.1 (Waller et al. 2003) via a 19-residue sequence that forms an amphipathic helix, and RESA binds to spectrin via a 48-residue domain in the RESA protein (Pei et al. 2007a). KAHRP binds to spectrin, and the binding domain of the parasite protein appears to be in the N-terminal region (Pei et al. 2005). The binding of malaria proteins like PfEMP3 can disrupt interactions between erythrocyte proteins (Pei et al. 2007b). MESA binds to the same domain of protein 4.1 as glycophorin C and p55, interfering with the binding of p55. Likely some of the effects of malaria on red cell morphology and membrane mechanical properties are mediated by this type of interruption of host protein interactions (Waller et al. 2003). Identification of binding domains between parasite proteins and the erythrocyte skeleton that appears essential for parasite growth and virulence may lead to the development of novel therapeutics to interfere with these interactions, a previously unexplored means of controlling malarial disease.

Protein phosphorylation also plays a critical role in regulating the mechanical stability of normal erythrocytes and the maintenance of their deformability (Manno et al. 1995). Levels of phosphorylation of erythrocyte membrane skeletal proteins also are affected by malaria infection (Murray and Perkins 1989). In *P. falciparum-infected* erythrocytes, there is a marked increase in phosphorylation of protein 4.1 (Lustigman et al. 1990). As phosphorylation of protein 4.1 inhibits protein 4.1-mediated spectrin-actin interaction (Ling et al. 1988), the parasite-induced phosphorylation could reduce membrane mechanical stability and contribute significantly to disease pathogenesis; phosphorylation of band 3 is also measurably increased (Chishti et al. 1994). The enzymes responsible for the phosphorylation of malaria proteins have not been identified. Among the proteins exported to the erythrocyte skeleton is falciparum-exported serine/threonine kinase (FEST) (Kun et al. 1997); its association with the MS in parasitized erythrocytes provides tantalizing evidence for an important role in phosphorylation of a cytoskeletal proteins in infected erythrocytes. In summary, Plasmodium alters the erythrocyte MS architecture in order to sustain within the host. Details of the invasion and its consequence is described by Mohandas and coworker in their recent review (Mohandas and An 2012) could be of further interest.

Conclusion

Extensive studies on erythroid MS from both normal and diseased individuals by several biochemical, biophysical, molecular biological and proteomics approaches have facilitated the development of structure and function of MS and also detailed molecular insights into some of its disorders like HS and HE. Parasite induced alteration of MS is also a newly emerging field of potential interest. Molecular analysis of the spectrin-based MS of the mammalian erythrocyte has provided us with a set of proteins with miscellaneous roles in organization and survival of cells. Characterization of genes and alternatively spliced variants combined with completion of the genome sequencings now are providing the first glimpses of the full extent of physiological roles of proteins in MS. At the same time, while the MS defines red cell shape, deformability and integrity, the mechanisms used to achieve these properties are still not well understood. Furthermore, although many protein components of the membrane and membrane skeleton have been defined, there are critical gaps and inaccuracies in this knowledge. Possibly, the protein composition is not yet complete, the reported stoichiometries (Mankelow et al. 2012) are probably not correct and many critical protein–protein interactions are missing. Even reconstitution of major components of the membrane skeleton, such as the actin-based junctional complex is still not possible. These deliberations imply that probably we have just arrived at the end of the beginning in terms of understanding the fundamental roles of MS and in applying this knowledge to practice.

Acknowledgements Avik Basu acknowledges a Senior Research Fellowship from Department of Atomic Energy (DAE), India. Authors also acknowledge MSACR project of DAE for funding and Dr. Sumanta Basu for initial artwork used in illustrations.

References

Agre P, Orringer EP, Bennett V (1982) Deficient red-cell spectrin in severe, recessively inherited spherocytosis. N Engl J Med 306:1155–1161

Agre P, Casella JF, Zinkham WH, Mcmillan C, Bennett V (1985) Partial deficiency of erythrocyte spectrin in hereditary spherocytosis. Nature 314:380–383

Beck KA, Nelson WJ (1998) A spectrin membrane skeleton of the Golgi complex. Biochim Biophys Acta 1404:153–160

Bennett V (1979) Immunoreactive forms of human-erythrocyte ankyrin are present in diverse cells and tissues. Nature 281:597–599

Bennett V (1985) The membrane skeleton of human erythrocytes and its implications for more complex cells. Annu Rev Biochem 54:273–304

Bennett V, Stenbuck PJ (1979a) Identification and partial-purification of ankyrin, the high-affinity membrane attachment site for human-erythrocyte spectrin. J Biol Chem 254:2533–2541

Bennett V, Stenbuck PJ (1979b) Membrane attachment protein for spectrin is associated with band-3 in human-erythrocyte membranes. Nature 280:468–473

Bennett V, Stenbuck PJ (1980) Association between ankyrin and the cytoplasmic domain of band-3 isolated from the human-erythrocyte membrane. J Biol Chem 255:6424–6432

Bennett V, Davis J, Fowler WE (1982) Brain spectrin, a membrane-associated protein related in structure and function to erythrocyte spectrin. Nature 299:126–131
Chishti AH, Maalouf GJ, Marfatia S, Palek J, Wang W, Fisher D, Liu SC (1994) Phosphorylation of protein-4.1 in plasmodium-falciparum-infected human red-blood-cells. Blood 83:3339–3345
Coetzer T, Lawler J, Prchal JT, Palek J (1987) Molecular determinants of clinical expression of hereditary elliptocytosis and pyropoikilocytosis. Blood 70:766–772
Coetzer T, Palek J, Lawler J, Liu SC, Jarolim P, Lahav M, Prchal JT, Wang W, Alter BP, Schewitz G, Mankad V, Gallanello R, Cao A (1990) Structural and functional-heterogeneity of alpha-spectrin mutations involving the spectrin heterodimer self-association site—relationships to hematologic expression of homozygous hereditary elliptocytosis and hereditary pyropoikilocytosis. Blood 75:2235–2244
Coetzer TL, Sahr K, Prchal J, Blacklock H, Peterson L, Koler R, Doyle J, Manaster J, Palek J (1991) Four different mutations in codon 28 of alpha-spectrin are associated with structurally and functionally abnormal spectrin alpha-i/74 in hereditary elliptocytosis. J Clin Invest 88: 743–749
Conboy JG, Chasis JA, Winardi R, Tchernia G, Kan YW, Mohandas N (1993) An isoform-specific mutation in the protein 4.1 gene results in hereditary elliptocytosis and complete deficiency of protein 4.1 in erythrocytes but not in nonerythroid cells. J Clin Invest 91:77–82
Cooke BM, Mohandas N, Coppel RL (2001) The malaria-infected red blood cell: structural and functional changes. Adv Parasitol 50:1–86
Cooke BM, Mohandas N, Coppel RL (2004) Malaria and the red blood cell membrane. Semin Hematol 41:173–188
Cooke BM, Buckingham DW, Glenister FK, Fernandez KM, Bannister LH, Marti M, Mohandas N, Coppel RL (2006) A Maurer's cleft-associated protein is essential for expression of the major malaria virulence antigen on the surface of infected red blood cells. J Cell Biol 172:899–908
Corbett JD, Agre P, Palek J, Golan DE (1994) Differential control of band-3 lateral and rotational mobility in intact red-cells. J Clin Invest 94:683–688
Da Costa L, Galimand J, Fenneteau O, Mohandas N (2013) Hereditary spherocytosis, elliptocytosis, and other red cell membrane disorders. Blood Rev 27:167–178
Dalla Venezia N, Maillet P, Morle L, Roda L, Delaunay J, Baklouti F (1998) A large deletion within the protein 4.1 gene associated with a stable truncated mRNA and an unaltered tissue-specific alternative splicing. Blood 91:4361–4367
del Giudice EM, Lombardi C, Francese M, Nobili B, Conte ML, Amendola G, Cutillo S, Iolascon A, Perrotta S (1998) Frequent de novo monoallelic expression of beta-spectrin gene (SPTB) in children with hereditary spherocytosis and isolated spectrin deficiency. Br J Haematol 101: 251–254
del Giudice EM, Nobili B, Francese M, D'Urso L, Iolascon A, Eber S, Perrotta S (2001) Clinical and molecular evaluation of non-dominant hereditary spherocytosis. Br J Haematol 112: 42–47
Delaunay J, Dhermy D (1993) Mutations involving the spectrin heterodimer contact site—clinical expression and alterations in specific function. Semin Hematol 30:21–33
delGiudice EM, Hayette S, Bozon M, Perrotta S, Alloisio N, Vallier A, Iolascon A, Delaunay T, Morle L (1996) Ankyrin Napoli: A de novo deletional frameshift mutation in exon 16 of ankyrin gene (ANK1) associated with spherocytosis. Br J Haematol 93:828–834
Demiralp DO, Peker S, Turgut B, Akar N (2012) Comprehensive identification of erythrocyte membrane protein deficiency by 2D gel electrophoresis based proteomic analysis in hereditary elliptocytosis and spherocytosis. Proteomics Clin Appl 6:403–411
Derick LH, Liu SC, Chishti AH, Palek J (1992) Protein immunolocalization in the spread erythrocyte-membrane skeleton. Eur J Cell Biol 57:317–320
Dhermy D, Galand C, Bournier O, Boulanger L, Cynober T, Schismanoff PO, Bursaux E, Tchernia G, Boivin P, Garbarz M (1997) Heterogenous band 3 deficiency in hereditary spherocytosis related to different band 3 gene defects. Br J Haematol 98:32–40 (vol 99, pg 474, 1997)
Discher D, Parra M, Conboy JG, Mohandas N (1993) Mechanochemistry of the alternatively spliced spectrin-actin binding domain in membrane skeletal protein-4.1. J Biol Chem 268:7186–7195

Eber S, Lux SE (2004) Hereditary spherocytosis—defects in proteins that connect the membrane skeleton to the lipid bilayer. Semin Hematol 41:118–141

Fowler VM (1990) Tropomodulin—a cytoskeletal protein that binds to the end of erythrocyte tropomyosin and inhibits tropomyosin binding to actin. J Cell Biol 111:471–482

Fowler VM, Bennett V (1984) Erythrocyte-membrane tropomyosin—purification and properties. J Biol Chem 259:5978–5989

Gallagher PG (2004) Hereditary elliptocytosis: spectrin and protein 4.1R. Semin Hematol 41:142–164

Gallagher PG, Petruzzi MJ, Weed SA, Zhang ZS, Marchesi SL, Mohandas N, Morrow JS, Forget BG (1997) Mutation of a highly conserved residue of beta I spectrin associated with fatal and near-fatal neonatal hemolytic anemia. J Clin Invest 99:267–277

Gardner K, Bennett V (1987) Modulation of spectrin actin assembly by erythrocyte adducin. Nature 328:359–362

Giorgi M, Cianci CD, Gallagher PG, Morrow JS (2001) Spectrin oligomerization is cooperatively coupled to membrane assembly: a linkage targeted by many hereditary hemolytic anemias? Exp Mol Pathol 70:215–230

Greenwood BM, Fidock DA, Kyle DE, Kappe SHI, Alonso PL, Collins FH, Duffy PE (2008) Malaria: progress, perils, and prospects for eradication. J Clin Invest 118:1266–1276

Hanspal M, Yoon SH, Yu H, Hanspal JS, Lambert S, Palek J, Prchal JT (1991) Molecular-basis of spectrin and ankyrin deficiencies in severe hereditary spherocytosis—evidence implicating a primary defect of ankyrin. Blood 77:165–173

Hassoun H, Vassiliadis JN, Murray J, Yi SJ, Hanspal M, Ware RE, Winter SS, Chiou SS, Palek J (1995) Molecular basis of spectrin deficiency in beta spectrin durham—a deletion within beta spectrin adjacent to the ankyrin-binding site precludes spectrin attachment to the membrane in hereditary spherocytosis. J Clin Invest 96:2623–2629

Husainchishti A, Faquin W, Wu CC, Branton D (1989) Purification of erythrocyte dematin (protein-4.9) reveals an endogenous protein-kinase that modulates actin-bundling activity. J Biol Chem 264:8985–8991

Ipsaro JJ, Harper SL, Messick TE, Marmorstein R, Mondragon A, Speicher DW (2010) Crystal structure and functional interpretation of the erythrocyte spectrin tetramerization domain complex. Blood 115:4843–4852

Kahana E, Pinder JC, Smith KS, Gratzer WB (1992) Fluorescence quenching of spectrin and other red-cell membrane cytoskeletal proteins—relation to hydrophobic binding-sites. Biochem J 282:75–80

Kanzaki A, Hayette S, Morle L, Inoue F, Matsuyama R, Inoue T, Yawata A, Wada H, Vallier A, Alloisio N, Yawata Y, Delaunay J (1997) Total absence of protein 4.2 and partial deficiency of band 3 in hereditary spherocytosis. Br J Haematol 99:522–530

Khan AA, Hanada T, Mohseni M, Jeong JJ, Zeng LX, Gaetani M, Li DH, Reed BC, Speicher DW, Chishti AH (2008) Dematin and adducin provide a novel link between the spectrin cytoskeleton and human erythrocyte membrane by directly interacting with glucose transporter-1. J Biol Chem 283:14600–14609

Koshino I, Mohandas N, Takakuwa Y (2012) Identification of a novel role for dematin in regulating red cell membrane function by modulating spectrin-actin interaction. J Biol Chem 287: 35244–35250

Kuhlman PA, Hughes CA, Bennett V, Fowler VM (1996) A new function for adducin. Calicium calmodulin-regulated capping of the barbed ends of actin filaments. J Biol Chem 271: 7986–7991

Kun JFJ, Hibbs AR, Saul A, McColl DJ, Coppel RL, Anders RF (1997) A putative Plasmodium falciparum exported serine/threonine protein kinase. Mol Biochem Parasitol 85:41–51

Lambert S, Conboy J, Zail S (1988) A molecular study of heterozygous protein 4.1 deficiency in hereditary elliptocytosis. Blood 72:1926–1929

Lazarides E, Woods C (1989) Biogenesis of the red blood cell membrane-skeleton and the control of erythroid morphogenesis. Annu Rev Cell Biol 5:427–452

Ling E, Danilov YN, Cohen CM (1988) Modulation of red-cell band-4.1 function by camp-dependent kinase and protein kinase-c phosphorylation. J Biol Chem 263:2209–2216
Liu SC, Derick LH, Palek J (1987) Visualization of the hexagonal lattice in the erythrocyte membrane skeleton. J Cell Biol 104:527–536
Liu SC, Derick LH, Palek J (1993) Dependence of the permanent deformation of red-blood-cell membranes on spectrin dimer tetramer equilibrium—implication for permanent membrane deformation of irreversibly sickled cells. Blood 81:522–528
Lustigman S, Anders RF, Brown GV, Coppel RL (1990) The mature-parasite-infected erythrocyte surface-antigen (Mesa) of Plasmodium-falciparum associates with the erythrocyte-membrane skeletal protein, band-4.1. Mol Biochem Parasitol 38:261–270
Macpherson GG, Warrell MJ, White NJ, Looareesuwan S, Warrell DA (1985) Human cerebral malaria—a quantitative ultrastructural analysis of parasitized erythrocyte sequestration. Am J Pathol 119:385–401
Mankelow TJ, Satchwell TJ, Burton NM (2012) Refined views of multi-protein complexes in the erythrocyte membrane. Blood Cells Mol Dis 49:1–10
Manno S, Takakuwa Y, Nagao K, Mohandas N (1995) Modulation of erythrocyte-membrane mechanical function by beta-spectrin phosphorylation and dephosphorylation. J Biol Chem 270:5659–5665
Marchesi VT (1985) Stabilizing infrastructure of cell-membranes. Annu Rev Cell Biol 1:531–561
Marfatia SM, Lue RA, Branton D, Chishti AH (1994) In-vitro binding-studies suggest a membrane-associated complex between erythroid P55, protein-4.1, and glycophorin-C. J Biol Chem 269:8631–8634
Mcguire M, Smith BL, Agre P (1988) Distinct variants of erythrocyte protein-4.1 inherited in linkage with elliptocytosis and Rh-type in 3 white families. Blood 72:287–293
Michalak K, Bobrowska M, Sikorski AF (1993) Interaction of bovine erythrocyte spectrin with aminophospholipid liposomes. Gen Physiol Biophys 12:163–170
Miller LH, Baruch DI, Marsh K, Doumbo OK (2002) The pathogenic basis of malaria. Nature 415:673–679
Mohandas N, An X (2012) Malaria and human red blood cells. Med Microbiol Immunol 201:593–598
Mohandas N, Chasis JA (1993) Red-blood-cell deformability, membrane material properties and shape—regulation by transmembrane, skeletal and cytosolic proteins and lipids. Semin Hematol 30:171–192
Mohandas N, Gallagher PG (2008) Red cell membrane: past, present, and future. Blood 112:3939–3948
Moriniere M, Ribeiro L, Dalla Venezia N, Deguillien M, Maillet P, Cynober T, Delhommeau F, Almeida H, Tamagnini G, Delaunay J, Baklouti F (2000) Elliptocytosis in patients with C-terminal domain mutations of protein 4.1 correlates with encoded messenger RNA levels rather than with alterations in primary protein structure. Blood 95:1834–1841
Murray MC, Perkins ME (1989) Phosphorylation of erythrocyte-membrane and cytoskeleton proteins in cells infected with Plasmodium-falciparum. Mol Biochem Parasitol 34:229–236
Ohanian V, Wolfe LC, John KM, Pinder JC, Lux SE, Gratzer WB (1984) Analysis of the ternary interaction of the red-cell membrane skeletal protein-spectrin, protein-actin, and protein 4.1. Biochemistry 23:4416–4420
Palek J, Jarolim P (1993) Cellular molecular-biology of Rbc membrane.4. Clinical expression and laboratory detection of red-blood-cell membrane-protein mutations. Semin Hematol 30: 249–283
Palek J, Lambert S (1990) Genetics of the red-cell membrane skeleton. Semin Hematol 27: 290–332
Pei XH, An XL, Guo XH, Tarnawski M, Coppel R, Mohandas N (2005) Structural and functional studies of interaction between Plasmodium falciparum knob-associated histidine-rich protein (KAHRP) and erythrocyte spectrin. J Biol Chem 280:31166–31171

Pei XH, Guo XH, Coppel R, Bhattacharjee S, Haldar K, Gratzer W, Mohandas N, An XL (2007a) The ring-infected erythrocyte surface antigen (RESA) of Plasmodium falciparum stabilizes spectrin tetramers and suppresses further invasion. Blood 110:1036–1042

Pei XH, Guo XH, Coppel R, Mohandas N, An XL (2007b) Plasmodium falciparum erythrocyte membrane protein 3 (PfEMP3) destabilizes erythrocyte membrane skeleton. J Biol Chem 282:26754–26758

Peker S, Akar N, Demiralp DO (2012) Proteomic identification of erythrocyte membrane protein deficiency in hereditary spherocytosis. Mol Biol Rep 39:3161–3167

Polprasert C, Chiangjong W, Thongboonkerd V (2012) Marked changes in red cell membrane proteins in hereditary spherocytosis: a proteomics approach. Mol Biosyst 8:2312–2322

Randon J, delGiudice EM, Bozon M, Perrotta S, DeVivo M, Iolascon A, Delaunay J, Morle L (1997) Frequent de novo mutations of the ANK1 gene mimic a recessive mode of transmission in hereditary spherocytosis: three new ANK1 variants: Ankyrins Bari, Napoli II and Anzio. Br J Haematol 96:500–506

Rebuck JW, Van Slyck EJ (1968) An unsuspected ultrastructural fault in human elliptocytes. Am J Clin Pathol 49:19–25

Saha S, Ramanathan R, Basu A, Banerjee D, Chakrabarti A (2011) Elevated levels of redox regulators, membrane-bound globin chains, and cytoskeletal protein fragments in hereditary spherocytosis erythrocyte proteome. Eur J Haematol 87:259–266

Sato SB, Ohnishi S (1983) Interaction of a peripheral protein of the erythrocyte-membrane, band-4.1, with phosphatidylserine-containing liposomes and erythrocyte inside-out vesicles. Eur J Biochem 130:19–25

Shen BW, Josephs R, Steck TL (1986) Ultrastructure of the intact skeleton of the human-erythrocyte membrane. J Cell Biol 102:997–1006

Sherman IW (1985) Membrane-structure and function of malaria parasites and the infected erythrocyte. Parasitology 91:609–645

Sikorski AF, Michalak K, Bobrowska M (1987) Interaction of spectrin with phospholipids—quenching of spectrin intrinsic fluorescence by phospholipid suspensions. Biochim Biophys Acta 904:55–60

Singer SJ, Nicolson GL (1972) The fluid mosaic model of the structure of cell membranes. Science 175:720–731

Takakuwa Y, Tchernia G, Rossi M, Benabadji M, Mohandas N (1986) Restoration of normal membrane stability to unstable protein-4.1-deficient erythrocyte-membranes by incorporation of purified protein-4.1. J Clin Invest 78:80–85

Tse WT, Lux SE (1999) Red blood cell membrane disorders. Br J Haematol 104:2–13

Tyler JM, Reinhardt BN, Branton D (1980) Associations of erythrocyte-membrane proteins—binding of purified bands 2.1 and 4.1 to spectrin. J Biol Chem 255:7034–7039

Ungewickell E, Gratzer W (1978) Self-association of human spectrin—thermodynamic and kinetic study. Eur J Biochem 88:379–385

Ungewickell E, Bennett PM, Calvert R, Ohanian V, Gratzer WB (1979) Invitro formation of a complex between cytoskeletal proteins of the human-erythrocyte. Nature 280:811–814

Waller KL, Nunomura W, An XL, Cooke BM, Mohandas N, Coppel RL (2003) Mature parasite-infected erythrocyte surface antigen (MESA) of Plasmodium falciparum binds to the 30-kDa domain of protein 4.1 in malaria-infected red blood cells. Blood 102:1911–1914

Weber A, Pennise CR, Babcock GG, Fowler VM (1994) Tropomodulin caps the pointed ends of actin-filaments. J Cell Biol 127:1627–1635

Wichterle H, Hanspal M, Palek J, Jarolim P (1996) Combination of two mutant alpha spectrin alleles underlies a severe spherocytic hemolytic anemia. J Clin Invest 98:2300–2307

Wickham ME, Rug M, Ralph SA, Klonis N, McFadden GI, Tilley L, Cowman AF (2001) Trafficking and assembly of the cytoadherence complex in Plasmodium falciparum-infected human erythrocytes. EMBO J 20:5636–5649

Woods CM, Lazarides E (1986) Spectrin assembly in avian erythroid development is determined by competing reactions of subunit homooligomerization and hetero-oligomerization. Nature 321:85–89

Yoshino H, Marchesi VT (1984) Isolation of spectrin subunits and reassociation in vitro—analysis by fluorescence polarization. J Biol Chem 259:4496–4500
Yu J, Steck TL (1975) Isolation and characterization of band-3, predominant polypeptide of human erythrocyte-membrane. J Biol Chem 250:9170–9175
Yu J, Fischman DA, Steck TL (1973) Selective solubilization of proteins and phospholipids from red blood cell membranes by nonionic detergents. J Supramol Struct 1:233–248

Chapter 5
Membrane Rafts in the Erythrocyte Membrane: A Novel Role of MPP1p55

Aleksander F. Sikorski, Joanna Podkalicka, Walis Jones, and Agnieszka Biernatowska

Abbreviations

a ZO1	A mammalian zonula occludens protein
DlgA	A Drosophila disc large tumor suppressor
DRM	Detergent resistant membrane
FLIM Fluorescence lifetime imaging L27	Lin-2 and Lin-7 receptor targeting proteins
ld	Liquid disordered
lo	Liquid ordered
MAGUK	Membrane associated guanylate kinase
MPP1	Membrane palmitoylated protein-1
PC	Phosphatidylcholine
PDZ PSD	Post synaptic density protein
PE	Phosphatidylethanolamine
PS	Phosphatidylserine
SH3	Src homology 3 domain
SM	Sphingomyelin

A.F. Sikorski (✉) • J. Podkalicka • A. Biernatowska
Laboratory of Cytobiochemistry, Faculty of Biotechnology, University of Wrocław, ul. F. Joliot-Curie 14a, 50383 Wrocław, Poland
e-mail: afsbc@ibmb.uni.wroc.pl

W. Jones
BioPharm Enterprises Limited, Tanyfron 65 Gwaun Afan Cwmafan, Port Talbot, Neath Port Talbot, Wales SA12 9EJ, UK

A. Chakrabarti, A. Surolia (eds.), *Biochemical Roles of Eukaryotic Cell Surface Macromolecules*, Advances in Experimental Medicine and Biology 842,
DOI 10.1007/978-3-319-11280-0_5

Introduction

Eukaryotic cell membranes are organized into functional lipid and protein domains, the most widely studied being membrane rafts. They are enriched in sphingolipids and cholesterol and contain several types of membrane proteins such as stomatin and flotillins, which are their "permanent" residents, along with several others, such as receptors, including growth factor receptors. Membrane rafts organize receptors and their downstream molecules to regulate a number of intracellular signaling pathways (for review see (Grzybek et al. 2005; Pike 2006; Hancock 2006; Simons and Sampaio 2011)). Important post-translational modifications, such as palmitoylation, addition of GPI anchors or sterol molecules, regulate raft-affinity of the majority of raft proteins.

Lateral interactions of cholesterol with membrane raft lipids seem to be crucial for maintaining these microdomains in the 'lo' (liquid-ordered) state, which is characterized by decreased conformational (trans-gauche) freedom and, consequently, reduced "fluidity", compared to the bulk cholesterol-poor membrane which exists in the 'ld' (liquid disordered) state. However, unlike the gel phase in artificial lipid systems they are characterized by similar rotational and lateral (translational) mobility to the bulk membrane (Ipsen et al. 1987; Simons and Vaz 2004). Rafts are enriched not only in cholesterol and sphingolipids, but also glycolipids and specific inner-layer phospholipid PE and PS species. Data on the lipidomics of the DRM reveal that, in addition to an abundance of sphingomyelin and cholesterol, they contain other phospholipids that mostly contain fully saturated or monounsaturated acyl chains. Predominant among these are the phosphatidylethanolamine glycerophospholipids and plasmalogens (Macdonald and Pike 2005; Pike et al. 2005; Koumanov et al. 2005; Brugger et al. 2006). Phosphatidylserine, which is a relatively minor membrane component, is three times more prevalent in the DRM than in the bulk volume of the plasma membrane, while phosphatidylinositols are rather diminished within the DRM, as are phosphatidylcholine species. PEs occur in the membrane predominantly as sn-1 saturated, sn-2 unsaturated glycerophospholipids, and recent data show that some DRM preparations are enriched in 1-stearoyl-2-linoleoyl-sn-glycero-3-phosphoethanolamine (SLPE), regardless of the method of isolation (Pike et al. 2005). Our own data indicate that this PE interacts with cholesterol comparably to SM, while dipalmitoyl-PE does not bind cholesterol (Grzybek et al. 2009). This suggests the importance of the structure of acyl chains of particular phospholipids (which, incidentally not all are saturated) and also explains the background of the mechanism by which inner-layer phospholipids participate in membrane rafts.

Membrane rafts contain several specific sets of membrane proteins (Brown and London 2000), which include membrane proteins belonging to the SPFH family (stomatin/prohibitin/flotillin/HflK), such as raft scaffold proteins flotillin-1 and -2, and stomatin or stomatin-like protein. These proteins share a common feature in that they associate with the raft domains, possibly through cholesterol-binding (Epand et al. 2005; Epand 2006; Browman et al. 2007) or oligomerization (Browman et al. 2007). Apart from these proteins, and other palmitoylated and transmembrane proteins, membrane rafts also include GPI-anchored proteins and proteins involved in signal transduction, such as, by way of example, tyrosine kinases of the src family, Gα subunits of heterotrimeric G proteins, endothelial nitric oxide synthase and hedgehog protein. Many of the raft proteins are modified with saturated acyl (palmitoyl) chains, or with cholesterol, which is thought to facilitate interactions with lo-domains (Wang et al. 2000). The mechanism(s) by which the proteins partition into membrane rafts may involve preferred solubility in the ordered domains and/or chemical affinity for raft lipids, as exemplified by the cholesterol-binding properties of some of the proteins (Fantini and Barrantes 2013). Another example includes a structural protein motif recognizing sphingolipids or specific glycolipids such as gangliosides (Hakomori 2002; Mahfoud et al. 2002).

It has long been known that rafts are engaged in cellular signal transduction pathways by hosting a number of receptors and their associate adapter proteins, and that they also facilitate signaling switches during the activation of the respective pathways. These proteins include receptor tyrosine kinases (EGF-R, IGF-1, c-kit), non-receptor kinases (e.g. src kinases: Src, Lck, Hck, Fyn, Blk, Lyn, Fgr, Yes, and Yrk (Boggon and Eck 2004)), serpentine (G-protein-linked seven transmembrane domain) receptors (Pike 2003), sigma receptors (Aydar et al. 2004) as well as heterotrimeric and monomeric G-proteins and other adaptor proteins. Also pro- or anti-apoptotic signaling elements have been found in raft domains, as recently reviewed (Hryniewicz-Jankowska et al. 2014). Of particular interest are the potential roles of membrane rafts as signaling platforms in neoplasia (Staubach and Hanisch 2011) and their possible usage as targets for anticancer drugs (Hryniewicz-Jankowska et al. 2014)

Erythrocyte Membrane Rafts

The existence of lateral heterogeneity within the erythrocyte membrane (Karnovsky et al. 1982; Schroeder et al. 1995; Rodgers and Glaser 1991) was known even before the formulation of the raft hypothesis by Simons and Ikonen (1997). As in the case with other cell types, the first attempts to characterize the lateral heterogeneity of erythrocyte membranes involved the extraction of membranes with the nonionic detergent, Triton X-100 (Salzer and Prohaska 2001). Subsequent freeze-fracture electron microscopy studies showed that the DRM fraction derived from human and ruminant erythrocytes were devoid of intramembrane particles from the inner fracture plane (Quinn et al. 2005), which is in agreement with the notion that transmembrane proteins are excluded from highly-ordered regions of the bilayer (McIntosh et al. 2003). Synchrotron X-ray diffraction studies indicated that d-spacings obtained for multilamellar stacks and vesicular suspensions of raft membranes (DRM) were, on average, more than 0.5 nm greater than corresponding arrangements of the original erythrocyte membranes from which they were obtained (Quinn et al. 2005).

Salzer and Prohaska provided the first systematic analysis of the protein composition of the DRM fraction of the erythrocyte membrane (Salzer and Prohaska 2001). Prominent components of this fraction were flotillins-1 and -2, typical components of raft-fractions in other cell types (Bickel et al. 1997; Lang et al. 1998). They also found variable amounts of membrane-skeleton and membrane-skeleton adapter proteins, such as spectrin, actin and proteins 4.1 and 4.2. The major integral protein, erythrocyte anion-exchanger, was absent from this fraction, as was glycophorin C.

Kamata et al. provided functional evidence for the presence of membrane rafts in erythrocyte membranes by showing inactivation of the Gsα-mediated signal transduction pathway by reversible disruption of rafts by lidocaine treatment (2008).

Stomatin, Flotillins and MPP1 Are the Stable Components of Erythrocyte Rafts

Among the constant protein components of erythrocyte membrane rafts, stomatin, flotillin-1 and flotillin-2 seem to be the major components playing an important structural role in their organization. They are considered fundamental marker-proteins of these structures. Although MPP1 is not commonly recognized as a membrane raft marker protein, our observations indicate that, at least in erythrocyte and erythroid cells, MPP1 is consistently found in DRM preparations of these cells, as described below.

Stomatin

Stomatin was first described as an integral protein of the erythrocyte membrane and was subsequently found to be widely expressed in various tissues and cells (Hiebl-Dirschmied et al. 1991a, b). Stomatin is absent in the erythrocyte membrane of patients suffering from a specific form of hemolytic anemia called stomatocytosis (for a review see e.g. (Boguslawska et al. 2010)). The protein consists of 287 amino acid residues, characterized by the presence of a highly charged 24-residue long N-terminus. This is followed by a 29-residue hydrophobic hairpin-domain, which is highly conserved amongst the close stomatin homologue proteins SLP-3, SLP-1, podocin and MEC-2 (Kadurin et al. 2009) and a large C-terminal region containing 234 residues. Stomatin typically has an unusual topology, with the afore-mentioned "hairpin domain" located within the lipid bilayer and the N- and C-termini facing the cytosol. There is data suggesting that a small portion of stomatin is expressed as a transmembrane protein, with the C-terminal domain being N-glycosylated and exposed to the cell surface (Kadurin et al. 2009). Palmitoylation of cysteine residues Cys-29 and Cys-86 further increases the affinity of stomatin for the membrane (Snyers et al. 1999).

Stomatin domain (SPFH, in mouse stomatin residues 86–213) has a mixed alpha/beta-fold structure. The N-terminal part forms an antiparallel curved beta-sheet (beta strands 1a, 1b, 2 and 3). Helices alpha-2 and alpha-4 extend in parallel and occupy the groove of the sheet, while the short alpha-1 and alpha-3 helices are oriented perpendicularly at both ends of the beta-sheet. The N- and C-termini are located at opposing sides of the molecule (Brand et al. 2012). Mouse stomatin forms a banana-shaped dimer via the four residues located at the C-terminal position of beta-3, in contrast to its archeal homologue from *Pyrococcus horikoshii*, which is trimeric (Yokoyama et al. 2008).

There is substantial evidence that stomatin forms higher-order oligomers; one of the crystal forms was found to form a ring-like oligomer 8 nm in diameter (Brand et al. 2012). Others had earlier found that erythrocyte stomatin forms oligomers of 9–12 mers, or even structures that were twice as large, reporting the presence of a short hydrophobic sequence (residues 264STIVFPLPI272) within the C-terminal region of stomatin that was responsible for oligomer formation (Umlauf et al. 2006).

Flotillins

Flotillins are also members of the SPFH-protein family, containing a common structural motif known as the SPFH domain (see above). Flotillin-1 is alternatively known as reggie-2, whilst flotillin-2 is known as reggie-1. Flotillin-1 has a hydrophobic sequence located within the C-terminal part of the SPFH domain (residues 134–151) in addition to the N-terminal hydrophobic stretch (residues 10–36). The N-terminal domain is considered as being important in the interaction of flotillin-1 with the membrane raft-domain and the C-terminal hydrophobic sequence is thought to be responsible for localization of flotillin-1 in the plasma membrane (Liu et al. 2005). SPFH domains in both flotillins (flotillin-1: residues 36–179, and flotillin-2: residues 41–183) are followed by another flotillin domain comprised of three coiled-coil stretches rich in EA residues (Solis et al. 2007). The C-terminal part of both molecules contains a PDZ3 domain (Chi et al. 2012). These proteins share 48 % structural identity and 71 % sequence homology. Flotillin-1 contains a palmitoylation site which is thought to be important for its membrane localization in kidney cells (Morrow et al. 2002). However, separate data indicates that palmitoylation at this site is less important in defining membrane localization than is the presence of the hydrophobic stretches located in this protein (Liu et al. 2005).

Flotillin-2 is myristoylated on the G2 residue, which is unique within the SPFH protein family members, and is palmitoylated on C4, 19 and 20. These post-translational modifications seem indispensable for membrane binding of flotillin-2 (Neumann-Giesen et al. 2004).

Cross-linking studies have shown that flotillins form rather stable homo- and hetero-tetramers of flotillin-1 and -2 in membranes and that oligomerization is dependent upon the C-terminal part of the molecules. Deletion studies have indicated that the presence of coiled-coil-2 and partially-coiled-coil-1 domains are responsible for tetramerization of flotillin-2. Moreover, the stability of flotillin-1 in the cell depends on the co-presence of flotillin-2 (Solis et al. 2007).

MPP1

MPP1 (p55) was first cloned by Ruff et al. (1991) and identified as a major palmitoylation substrate in the erythrocyte membrane. It is a member of the MAGUK (membrane associated guanylate kinase) family of proteins (te Velthuis et al. 2007; de Mendoza et al. 2010). It partially fulfills the criteria for scaffolding proteins (Pan et al. 2012), in that it contains several functional domains which are potentially responsible for simultaneous binding of regulatory and skeletal proteins and is, therefore, an important protein of the membrane skeleton ternary complex (Marfatia et al. 1995). All MAGUK family proteins share an enzymatically inactive guanylate kinase domain that mediates protein–protein interactions and intramolecular interactions with the SH3 domain and 0–1 unit of SH3 domain; 0–6 units of the PDZ domain, and 0–2 units of L27 (found in Lin-2 and Lin-7 receptor targeting proteins) domains. Different members of this family have ww, card, ZU5 or protein kinase domains. MPP1 also shares single GUK, SH3 and PDZ domains, as well as a D5/Hook/I3 domain which is responsible for protein 4.1R binding (Marfatia et al. 1997; Seo et al. 2009a). The best known role of MPP1 is its participation in the junctional complex formation of the erythrocyte membrane skeleton (for a current review see (Machnicka et al. 2012)). In this role, the PDZ-domain of MPP1 interacts with the cytoplasmic domain of glycophorin C, while the central region (D5-domain) is responsible for the interaction with protein 4.1R (Seo et al. 2009a, b). This interaction markedly strengthens protein 4.1-glycophorin C binding (Hemming et al. 1995). The functional role of MPP1 palmitoylation is currently not well understood, but one of the apparent functional aspects of this post-translational modification is discussed below.

The role of MPP1 in non-erythroid cells is understood rather poorly. Mburu et al. (2006) have shown that MPP1 forms a complex with whirlin, the protein which binds to the Usher protein network in the cochlea and the Crumbs network in the retina, by direct association with USH2A (usherin) and VLGR1 (a member of the 7-transmembrane receptor G-protein, which binds calcium and is expressed in the central nervous system). Mutations in this gene are associated with Usher syndrome 2 and familial febrile seizures. MPP1 binds also to MPP5, another member of the MPP subfamily of proteins (Gosens et al. 2007).

One of the established physiological roles of MPP1 is its engagement in the regulation of neutrophil polarity. Using a MPP1 knockout (p55$^{-/-}$) mouse model, Quinn et al. (2009) showed that, upon agonist-stimulation of neutrophils, MPP1 is rapidly recruited to the leading edge. Neutrophils of knockout mouse do not migrate efficiently *in vitro* and form multiple pseudopods upon chemotactic stimulation, in contrast to normal mouse neutrophils, which form a single pseudopod at the cell front required for efficient chemotaxis. Phosphorylation of Akt is decreased in these cells upon stimulation with chemoattractant, and this appears to be mediated by a PI3Kγ kinase-independent mechanism.

Visualization of Erythrocyte Membrane Rafts

Several techniques have been used to visualize membrane rafts *in situ* within living cells, including such methodologies as homo- and hetero-transfer, fluorescence polarization anisotropy, FCS, total internal reflection (TIRF), single-molecule microscopy and the super-resolution microscopical techniques of stimulated emission depletion (STED), photo-activated localization microscopy (PALM), and stochastic optical reconstruction (STORM). All of these techniques (and combinations of them) support the hypothesis of lateral membrane heterogeneity and the existence of nano-domains [for reviews, see: (Jacobson et al. 2007; He and Marguet 2011)]. Among others erythrocyte membranes were also a subject of several experimental efforts.

Mikhalyov and Samsonov (2011) used the fluorescent probe N-(BODIPY®-FL-propionyl)-neuraminosyl-GM1 (BODIPY-GM1) to detect rafts in erythrocyte membranes via fluorescence video-microscopy. Their observations indicated that the probe was uniformly distributed over the plasma membrane at 23 °C. However, a partial phase-separation of the probe was observed at 4 °C, detectable as discrete bright spots whose dimensions were comparable to the limit of resolution of the technique. They also observed red-shifted fluorescence of the probe, characteristic of high local concentrations of the BODIPY fluorophore. The shift was greatest at 4 °C, and the smallest at 37 °C and was eliminated by treatment of the erythrocytes with methyl-β-cyclodextrin. They also observed that distinct GM1 patches distributed over the entire membrane at both 23 °C and at 37 °C in erythrocytes stained with Alexa FL 647 cholera toxin subunit B conjugate (CTB-A647). The authors conclude that rafts could be detected in erythrocytes based on fluorometry and fluorescence microscopy methods.

The combined FCS (scanning fluorescence correlation spectroscopy) and Laurdan GP (generalized polarization) can be used to measure the fluctuations of the membrane packing microheterogeneity. Such studies have facilitated the measurement of generalized polarization fluctuations in the plasma membrane of intact rabbit erythrocytes and Chinese hamster ovary cells, indicative of the existence of tightly packed micro-domains moving within a more fluid background phase. These structures, which are characterized by different lipid packing, have different properties: 1: fast fluctuations with a small amplitude, corresponding to the domains of ~50 nm in diameter, 2: slow fluctuations, with a large amplitude, corresponding to domains of ~150 nm in diameter and 3: slow fluctuations with a small amplitude, corresponding to the large domains of diameter within a range of 150–600 nm (Sanchez et al. 2012). According to these authors, the small size and characteristic high-lipid packing indicate that these micro-domains have properties that have been proposed for lipid rafts.

The data presented above, however, do not reach the resolution achieved in the observation of other living cells through using STED far-field fluorescence nanoscopy, where raft components were found trapped transiently (~10–20 ms) in cholesterol-mediated complexes of areas <20 nm in diameter (Eggeling et al. 2009).

Role of Erythrocyte Membrane Rafts in Signaling

As was previously mentioned, one of the most recognized physiological functions of membrane rafts in living cells is the functional organization of signal-transduction platforms. Although the erythrocyte is a specialized cell in which not many signal transduction events can be observed, there are several indications that those signaling pathways do exist and are also raft-dependent in these cells.

Murphy et al. (2006) demonstrated that, by using a peptide that inhibits Gs-receptor interaction, erythrocyte Gs is functional and that propranolol, an antagonist of G protein–coupled β-adrenergic receptors, blocked Gs activity in erythrocytes. This drug was proposed for use in malarial infection treatment (Murphy et al. 2004, 2006, 2007).

One of the best examples of a role of membrane rafts in cell signaling in erythrocytes is the above-mentioned work demonstrating that disruption of rafts in erythrocyte membranes via lidocaine treatment, which occurs without altering membrane cholesterol content, suppressed Gsα-mediated signal transduction. This signaling pathway can be triggered via the activation of the adenosine receptor by the agonist, 5′-*N*-ethylcarboxamidoadenosine. Furthermore, they showed that raft-mediated signal transduction resulted in elevated cAMP levels, leading subsequently to the phosphorylation of adducin, a membrane skeletal protein. Signal transduction was restored upon removal of lidocaine from the erythrocyte membrane (Kamata et al. 2008). These authors (Koshino and Takakuwa 2009) showed that disruption of erythrocyte membrane-rafts with lidocaine also prevented *Plasmodium* invasion of erythrocytes in culture.

Another signaling pathway that has been discovered in erythrocytes is an apoptotic signal transduction mechanism. Mature red cells contain Fas, FasL, FADD (Fas-associated death domain), caspase 8, and caspase 3. Circulating, aged and oxidatively-stressed erythrocytes showed co-localization of Fas with the raft marker proteins Gsα and CD59. Aged red cells had significantly lower aminophospholipid translocase activity. These events were independent of calpain, but dependent on reactive oxygen species (ROS). Upon inhibition of ROS generation by treatment of cells with *t*-butyl hydroperoxide, Fas did not translocate into rafts and neither formation of a Fas-associated signaling complex nor caspase activation was observed. This indicates that translocation of Fas into rafts was the trigger for the signal leading to caspase 3 activation (Mandal et al. 2005). It was shown that administration of *N*-acetyl cysteine in combination with atorvastatin, an inhibitor of cholesterol synthesis, proved to be more effective than either of the drugs alone towards the rectification of arsenic-mediated early erythrocyte death. Cholesterol depletion via atorvastatin treatment, along with reduced glutathione (GSH) repletion via *N*-acetyl cysteine treatment blocked Fas activation, leading to premature death of erythrocytes (Biswas et al. 2011). Similarly, mouse erythrocytes exposed to lead ions (Pb^{2+}) undergo FAS-activated apoptotic death, with FAS observed to translocate to the DRM fraction in such cells (Mandal et al. 2012).

Erythrocyte Rafts and Pathology

Although numerous reports exist on the importance of rafts in many pathologies, such as neuropathological disorders and neoplasia, there is sparse data on red cell pathologies in which membrane rafts are reported to play an important role.

As one of the main protein components of erythrocyte membrane rafts is stomatin, the behavior of this protein in stomatocytosis, particularly in overhydrated stomatocytosis, was studied (Wilkinson et al. 2008). In this rare, dominantly-inherited, hemolytic anemia, erythrocytes show stomatocytic morphology, deficiency of stomatin in the membrane and a major leak of the monovalent cations Na^+ and K^+ across the membrane (Lande et al. 1982). However, the *STOM* gene encoding stomatin is almost certainly not mutated in this pathology. Rather, it appears that there is probably a trafficking defect in the developing erythroid cells, so that stomatin is obstructed within the secretory pathway and is not able to be incorporated into the plasma membrane of the mature erythrocytes (Fricke et al. 2005). When the DRM fraction was obtained from the erythrocytes of such patients, it was found to contain a reduced amount of stomatin, with the total membrane stomatin content reduced by more than 50 %. At the same time, the other raft proteins, such as flotillin-1 and -2, and Glut-1, were present in the DRM fraction in normal quantities. According to the authors, the DRM of patient erythrocytes contained decreased amounts of DRM-associated actin and tropomodulin (Wilkinson et al. 2008). The authors suggest that the raft tropomodulin, seen in normal erythrocytes, may act as a reservoir of available tropomodulin accessible for control of actin turnover. A fuller answer to this question still awaits further studies. As our own experience is concerned (see below), DRM isolated from human erythrocytes under alkaline conditions, or in the presence of latrunculin, contained normal quantities of flotillins and stomatin, but they contained only traces of actin or spectrin (Lach et al. 2012). This might be an indication of really weak raft-membrane skeleton interactions.

In some of the cases of hereditary hereditart spherocytosis, erythrocyte membranes were found to have membrane raft proteins flotillin-1 and -2 (Margetis et al. 2007) mildly reduced (75–90 %) and were observed to contain sorcin. Sorcin is a 22 kDa (band 8) protein, which is found in nanovesicles released from normal erythrocytes in the presence of high concentrations of Ca^{2+} (Allan et al. 1980; Salzer et al. 2002).

During our studies on hereditary spherocytoses, we found a family whose two male siblings (brothers) displayed symptoms of hemolytic anemia that did not fit to the traditional characteristics of the known disease. The parents of the brothers and their sister, along with the family members on both sides of the siblings' parents, were without symptoms of hemolytic anemia. As the patients showed the presence of stomatocytes, their Na^+/K^+ equilibrium was tested, along with their red blood cell membrane protein profile via SDS-PAGE. However, no changes were noticed in either of these parameters, compared to the unaffected individuals tested. Further, no protein acyl transferase (PAT) activity was detectable in their erythrocytes, which was caused by an absence of membrane DHHC17 protein, while near-normal transcript levels of DHHC17 gene were present in the reticulocytes of the two patients.

DHHC17 is a member of the protein acyltransferase (PAT) family of genes, of which 23 are present in the human genome (Korycka et al. 2012), but only one form is expressed in reticulocytes (Lach et al. 2012).

MPP1 and Its Palmitoylation Play a Crucial Role in Lateral Membrane Organization in Erythroid Cells

Lack of palmitoylation activity was found to lead to marked changes in lateral membrane organization, as revealed by a marked decrease in the DRM fraction of the erythrocyte membrane (see above). It was found that the mechanism underlying these changes involves MPP1, which is the major palmitoylation substrate of the erythrocyte membrane. It was confirmed that this protein was absent from the DRM fraction of erythrocytes of both of the patients (in the case of above-described hemolytic anemia) or from DRM erythrocyte fractions derived from healthy control individuals in which palmitoylation was inhibited by treatment with the potent PAT inhibitor, 2-BrP (2-bromo palmitate), first by MS/MS analysis (Grzybek and Sikorski –unpublished data) and then, by immunoblotting (Lach et al. 2012).

Separately, DHHC17-directed siRNA treatment was performed on reticulocytes isolated from human umbilical cord blood. Here, Western Blot analysis of sucrose density gradient fractions using anti-MPP1 antibodies indicated that the DHHC17 siRNA-transfected reticulocyte DRM fraction contained much less MPP1 than the same fraction obtained from reticulocytes transfected with control RNA of a scrambled sequence. A decreased amount of DRM was also observed when erythrocyte ghosts were resealed with a specific anti-MPP1 antibody.

The possibility of a mutation in the *MPP1* gene seemed very likely, as this gene locus is located in the X chromosome, which would fit the pattern of inheritance of the studied family, where only male individuals were affected. However, only a silent mutation (ACG_ACT; T85T) was identified in these patients. No mutations were found in the nucleotide sequences coding for the cytoplasmic-domain of glycophorin C (*GYPC*), or in the coding sequence of the *EPB4.1R* gene (coding for protein 4.1). Moreover, sequencing of the stomatin transcript, which was present at the normal level, did not reveal mutations or polymorphisms.

The change in membrane solubility in 1 % Triton X-100 solution cannot be an effect of differences in lipid composition, as both the TLC and quantitative analysis of lipids in lipid extracts from erythrocyte ghosts showed no significant variations among the major lipid classes (including cholesterol content) between control healthy individuals and patients. Therefore, we concluded that (palmitoylated) MPP1 might play a crucial role in lateral membrane organization in the human erythrocyte, which is connected to the unique pathology of this cell.

Moreover, further studies performed on HEL cells (derived from erythroid precursors) showed a similar effect of palmitoylation inhibition on the DRM fraction.

When a stable HEL cell-line with a silenced *MPP1* gene was used for experiments, the same effect was observed. Moreover, the FLIM experiments using di-4 probe performed on normal and 2-BrP-treated erythrocytes, normal and 2-BrP-treated HEL cells and HEL cells with stably-silenced *MPP1* expression, demonstrated significant decrease in membrane-order upon inhibition of palmitoylation, or decrease in cellular MPP1 level (Biernatowska et al. 2013; Lach et al. 2012). Similar conclusions were drawn from studies on giant plasma membrane vesicles derived from HEL cells, performed by using several advanced biophysical techniques in parallel (Podkalicka, Grzybek et al. – to be published).

Furthermore, *MPP1*-knockdown significantly affects the activation of MAP-kinase signaling via raft-dependent tyrosine kinase receptors, indicating the importance of MPP1 for lateral membrane organization (Biernatowska et al. 2013).

In conclusion, palmitoylation of MPP1 appears to be at least one of the mechanisms controlling lateral organization of the cell membrane. Thus, these studies point to a new role for MPP1, and present a novel linkage between membrane-raft organization and protein palmitoylation.

An important question that remains is how MPP1 affects the organization of the membrane domains. Our hypothesis is that, upon palmitoylation, the affinity of MPP1 for membrane-skeleton binding decreases and MPP1 becomes freely available for binding to the pre-existing nanoclusters or 'unstable rafts elements' within the membrane, namely, protein-cholesterol/lipid complexes corresponding to cholesterol-depletion sensitive, short-lived (<0.1 ms) nanoclusters (<10 nm in diameter), observed previously by others (Fujita and Kinoshita 2010; Mayor and Rao 2004; Sharma et al. 2004). This suggestion is based on the observations that, in normal erythrocytes and in HEL cells, the unpalmitoylated MPP1 remains in the high-density "skeletal" fraction within the density-gradient profile of the DRM. Binding of palmitoylated MPP1 to the pre-existing nanoclusters induces their fusion into nanodomains and stabilizes them as membrane "rafts" (resting-state rafts), which are larger (~20 nm) in diameter (Eggeling et al. 2009), more stable, and detergent-resistant domains, which become functional. This is consistent with the model proposed by Hancock (2006). Since palmitoylation has been reported previously to be involved in the regulation of membrane-protein clustering, such as tetraspanin (Berditchevski et al. 2002; Hemler 2005) or $GABA_A$ receptors (Rathenberg et al. 2004), and has also been shown to promote oligomerization of certain proteins (Charollais and Van Der Goot 2009; Feig et al. 2007), we postulate that palmitoylated MPP1 in erythroid cells plays a role in raft protein(s) oligomerization.

It should be stressed, however, that the MPP1-based mechanism of raft organization is not the only possible mechanism driving lateral membrane organization. For example, a similar function was recently suggested for the palmitoylation of Rac1 in COS-7 and MEF cells (Navarro-Lerida et al. 2012), and these authors implicate a role for the actin cytoskeleton in the mechanism of raft clustering in these cells.

The Involvement of Actin in Erythrocyte Membrane Rafts Organization?

Substantial indications are available in the literature on the relationship of domain formation and stabilization with cortical actin (Goswami et al. 2008; Hummel et al. 2011). In addition, MPP1 is a membrane-skeleton protein that participates in the formation of "vertical" linkages between the membrane-skeleton and the membrane bilayer containing integral proteins (for recent reviews, see (Machnicka et al. 2012, 2014)). It is known that the DRM fraction from erythrocyte membranes contains both spectrin and actin, in addition to MPP1 and raft marker proteins such as flotillins, stomatin and other GPI-anchored proteins. It could be speculated that the presence of MPP1 in raft domains could arise as a result of spectrin-actin complex formation. Also, direct interactions of membrane skeletal proteins with membrane-lipids have been implied by numerous works (for reviews, see e.g. (Hanus-Lorenz et al. 2001; Boguslawska et al. 2014). Such data might contribute to establishing the link between detergent-resistant membranes and MPP1. However, the above-mentioned simple experiments on the isolation of a DRM fraction under alkaline conditions, i.e. in the presence of Na_2CO_3, or from latrunculin-treated erythrocyte ghosts, rather exclude this possibility (Lach et al. 2012).

Conclusions

From the above discussion, several conclusions can be drawn. The most important and, probably, the most obvious is that membrane rafts in the mature erythrocyte are not the remnants from erythroid precursor cell(s), but are real and dynamic domains functioning within the membrane, helping to fulfil physiological roles, of which only a few of them are just now being recognized.

The next conclusion is the presence and role of the MPP1 protein in these domains. It has been known for some time that palmitoylation drives proteins into the raft-domain in non-erythroid cells (Levental et al. 2009; Yang et al. 2010; Barnes et al. 2006; Babina et al. 2014), but the role of MPP1 in this remained unknown until recently, although it was well known that this was a major palmitoylation target in erythrocytes (Maretzki et al. 1990; Ruff et al. 1991). This fact was noticed in early studies, explaining the property of tighter binding of MPP1 to the membrane, even after extraction at pH 11 (Ruff et al. 1991). However, the most important conclusion stems from the fact that (palmitoylated) MPP1 functions as an organizer of membrane rafts in erythroid cells. We postulate that the binding of palmitoylated MPP1 to the pre-existing nano-assemblies induces their fusion into nano-domains and stabilizes them as larger (~20 nm diameter) "resting state rafts" that are more stable and detergent-resistant domains which assume a level of functional organization.

The above conclusion is strengthened by the fact that a lack of MPP1 palmitoylation, resulting from an absence of DHHC17, the only erythroid palmitoyltransferase, leads to the observed pathology. However, the molecular mechanism by which this occurs is still to be identified.

Therefore, erythrocyte membrane understanding will increasingly focus on what components and conditions are necessary to fulfill the functional roles of membrane domains, such as raft assemblies, and the mechanisms by which these are effected. In this regard, MPP1, particularly its altered behavior in the free and in the palmitoylated states, now becomes an increasingly important component and target in the regulation, maintenance and function of the normal erythrocyte membrane.

References

Allan D, Thomas P, Limbrick AR (1980) The isolation and characterization of 60 nm vesicles ('nanovesicles') produced during ionophore A23187-induced budding of human erythrocytes. Biochem J 188(3):881–887

Aydar E, Palmer CP, Djamgoz MB (2004) Sigma receptors and cancer: possible involvement of ion channels. Cancer Res 64(15):5029–5035. doi:10.1158/0008-5472.CAN-03-2329

Babina IS, McSherry EA, Donatello S, Hill AD, Hopkins AM (2014) A novel mechanism of regulating breast cancer cell migration via palmitoylation-dependent alterations in the lipid raft affiliation of CD44. Breast Cancer Res 16(1):R19. doi:10.1186/bcr3614

Barnes NC, Powell MS, Trist HM, Gavin AL, Wines BD, Hogarth PM (2006) Raft localisation of FcgammaRIIa and efficient signaling are dependent on palmitoylation of cysteine 208. Immunol Lett 104(1–2):118–123. doi:10.1016/j.imlet.2005.11.007

Berditchevski F, Odintsova E, Sawada S, Gilbert E (2002) Expression of the palmitoylation-deficient CD151 weakens the association of alpha 3 beta 1 integrin with the tetraspanin-enriched microdomains and affects integrin-dependent signaling. J Biol Chem 277(40): 36991–37000

Bickel PE, Scherer PE, Schnitzer JE, Oh P, Lisanti MP, Lodish HF (1997) Flotillin and epidermal surface antigen define a new family of caveolae-associated integral membrane proteins. J Biol Chem 272(21):13793–13802

Biernatowska A, Podkalicka J, Majkowski M, Hryniewicz-Jankowska A, Augoff K, Kozak K, Korzeniewski J, Sikorski AF (2013) The role of MPP1/p55 and its palmitoylation in resting state raft organization in HEL cells. Biochim Biophys Acta 1833(8):1876–1884. doi:10.1016/j.bbamcr.2013.03.009

Biswas D, Sen G, Sarkar A, Biswas T (2011) Atorvastatin acts synergistically with N-acetyl cysteine to provide therapeutic advantage against Fas-activated erythrocyte apoptosis during chronic arsenic exposure in rats. Toxicol Appl Pharmacol 250(1):39–53. doi:10.1016/j.taap.2010.10.002

Boggon TJ, Eck MJ (2004) Structure and regulation of Src family kinases. Oncogene 23(48):7918–7927. doi:10.1038/sj.onc.1208081

Boguslawska DM, Machnicka B, Sikorski AF (2010) Hereditary stomatocytoses—diagnostic problems and their molecular basis. Pol Merkur Lekarski 29(170):119–124

Boguslawska DM, Machnicka B, Hryniewicz-Jankowska A, Czogalla A (2014) Spectrin and phospholipids—the current picture of their fascinating interplay. Cell Mol Biol Lett 19(1):158–179. doi:10.2478/s11658-014-0185-5

Brand J, Smith ES, Schwefel D, Lapatsina L, Poole K, Omerbasic D, Kozlenkov A, Behlke J, Lewin GR, Daumke O (2012) A stomatin dimer modulates the activity of acid-sensing ion channels. EMBO J 31(17):3635–3646. doi:10.1038/emboj.2012.203

Browman DT, Hoegg MB, Robbins SM (2007) The SPFH domain-containing proteins: more than lipid raft markers. Trends Cell Biol 17(8):394–402. doi:10.1016/j.tcb.2007.06.005

Brown DA, London E (2000) Structure and function of sphingolipid- and cholesterol-rich membrane rafts. J Biol Chem 275(23):17221–17224. doi:10.1074/jbc.R000005200

Brugger B, Glass B, Haberkant P, Leibrecht I, Wieland FT, Krausslich HG (2006) The HIV lipidome: a raft with an unusual composition. Proc Natl Acad Sci U S A 103(8):2641–2646. doi:10.1073/pnas.0511136103

Charollais J, Van Der Goot FG (2009) Palmitoylation of membrane proteins (Review). Mol Membr Biol 26(1):55–66

Chi CN, Haq SR, Rinaldo S, Dogan J, Cutruzzola F, Engstrom A, Gianni S, Lundstrom P, Jemth P (2012) Interactions outside the boundaries of the canonical binding groove of a PDZ domain influence ligand binding. Biochemistry 51(44):8971–8979. doi:10.1021/bi300792h

de Mendoza A, Suga H, Ruiz-Trillo I (2010) Evolution of the MAGUK protein gene family in premetazoan lineages. BMC Evol Biol 10:93. doi:10.1186/1471-2148-10-93

Eggeling C, Ringemann C, Medda R, Schwarzmann G, Sandhoff K, Polyakova S, Belov VN, Hein B, von Middendorff C, Schonle A, Hell SW (2009) Direct observation of the nanoscale dynamics of membrane lipids in a living cell. Nature 457(7233):1159–1162. doi:10.1038/nature07596

Epand RM (2006) Cholesterol and the interaction of proteins with membrane domains. Prog Lipid Res 45(4):279–294. doi:10.1016/j.plipres.2006.02.001

Epand RM, Sayer BG, Epand RF (2005) Caveolin scaffolding region and cholesterol-rich domains in membranes. J Mol Biol 345(2):339–350. doi:10.1016/j.jmb.2004.10.064

Fantini J, Barrantes FJ (2013) How cholesterol interacts with membrane proteins: an exploration of cholesterol-binding sites including CRAC, CARC, and tilted domains. Front Physiol 4:31. doi:10.3389/fphys.2013.00031

Feig C, Tchikov V, Schutze S, Peter ME (2007) Palmitoylation of CD95 facilitates formation of SDS-stable receptor aggregates that initiate apoptosis signaling. EMBO J 26(1):221–231

Fricke B, Parsons SF, Knopfle G, von During M, Stewart GW (2005) Stomatin is mis-trafficked in the erythrocytes of overhydrated hereditary stomatocytosis, and is absent from normal primitive yolk sac-derived erythrocytes. Br J Haematol 131(2):265–277. doi:10.1111/j.1365-2141.2005.05742.x

Fujita M, Kinoshita T (2010) Structural remodeling of GPI anchors during biosynthesis and after attachment to proteins. FEBS Lett 584(9):1670–1677, S0014-5793(09)00871-0 [pii]10.1016/j.febslet.2009.10.079

Gosens I, van Wijk E, Kersten FF, Krieger E, van der Zwaag B, Marker T, Letteboer SJ, Dusseljee S, Peters T, Spierenburg HA, Punte IM, Wolfrum U, Cremers FP, Kremer H, Roepman R (2007) MPP1 links the Usher protein network and the Crumbs protein complex in the retina. Hum Mol Genet 16(16):1993–2003. doi:10.1093/hmg/ddm147

Goswami D, Gowrishankar K, Bilgrami S, Ghosh S, Raghupathy R, Chadda R, Vishwakarma R, Rao M, Mayor S (2008) Nanoclusters of GPI-anchored proteins are formed by cortical actin-driven activity. Cell 135(6):1085–1097. doi:10.1016/j.cell.2008.11.032

Grzybek M, Kozubek A, Dubielecka P, Sikorski AF (2005) Rafts—the current picture. Folia Histochem Cytobiol 43(1):3–10

Grzybek M, Kubiak J, Lach A, Przybylo M, Sikorski AF (2009) A raft-associated species of phosphatidylethanolamine interacts with cholesterol comparably to sphingomyelin. A Langmuir-Blodgett monolayer study. PLoS One 4(3):e5053

Hakomori SI (2002) The glycosynapse. Proc Natl Acad Sci U S A 99(1):225–232. doi:10.1073/pnas.012540899

Hancock JF (2006) Lipid rafts: contentious only from simplistic standpoints. Nat Rev Mol Cell Biol 7(6):456–462. doi:10.1038/nrm1925

Hanus-Lorenz B, Hryniewicz-Jankowska A, Leluk J, Lorenz M, Skala J, Sikorski AF (2001) Spectrin motifs are detected in plant and yeast genomes. Cell Mol Biol Lett 6(2):207

He HT, Marguet D (2011) Detecting nanodomains in living cell membrane by fluorescence correlation spectroscopy. Annu Rev Phys Chem 62:417–436. doi:10.1146/annurev-physchem-032210-103402

Hemler ME (2005) Tetraspanin functions and associated microdomains. Nat Rev Mol Cell Biol 6(10):801–811

Hemming NJ, Anstee DJ, Staricoff MA, Tanner MJ, Mohandas N (1995) Identification of the membrane attachment sites for protein 4.1 in the human erythrocyte. J Biol Chem 270(10): 5360–5366

Hiebl-Dirschmied CM, Adolf GR, Prohaska R (1991a) Isolation and partial characterization of the human erythrocyte band 7 integral membrane protein. Biochim Biophys Acta 1065(2):195–202

Hiebl-Dirschmied CM, Entler B, Glotzmann C, Maurer-Fogy I, Stratowa C, Prohaska R (1991b) Cloning and nucleotide sequence of cDNA encoding human erythrocyte band 7 integral membrane protein. Biochim Biophys Acta 1090(1):123–124

Hryniewicz-Jankowska A, Augoff K, Biernatowska A, Podkalicka J, Sikorski AF (2014) Membrane rafts as a novel target in cancer therapy. Biochim Biophys Acta 1845(2):155–165. doi:10.1016/j.bbcan.2014.01.006

Hummel I, Klappe K, Ercan C, Kok JW (2011) Multidrug resistance-related protein 1 (MRP1) function and localization depend on cortical actin. Mol Pharmacol 79(2):229–240. doi:10.1124/mol.110.069013

Ipsen JH, Karlstrom G, Mouritsen OG, Wennerstrom H, Zuckermann MJ (1987) Phase equilibria in the phosphatidylcholine-cholesterol system. Biochim Biophys Acta 905:162–172

Jacobson K, Mouritsen OG, Anderson RG (2007) Lipid rafts: at a crossroad between cell biology and physics. Nat Cell Biol 9(1):7–14. doi:10.1038/ncb0107-7

Kadurin I, Huber S, Grunder S (2009) A single conserved proline residue determines the membrane topology of stomatin. Biochem J 418(3):587–594. doi:10.1042/BJ20081662

Kamata K, Manno S, Ozaki M, Takakuwa Y (2008) Functional evidence for presence of lipid rafts in erythrocyte membranes: Gsalpha in rafts is essential for signal transduction. Am J Hematol 83(5):371–375. doi:10.1002/ajh.21126

Karnovsky MJ, Kleinfeld AM, Hoover RL, Klausner RD (1982) The concept of lipid domains in membranes. J Cell Biol 94(1):1–6

Korycka J, Lach A, Heger E, Boguslawska DM, Wolny M, Toporkiewicz M, Augoff K, Korzeniewski J, Sikorski AF (2012) Human DHHC proteins: a spotlight on the hidden player of palmitoylation. Eur J Cell Biol 91(2):107–117. doi:10.1016/j.ejcb.2011.09.013

Koshino I, Takakuwa Y (2009) Disruption of lipid rafts by lidocaine inhibits erythrocyte invasion by Plasmodium falciparum. Exp Parasitol 123(4):381–383. doi:10.1016/j.exppara.2009.08.019

Koumanov KS, Tessier C, Momchilova AB, Rainteau D, Wolf C, Quinn PJ (2005) Comparative lipid analysis and structure of detergent-resistant membrane raft fractions isolated from human and ruminant erythrocytes. Arch Biochem Biophys 434(1):150–158. doi:10.1016/j.abb.2004.10.025

Lach A, Grzybek M, Heger E, Korycka J, Wolny M, Kubiak J, Kolondra A, Boguslawska DM, Augoff K, Majkowski M, Podkalicka J, Kaczor J, Stefanko A, Kuliczkowski K, Sikorski AF (2012) Palmitoylation of MPP1 (membrane-palmitoylated protein 1)/p55 is crucial for lateral membrane organization in erythroid cells. J Biol Chem 287(23):18974–18984. doi:10.1074/jbc.M111.332981

Lande WM, Thiemann PV, Mentzer WC Jr (1982) Missing band 7 membrane protein in two patients with high Na, low K erythrocytes. J Clin Invest 70(6):1273–1280

Lang DM, Lommel S, Jung M, Ankerhold R, Petrausch B, Laessing U, Wiechers MF, Plattner H, Stuermer CA (1998) Identification of reggie-1 and reggie-2 as plasmamembrane-associated proteins which cocluster with activated GPI-anchored cell adhesion molecules in non-caveolar micropatches in neurons. J Neurobiol 37(4):502–523

Levental I, Byfield FJ, Chowdhury P, Gai F, Baumgart T, Janmey PA (2009) Cholesterol-dependent phase separation in cell-derived giant plasma-membrane vesicles. Biochem J 424(2):163–167. doi:10.1042/BJ20091283

Liu J, Deyoung SM, Zhang M, Dold LH, Saltiel AR (2005) The stomatin/prohibitin/flotillin/HflK/C domain of flotillin-1 contains distinct sequences that direct plasma membrane localization and protein interactions in 3T3-L1 adipocytes. J Biol Chem 280(16):16125–16134. doi:10.1074/jbc.M500940200

Macdonald JL, Pike LJ (2005) A simplified method for the preparation of detergent-free lipid rafts. J Lipid Res 46(5):1061–1067. doi:10.1194/jlr.D400041-JLR200
Machnicka B, Grochowalska R, Boguslawska DM, Sikorski AF, Lecomte MC (2012) Spectrin-based skeleton as an actor in cell signaling. Cell Mol Life Sci 69(2):191–201. doi:10.1007/s00018-011-0804-5
Machnicka B, Czogalla A, Hryniewicz-Jankowska A, Boguslawska DM, Grochowalska R, Heger E, Sikorski AF (2014) Spectrins: a structural platform for stabilization and activation of membrane channels, receptors and transporters. Biochim Biophys Acta 1838(2):620–634. doi:10.1016/j.bbamem.2013.05.002
Mahfoud R, Garmy N, Maresca M, Yahi N, Puigserver A, Fantini J (2002) Identification of a common sphingolipid-binding domain in Alzheimer, prion, and HIV-1 proteins. J Biol Chem 277(13):11292–11296. doi:10.1074/jbc.M111679200
Mandal D, Mazumder A, Das P, Kundu M, Basu J (2005) Fas-, caspase 8-, and caspase 3-dependent signaling regulates the activity of the aminophospholipid translocase and phosphatidylserine externalization in human erythrocytes. J Biol Chem 280(47):39460–39467. doi:10.1074/jbc.M506928200
Mandal S, Mukherjee S, Chowdhury KD, Sarkar A, Basu K, Paul S, Karmakar D, Chatterjee M, Biswas T, Sadhukhan GC, Sen G (2012) S-allyl cysteine in combination with clotrimazole downregulates Fas induced apoptotic events in erythrocytes of mice exposed to lead. Biochim Biophys Acta 1820(1):9–23. doi:10.1016/j.bbagen.2011.09.019
Maretzki D, Mariani M, Lutz HU (1990) Fatty acid acylation of membrane skeletal proteins in human erythrocytes. FEBS Lett 259(2):305–310
Marfatia SM, Leu RA, Branton D, Chishti AH (1995) Identification of the protein 4.1 binding interface on glycophorin C and p55, a homologue of the Drosophila discs-large tumor suppressor protein. J Biol Chem 270(2):715–719
Marfatia SM, Morais-Cabral JH, Kim AC, Byron O, Chishti AH (1997) The PDZ domain of human erythrocyte p55 mediates its binding to the cytoplasmic carboxyl terminus of glycophorin C. Analysis of the binding interface by in vitro mutagenesis. J Biol Chem 272(39):24191–24197
Margetis P, Antonelou M, Karababa F, Loutradi A, Margaritis L, Papassideri I (2007) Physiologically important secondary modifications of red cell membrane in hereditary spherocytosis-evidence for in vivo oxidation and lipid rafts protein variations. Blood Cells Mol Dis 38(3):210–220. doi:10.1016/j.bcmd.2006.10.163
Mayor S, Rao M (2004) Rafts: scale-dependent, active lipid organization at the cell surface. Traffic 5(4):231–240
Mburu P, Kikkawa Y, Townsend S, Romero R, Yonekawa H, Brown SD (2006) Whirlin complexes with p55 at the stereocilia tip during hair cell development. Proc Natl Acad Sci U S A 103(29):10973–10978. doi:10.1073/pnas.0600923103
McIntosh TJ, Vidal A, Simon SA (2003) Sorting of lipids and transmembrane peptides between detergent-soluble bilayers and detergent-resistant rafts. Biophys J 85(3):1656–1666. doi:10.1016/S0006-3495(03)74595-0
Mikhalyov I, Samsonov A (2011) Lipid raft detecting in membranes of live erythrocytes. Biochim Biophys Acta 1808(7):1930–1939. doi:10.1016/j.bbamem.2011.04.002
Morrow IC, Rea S, Martin S, Prior IA, Prohaska R, Hancock JF, James DE, Parton RG (2002) Flotillin-1/reggie-2 traffics to surface raft domains via a novel golgi-independent pathway. Identification of a novel membrane targeting domain and a role for palmitoylation. J Biol Chem 277(50):48834–48841. doi:10.1074/jbc.M209082200
Murphy SC, Samuel BU, Harrison T, Speicher KD, Speicher DW, Reid ME, Prohaska R, Low PS, Tanner MJ, Mohandas N, Haldar K (2004) Erythrocyte detergent-resistant membrane proteins: their characterization and selective uptake during malarial infection. Blood 103(5):1920–1928. doi:10.1182/blood-2003-09-3165
Murphy SC, Harrison T, Hamm HE, Lomasney JW, Mohandas N, Haldar K (2006) Erythrocyte G protein as a novel target for malarial chemotherapy. PLoS Med 3(12):e528. doi:10.1371/journal.pmed.0030528

Murphy SC, Fernandez-Pol S, Chung PH, Prasanna Murthy SN, Milne SB, Salomao M, Brown HA, Lomasney JW, Mohandas N, Haldar K (2007) Cytoplasmic remodeling of erythrocyte raft lipids during infection by the human malaria parasite Plasmodium falciparum. Blood 110(6):2132–2139. doi:10.1182/blood-2007-04-083873

Navarro-Lerida I, Sanchez-Perales S, Calvo M, Rentero C, Zheng Y, Enrich C, Del Pozo MA (2012) A palmitoylation switch mechanism regulates Rac1 function and membrane organization. EMBO J 31(3):534–551. doi:10.1038/emboj.2011.446emboj2011446 [pii]

Neumann-Giesen C, Falkenbach B, Beicht P, Claasen S, Luers G, Stuermer CA, Herzog V, Tikkanen R (2004) Membrane and raft association of reggie-1/flotillin-2: role of myristoylation, palmitoylation and oligomerization and induction of filopodia by overexpression. Biochem J 378(Pt 2):509–518. doi:10.1042/BJ20031100

Pan CQ, Sudol M, Sheetz M, Low BC (2012) Modularity and functional plasticity of scaffold proteins as p(l)acemakers in cell signaling. Cell Signal 24(11):2143–2165. doi:10.1016/j.cellsig.2012.06.002

Pike LJ (2003) Lipid rafts: bringing order to chaos. J Lipid Res 44(4):655–667. doi:10.1194/jlr.R200021-JLR200

Pike LJ (2006) Rafts defined: a report on the Keystone symposium on lipid rafts and cell function. J Lipid Res 47(7):1597–1598. doi:10.1194/jlr.E600002-JLR200

Pike LJ, Han X, Gross RW (2005) Epidermal growth factor receptors are localized to lipid rafts that contain a balance of inner and outer leaflet lipids: a shotgun lipidomics study. J Biol Chem 280(29):26796–26804. doi:10.1074/jbc.M503805200

Quinn PJ, Tessier C, Rainteau D, Koumanov KS, Wolf C (2005) Structure and thermotropic phase behaviour of detergent-resistant membrane raft fractions isolated from human and ruminant erythrocytes. Biochim Biophys Acta 1713(1):5–14. doi:10.1016/j.bbamem.2005.04.013

Quinn BJ, Welch EJ, Kim AC, Lokuta MA, Huttenlocher A, Khan AA, Kuchay SM, Chishti AH (2009) Erythrocyte scaffolding protein p55/MPP1 functions as an essential regulator of neutrophil polarity. Proc Natl Acad Sci U S A 106(47):19842–19847. doi:10.1073/pnas.0906761106

Rathenberg J, Kittler JT, Moss SJ (2004) Palmitoylation regulates the clustering and cell surface stability of GABAA receptors. Mol Cell Neurosci 26(2):251–257

Rodgers W, Glaser M (1991) Characterization of lipid domains in erythrocyte membranes. Proc Natl Acad Sci U S A 88(4):1364–1368

Ruff P, Speicher DW, Husain-Chishti A (1991) Molecular identification of a major palmitoylated erythrocyte membrane protein containing the src homology 3 motif. Proc Natl Acad Sci U S A 88(15):6595–6599

Salzer U, Prohaska R (2001) Stomatin, flotillin-1, and flotillin-2 are major integral proteins of erythrocyte lipid rafts. Blood 97(4):1141–1143

Salzer U, Hinterdorfer P, Hunger U, Borken C, Prohaska R (2002) Ca(++)-dependent vesicle release from erythrocytes involves stomatin-specific lipid rafts, synexin (annexin VII), and sorcin. Blood 99(7):2569–2577

Sanchez SA, Tricerri MA, Gratton E (2012) Laurdan generalized polarization fluctuations measures membrane packing micro-heterogeneity in vivo. Proc Natl Acad Sci U S A 109(19):7314–7319. doi:10.1073/pnas.1118288109

Schroeder F, Woodford JK, Kavecansky J, Wood WG, Joiner C (1995) Cholesterol domains in biological membranes. Mol Membr Biol 12(1):113–119

Seo PS, Jeong JJ, Zeng L, Takoudis CG, Quinn BJ, Khan AA, Hanada T, Chishti AH (2009a) Alternatively spliced exon 5 of the FERM domain of protein 4.1R encodes a novel binding site for erythrocyte p55 and is critical for membrane targeting in epithelial cells. Biochim Biophys Acta 1793(2):281–289. doi:10.1016/j.bbamcr.2008.09.012

Seo PS, Quinn BJ, Khan AA, Zeng L, Takoudis CG, Hanada T, Bolis A, Bolino A, Chishti AH (2009b) Identification of erythrocyte p55/MPP1 as a binding partner of NF2 tumor suppressor protein/Merlin. Exp Biol Med 234(3):255–262. doi:10.3181/0809-RM-275

Sharma P, Varma R, Sarasij RC, Ira, Gousset K, Krishnamoorthy G, Rao M, Mayor S (2004) Nanoscale organization of multiple GPI-anchored proteins in living cell membranes. Cell 116(4):577–589

Simons K, Ikonen E (1997) Functional rafts in cell membranes. Nature 387(6633):569–572. doi:10.1038/42408
Simons K, Sampaio JL (2011) Membrane organization and lipid rafts. Cold Spring Harb Perspect Biol 3(10):a004697. doi:10.1101/cshperspect.a004697
Simons K, Vaz WL (2004) Model systems, lipid rafts, and cell membranes. Annu Rev Biophys Biomol Struct 33:269–295. doi:10.1146/annurev.biophys.32.110601.141803
Snyers L, Umlauf E, Prohaska R (1999) Cysteine 29 is the major palmitoylation site on stomatin. FEBS Lett 449(2–3):101–104
Solis GP, Hoegg M, Munderloh C, Schrock Y, Malaga-Trillo E, Rivera-Milla E, Stuermer CA (2007) Reggie/flotillin proteins are organized into stable tetramers in membrane microdomains. Biochem J 403(2):313–322. doi:10.1042/BJ20061686
Staubach S, Hanisch FG (2011) Lipid rafts: signaling and sorting platforms of cells and their roles in cancer. Expert Rev Proteomics 8(2):263–277. doi:10.1586/epr.11.2
te Velthuis AJ, Admiraal JF, Bagowski CP (2007) Molecular evolution of the MAGUK family in metazoan genomes. BMC Evol Biol 7:129. doi:10.1186/1471-2148-7-129
Umlauf E, Mairhofer M, Prohaska R (2006) Characterization of the stomatin domain involved in homo-oligomerization and lipid raft association. J Biol Chem 281(33):23349–23356. doi:10.1074/jbc.M513720200
Wang TY, Leventis R, Silvius JR (2000) Fluorescence-based evaluation of the partitioning of lipids and lipidated peptides into liquid-ordered lipid microdomains: a model for molecular partitioning into "lipid rafts". Biophys J 79(2):919–933. doi:10.1016/S0006-3495(00)76347-8
Wilkinson DK, Turner EJ, Parkin ET, Garner AE, Harrison PJ, Crawford M, Stewart GW, Hooper NM (2008) Membrane raft actin deficiency and altered Ca2+-induced vesiculation in stomatin-deficient overhydrated hereditary stomatocytosis. Biochim Biophys Acta 1778(1):125–132. doi:10.1016/j.bbamem.2007.09.016
Yang W, Di Vizio D, Kirchner M, Steen H, Freeman MR (2010) Proteome scale characterization of human S-acylated proteins in lipid raft-enriched and non-raft membranes. Mol Cell Proteomics 9(1):54–70. doi:10.1074/mcp.M800448-MCP200
Yokoyama H, Fujii S, Matsui I (2008) Crystal structure of a core domain of stomatin from Pyrococcus horikoshii illustrates a novel trimeric and coiled-coil fold. J Mol Biol 376(3): 868–878. doi:10.1016/j.jmb.2007.12.024

Chapter 6
Immuno-Modulatory Role of Porins: Host Immune Responses, Signaling Mechanisms and Vaccine Potential

Sanica C. Sakharwade, G.V.R. Krishna Prasad, and Arunika Mukhopadhaya

Introduction

Our immune system is a complex network of defense mechanisms which provides protection against a vast array of pathogens. The immune system responds to infectious microbes by triggering two branches: the innate immune system and the adaptive immune system. The innate immune system plays a crucial role during early stages of infection. Innate immune cells, like macrophages and granulocytes respond to invading pathogens by producing pro-inflammatory cytokines and chemokines leading to inflammation and killing of pathogens either directly or indirectly by activation of adaptive immune cells. The adaptive immune system combats infections effectively starting from 4 to 5 days of the infection with the help of B cell and T cell mediated responses. Sometimes, the adaptive immune system might fail to protect against the invading microbes. Some bacteria evade this defense by manipulating the immune system to establish themselves inside the human host and cause disease. At certain times, unregulated production of cytokines results in septic shock leading to multiple organ failure and ultimately death. Therefore, it is important to understand the role of pathogens in the context of host-immunomodulation.

Gram-negative pathogens pose a significant health risk to humans worldwide. The outer membrane of gram-negative bacteria contains two major components; lipopolysaccharide (LPS) and outer membrane proteins (OMPs). Among these, the bacterial endotoxin LPS is one of the well-studied immuno-stimulatory components of the bacteria and is known to cause inflammation and sepsis when present in excess.

S.C. Sakharwade • G.V.R.K. Prasad • A. Mukhopadhaya, Ph.D. (✉)
Department of Biological Sciences, Indian Institute of Science Education and Research (IISER) Mohali, Sector 81, SAS Nagar, Manauli 140306, Punjab, India
e-mail: arunika@iisermohali.ac.in

A. Chakrabarti, A. Surolia (eds.), *Biochemical Roles of Eukaryotic Cell Surface Macromolecules*, Advances in Experimental Medicine and Biology 842,
DOI 10.1007/978-3-319-11280-0_6

About one-third of the genome of gram-negative bacteria encode for OMPs (Koebnik et al. 2000). Porins are a type of OMPs that form transport channels across the membrane. They form beta barrel structures and have several roles. They act as receptors for phages and complement proteins, they can mediate antibiotic resistance (Achouak et al. 2001), anti-microbial peptide resistance (Galdiero et al. 2012), bile resistance (Wibbenmeyer et al. 2002; Hung and Mekalanos 2005) and also can act as adhesins (Duperthuy et al. 2010, 2011). All these properties suggest that they also have the ability to act as virulence factors. In fact, in *Vibrio splendidus*, one of the porins, OmpU helps in invasion of the host cells (Duperthuy et al. 2011). In *Serratia marscecens*, the opportunistic pathogen, antibiotic resistance is mainly imparted by porins (Ruiz et al. 2003). In *Pseudomonas aeruginosa* and *Neisseria gonorrhoeae*, porins trigger apoptosis of the host cells (Buommino et al. 1999; Muller et al. 2000).

Over the past two decades, accumulating evidence suggests that porins have immuno-modulatory properties. They can act as PAMPs (pathogen associated molecular patterns) and can be recognized by PRRs (pattern recognition receptors) present on host cells of mainly immune origin (Achouak et al. 2001; Galdiero et al. 2012). This review gives an overview of how porins modulate the host innate and adaptive responses, activate various signaling pathways and how they can be used as vaccines or adjuvants against various gram-negative bacterial infections.

Role of Porins in Modulation of Immune Responses

Innate Immune Responses

The innate immune system is the first line of defense against pathogen intrusion. It consists of various barriers like mechanical, chemical and physiological barriers, humoral factors and finally the inflammatory responses. The innate immune cells consist of mainly, monocytes, macrophages, dendritic cells (DCs), natural killer cells, mast cells and granulocytes such as neutrophils, basophils and eosinophils. These various cell types have specific functions that together mount an immune response towards detection and clearance of the pathogen from the host system. The innate immune responses however, do not induce memory generation and hence do not provide any additional protection upon re-challenge by the same pathogen.

Numerous gram-negative bacterial porins stimulate the production of pro-inflammatory cytokines. TNFα and IL-1β act on endothelial cells causing dilatation of vessels and hence, initiate the inflammatory process. Both these cytokines can signal the hypothalamus to induce fever. TNFα can also act on hepatocytes, along with IL-6 to induce the acute phase response. IL-12, a cytokine secreted by macrophages and dendritic cells, is involved in the differentiation of T cells. Nitric oxide is a reactive intermediate formed during phagocytosis and is toxic in nature. Porins from many gram-negative bacteria such as *Fusobacterium nucleatum*, *Haemophilus influenzae*, *Helicobater pylori*, *Neisseria*, *Salmonella* and *Shigella* spp. induce pro-inflammatory cytokines like TNFα, IL-6 and IL-1β secretion in a variety of cell lines and primary cells, such as monocytes and macrophages of human and mouse origin (Table 6.1) (Tufano et al. 1994, 1995; Toussi et al. 2012; Galdiero et al. 2001a, b, 2004; Vitiello et al. 2004, 2011; Liu et al. 2010; Singleton et al. 2005; Massari et al. 2006; Al-Bader et al. 2004; Moreno-Eutimio et al. 2013, Galdiero et al. 2005, 2006a; Ray et al. 2003; Biswas et al. 2007; Elena et al. 2009; Pore et al. 2012). *Vibrio cholerae* OmpU stimulates monocytes and macrophages to produce TNFα and IL-6 (Sakharwade et al. 2013). Omp16 of *Brucella abortus*, an outer membrane protein lipid anchor induces TNFα and IL-12 in mouse derived macrophages (Pasquevich et al. 2010). *Pasteurella multocida* porin and *Shigella* porins induce IL-12 secretion in mouse peritoneal macrophages and HEK cells (Iovane et al. 1998; Ray et al. 2003). PorA of *Neisseria meningitidis* induces IL-12 secretion in human PBMCs (peripheral blood mononuclear cells) derived DCs (Al-Bader et al. 2004). *Salmonella* porins, OmpA of *Shigella flexineri* and OmpU of *V. cholerae* are able to stimulate nitric oxide production in macrophages of mouse origin (Sakharwade et al. 2013; Pore et al. 2012; Gupta et al. 1999). In contrast to the reports demonstrating porin-induced pro-inflammatory responses, *Salmonella* porins which are pro-inflammatory in nature can also induce the production of IL-10, a potent immune-suppressive cytokine in human cell line and mouse primary cells (Galdiero et al. 2005).

Porins can also induce secretion of chemokines such as MIP-1α, MIP-1β, RANTES and IL-8. IL-8 is a potent neutrophil chemo-attractant factor. It promotes angiogenesis and phagocytosis. Chemokines like MIP-1α and MIP1β act on granulocytes and lead to acute inflammation and increase infiltration of neutrophils at the site of infection. They also aid in release of pro-inflammatory cytokines from macrophages. RANTES recruits T cells, eosinophils, basophils and leukocytes to inflammatory sites. OmpU deleted strain of *V. cholerae* showed decreased production of pro-inflammatory cytokines along with chemokines (Bandyopadhaya et al. 2007b, 2009; Sarkar et al. 2012). PorA of *N. meningitidis* and *Shigella* porins induce strong chemokine response in human PBMCs derived DCs and mouse peritoneal macrophages respectively (Al-Bader et al. 2004; Ray et al. 2003; Biswas et al. 2007).

Further, several studies on how porins affect neutrophil functions have been carried out. *Nesserial* porins are able to inhibit chemokine induced actin polymerization as well as degranulation in human PBMCs derived neutrophils (Bjerknes et al. 1995). Further, meningococcal porins down-regulate complement receptors CD35 and CD11b on neutrophils, but increase their oxidative burst capacity (Bjerknes et al. 1995). However, PorB of *N. gonorrhoeae* down-regulates oxidative burst and inhibits granule fusion with plasma and phagosomal membranes (Lorenzen et al. 2000). *Salmonella* Typhimurium porins induce the production of platelet-activating factor by human neutrophils (Tufano et al. 1992) as well as cause leukocyte transmigration *in vitro* (Galdiero et al. 1999). *Pasteurella haemolytica* porins decrease phagocytic index and intracellular killing capacity of bovine polymorphonuclear leukocytes (Galdiero et al. 1998b). *Klebsiella pneumoniae* OmpK35 and OmpK36 may affect neutrophil phagocytosis as deletion mutants caused an

Table 6.1 Innate immune responses initiated by porins

Organism	Porin	Model	Signaling mediators implicated	Inflammatory response	References
Brucella abortus	Omp16	Mouse macrophages		TNFα, IL-12	Pasquevich et al. (2010)
Fusobacterium nucleatum	FomA	HEK	TLR2 dependent NFκB activation	IL-8	Toussi et al. (2012)
Haemophilus influenza	Hib	Mice	TLR2-MyD88 expression along with CD14	TNFα, IL-6	Galdiero et al. (2004)
		THP-1		TNFα, IL-6	
		HEK		IL-8	
	P2	U87-MG		IL-6	Vitiello et al. (2011)
		U937	MEK1-MEK2-MAPK pathway		Galdiero et al. (2003c)
	Hib	Rat brain		TNFα, IL-1α, MIP-2	Galdiero et al. (2001a)
Helicobacter pylori	30 kDa porin	Human PBMCs derived monocytes		TNFα, IL-6, IL-8	Tufano et al. (1994)
Klebsiella pneumoniae	35 and 36 kDa	Human PMNs		Induction of opsonizing antibodies	Alcantar-Curiel et al. (2000)
Neiserria lactamica	PorB	BEAS-2B		IL-8	Liu et al. (2010)
		Detroit 562 cells			
Neiserria meningitidis	PorB	HEK 293	TLR2/TLR1 mediated NFκB activation	IL-8	Massari et al. (2006)
		Mouse splenic dendritic cells		IL-6	Singleton et al. (2005)
	PorA	Human monocyte derived dendritic cells		IL-8, RANTES, MIP-1α, MIP-1β,	Al-Bader et al. (2004)
				TNFα, IL-6, IL-12p40	
Pasteurella haemolytica	LAP	Calves and sheep lungs		Acute bronchopneumonia	Brogden et al. (1995)

Pasteurella multocida	37.5 kDa porin	Murine splenocytes		IL-1α, IL-6, TNF-α, IFN-γ	Iovane et al. (1998)
				IL-1α, IL-6, TNF-α, IFN-γ, IL-12p40	
Pseudomonas aeruginosa	38 kDa	Mice		Fibroblast increase; collagen edema	Baroni et al. (2001)
	38 kDa	Human PBMCs		TNFα	Cusumano et al. (1997)
				IL-6	
***Salmonella* Typhi**	OmpS1	Mouse bone marrow derived macrophages		TNFα, IL-6, IL-10	Moreno-Eutimio et al. (2013)
	OmpS2	Mouse spleen and lymph node		Increased CD40 expression by OmpS2	
***Salmonella* Typhimurium**	34 and 36 kDa porins	Mouse		TNFα, IL-6	Tufano et al. (1995)
		RAW 264.7	p38 and JNK MAPK activation	Nitric oxide	Vitiello et al. (2008a)
		Mouse gut macrophages		Nitric oxide	Gupta et al. (1999)
		U937 monocytes	Raf-1-MEK1-MEK2-MAPK pathway leading to activation of NFκB and AP-1 activation	TNF-a, IL-1β, IL-6, IL-10 and IL-1β	Galdiero et al. (2002, 2003a, 2005, 2006a), Finamore et al. (2009)
			PTK induced PKA and PKC activation		
		THP-1	PTK and p38 MAPK leading to activation of NFκB and AP-1 activation	TNFα, IL-6, IL-8	Vitiello et al. (2004), Galdiero et al. (2001b)

(continued)

Table 6.1 (continued)

Organism	Porin	Model	Signaling mediators implicated	Inflammatory response	References
Shigella dysenteriae	MOMP	Mouse peritoneal macrophages	TLR2/6 mediated MyD88 dependent NFκB activation	TNFα, IL-12	Ray et al. (2003), Biswas et al. (2007)
				MIP-1α, MIP-1β, RANTES	
				Increase in CD80, MHC-II, CD40 expression	
		Int407	TLR2/6 mediated MyD88 dependent NFκB activation	MIP-1α, MCP-1, IL-8	Mukherjee et al. (2014)
Shigella flexneri	38 and 40 kDa	Caco-2	Up-regulation of TLR2 and MyD88	TNFα, IL-8, IL-1β	Elena et al. (2009)
	38 kDa	Mouse peritoneal macrophages		MIP-1α, MIP-1β and RANTES	Biswas et al. (2007)
				TNFα, IL-12	Biswas et al. (2007), Ray et al. (2003)
	OmpA	HEK 293		IL-6, IL-12p70, and IL-1β	Pore et al. (2012)
		Mouse peritoneal macrophages		Nitric oxide	
				Nitric oxide	
		RAW 264.7		Increase in CD80 and MHC-II expression	
Vibrio cholerae	OmpU	RAW 264.7		Nitric oxide, TNFα, IL-6	Sakharwade et al. (2013)
		THP-1, human PBMCs		TNFα, IL-6	
	OmpU deleted strain	Int407		IL-1α, IL-6, MCP-1	Bandyopadhaya et al. (2007a)
				TNF-a, IL-6, IL-1α	Bandyopadhaya et al. (2009)
				IL-8	Sarkar et al. (2012)

Cell lines: BEAS-2B (human airway epithelial cell line), Caco-2 (human epithelial colorectal adenocarcinoma cell line), Detroit 562 cells (human pharynx carcinoma cell line), HEK (human epithelial kidney cell line), Int407 (human intestinal epithelial cell line), RAW 264.7 (murine macrophage cell line), THP-1 (human monocytic cell line), U87-MG (human astroglioma cells), U937 (human leukemic monocytic cell line) increase in phagocytosis capacity (Tsai et al. 2011). *H. pylori* porins decrease chemotaxis ability of human neutrophils and can interfere with intracellular killing (Tufano et al. 1994).

The complement system falls under the humoral branch of the innate immune system. This system consists of plasma proteins that interact with each other and ultimately lead to opsonization of pathogens or induction of several inflammatory responses. There are three pathways (classical, mannan binding lectin and alternative pathways) by which the complement system is activated which subsequently converge at C3 convertase enzyme and formation of membrane attack complex (MAC) that leads to killing of target microbes. C3b can opsonize microbes by binding to complement receptors on phagocytes. C3a, C4a and C5a can recruit phagocytes to inflammatory sites. Most porins activate the classical pathway by binding to C1q.

Porins from *Salmonella minnesota* bind to C1q (Latsch et al. 1990). *N. gonorrhoeae* Por1B binds to C3b and C4 (Lewis et al. 2008) and Por1A and Por1B bind to C4 binding protein as well (Ram et al. 2001). Similarly, *N. meningitidis* OMPs activate the complement system (Bjerre et al. 2002). *S.* Typhimurium porins activate the classical complement pathway as measured by consumption of C1s, C3 and C4 in human or guinea pig serum (Galdiero et al. 1984). *K. pneumoniae* OmpK36 also activates the classical complement pathway *in vivo* by binding to C1q and leads to deposition of C3, C5-9 (MAC) components on the porin (Alberti et al. 1993, 1996). Similarly, *Aeromonas hydrophila* 39 kDa porin and *Aeromonas salmonicida* 40 kDa porin activate the classical pathway in an antibody independent manner by binding to C1q (Merino et al. 1998, 2005). The MOMP (major outer membrane protein) of *Legionella pneumophila* binds to C3 and cause phagocytosis of MOMP incorporated vesicles by human monocytes (Bellinger-Kawahara and Horwitz 1990).

All these facts have led to the understanding that porins are able to induce pro-inflammatory cytokine and chemokine responses as studied *in vitro* and *in vivo* in both mouse and human cells. Some porins also have the ability to induce production of cytokines involved in activation of cells important for innate immune responses or adaptive immune responses. Further, porins are able to interfere with neutrophil function as well as activate the complement system. Interestingly, observation from our laboratory revealed that *V. cholerae* OmpU is able to down-regulate LPS mediated effects, although it is pro-inflammatory in nature.

Adaptive Immune Response

The adaptive response starts later as the infection progresses; it is specific, more potent than the innate immune responses and is associated with memory induction. The adaptive immune response is initiated with the help of signals generated by the innate immune cells which can activate lymphocytes; the T and B cells. Antigen presenting cells such as, macrophages and dendritic cells, present endocytosed antigens to $CD4^+$ and $CD8^+$ T cells and activate them. Activated T helper cells (Th) further help in B cell mediated antibody responses. T cells require certain signals from antigen presenting cells (APCs) in order to be activated optimally. The interaction of antigen presented by MHC molecules on APCs and certain co-stimulatory molecules like B7 (CD80 and CD86), CD40 etc. with their respective receptors present on T cells, trigger their activation. Cytokines also play an important part in differentiation of T cells. In presence of cytokines such as IL-12 and IFNγ, $CD4^+$ T cells differentiate towards Th1 type and in the presence of IL-4, $CD4^+$ T cells differentiate towards Th2 type. Th1 cells activate macrophages and differentiation of B cells, followed by antibody production, shaping the immune responses towards cell-mediated immunity (Fig. 6.1). On the other hand, Th2 polarization is required for humoral immunity and hyper-sensitivity.

Some porins can modulate adaptive-responses of the host. Porins of *Shigella*, *Salmonella* and *Neisseria* species affect antigen presenting cells such as macrophages and dendritic cells in numerous ways. *Shigella* porins induce expression of CD40 and CD80 co-stimulatory molecules as well as MHC-II molecules on macrophages of mouse origin (Elena et al. 2009; Pore et al. 2012). *Salmonella* porins induce expression of CD40 and CD86 on dendritic cells of mouse origin (Moreno-Eutimio et al. 2013; Cervantes-Barragan et al. 2009). *N. meningitidis* PorA and PorB increase expression of co-stimulatory molecules along with MHC-II molecules in human PBMCs derived DCs and mouse splenic DCs respectively (Al-Bader et al. 2004; Singleton et al. 2005). OmpA porin of *Acinetobacter baumannii* can stimulate mouse bone marrow derived dendritic cells to secrete IL-12 along with

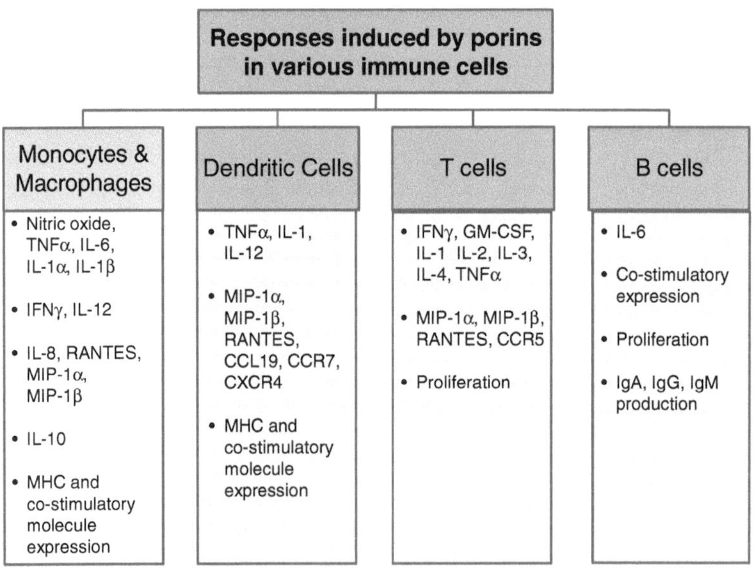

Fig. 6.1 Different immune responses elicited by gram negative porins in different types of immune cells

increased surface expression of co-stimulatory moleculesas well as maturation of dendritic cells and can polarizes T cells towards Th1 type response (Lee et al. 2007). *S.* Typhimurium porins induce Th1 and Th2 differentiation of T cells (Galdiero et al. 1998a). Many *S.* Typhimurium porins have been studied for the effect on B cell responses. Collectively, these porins can induce co-stimulatory molecule expression of B cells of mouse and human origin (Cervantes-Barragan et al. 2009; Galdiero et al. 2003b). Also, they generate IgM and IgG antibody responses (Secundino et al. 2006; Gil-Cruz et al. 2009). Similarly, porins of *N. meningitidis* and *N. gonorrheae* cause generation of IgM responses and induce CD86 expression in mouse splenic B cells (Snapper et al. 1997; Wetzler et al. 1996). *S. dysenteriae* MOMP can induce co-stimulatory molecule expression in mouse peritoneal B1 and B2 cells and also generate IgM, IgA and IgG responses (Ray et al. 2004; Ray and Biswas 2005). The *Helicobacter pylori* 30 kDa porin induces IFNγ, GM-CSF, IL-3 and IL-4 secretion in lymphocytes derived from human (Tufano et al. 1994) which increase proliferation of mast cells, decrease IFNγ secretion by macrophages, induce class switching in B cell and differentiation of Th2 cells.

So far, the literature indicates that various gram-negative porins have the capacity to induce adaptive immune responses. Porins can provide signal for Th1 or Th2 differentiation as well as B cell activation, class switching phenomenon and affinity maturation.

Signaling Cascades Initiated by Porins

Identification of pathogenic and non-pathogenic organisms by innate immune cells occurs upon recognition of various PAMPs by PRRs (Medzhitov and Janeway 1997; Kumar et al. 2012; Kawai and Akira 2009). PRRs, then initiate intracellular signaling pathways that lead to recruitment of phagocytic cells, monocytes to the site of infection and activation of innate and adaptive immunity (Medzhitov 2007).

TLRs are one of the major type of PRRs present on immune cells (Akira and Takeda 2004; Kaisho and Akira 2001; Armant and Fenton 2002). Upon binding to specific microbial components, TLRs trigger intracellular signaling cascades that result in production of inflammatory cytokines and chemokines from several immune cells (Akira and Takeda 2004; West et al. 2006; Mogensen 2009) (Fig. 6.2). These inflammatory cytokines can induce dendritic cell maturation which is characterized by up-regulation of co-stimulatory molecules and altered expression of chemokine receptors on the surface of DCs. Thus, TLR mediated DC maturation acts a link between innate and adaptive immunity (Akira et al. 2001).

TLR signaling further activates transcription factors such as, NFκB and AP-1 (Bell et al. 2003; Kawai and Akira 2005; Karin and Greten 2005). Briefly, TLRs upon binding to specific ligands interact with intracellular TIR domain containing adaptor molecule MyD88. MyD88 then recruits IRAK1 (IL-1 receptor associated kinase 1) which then forms a complex with IRAK4 or IRAK2. Phosphorylated IRAK1 recruits TNF receptor associated factor 6 (TRAF6) and E2 ubiquitin conjugating enzyme 13 (UBC13). TRAF6 and UBC13 poly-ubiquitinylate IRAK1 and TRAF6, leading to activation of MAPK and NFκB pathways.

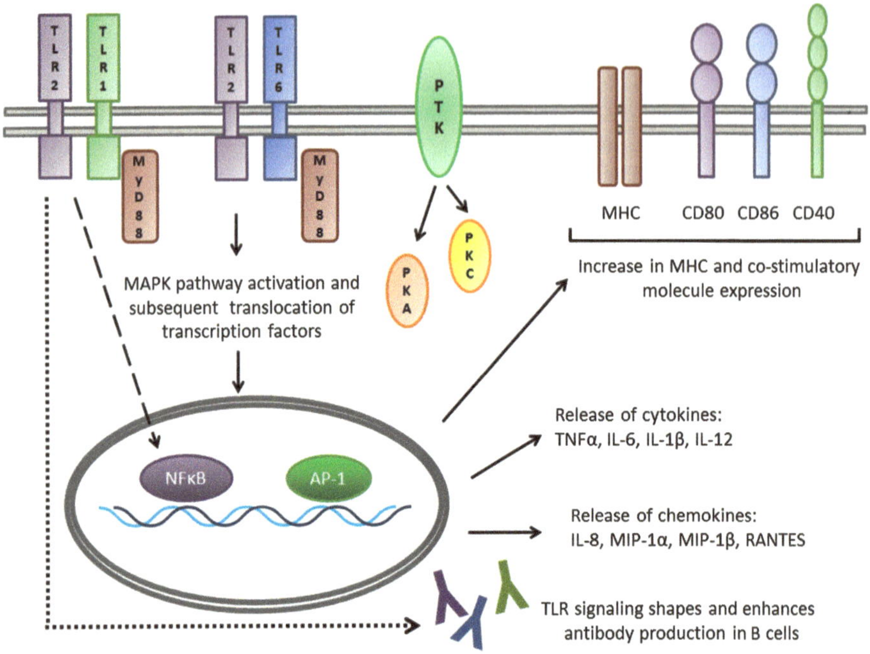

Fig. 6.2 Gram negative porins can stimulate TLR pathway and result into cell surface expression of molecules of immune importance and release of pro-inflammatory cytokines, chemokines and antibody secretion by B cells

Several studies suggest that porins act as PAMPs as they interact with TLRs and initiate down-stream signaling. Porins of various gram-negative bacteria mediate signaling via TLR pathway (Table 6.1). Data from the studies on *Shigella*, *Neisseria* and *Heamophilus* spp. indicate that these porins are recognized by TLR2 predominantly heterodimerized with either TLR1 or TLR6. Upon binding to porins, TLR2/TLR1 or TLR2/TLR6 activate NFκB or AP-1 via MyD88 dependent pathway leading to translocation of the nuclear factors into the nucleus and transcription of pro-inflammatory cytokine genes and chemokine genes mediated by various cytosolic signaling cascades (Massari et al. 2002, 2006; Banerjee et al. 2008; Ray and Biswas 2005; Biswas et al. 2007; Singleton et al. 2005). *S.* Typhimurium porins induce phosphorylation of protein tyrosine kinases (PTK), protein kinase A (PKA) and protein kinase C (PKC) in U937 monocytic cell line and also activate transcription factors AP-1 and NFκB by Raf-1-MEK1/2-MAPK pathway (Galdiero et al. 2002, 2003a). Inhibitor studies suggest that p38 MAPK is mainly involved in transcription factor activation. Studies on *Neisseria* spp. porins are implicated in TLR mediated NFκB activation which occurs via Raf-1-MEK1/2-MAPK pathway; (Massari et al. 2003; MacLeod et al. 2008). *H. influenzae* porin P2 and porins of *Salmonella* and *Neisseria* also activate the MAPK pathway (Galdiero et al. 2002, 2003c; Vitiello et al. 2004; MacLeod et al. 2008). The three-dimensional structure model of porin P2 constructed on the basis of crystal structure of *K. pneumoniae* OmpK36 and *Escherichia coli* PhoE and OmpF predict that the domains of surface exposed loops are involved in activation of signal transduction pathway (Table 6.1). In particular,

Table 6.2 Adaptive responses initiated by porins

Cell type	Organism	Porin	Model	Effect on adaptive responses	References
Dendritic cells	*Acinetobacter baumannii*	OmpA	Bone marrow	Increase in CD40, CD54, CD80 and CD86	Lee et al. (2007)
				Production of IL-12	
				Augmentation of syngeneic and allogeneic immunostimulatory capacity	
	Neiserria meningitidis	PorA	Human monocyte derived dendritic cells	Increase in expression of MHC-II, CD40, CD54, CD80, CD86	Al-Bader et al. (2004)
				Decrease in receptor mediated endocytosis	
				Induction of allostimulatory activity	
				Induction of $CD4^+$ T cell proliferation	
		PorB	Mouse spleen	Increase in CD86 and MHC molecules	Singleton et al. (2005)
				Induction of allostimulatory activity	
				Activation of antigen specific T cells	
	Salmonella Typhi	OmpS1, OmpS2	Mouse bone marrow derived DCs	TNFα, IL-6, IL-10 expression	Moreno-Eutimio et al. (2013)
			Mouse spleen and lymph nodes	Increased in MHC-II and CD40 expression	
	Salmonella Typhimurium	OmpC and OmpF	Mouse spleen	Increase in CD86 and CD40 expression	Cervantes-Barragan et al. (2009)
	Shigella dysenteriae	MOMP incorporated liposome	Mouse spleen	Up-regulation of MHC class II, CD40, CD80 and ICAM-1	Banerjee et al. (2008)
				Increase in expression of TNFα, IL-12, MIP-1α, MIP-1β, RANTES, CCL19, CCR7, CXCR4	

(continued)

Table 6.2 (continued)

Cell type	Organism	Porin	Model	Effect on adaptive responses	References
T cells	*Acinetobacter baumannii*	OmpA	Mouse spleen	Induction of IFNγ	Lee et al. (2007)
	Helicobacter pylori	30 kDa porin	Human PBMCs derived lymphocytes	IFNγ, GM-CSF, IL-3, IL-4	Tufano et al. (1994)
	Salmonella Typhimurium	34 and 36 kDa	Mouse spleen	IFNγ and IL-4 release by $CD4^+$ T cells purified from immunized mice	Galdiero et al. (1998a)
	Shigella dysenteriae	MOMP	Mouse spleen	Induction of IL-2, CD25, MIP-1α, MIP-1β, RANTES, CCR5	Biswas et al. (2008, 2009)
				Release of TNFα, IFNγ	
	Shigella flexneri	OmpA	Mouse spleen	Release of IL-2, IFNγ	Pore et al. (2012)
				Induction of MIP-1α, MIP-1β, RANTES, CCR5	

B cells	*Fusobacterium nucleatum*	FomA	Mouse spleen	IL-6 production	Toussi et al. (2012)
				Increase in CD86 and MHC-II expression	
	Neiserria gonorrheae	PIA, PIB	Mouse spleen and lymph node	Increase in CD86 expression	Wetzler et al. (1996)
			Mouse spleen	Increase in IgM secretion	Snapper et al. (1997)
		PorB		Increase in B cell proliferation	Wetzler et al. (1996)
				Increase in T cell co-stimulatory activity	
	Neiserria meningitidis	C1, C3	Mouse spleen and lymph node	Increase in CD86 expression	Wetzler et al. (1996)
		C3	Mouse spleen	Increase in IgM secretion	Snapper et al. (1997)
		PorB	Mouse spleen	Increase in CD86 expression	MacLeod et al. (2008)
				Increase in proliferation	
				PTK-MAPK pathway involved	
	Salmonella Typhimurium	OmpD	Mouse spleen	Increase in IgM secretion	Gil-Cruz et al. (2009)
				Induction of B cells showing B1b markers	
		OmpC	Mouse spleen	Increase in IgG2b, IgG1, IgG2a production	Secundino et al. (2006)
		OmpC and OmpF	Mouse spleen	Increase in MHC-II, CD86 and CD40 expression	Cervantes-Barragan et al. (2009)
		34 and 36 kDa	Human PBMCs	Increase in CD80 and CD86 expression	Galdiero et al. (2003b)
	Shigella dysenteriae	MOMP	Mouse peritoneal cavity B2 cells	Increase in of expression of CD86, IgA, IgM, IgG2a	Ray and Biswas (2005)
			Mouse peritoneal B1 cells	Up-regulation of CD80–CD86, IgA, IgM	Ray et al. (2004)
				TLR2/6-MyD88 involvement	

synthetic peptide corresponding to surface exposed loops L5, L6 and L7 activate JNK and p38 MAPK similarly as the intact protein with L7 being the most active peptide (Galdiero et al. 2003c). Further studies on L7 showed that only six amino acids contribute to the overall activity and induction of TNFα and IL-6 production (Galdiero et al. 2006b).

Porins as Vaccine Candidates

For an agent to be used as a good vaccine, it must be highly immunogenic and a major protective antigen. It is desirable for a vaccine candidate to drive the $CD4^{+}$ T cell responses towards Th1 to ensure both humoral and cell mediated immunity against pathogens. Activated Th1 cells aid in reduction and clearance of pathogens (intra-cellular and extra-cellular) by secreting IFN-γ, TNFα, IL-2 and IL-3 and help in activation and differentiation of B cells, $CD8^{+}$ T cells and macrophages. In certain cases, like anti-parasite responses, Th2 differentiation is important. Th2 cells produce IL-4, IL-5, IL-13, IL-6 and IL-10 and mainly support B cell activation and differentiation. $CD8^{+}$ T cells clear intra-cellular pathogens by killing infected cells or by releasing cytokines that would help in the process. Antibodies produced by B cells can bind to the enzymatic active sites of toxins or prevent their diffusion, neutralize viral replication, promote phagocytosis of extracellular bacteria by opsonization and can activate the complement cascade. IgM followed by IgG antibodies appear a few days after immunization. B cell maturation is associated with two major events: Ig class-switch recombination from IgM towards IgG, IgA or IgE, and maturation of the affinity of B cells for their specific antigen.

Porins have been widely studied for their capacity to act as adjuvants or as potential vaccines in various animal models (Table 6.3). *A. hydrophila* is a gram-negative organism that is pathogenic in fish, amphibian and humans as well. Administration of OmpF of *A. hydrophila* leads to increased IgG expression in mouse model along with increased lymphocyte proliferation and T cell activation *in vitro* (Yadav et al. 2014). OmpTS of *A. hydrophila* was highly immunogenic in fish model *Labeo rohita* and Omp48 immunized fish showed survival against *A. hydrophila* and *Edwardsiella tarda* infections (Khushiramani et al. 2007, 2014). *Borrelia burgdorferi* is a spirochete that causes Lyme's disease. One of its porins, Oms66 showed protection against infection in immunized mice (Exner et al. 2000). Mice immunized with Omp16 from *B. abortus*, that causes brucellosis, showed protection from infection. Further, Omp16 was able to activate dendritic cells and induce IFNγ secretion from mouse splenic T cells and induce foot pad swelling (Pasquevich et al. 2010). *Burkholderia pseudomallei* infects both animals and humans causing melioidosis, which has a mortality rate of 20–50 % in humans even with treatment. Mice immunized with OmpA of *B. pseudomallei* showed protection against infection (Hara et al. 2009). *Chlamydia trachomatis* is an obligate intracellular pathogen that causes urethritis, proctitis, trachoma, infertility and is the single most infectious agent associated with blindness. Administration of *C. trachomatis* MOMP in mice

Table 6.3 Porins and outer membrane proteins as vaccine candidates or adjuvants

Organism	Porin	Administered as (route)	Model	Effect observed	References
Acinetobacter baumannii	Outer membrane vesicles	Intra-muscular	Mouse	Protection against two *A. baumannii* strains	McConnell et al. (2011)
		With alum as adjuvant			
Aeromonas hydrophila	OmpF	Intra-peritoneal	Mouse	Increased IgG expression	Yadav et al. (2014)
		With Freund's adjuvant		Increased lymphocyte proliferation *in vitro*	
				IL-4 and IFNγ secretion by T cells *in vitro*	
	Omp48	Intra-muscular	*Labeo rohita*	Immunized fish showed increased survival against *A. hydrophila* and *E. tarda*	Khushiramani et al. (2014)
	OmpTS	Intra-peritoneal	*Labeo rohita*	Highly immunogenic in fish	Khushiramani et al. (2007)
		With Freund's adjuvant			
Borrelia burgdorferi	Oms66	Skin implantation	Mouse	Protection against infection	Exner et al. (2000)
Brucella abortus	Omp16	Intra-peritoneal and oral	Mouse	Induction of IFNγ by splenic T cells	Pasquevich et al. (2010)
		Oral		Foot pad swelling	
		Intra-venous		Activation of dendritic cells	
		Plant derived Omp16; Oral		Protection against infection	
Burkholderia pseudomallei	OmpA	Intra-peritoneal	Mouse	Protection against *B. pseudomallei* infection	Hara et al. (2009)

(continued)

Table 6.3 (continued)

Organism	Porin	Administered as (route)	Model	Effect observed	References
Chlamydia trachomatis	MOMP	Intra-muscular MOMP DNA priming with ISCOM	Mouse	Stronger delayed-type hypersensitivity	Dong-Ji et al. (2000)
				IFNγ and IgA production	
				Protection against lung challenge	
		Sub-cutaneous administration of sonicated and vortexed MOMP with Freund's adjuvant	Mouse	Increased IgG and IgA production by sonicated MOMP	Pal et al. (2001)
				Increased T cell responses (proliferation, IL-4 and IFNγ production) by vortexed MOMP	
				Sonicated MOMP immunization showed protective effect upon genital challenge	
		Intra-venous adoptive immunization with antigen-pulsed DCs	Mouse	Proliferation of infection-sensitized $CD4^+$ T	Shaw et al. (2002)
				IL-12 and IFNγ secretion	
				No protection against infection	
		Intra-venous adoptive transfer of adenoviral transfected DC	Mouse	Increased expression of MHC-II, CD80 on DC	Lu et al. (2010)
				Increased IL-12 secretion	
				Protection against genital tract challenge infection	
	MOMP	Intra-muscular	*Macaca mulatta*	High amount of IgG and IgA present in plasma, vaginal washes, tears, saliva, and stool	Cheng et al. (2011)
	Subunit vaccine	With CpG-2395 and Montanide ISA 720 VG as adjuvants			
				Lymphoproliferative response	
				TNFα, IFNγ production	
	Porin B and Porin D in *V. cholerae* ghosts	Intra-muscular	Mouse	Cross protection against *C. muridarum*, *C. trachomatis* and *V. cholerae*	Eko et al. (2011)
	Multi subunit vaccine			Cross-reactive chlamydial specific genital mucosal Th1/Th2 cytokine responses and IgA and IgG2a antibody responses	

Fusobacterium nucleatum	FomA	Subcutaneous	Mouse	Increased IgG and IgM production	Toussi et al. (2012)
		With OVA/alum as adjuvant		IL-10 and IL-6 expression	
Neiserria lactamica	PorB	Sub-cutaneous	Mouse	High titers of IgG1 and IgG2b	Liu et al. (2008)
		With OVA as adjuvant		Production of IL-4, IL-10,	
				IL-12 and INF-γ	
Neisseria gonorrhoeae	Class I	Nasal	Mouse	Clearance of vaginal infection	Plante et al. (2000)
	PorB	Intra-muscular	Mouse	Th1 responses	Zhu et al. (2004)
	DNA vaccine	Epidermal particle bombardment		Th2 responses	
Neisseria meningitidis	PorA in liposomes	Sub-cutaneous	Mouse	Anti-sera showed bactericidal activity	Christodoulides et al. (1998)
		With alum as adjuvant			
		Intra-peritoneal	Mouse	Anti-sera showed bactericidal activity	Humphries et al. (2004)
	PorB	Intra-peritoneal	Mouse	Anti-sera showed bactericidal activity	Wright et al. (2002)
		With alum as adjuvant			
		Intra-nasal	Mouse	Protective effect as adjuvant against *Francisella tularensis* infection	Chiavolini et al. (2008)
Pseudomonas aeruginosa	OmpF epitopes expressed in CPMV	Sub-cutaneous	Mouse	IgG2a response	Brennan et al. (1999)
		With Freund's and QuilA adjuvants			
	Recombinant OprF and OmpI fusion proteins	Intra-peritoneal	Mouse	Protection upon challenge with *P. aeruginosa*	von Specht et al. (1995)
		With alum as adjuvant			
	OprF	Intra-dermal	Mouse	IgG1 response	Price et al. (2001)
	Expressed in plasmid vector	(Gene gun)		Protection to pulmonary infection by *P. aeruginosa*	

(continued)

Table 6.3 (continued)

Organism	Porin	Administered as (route)	Model	Effect observed	References
***Salmonella* Typhi**	OmpC, OmpF and OmpA	Intra-peritoneal	Mouse	Administration of two or three porins retard lethal effect; None were protective on their own	Toobak et al. (2013)
		With Freund's adjuvant			
	OmpS1, OmpS2	Intra-peritoneal	Mouse	Protection against *S.* Typhi infection	Moreno-Eutimio et al. (2013)
		With OVA adjuvant		Act as adjuvants	
	34, 36 kDa	Intra-peritoneal	Mouse	Protection against *S.* Typhi infection	Isibasi et al. (1992)
	34, 35, 36 kDa	Intra-peritoneal	Mouse	Protection against *S.* Typhi infection	Singh et al. (1999)
	OmpA	Intra-peritoneal	Mouse	Protection against *S.* Typhi infection	Isibasi et al. (1988)
***Salmonella* Typhimurium**	34, 35, 36 kDa	Sub-cutaneous	Mouse	Protection against *S.* Typhimurium and *S. enterditis*	Tabaraie et al. (1994)
	LT2	Intra-peritoneal	Mouse	Protection against *S.* Typhimurium challenge	Matsui and Arai (1990)
Treponema pallidum	Outer membrane vesicle	Sub-cutaneous	Mouse	Anti-OMV serum possessed complement-dependent treponemicidal activity	Blanco et al. (1999)
				Antibody response against TROMPs	
***Vibrio* strains**	OmpK	Intra-peritoneal	*Epinephelus coioides*	Tolerance to infection by pathogenic strains	Li et al. (2010b)
***V. harveyi* (11)**		With Freund's incomplete adjuvant			
***V. alginolyticus* (6)**					
***V. parahaemolyticus* (2)**					

Vibrio anguillarum	Omp38 DNA construct	Oral	*Lates calcarifer*	Moderate protection against *V. anguillarum* infection	Rajesh Kumar et al. (2008)
Vibrio alginolyticus	OmpW	Intra-peritoneal	*Larimichthys crocea*	Resistance against *V. alginolyticus* infection	Qian et al. (2007)
	OmpU	Intra-peritoneal	*Lutjanus erythropterus*	Protection against *V. alginolyticus* infection	Cai et al. (2013)
Vibrio cholerae	22, 30, 42, 43 kDa OMPs anti-sera	Ileal loop challenge	Rabbit	Reduced *V. cholerae*-induced fluids secretion in ileal loop	Das et al. (1998)
	Outer membrane vesicle	Intra-peritoneal	Mouse	Protection against *V. cholerae* infection	Schild et al. (2008); Leitner et al. (2013)
Vibrio harveyi	VhhP2	Oral and intra-peritoneal	*Paralichthys olivaceus*	Immunoprotection against *V. harveyi* infection	Sun et al. (2009)
	OmpK	Intra-peritoneal	*Epinephelus coioides*	Protection against *V. harveyi* infection	Ningqiu et al. (2008)
		With Freund's incomplete adjuvant			
		Intra-peritoneal	*Pseudosciaena crocea*	Protection against *V. harveyi* infection	Zhang et al. (2007)
Vibrio parahaemolyticus	VP1061, VP2850	Intra-peritoneal	*Carassius carassius*	Cross-protective immune reaction against the infections of *V. alginolyticus*, *A. hydrophila*, and *P. fluorescens*	Li et al. (2010a)
		With Freund's incomplete adjuvant	Mouse		
	OmpW, OmpV, OmpU and OmpK	Intra-peritoneal	*Larimichthys crocea*	Protection against *V. parahaemolyticus* infection	Mao et al. (2007)

induced IgG, IgM responses; T cells responses, co-stimulatory molecule expression in dendritic cells along with IL-12 secretion (Dong-Ji et al. 2000; Pal et al. 2001; Shaw et al. 2002). MOMP immunized mice showed protection against genital challenge by the bacteria (Lu et al. 2010). MOMP subunit vaccine administered in rhesus macaques also showed similar T and B cell responses (Cheng et al. 2011). Other porins of *C. trachomatis*, Porin B and Porin D, when administered to mice in the form of *V. cholerae* ghosts, induced cross-reactive chlamydial specific genital mucosal T and B cell responses (Eko et al. 2011). FomA of *F. nulceatum*, bacteria involved in periodontal disease, leads to IgG and IgM antibody production along with IL-6 and IL-10 secretion upon treatment in mice (Toussi et al. 2012). *Neiserria lactamica* is a commensal found in infants that can cause pneumonia in children. PorB induces high levels of IgA and IgM antibody responses along with IL-4, IL-12, IL-10 and IFNγ production in mice (Liu et al. 2008). Similarly, PorB of *N. gonorrhoeae*, the bacteria that causes gonorrhea, induces Th1 and Th2 type responses in mice (Zhu et al. 2004). Mice immunized with Class I porins of *N. gonorrhoeae*, showed reduction in vaginal infection (Plante et al. 2000). PorB of *N. meningitidis*, the causal bacteria of meningococcal disease, showed a protective response against *Francisella tularensis* infection (Chiavolini et al. 2008). Further, antisera of *N. meningitidis* PorB immunized mice showed bactericidal activity (Wright et al. 2002). Similarly, antisera of *N. meningitidis* PorA incorporated liposome immunized mice showed bactericidal activity (Christodoulides et al. 1998; Humphries et al. 2004). *P. aeruginosa* is an opportunistic pathogen that colonizes in the lungs, kidneys and urinary tract. *P. aeruginosa* OmpF epitopes induced IgG2a response in mice whereas OprF induced IgG1 response in mice (Brennan et al. 1999). Mice immunized with OprF and OmpI fusion proteins or OprF only were able to resist infection (von Specht et al. 1995; Price et al. 2001). *Salmonella* causes food poisoning that is characterized by enteritis and diarrhea, leading to typhoid. Administration of various *Salmonella* Typhi porins offer protection against infection in mice. OmpA, OmpC, OmpF, OmpS1 and OmpS2 have been studied in this regard (Toobak et al. 2013; Moreno-Eutimio et al. 2013; Isibasi et al. 1988, 1992; Singh et al. 1999). OmpS1 and OmpS2 also show adjuvant properties (Moreno-Eutimio et al. 2013). Various *S.* Typhimurium porins immunized mice have also shown protection to infection (Tabaraie et al. 1994; Matsui and Arai 1990). Outer membrane vesicles (OMVs) of *Treponema pallidum*, the causative agent of syphilis, when administered to mice showed an antibody response against outer membrane proteins. Anti-OMV serum showed complement dependent treponemicidal activity (Blanco et al. 1999). OMVs of *A. baumanii* immunized mice showed protection against two strains of *A. baumanii* (McConnell et al. 2011). Similarly, administration of *V. cholerae* OMVs showed protection against cholera in mice (Schild et al. 2008; Leitner et al. 2013). Anti-sera against 22, 30, 42 and 43 kDa *V. cholerae* OMPs reduced *V. cholerae* induced fluid secretion in ileal loop model in rabbits (Das et al. 1998). Many *Vibrio spp.* like *V. anguillarum*, *V. harveyi*, *V. alginolyticus* and *V. parahaemolyticus* affect fish and other marine animals. Sea food contaminated with *V. parahaemolyticus* can cause gastroenteritis in humans. *V. anguillarum* Omp38 immunized Asian seabass showed moderate protection against infection

(Rajesh Kumar et al. 2008). Similarly, immunization with OmpW and OmpU of *V. alginolyticus* showed protection upon challenge with the bacteria in large yellow croaker and crimson snapper respectively (Qian et al. 2007; Cai et al. 2013). Immunization of fish with certain *V. harveyi* outer membrane proteins was able to protect fish from infection. Vhhp2 administration in olive flounder and OmpK immunization of large yellow croaker and orange spotted grouper had a protective effect against *V. harveyi* infection in the fishes (Sun et al. 2009; Ningqiu et al. 2008; Zhang et al. 2007). Large yellow croaker fish immunized with *V. parahaemolyticus* OmpW, OmpV, OmpU and OmpK showed protection against infection (Mao et al. 2007). Further, immunization of crucian carp and mice with *V. parahaemolyticus* VP1061 and VP2850 proteins induced a cross protective effect against *V. alginolyticus*, *A. hydrophila* and *Pseudomonas fluorescens* (Li et al. 2010a). OmpK, a homologous protein of the *Vibrio* species was administered to orange spotted grouper. Fish immunized by OmpK were able to survive infection from various strains of *V. harveyi*, *V. alginolyticus* and *V. parahaemolyticus* (Li et al. 2010b).

In sum, the above studies highlight the role of various gram-negative bacterial porins and few outer membrane proteins as vaccine candidates. Porins are able to stimulate T cell and B cell responses as well as offer protection against various gram-negative bacterial infections.

Conclusion

Outer membrane proteins are crucial for maintaining bacterial structure and homeostasis. These proteins are also important for gram-negative bacterial pathogenesis as they modulate host immune responses. Porins, a class of outer membrane proteins induce inflammatory responses in a range of host cells. They can also activate dendritic cells, T cells and B cells as well as shape adaptive immune responses. The signaling cascades activated by various porins have been delineated and their characterization has added to our knowledge on how they modulate host cell responses. Further, multiple porins have been reported for their vaccine potential and are undergoing further studies for their use as vaccines or adjuvants.

A number of patents have been filed since 2005 for the use of porins and outer membrane proteins of various gram-negative bacteria as vaccines or adjuvants. The use of *Salmonella* spp. OmpC and OmpF as adjuvant for influenza vaccine show improved immune response as compared to administration of influenza vaccine alone (Leclerc and Lopez 2010). Class 1 porins of *N. meningitidis* show significant immune stimulating capability and has the potential to be used as a vaccine for meningitidis (Seid et al. 2006; Paradiso et al. 2007; Van et al. 2007; Granoff et al. 2013). OmpK36 and its homologues from *K. pneumoniae*, *S.* Typhi, or *E. coli* open up a prospective in the diagnosis, treatment and prevention of *enterobacteriaceae* infection (Siu et al. 2013). The MOMP of *H. influenzae* and *C. trachomatis* show protective response against influenza/otitis media and Chlamydiophilia infections respectively (Berthet et al. 2011; Stephens and Kawa 2011). Besides porins, surface protein of *Moraxella catarrhalis* and outer membrane vesicles of *V. cholerae* were successfully tested as vaccines (Chen et al. 2005; Camilli et al. 2014). All these studies highlight the necessity to examine porins and other outer membrane bacterial components for their adjuvant capacity and vaccine potential.

In conclusion, porins have emerged to have many more functions than previously believed and have the potential to be used for diagnosis and treatment of various gram-negative bacterial infections.

References

Achouak W, Heulin T, Pages JM (2001) Multiple facets of bacterial porins. FEMS Microbiol Lett 199(1):1–7. doi:S0378-1097(01)00127-6 [pii]

Akira S, Takeda K (2004) Toll-like receptor signalling. Nat Rev Immunol 4(7):499–511. doi:10.1038/nri1391, nri1391 [pii]

Akira S, Takeda K, Kaisho T (2001) Toll-like receptors: critical proteins linking innate and acquired immunity. Nat Immunol 2(8):675–680. doi:10.1038/90609, 90609 [pii]

Al-Bader T, Jolley KA, Humphries HE, Holloway J, Heckels JE, Semper AE, Friedmann PS, Christodoulides M (2004) Activation of human dendritic cells by the PorA protein of *Neisseria meningitidis*. Cell Microbiol 6(7):651–662. doi:10.1111/j.1462-5822.2004.00392.x, CMI392 [pii]

Alberti S, Marques G, Camprubi S, Merino S, Tomas JM, Vivanco F, Benedi VJ (1993) C1q binding and activation of the complement classical pathway by *Klebsiella pneumoniae* outer membrane proteins. Infect Immun 61(3):852–860

Alberti S, Marques G, Hernandez-Alles S, Rubires X, Tomas JM, Vivanco F, Benedi VJ (1996) Interaction between complement subcomponent C1q and the *Klebsiella pneumoniae* porin OmpK36. Infect Immun 64(11):4719–4725

Alcantar-Curiel MD, Garcia-Latorre E, Santos JI (2000) *Klebsiella pneumoniae* 35 and 36 kDa porins are common antigens in different serotypes and induce opsonizing antibodies. Arch Med Res 31(1):28–36. doi:S0188-4409(99)00083-1 [pii]

Armant MA, Fenton MJ (2002) Toll-like receptors: a family of pattern-recognition receptors in mammals. Genome Biol 3(8):reviews3011.1–reviews3011.6

Bandyopadhaya A, Sarkar M, Chaudhuri K (2007a) Human intestinal epithelial cell cytokine mRNA responses mediated by NF-kappaB are modulated by the motility and adhesion process of *Vibrio cholerae*. Int J Biochem Cell Biol 39(10):1863–1876. doi:10.1016/j.biocel.2007.05.005, S1357-2725(07)00147-1 [pii]

Bandyopadhaya A, Sarkar M, Chaudhuri K (2007b) Transcriptional upregulation of inflammatory cytokines in human intestinal epithelial cells following *Vibrio cholerae* infection. FEBS J 274(17):4631–4642. doi:10.1111/j.1742-4658.2007.05991.x, EJB5991 [pii]

Bandyopadhaya A, Bhowmick S, Chaudhuri K (2009) Activation of proinflammatory response in human intestinal epithelial cells following *Vibrio cholerae* infection through PI3K/Akt pathway. Can J Microbiol 55(11):1310–1318. doi:10.1139/w09-093, w09-093 [pii]

Banerjee P, Biswas A, Biswas T (2008) Porin-incorporated liposome induces Toll-like receptors 2- and 6-dependent maturation and type 1 response of dendritic cell. Int Immunol 20(12):1551–1563. doi:10.1093/intimm/dxn114, dxn114 [pii]

Baroni A, Gorga F, Baldi A, Perfetto B, Paoletti I, Russo A, Lembo L, Rossano F (2001) Histopathological features and modulation of type IV collagen expression induced by *Pseudomonas aeruginosa* lipopolysaccharide (LPS) and porins on mouse skin. Histol Histopathol 16(3):685–692

Bell JK, Mullen GE, Leifer CA, Mazzoni A, Davies DR, Segal DM (2003) Leucine-rich repeats and pathogen recognition in Toll-like receptors. Trends Immunol 24(10):528–533. doi:S1471490603002424 [pii]

Bellinger-Kawahara C, Horwitz MA (1990) Complement component C3 fixes selectively to the major outer membrane protein (MOMP) of *Legionella pneumophila* and mediates phagocytosis of liposome-MOMP complexes by human monocytes. J Exp Med 172(4):1201–1210

Berthet FXJ, Denoel P, Poolman J, Thonnard J, Bakaletz L (2011) *Haemophilus influenzae* outer membrane protein and use thereof in vaccination. Google patents

Biswas A, Banerjee P, Mukherjee G, Biswas T (2007) Porin of *Shigella dysenteriae* activates mouse peritoneal macrophage through Toll-like receptors 2 and 6 to induce polarized type I response. Mol Immunol 44(5):812–820

Biswas A, Banerjee P, Biswas T (2008) Priming of CD4+ T cells with porin of *Shigella dysenteriae* activates the cells toward type 1 polarization. Int Immunol 20(1):81–88. doi:10.1093/intimm/dxm122, dxm122 [pii]

Biswas A, Banerjee P, Biswas T (2009) Porin of *Shigella dysenteriae* directly promotes toll-like receptor 2-mediated CD4+ T cell survival and effector function. Mol Immunol 46(15):3076–3085. doi:10.1016/j.molimm.2009.06.006, S0161-5890(09)00422-2 [pii]

Bjerknes R, Guttormsen HK, Solberg CO, Wetzler LM (1995) Neisserial porins inhibit human neutrophil actin polymerization, degranulation, opsonin receptor expression, and phagocytosis but prime the neutrophils to increase their oxidative burst. Infect Immun 63(1):160–167

Bjerre A, Brusletto B, Mollnes TE, Fritzsonn E, Rosenqvist E, Wedege E, Namork E, Kierulf P, Brandtzaeg P (2002) Complement activation induced by purified *Neisseria meningitidis* lipopolysaccharide (LPS), outer membrane vesicles, whole bacteria, and an LPS-free mutant. J Infect Dis 185(2):220–228. doi:10.1086/338269, JID010768 [pii]

Blanco DR, Champion CI, Lewinski MA, Shang ES, Simkins SG, Miller JN, Lovett MA (1999) Immunization with *Treponema pallidum* outer membrane vesicles induces high-titer complement-dependent treponemicidal activity and aggregation of *T. pallidum* rare outer membrane proteins (TROMPs). J Immunol 163(5):2741–2746. doi:ji_v163n5p2741 [pii]

Brennan FR, Jones TD, Gilleland LB, Bellaby T, Xu F, North PC, Thompson A, Staczek J, Lin T, Johnson JE, Hamilton WD, Gilleland HE Jr (1999) *Pseudomonas aeruginosa* outer-membrane protein F epitopes are highly immunogenic in mice when expressed on a plant virus. Microbiology 145(Pt 1):211–220

Brogden KA, Ackermann MR, Debey BM (1995) *Pasteurella haemolytica* lipopolysaccharide-associated protein induces pulmonary inflammation after bronchoscopic deposition in calves and sheep. Infect Immun 63(9):3595–3599

Buommino E, Morelli F, Metafora S, Rossano F, Perfetto B, Baroni A, Tufano MA (1999) Porin from *Pseudomonas aeruginosa* induces apoptosis in an epithelial cell line derived from rat seminal vesicles. Infect Immun 67(9):4794–4800

Cai SH, Lu YS, Wu ZH, Jian JC (2013) Cloning, expression of *Vibrio alginolyticus* outer membrane protein-OmpU gene and its potential application as vaccine in crimson snapper, *Lutjanus erythropterus* Bloch. J Fish Dis 36(8):695–702. doi:10.1111/jfd.12036

Camilli A, Schild S, Nelson EJ (2014) Cholera vaccines. Google patents

Cervantes-Barragan L, Gil-Cruz C, Pastelin-Palacios R, Lang KS, Isibasi A, Ludewig B, Lopez-Macias C (2009) TLR2 and TLR4 signaling shapes specific antibody responses to *Salmonella* Typhi antigens. Eur J Immunol 39(1):126–135. doi:10.1002/eji.200838185

Chen D, VanDerMeid KR, McMichael JC, Barniak VL (2005) 74-kilodalton outer membrane protein from *Moraxella catarrhalis*. Google patents

Cheng C, Pal S, Bettahi I, Oxford KL, Barry PA, de la Maza LM (2011) Immunogenicity of a vaccine formulated with the *Chlamydia trachomatis* serovar F, native major outer membrane protein in a nonhuman primate model. Vaccine 29(18):3456–3464. doi:10.1016/j.vaccine.2011.02.057, S0264-410X(11)00292-1 [pii]

Chiavolini D, Weir S, Murphy JR, Wetzler LM (2008) *Neisseria meningitidis* PorB, a Toll-like receptor 2 ligand, improves the capacity of *Francisella tularensis* lipopolysaccharide to protect mice against experimental tularemia. Clin Vaccine Immunol 15(9):1322–1329. doi:10.1128/CVI.00125-08, CVI.00125-08 [pii]

Christodoulides M, Brooks JL, Rattue E, Heckels JE (1998) Immunization with recombinant class 1 outer-membrane protein from *Neisseria meningitidis*: influence of liposomes and adjuvants on antibody avidity, recognition of native protein and the induction of a bactericidal immune response against meningococci. Microbiology 144(Pt 11):3027–3037

Cusumano V, Tufano MA, Mancuso G, Carbone M, Rossano F, Fera MT, Ciliberti FA, Ruocco E, Merendino RA, Teti G (1997) Porins of *Pseudomonas aeruginosa* induce release of tumor necrosis factor alpha and interleukin-6 by human leukocytes. Infect Immun 65(5):1683–1687

Das M, Chopra AK, Cantu JM, Peterson JW (1998) Antisera to selected outer membrane proteins of *Vibrio cholerae* protect against challenge with homologous and heterologous strains of *V. cholerae*. FEMS Immunol Med Microbiol 22(4):303–308. doi:S0928-8244(98)00101-1 [pii]

Dong-Ji Z, Yang X, Shen C, Lu H, Murdin A, Brunham RC (2000) Priming with *Chlamydia trachomatis* major outer membrane protein (MOMP) DNA followed by MOMP ISCOM boosting enhances protection and is associated with increased immunoglobulin A and Th1 cellular immune responses. Infect Immun 68(6):3074–3078

Duperthuy M, Binesse J, Le Roux F, Romestand B, Caro A, Got P, Givaudan A, Mazel D, Bachere E, Destoumieux-Garzon D (2010) The major outer membrane protein OmpU of *Vibrio splendidus* contributes to host antimicrobial peptide resistance and is required for virulence in the oyster *Crassostrea gigas*. Environ Microbiol 12(4):951–963. doi:10.1111/j.1462-2920.2009.02138.x, EMI2138 [pii]

Duperthuy M, Schmitt P, Garzon E, Caro A, Rosa RD, Le Roux F, Lautredou-Audouy N, Got P, Romestand B, de Lorgeril J, Kieffer-Jaquinod S, Bachere E, Destoumieux-Garzon D (2011) Use of OmpU porins for attachment and invasion of *Crassostrea gigas* immune cells by the oyster pathogen *Vibrio splendidus*. Proc Natl Acad Sci U S A 108(7):2993–2998. doi:10.1073/pnas.1015326108, 1015326108 [pii]

Eko FO, Okenu DN, Singh UP, He Q, Black C, Igietseme JU (2011) Evaluation of a broadly protective *Chlamydia*-cholera combination vaccine candidate. Vaccine 29(21):3802–3810. doi:10.1016/j.vaccine.2011.03.027, S0264-410X(11)00379-3 [pii]

Elena G, Giovanna D, Brunella P, De Anna F, Alessandro M, Antonietta TM (2009) Proinflammatory signal transduction pathway induced by *Shigella flexneri* porins in caco-2 cells. Braz J Microbiol 40(3):701–713. doi:10.1590/S1517-838220090003000036, S1517-838220090003000036 [pii]

Exner MM, Wu X, Blanco DR, Miller JN, Lovett MA (2000) Protection elicited by native outer membrane protein Oms66 (p66) against host-adapted *Borrelia burgdorferi*: conformational nature of bactericidal epitopes. Infect Immun 68(5):2647–2654

Finamore E, Vitiello M, D'Isanto M, Galdiero E, Falanga A, Kampanaraki A, Raieta K, Galdiero M (2009) Evidence for IL-6 promoter nuclear activation in U937 cells stimulated with *Salmonella enterica* serovar Typhimurium porins. Eur Cytokine Netw 20(3):140–147. doi:10.1684/ecn.2009.0158, ecn.2009.0158 [pii]

Galdiero F, Tufano MA, Sommese L, Folgore A, Tedesco F (1984) Activation of complement system by porins extracted from *Salmonella* Typhimurium. Infect Immun 46(2):559–563

Galdiero M, De Martino L, Marcatili A, Nuzzo I, Vitiello M, Cipollaro de l'Ero G (1998a) Th1 and Th2 cell involvement in immune response to *Salmonella* Typhimurium porins. Immunology 94(1):5–13

Galdiero M, Palomba E, De L, Vitiello M, Pagnini P (1998b) Effects of the major *Pasteurella multocida* porin on bovine neutrophils. Am J Vet Res 59(10):1270–1274

Galdiero M, Folgore A, Molitierno M, Greco R (1999) Porins and lipopolysaccharide (LPS) from *Salmonella* Typhimurium induce leucocyte transmigration through human endothelial cells in vitro. Clin Exp Immunol 116(3):453–461

Galdiero M, D'Amico M, Gorga F, Di Filippo C, D'Isanto M, Vitiello M, Longanella A, Tortora A (2001a) *Haemophilus influenzae* porin contributes to signaling of the inflammatory cascade in rat brain. Infect Immun 69(1):221–227. doi:10.1128/IAI.69.1.221-227.2001

Galdiero M, D'Isanto M, Vitiello M, Finamore E, Peluso L (2001b) Porins from *Salmonella enterica* serovar Typhimurium induce TNF-alpha, IL-6 and IL-8 release by CD14-independent and CD11a/CD18-dependent mechanisms. Microbiology 147(Pt 10):2697–2704

Galdiero M, Vitiello M, Sanzari E, D'Isanto M, Tortora A, Longanella A, Galdiero S (2002) Porins from *Salmonella enterica* serovar Typhimurium activate the transcription factors activating protein 1 and NF-kappaB through the Raf-1-mitogen-activated protein kinase cascade. Infect Immun 70(2):558–568

Galdiero M, D'Isanto M, Vitiello M, Finamore E, Peluso L (2003a) Monocytic activation of protein tyrosine kinase, protein kinase A and protein kinase C induced by porins isolated from *Salmonella enterica* serovar Typhimurium. J Infect 46(2):111–119

Galdiero M, Pisciotta MG, Galdiero E, Carratelli CR (2003b) Porins and lipopolysaccharide from *Salmonella* Typhimurium regulate the expression of CD80 and CD86 molecules on B cells and macrophages but not CD28 and CD152 on T cells. Clin Microbiol Infect 9(11):1104–1111. doi:728 [pii]

Galdiero S, Capasso D, Vitiello M, D'Isanto M, Pedone C, Galdiero M (2003c) Role of surface-exposed loops of *Haemophilus influenzae* protein P2 in the mitogen-activated protein kinase cascade. Infect Immun 71(5):2798–2809

Galdiero M, Finamore E, Rossano F, Gambuzza M, Catania MR, Teti G, Midiri A, Mancuso G (2004) *Haemophilus influenzae* porin induces Toll-like receptor 2-mediated cytokine production in human monocytes and mouse macrophages. Infect Immun 72(2):1204–1209

Galdiero M, Tortora A, Damiano N, Vitiello M, Longanella A, Galdiero E (2005) Induction of cytokine mRNA expression in U937 cells by *Salmonella* Typhimurium porins is regulated by different phosphorylation pathways. Med Microbiol Immunol 194(1):13–23. doi:10.1007/s00430-003-0209-7

Galdiero M, Vitiello M, D'Isanto M, Raieta K, Galdiero E (2006a) STAT1 and STAT3 phosphorylation by porins are independent of JAKs but are dependent on MAPK pathway and plays a role in U937 cells production of interleukin-6. Cytokine 36(5–6):218–228. doi:10.1016/j.cyto.2006.12.003, S1043-4666(06)00345-0 [pii]

Galdiero S, Vitiello M, Amodeo P, D'Isanto M, Cantisani M, Pedone C, Galdiero M (2006b) Structural requirements for proinflammatory activity of porin P2 Loop 7 from *Haemophilus influenzae*. Biochemistry 45(14):4491–4501. doi:10.1021/bi052262p

Galdiero S, Falanga A, Cantisani M, Tarallo R, Della Pepa ME, D'Oriano V, Galdiero M (2012) Microbe-host interactions: structure and role of Gram-negative bacterial porins. Curr Protein Pept Sci 13(8):843–854. doi:CPPS-EPUB-20121210-11 [pii]

Gil-Cruz C, Bobat S, Marshall JL, Kingsley RA, Ross EA, Henderson IR, Leyton DL, Coughlan RE, Khan M, Jensen KT, Buckley CD, Dougan G, MacLennan IC, Lopez-Macias C, Cunningham AF (2009) The porin OmpD from nontyphoidal *Salmonella* is a key target for a protective B1b cell antibody response. Proc Natl Acad Sci U S A 106(24):9803–9808. doi:10.1073/pnas.0812431106, 0812431106 [pii]

Granoff DM, Aaberge IS, Haneberg B, Holst J, Raff H (2013) Combination meningitidis b/c vaccines. Google patents

Gupta S, Kumar D, Vohra H, Ganguly NK (1999) Involvement of signal transduction pathways in *Salmonella* Typhimurium porin activated gut macrophages. Mol Cell Biochem 194(1–2):235–243

Hara Y, Mohamed R, Nathan S (2009) Immunogenic *Burkholderia pseudomallei* outer membrane proteins as potential candidate vaccine targets. PLoS One 4(8):e6496. doi:10.1371/journal.pone.0006496

Humphries HE, Williams JN, Christodoulides M, Heckels JE (2004) Recombinant meningococcal PorA protein, expressed using a vector system with potential for human vaccination, induces a bactericidal immune response. Vaccine 22(11–12):1564–1569. doi:10.1016/j.vaccine.2003.09.042, S0264410X04000416 [pii]

Hung DT, Mekalanos JJ (2005) Bile acids induce cholera toxin expression in *Vibrio cholerae* in a ToxT-independent manner. Proc Natl Acad Sci U S A 102(8):3028–3033. doi:10.1073/pnas.0409559102, 0409559102 [pii]

Iovane G, Pagnini P, Galdiero M, Cipollaro de l'Ero G, Vitiello M, D'Isanto M, Marcatili A (1998) Role of *Pasteurella multocida* porin on cytokine expression and release by murine splenocytes. Vet Immunol Immunopathol 66(3–4):391–404. doi:S0165-2427(98)00183-4 [pii]

Isibasi A, Ortiz V, Vargas M, Paniagua J, Gonzalez C, Moreno J, Kumate J (1988) Protection against Salmonella Typhi infection in mice after immunization with outer membrane proteins isolated from Salmonella Typhi 9,12, d, Vi. Infect Immun 56(11):2953–2959

Isibasi A, Ortiz-Navarrete V, Paniagua J, Pelayo R, Gonzalez CR, Garcia JA, Kumate J (1992) Active protection of mice against *Salmonella* Typhi by immunization with strain-specific porins. Vaccine 10(12):811–813

Kaisho T, Akira S (2001) Toll-like receptors and their signaling mechanism in innate immunity. Acta Odontol Scand 59(3):124–130

Karin M, Greten FR (2005) NF-kappaB: linking inflammation and immunity to cancer development and progression. Nat Rev Immunol 5(10):749–759. doi:10.1038/nri1703, nri1703 [pii]

Kawai T, Akira S (2005) Pathogen recognition with Toll-like receptors. Curr Opin Immunol 17(4):338–344. doi:10.1016/j.coi.2005.02.007, S0952-7915(05)00079-8 [pii]

Kawai T, Akira S (2009) The roles of TLRs, RLRs and NLRs in pathogen recognition. Int Immunol 21(4):317–337. doi:10.1093/intimm/dxp017, dxp017 [pii]

Khushiramani R, Girisha SK, Karunasagar I (2007) Cloning and expression of an outer membrane protein ompTS of *Aeromonas hydrophila* and study of immunogenicity in fish. Protein Expr Purif 51(2):303–307. doi:10.1016/j.pep.2006.07.021, S1046-5928(06)00234-8 [pii]

Khushiramani RM, Maiti B, Shekar M, Girisha SK, Akash N, Deepanjali A, Karunasagar I (2014) Recombinant *Aeromonas hydrophila* outer membrane protein 48 (Omp48) induces a protective immune response against Aeromonas hydrophila and Edwardsiella tarda. Res Microbiol 163(4):286–291. doi:10.1016/j.resmic.2012.03.001, S0923-2508(12)00039-3 [pii]

Koebnik R, Locher KP, Van Gelder P (2000) Structure and function of bacterial outer membrane proteins: barrels in a nutshell. Mol Microbiol 37(2):239–253. doi:mmi1983 [pii]

Kumar H, Kawai T, Akira S (2012) Pathogen recognition by the innate immune system. Int Rev Immunol 30(1):16–34. doi:10.3109/08830185.2010.529976

Latsch M, Mollerfeld J, Ringsdorf H, Loos M (1990) Studies on the interaction of C1q, a subcomponent of the first component of complement, with porins from *Salmonella* Minnesota incorporated into artificial membranes. FEBS Lett 276(1–2):201–204. doi:0014-5793(90)80542-Q [pii]

Leclerc D, Lopez MCIR (2010) Compositions comprising salmonella porins and uses thereof as adjuvants and vaccines. Google patents

Lee JS, Lee JC, Lee CM, Jung ID, Jeong YI, Seong EY, Chung HY, Park YM (2007) Outer membrane protein A of *Acinetobacter baumannii* induces differentiation of CD4+ T cells toward a Th1 polarizing phenotype through the activation of dendritic cells. Biochem Pharmacol 74(1):86–97. doi:10.1016/j.bcp.2007.02.012, S0006-2952(07)00131-1 [pii]

Leitner DR, Feichter S, Schild-Prufert K, Rechberger GN, Reidl J, Schild S (2013) Lipopolysaccharide modifications of a cholera vaccine candidate based on outer membrane vesicles reduce endotoxicity and reveal the major protective antigen. Infect Immun 81(7):2379–2393. doi:10.1128/IAI.01382-12, IAI.01382-12 [pii]

Lewis LA, Ram S, Prasad A, Gulati S, Getzlaff S, Blom AM, Vogel U, Rice PA (2008) Defining targets for complement components C4b and C3b on the pathogenic neisseriae. Infect Immun 76(1):339–350. doi:10.1128/IAI.00613-07, IAI.00613-07 [pii]

Li H, Ye MZ, Peng B, Wu HK, Xu CX, Xiong XP, Wang C, Wang SY, Peng XX (2010a) Immunoproteomic identification of polyvalent vaccine candidates from *Vibrio parahaemolyticus* outer membrane proteins. J Proteome Res 9(5):2573–2583. doi:10.1021/pr1000219

Li N, Yang Z, Bai J, Fu X, Liu L, Shi C, Wu S (2010b) A shared antigen among *Vibrio* species: outer membrane protein-OmpK as a versatile Vibriosis vaccine candidate in Orange-spotted grouper (Epinephelus coioides). Fish Shellfish Immunol 28(5–6):952–956. doi:10.1016/j.fsi.2010.02.010, S1050-4648(10)00060-4 [pii]

Liu X, Wetzler LM, Massari P (2008) The PorB porin from commensal *Neisseria lactamica* induces Th1 and Th2 immune responses to ovalbumin in mice and is a potential immune adjuvant. Vaccine 26(6):786–796. doi:10.1016/j.vaccine.2007.11.080, S0264-410X(07)01431-4 [pii]

Liu X, Wetzler LM, Nascimento LO, Massari P (2010) Human airway epithelial cell responses to *Neisseria lactamica* and purified porin via Toll-like receptor 2-dependent signaling. Infect Immun 78(12):5314–5323. doi:10.1128/IAI.00681-10, IAI.00681-10 [pii]

Lorenzen DR, Gunther D, Pandit J, Rudel T, Brandt E, Meyer TF (2000) *Neisseria gonorrhoeae* porin modifies the oxidative burst of human professional phagocytes. Infect Immun 68(11): 6215–6222

Lu H, Wang H, Zhao HM, Zhao L, Chen Q, Qi M, Liu J, Yu H, Yu XP, Yang X, Zhao WM (2010) Dendritic cells (DCs) transfected with a recombinant adenovirus carrying chlamydial major outer membrane protein antigen elicit protective immune responses against genital tract challenge infection. Biochem Cell Biol 88(4):757–765. doi:10.1139/O10-011, o10-011 [pii]

MacLeod H, Bhasin N, Wetzler LM (2008) Role of protein tyrosine kinase and Erk1/2 activities in the Toll-like receptor 2-induced cellular activation of murine B cells by neisserial porin. Clin Vaccine Immunol 15(4):630–637. doi:10.1128/CVI.00435-07, CVI.00435-07 [pii]

Mao Z, Yu L, You Z, Wei Y, Liu Y (2007) Cloning, expression and immunogenicty analysis of five outer membrane proteins of *Vibrio parahaemolyticus* zj2003. Fish Shellfish Immunol 23(3):567–575. doi:10.1016/j.fsi.2007.01.004, S1050-4648(07)00010-1 [pii]

Massari P, Henneke P, Ho Y, Latz E, Golenbock DT, Wetzler LM (2002) Cutting edge: immune stimulation by neisserial porins is Toll-like receptor 2 and MyD88 dependent. J Immunol 168(4):1533–1537

Massari P, Ram S, Macleod H, Wetzler LM (2003) The role of porins in neisserial pathogenesis and immunity. Trends Microbiol 11(2):87–93

Massari P, Visintin A, Gunawardana J, Halmen KA, King CA, Golenbock DT, Wetzler LM (2006) Meningococcal porin PorB binds to TLR2 and requires TLR1 for signaling. J Immunol 176(4):2373–2380. doi:176/4/2373 [pii]

Matsui K, Arai T (1990) Protective immunities induced by porins from mutant strains of *Salmonella* Typhimurium. Microbiol Immunol 34(11):917–927

McConnell MJ, Rumbo C, Bou G, Pachon J (2011) Outer membrane vesicles as an acellular vaccine against *Acinetobacter baumannii*. Vaccine 29(34):5705–5710. doi:10.1016/j.vaccine.2011.06.001, S0264-410X(11)00861-9 [pii]

Medzhitov R (2007) Recognition of microorganisms and activation of the immune response. Nature 449(7164):819–826. doi:10.1038/nature06246, nature06246 [pii]

Medzhitov R, Janeway CA Jr (1997) Innate immunity: the virtues of a nonclonal system of recognition. Cell 91(3):295–298. doi:S0092-8674(00)80412-2 [pii]

Merino S, Nogueras MM, Aguilar A, Rubires X, Alberti S, Benedi VJ, Tomas JM (1998) Activation of the complement classical pathway (C1q binding) by mesophilic *Aeromonas hydrophila* outer membrane protein. Infect Immun 66(8):3825–3831

Merino S, Vilches S, Canals R, Ramirez S, Tomas JM (2005) A C1q-binding 40 kDa porin from *Aeromonas salmonicida*: cloning, sequencing, role in serum susceptibility and fish immunoprotection. Microb Pathog 38(5–6):227–237. doi:10.1016/j.micpath.2005.02.006, S0882-4010(05)00022-7 [pii]

Mogensen TH (2009) Pathogen recognition and inflammatory signaling in innate immune defenses. Clin Microbiol Rev 22(2):240–273. doi:10.1128/CMR.00046-08, Table of Contents. 22/2/240 [pii]

Moreno-Eutimio MA, Tenorio-Calvo A, Pastelin-Palacios R, Perez-Shibayama C, Gil-Cruz C, Lopez-Santiago R, Baeza I, Fernandez-Mora M, Bonifaz L, Isibasi A, Calva E, Lopez-Macias C (2013) *Salmonella* Typhi OmpS1 and OmpS2 porins are potent protective immunogens with adjuvant properties. Immunology 139(4):459–471. doi:10.1111/imm.12093

Mukherjee S, Sinha D, Ghosh AK, Biswas T (2014) Bacterial ligand stimulates TLR2-dependent chemokines of colon cell. Immunobiology 219(5):350–356. doi:10.1016/j.imbio.2013.12.002, S0171-2985(13)00210-6 [pii]

Muller A, Gunther D, Brinkmann V, Hurwitz R, Meyer TF, Rudel T (2000) Targeting of the pro-apoptotic VDAC-like porin (PorB) of *Neisseria gonorrhoeae* to mitochondria of infected cells. EMBO J 19(20):5332–5343. doi:10.1093/emboj/19.20.5332

Ningqiu L, Junjie B, Shuqin W, Xiaozhe F, Haihua L, Xing Y, Cunbin S (2008) An outer membrane protein, OmpK, is an effective vaccine candidate for *Vibrio harveyi* in Orange-spotted grouper (*Epinephelus coioides*). Fish Shellfish Immunol 25(6):829–833. doi:10.1016/j.fsi.2008.09.007, S1050-4648(08)00215-5 [pii]

Pal S, Theodor I, Peterson EM, de la Maza LM (2001) Immunization with the *Chlamydia trachomatis* mouse pneumonitis major outer membrane protein can elicit a protective immune response against a genital challenge. Infect Immun 69(10):6240–6247. doi:10.1128/IAI.69.10.6240-6247.2001

Paradiso PR, Seid RC, Poolman JT, Hoogerhout P, Wiertz EJHJ, Van DLP, Heckels JE, Clarke IN (2007) Meningococcal class 1 outer-membrane protein vaccine. Google patents
Pasquevich KA, Garcia Samartino C, Coria LM, Estein SM, Zwerdling A, Ibanez AE, Barrionuevo P, Oliveira FS, Carvalho NB, Borkowski J, Oliveira SC, Warzecha H, Giambartolomei GH, Cassataro J (2010) The protein moiety of *Brucella abortus* outer membrane protein 16 is a new bacterial pathogen-associated molecular pattern that activates dendritic cells in vivo, induces a Th1 immune response, and is a promising self-adjuvanting vaccine against systemic and oral acquired brucellosis. J Immunol 184(9):5200–5212. doi:10.4049/jimmunol.0902209, jimmunol.0902209 [pii]
Plante M, Jerse A, Hamel J, Couture F, Rioux CR, Brodeur BR, Martin D (2000) Intranasal immunization with gonococcal outer membrane preparations reduces the duration of vaginal colonization of mice by *Neisseria gonorrhoeae*. J Infect Dis 182(3):848–855. doi:10.1086/315801, JID991554 [pii]
Pore D, Mahata N, Chakrabarti MK (2012) Outer membrane protein A (OmpA) of *Shigella flexneri* 2a links innate and adaptive immunity in a TLR2-dependent manner and involvement of IL-12 and nitric oxide. J Biol Chem 287(15):12589–12601. doi:10.1074/jbc.M111.335554, M111.335554 [pii]
Price BM, Galloway DR, Baker NR, Gilleland LB, Staczek J, Gilleland HE Jr (2001) Protection against *Pseudomonas aeruginosa* chronic lung infection in mice by genetic immunization against outer membrane protein F (OprF) of P. aeruginosa. Infect Immun 69(5):3510–3515. doi:10.1128/IAI.69.5.3510-3515.2001
Qian R, Chu W, Mao Z, Zhang C, Wei Y, Yu L (2007) Expression, characterization and immunogenicity of a major outer membrane protein from *Vibrio alginolyticus*. Acta Biochim Biophys Sin (Shanghai) 39(3):194–200
Rajesh Kumar S, Ishaq Ahmed VP, Parameswaran V, Sudhakaran R, Sarath Babu V, Sahul Hameed AS (2008) Potential use of chitosan nanoparticles for oral delivery of DNA vaccine in Asian sea bass (*Lates calcarifer*) to protect from *Vibrio* (Listonella) *anguillarum*. Fish Shellfish Immunol 25(1–2):47–56. doi:10.1016/j.fsi.2007.12.004, S1050-4648(07)00222-7 [pii]
Ram S, Cullinane M, Blom AM, Gulati S, McQuillen DP, Monks BG, O'Connell C, Boden R, Elkins C, Pangburn MK, Dahlback B, Rice PA (2001) Binding of C4b-binding protein to porin: a molecular mechanism of serum resistance of *Neisseria gonorrhoeae*. J Exp Med 193(3):281–295
Ray A, Biswas T (2005) Porin of Shigella dysenteriae enhances Toll-like receptors 2 and 6 of mouse peritoneal B-2 cells and induces the expression of immunoglobulin M, immunoglobulin G2a and immunoglobulin A. Immunology 114(1):94–100. doi:10.1111/j.1365-2567.2004.02002.x, IMM2002 [pii]
Ray A, Chatterjee NS, Bhattacharya SK, Biswas T (2003) Porin of *Shigella dysenteriae* enhances mRNA levels for Toll-like receptor 2 and MyD88, up-regulates CD80 of murine macrophage, and induces the release of interleukin-12. FEMS Immunol Med Microbiol 39(3):213–219. doi:S0928824403002335 [pii]
Ray A, Karmakar P, Biswas T (2004) Up-regulation of CD80-CD86 and IgA on mouse peritoneal B-1 cells by porin of *Shigella dysenteriae* is Toll-like receptors 2 and 6 dependent. Mol Immunol 41(12):1167–1175. doi:10.1016/j.molimm.2004.06.007, S016158900400238X [pii]
Ruiz N, Montero T, Hernandez-Borrell J, Vinas M (2003) The role of *Serratia marcescens* porins in antibiotic resistance. Microb Drug Resist 9(3):257–264. doi:10.1089/107662903322286463
Sakharwade SC, Sharma PK, Mukhopadhaya A (2013) *Vibrio cholerae* porin OmpU induces pro-inflammatory responses, but down-regulates LPS-mediated effects in RAW 264.7, THP-1 and human PBMCs. PLoS One 8(9):e76583 [pii]
Sarkar M, Bhowmick S, Casola A, Chaudhuri K (2012) Interleukin-8 gene regulation in epithelial cells by *Vibrio cholerae*: role of multiple promoter elements, adherence and motility of bacteria and host MAPKs. FEBS J 279(8):1464–1473. doi:10.1111/j.1742-4658.2012.08539.x
Schild S, Nelson EJ, Camilli A (2008) Immunization with *Vibrio cholerae* outer membrane vesicles induces protective immunity in mice. Infect Immun 76(10):4554–4563. doi:10.1128/IAI.00532-08, IAI.00532-08 [pii]

Secundino I, Lopez-Macias C, Cervantes-Barragan L, Gil-Cruz C, Rios-Sarabia N, Pastelin-Palacios R, Villasis-Keever MA, Becker I, Puente JL, Calva E, Isibasi A (2006) *Salmonella* porins induce a sustained, lifelong specific bactericidal antibody memory response. Immunology 117(1):59–70. doi:10.1111/j.1365-2567.2005.02263.x, IMM2263 [pii]

Seid RC, Paradiso PR, Poolman JT, Hoogerhout P, Wiertz EJHJ, van der Ley P, Heckels JE, Clarke IN (2006) Meningococcal class 1 outer-membrane protein vaccine. Google patents

Shaw J, Grund V, Durling L, Crane D, Caldwell HD (2002) Dendritic cells pulsed with a recombinant chlamydial major outer membrane protein antigen elicit a CD4(+) type 2 rather than type 1 immune response that is not protective. Infect Immun 70(3):1097–1105

Singh M, Vohra H, Kumar L, Ganguly NK (1999) Induction of systemic and mucosal immune response in mice immunised with porins of *Salmonella* Typhi. J Med Microbiol 48(1):79–88

Singleton TE, Massari P, Wetzler LM (2005) Neisserial porin-induced dendritic cell activation is MyD88 and TLR2 dependent. J Immunol 174(6):3545–3550. doi:174/6/3545 [pii]

Siu LK, Chang FY, Lin YC, Fung CP, Liu YM, Chen JH, Tsai YK, Chong PCS, Leng CH, Liu SJ (2013) Use of outer membrane porin k36 protein (ompk36) in treatment/prevention/diagnosis of enterobacteriaceae infection. Google patents

Snapper CM, Rosas FR, Kehry MR, Mond JJ, Wetzler LM (1997) Neisserial porins may provide critical second signals to polysaccharide-activated murine B cells for induction of immunoglobulin secretion. Infect Immun 65(8):3203–3208

Stephens RS, Kawa D (2011) Porin B (PorB) as a therapeutic target for prevention and treatment of infection by *Chlamydia*. Google patents

Sun K, Zhang WW, Hou JH, Sun L (2009) Immunoprotective analysis of VhhP2, a Vibrio harveyi vaccine candidate. Vaccine 27(21):2733–2740. doi:10.1016/j.vaccine.2009.03.012, S0264-410X(09)00405-8 [pii]

Tabaraie B, Sharma BK, Sharma PR, Sehgal R, Ganguly NK (1994) Evaluation of Salmonella porins as a broad spectrum vaccine candidate. Microbiol Immunol 38(7):553–559

Toobak H, Rasooli I, Talei D, Jahangiri A, Owlia P, Darvish Alipour Astaneh S (2013) Immune response variations to *Salmonella enterica* serovar Typhi recombinant porin proteins in mice. Biologicals 41(4):224–230. doi:10.1016/j.biologicals.2013.05.005, S1045-1056(13)00058-4 [pii]

Toussi DN, Liu X, Massari P (2012) The FomA porin from *Fusobacterium nucleatum* is a Toll-like receptor 2 agonist with immune adjuvant activity. Clin Vaccine Immunol 19(7):1093–1101. doi:10.1128/CVI.00236-12, CVI.00236-12 [pii]

Tsai YK, Fung CP, Lin JC, Chen JH, Chang FY, Chen TL, Siu LK (2011) *Klebsiella pneumonia*e outer membrane porins OmpK35 and OmpK36 play roles in both antimicrobial resistance and virulence. Antimicrob Agents Chemother 55(4):1485–1493. doi:10.1128/AAC.01275-10, AAC.01275-10 [pii]

Tufano MA, Tetta C, Biancone L, Iorio EL, Baroni A, Giovane A, Camussi G (1992) *Salmonella* Typhimurium porins stimulate platelet-activating factor synthesis by human polymorphonuclear neutrophils. J Immunol 149(3):1023–1030

Tufano MA, Rossano F, Catalanotti P, Liguori G, Capasso C, Ceccarelli MT, Marinelli P (1994) Immunobiological activities of *Helicobacter pylori* porins. Infect Immun 62(4):1392–1399

Tufano MA, Caralanotti P, Capasso C, De Paolis P, Ranieri M, Rossano F (1995) Compartmentalization of intravesical and systemic interleukin-6 and tumor necrosis factor α in mice stimulated with porins and lipopolysaccharide from *Salmonella* Typhimurium. J Endotoxin Res 2(5):359–364

Van DERLEYPA, Poolman JT, Hoogerhout P (2007) Immunogenic meningococcal lps and outer membrane vesicles and vaccine there from. Google patents

Vitiello M, D'Isanto M, Galdiero M, Raieta K, Tortora A, Rotondo P, Peluso L (2004) Interleukin-8 production by THP-1 cells stimulated by *Salmonella enterica* serovar Typhimurium porins is mediated by AP-1, NF-kappaB and MAPK pathways. Cytokine 27(1):15–24. doi:10.1016/j.cyto.2004.03.010, S104346660400095X [pii]

Vitiello M, D'Isanto M, Finamore E, Ciarcia R, Kampanaraki A, Galdiero M (2008a) Role of mitogen-activated protein kinases in the iNOS production and cytokine secretion by *Salmonella enterica* serovar Typhimurium porins. Cytokine 41(3):279–285

Vitiello M, Galdiero S, D'Isanto M, D'Amico M, Di Filippo C, Cantisani M, Galdiero M, Pedone C (2008b) Pathophysiological changes of gram-negative bacterial infection can be reproduced by a synthetic peptide mimicking loop L7 sequence of *Haemophilus influenzae* porin. Microbes Infect 10(6):657–663. doi:10.1016/j.micinf.2008.03.002, S1286-4579(08)00082-8 [pii]

Vitiello M, Finamore E, Cantisani M, Bevilacqua P, Incoronato N, Falanga A, Galdiero E, Galdiero M (2011) P2 porin and loop L7 from *Haemophilus influenzae* modulate expression of IL-6 and adhesion molecules in astrocytes. Microbiol Immunol 55(5):347–356. doi:10.1111/j.1348-0421.2011.00318.x

von Specht BU, Knapp B, Muth G, Broker M, Hungerer KD, Diehl KD, Massarrat K, Seemann A, Domdey H (1995) Protection of immunocompromised mice against lethal infection with *Pseudomonas aeruginosa* by active or passive immunization with recombinant *P. aeruginosa* outer membrane protein F and outer membrane protein I fusion proteins. Infect Immun 63(5):1855–1862

West AP, Koblansky AA, Ghosh S (2006) Recognition and signaling by toll-like receptors. Annu Rev Cell Dev Biol 22:409–437

Wetzler LM, Ho Y, Reiser H (1996) Neisserial porins induce B lymphocytes to express costimulatory B7-2 molecules and to proliferate. J Exp Med 183(3):1151–1159

Wibbenmeyer JA, Provenzano D, Landry CF, Klose KE, Delcour AH (2002) *Vibrio cholerae* OmpU and OmpT porins are differentially affected by bile. Infect Immun 70(1):121–126

Wright JC, Williams JN, Christodoulides M, Heckels JE (2002) Immunization with the recombinant PorB outer membrane protein induces a bactericidal immune response against *Neisseria meningitidis*. Infect Immun 70(8):4028–4034

Yadav SK, Sahoo PK, Dixit A (2014) Characterization of immune response elicited by the recombinant outer membrane protein OmpF of *Aeromonas hydrophila*, a potential vaccine candidate in murine model. Mol Biol Rep 41(3):1837–1848. doi:10.1007/s11033-014-3033-9

Zhang C, Yu L, Qian R (2007) Characterization of OmpK, GAPDH and their fusion OmpK-GAPDH derived from *Vibrio harveyi* outer membrane proteins: their immunoprotective ability against vibriosis in large yellow croaker (*Pseudosciaena crocea*). J Appl Microbiol 103(5):1587–1599. doi:10.1111/j.1365-2672.2007.03386.x, JAM3386 [pii]

Zhu W, Thomas CE, Sparling PF (2004) DNA immunization of mice with a plasmid encoding *Neisseria gonorrhea* PorB protein by intramuscular injection and epidermal particle bombardment. Vaccine 22(5–6):660–669. doi:S0264410X03006479 [pii]

Chapter 7
Vibrio cholerae Cytolysin: Structure–Function Mechanism of an Atypical β-Barrel Pore-Forming Toxin

Anand Kumar Rai and Kausik Chattopadhyay

Introduction

Every living cell is surrounded by a plasma membrane composed of phospholipid bilayer. Plasma membranes allow selective passage of solvents, ions, small molecules, and macromolecules into and out of the cells, while at the same time preventing unwanted exchange of substances between the cells and the outside environment. Thus, plasma membranes serve an extremely crucial function in terms of acting as the selective permeability barrier for the living cells. Virulence mechanisms of many pathogenic bacteria involve breaching of this permeability barrier function of the target host cell membranes by employing a specialized class of toxins, known as pore-forming toxins (PFTs) (Alouf and Popoff 2006).

PFTs represent a unique class of membrane-damaging proteins that act to kill their target cells by forming pores in the cell membranes. PFTs are found in a wide array of organisms starting from bacteria to humans (Dunstone and Tweten 2012; Voskoboinik and Trapani 2006; Voskoboinik et al. 2006; McCormack et al. 2013), and are implicated in diverse biological functions that include bacterial pathogenesis processes (Alouf and Popoff 2006) as well as vertebrate immune responses (Voskoboinik et al. 2006; McCormack et al. 2013; Kondos et al. 2010; Voskoboinik and Trapani 2006). Some of the classic examples of the PFTs include the membrane attack complexes generated by the complement cascade of the vertebrate innate immune system (Borsos et al. 1964; Humphrey and Dourmashkin 1969; Bhakdi and Tranum-Jensen 1978; Mayer 1972), and the perforin protein produced by the cyto-

A.K. Rai • K. Chattopadhyay (✉)
Centre for Protein Science, Design and Engineering,
Department of Biological Sciences, Indian Institute of Science Education and Research (IISER), Mohali, Sector 81, Knowledge City, SAS Nagar, Mohali, Manauli 140306, Punjab, India
e-mail: kausik@iisermohali.ac.in

A. Chakrabarti, A. Surolia (eds.), *Biochemical Roles of Eukaryotic Cell Surface Macromolecules*, Advances in Experimental Medicine and Biology 842,
DOI 10.1007/978-3-319-11280-0_7

toxic T lymphocytes of the adaptive immune system (Dennert and Podack 1983; Podack and Dennert 1983; Blumenthal et al. 1984; Law et al. 2010). PFTs are known to be important components of the venoms of insects (Smith et al. 1994), snakes (Wu et al. 2012) and sea anemones (Kristan et al. 2009; Tejuca et al. 2009).

As mentioned above, PFTs are implicated in the virulence mechanisms of a wide array of bacterial pathogens (Los et al. 2013). In fact, one third of the bacterial toxins belong to the PFT family (Alouf 2001). In their generalized mode of action, bacterial PFTs are commonly generated as water-soluble molecules, which upon interacting with the target cell membranes convert into membrane-inserted water-filled pore(s) (van der Goot 2003; Parker and Feil 2005; Gonzalez et al. 2008; Iacovache et al. 2008, 2010). Depending on the specific PFT under consideration, membrane-inserted pore diameters can vary in the range from less than 1 nm up to 50 nm (Rosado et al. 2008; Parker and Feil 2005). PFT-mediated pore formation acts to punch holes in the target cell membrane, which in turn allow free diffusion of solvents, ions and other substances. In some cases, membrane pores formed by the PFT molecules are used by the pathogenic bacteria to inject toxic substances into the target host cells (Young and Collier 2007). Many bacterial pathogens, which survive within the subcellular organelles, are known to employ PFTs to rupture the organelle membranes for transmission into the cytoplasm toward exerting their pathogenesis process (Hamon et al. 2012; Schnupf and Portnoy 2007).

Mode of actions of the PFT family of proteins highlights a remarkable dimorphic nature of their structural property. With a unique primary amino acid sequence, PFTs are capable of adopting two distinct structural forms that can be accommodated into two discrete physicochemical environments: the hydrophilic environment of the aqueous phase and the hydrophobic environment of the membrane lipid bilayer. Such a property of the PFT family of molecules makes them unique models to address questions concerning the dynamics of protein structure and folding (Heuck et al. 2001; Chattopadhyay and Banerjee 2003).

Based on the structural and functional considerations, PFTs can be grouped into a number of distinct subclasses (Alouf and Popoff 2006). The most common way of classifying the PFT family members is on the basis of the structural mechanism used for the membrane pore formation process. Accordingly, PFTs are broadly classified into two structural subfamilies: α-PFTs and β-PFTs (Iacovache et al. 2010) (Fig. 7.1). α-PFTs employ α-helices to generate the transmembrane pore structures. Pore-forming Colicins (Wiener et al. 1997) and Cytolysin A (Mueller et al. 2009) produced by *Escherichia coli* and closely related bacteria are the archetypical examples in the α-PFT category. Members in the β-PFT subfamily are known to form transmembrane pores composed of β-strand-rich motifs, commonly termed as β-barrel structure (Heuck et al. 2001; Prevost et al. 2001; Menestrina et al. 2001). *Staphylococcus aureus* α-hemolysin is one of the most well-studied β-PFT family members (Song et al. 1996).

In the present review, we will discuss the structure–function relationship of *Vibrio cholerae* cytolysin, one of the prominent members in the β-PFT family of bacterial protein toxins.

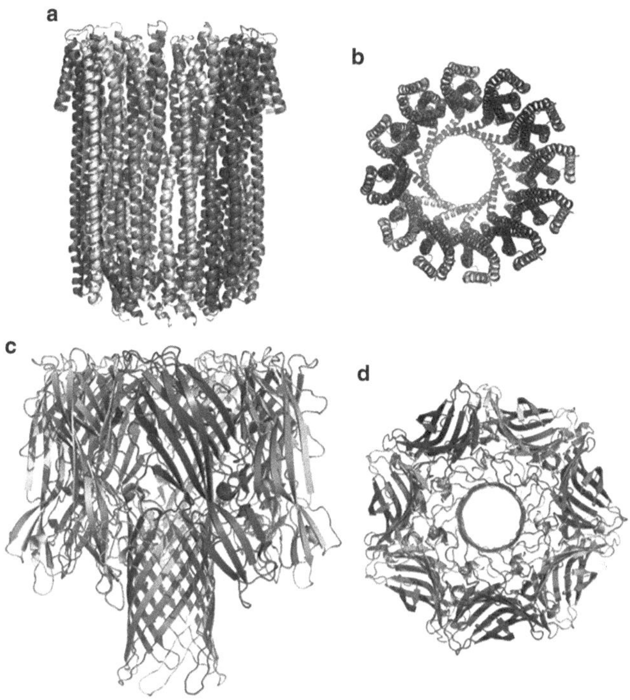

Fig. 7.1 Structural classification of α-PFTs and β-PFTs. (**a**, **b**) Oligomeric pore structure of Cytolysin A from *E. coli* (PDB: 2WCD) is shown as an example of α-PFT. (**c**, **d**) Oligomeric pore structure of the *S. aureus* α-hemolysin (PDB: 7AHL) represents the classic example of β-PFT. (**a**) and (**c**) show the *side views* of the pore structures, while *top views* of the pores are shown in (**b**) and (**d**). Structural models are visualized using the program PyMOL [DeLano WL, The PyMOL Molecular Graphics System (2002) found online (http://pymol.org)]

Vibrio cholerae Cytolysin

Vibrio cholerae cytolysin (VCC) is a membrane-damaging cytolytic/cytotoxic protein produced by many pathogenic strains of the Gram negative bacteria *V. cholerae*, the causative agent of severe diarrheal disease cholera (Kaper et al. 1995). VCC shows potent lytic activity against variety of erythrocytes and mammalian cells (Honda and Finkelstein 1979; Goldberg and Murphy 1984; Richardson et al. 1986; McCardell et al. 1985; Saka et al. 2008; Mitra et al. 2000; Coelho et al. 2000; Mukherjee et al. 2008; Figueroa-Arredondo et al. 2001; Zitzer et al. 1997a; Chakraborty et al. 2011). It is also shown to possess enterotoxic activity in terms of triggering bloody fluid accumulation in the rabbit ligated ileal loops (Ichinose et al. 1987). Based on such observations, VCC has been considered as a potential virulence factor of *V. cholerae*. VCC is particularly implicated in the pathogenesis process of those strains that lack 'cholera toxin', the major virulence factor responsible for causing the massive dehydrating diarrhea during *V. cholerae* infection (Saka et al. 2008; Kaper et al. 1995).

VCC is encoded by the *hlyA* gene in *V. cholerae* (Goldberg and Murphy 1984; Yamamoto et al. 1990; Rader and Murphy 1988). VCC is synthesized as an ~82 kDa molecule, Pre-Pro-VCC (Yamamoto et al. 1990). In the process of secretion, the N-terminal 25-residue signal peptide is cleaved to generate an inactive precursor of the toxin (~79.5 kDa), termed as Pro-VCC. Subsequently, ~15 kDa N-terminal region from Pro-VCC is proteolytically removed resulting in the generation of the cytolytically active mature form of VCC (Nagamune et al. 1996). Proteolytic maturation of VCC is believed to be mediated by the HA/protease, which represents the major extracellular proteolytic activity of *V. cholerae* (Nagamune et al. 1996). Conversion of Pro-VCC into the mature form of the toxin can also be achieved in vitro by other proteases like trypsin, chymotrypsin, and subtilisin (Nagamune et al. 1996). It has also been shown that the proteolytic activation of Pro-VCC can be triggered by the action of the proteases present on the surface of the target eukaryotic cells as well (Valeva et al. 2004).

Active form of VCC, in its purified form, has been shown to trigger lysis of erythrocytes, and other eukaryotic cells by forming transmembrane oligomeric pores of 1–2 nm diameters (Ikigai et al. 1996). Membrane pore-forming activity of VCC could also be mimicked in the lipid bilayer of model membranes (Ikigai et al. 1997). Apart from of its pore-forming cytolytic activity, VCC shows a potent lectin-like activity in interacting with complex glycoproteins and glycolipids containing terminal β1-galactosyl moiety (Saha and Banerjee 1997). Structural studies have characterized VCC as a β-PFT molecule, and suggest that it would follow the overall scheme of the generalized β-PFT mechanism (De and Olson 2011; Olson and Gouaux 2005; Valeva et al. 2005). However, VCC differs from the archetypical members of the β-PFT family in several aspects, particularly in the intricate details of the structure-function mechanism(s) associated with its membrane pore formation process. In the subsequent sections, we will discuss our current understanding regarding the structure-function relationship of VCC, in the context of its role as a β-PFT family of bacterial protein toxins.

Structural Features of VCC

As stated earlier, VCC is secreted as water-soluble monomeric precursor Pro-VCC, which upon removal of the Pro-domain gets converted into the mature active form of the toxin (Nagamune et al. 1996). Mature form of VCC causes lysis of its target cells by forming transmembrane oligomeric pores (Ikigai et al. 1996; Zitzer et al. 1995). High resolution three-dimensional structure has been determined for the water-soluble, monomeric precursor form Pro-VCC (Olson and Gouaux 2005) (Fig. 7.2). Structure of the oligomeric pore state of VCC has also been elucidated recently (De and Olson 2011) (Fig. 7.2). Structural studies have confirmed that VCC is indeed a β-PFT family of toxin, and it acts by forming transmembrane heptameric β-barrel pores in the membrane lipid bilayer. Consistent with the structural organization of the archetypical β-PFT pores, transmembrane heptameric pore

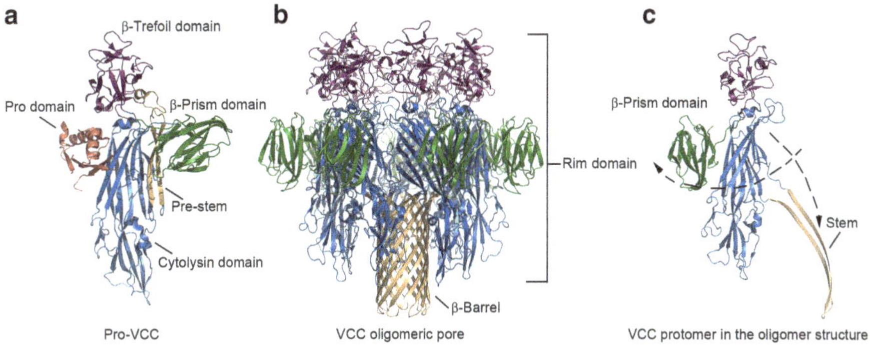

Fig. 7.2 Structures of the monomeric and oligomeric pore form of VCC. (**a**) Structure of the monomeric precursor form Pro-VCC (PDB: 1XEZ). (**b**) Heptameric β-barrel pore structure of VCC (PDB: 3O44). (**c**) VCC protomer in the oligomeric pore structure. Individual domains are marked. Structural rearrangements of the stem loop and the β-Prism domain are indicated in (**c**). Structural models are visualized using the program PyMOL [DeLano WL, The PyMOL Molecular Graphics System (2002) found online (http://pymol.org)]

complex of VCC depicts a mushroom-shaped architecture, which is grossly divided into two parts: (a) transmembrane β-barrel segment, and (b) membrane-proximal rim domain (Fig. 7.2). Analysis of the structures, however, highlights several unique features that are not commonly documented in the archetypical β-PFT family of bacterial protein toxins. Consistent with the conventional β-PFT molecular structures, VCC contains a central scaffold of cytolysin domain that constitutes the core structure of the mushroom-shaped oligomeric pore complex (Fig. 7.2). The cytolysin domain also encompasses the pore-forming motif of the toxin. In addition to the cytolysin domain, however, VCC documents presence of three additional domains that are not commonly found in the classical β-PFT molecules: an N-terminal Pro-domain present in the Pro-VCC precursor form of the toxin, and two lectin-like domains located at the C-terminal side of cytolysin domain (Fig. 7.2).

Cytolysin domain: VCC harbors a central ~325 residue cytolysin domain (Olson and Gouaux 2005; De and Olson 2011) that resembles overall architecture of the cytolysin domains present in the prototype β-PFTs like *S. aureus* α-hemolysin (Song et al. 1996). Consistent with the structural features of the transmembrane oligomeric pores of the conventional β-PFT molecules, cytolysin domain of VCC provides the central scaffold of the pore complex. VCC forms mushroom-shaped heptameric transmembrane pore structures that can be divided grossly into two parts: (a) transmembrane segment that creates the β-barrel pore, and, (b) membrane-proximal rim-domain that remains sited onto the membrane surface. The transmembrane β-barrel segment of the VCC pore complex is constituted solely from the contribution of the cytolsyin domain. Moreover, majority of the rim-domain is also constructed by the structural components of the cytolysin domain.

This cytolysin domain encompasses the 42-residue long, pore-forming segment of VCC that contributes to the generation of the transmembrane β-barrel pore structure (Fig. 7.2). In the water-soluble monomeric state of the protein, this region assumes two-strand β-sheet structure, and remains compactly packed against the cytolysin domain, in the form of a so called 'pre-stem' configuration. The interactions that keep the pre-stem loop packed within the cytolysin domain include a range of polar and non-polar/hydrophobic residues. In the process of oligomeric pore formation, the 'pre-stem' loop from each of the toxin protomers undergoes distinct conformational change to attain a so called 'stem' configuration, and gets inserted into the membrane lipid bilayer in a synchronous manner. In this way, stem region from each toxin subunit contributes two β-strands toward generation of the stem region of the transmembrane β-barrel pore. In the transmembrane oligomeric pore structure, stem regions from the neighboring protomers make extensive interactions between each other, and thus presumably contribute toward remarkable stability of the oligomeric pore states (SDS-stability, resistance toward proteolysis etc.) (Ray et al. 2003).

While part of the VCC cytolysin domain contributes toward generation of the transmembrane segments, rest of it constitutes the membrane-proximal rim-domain of the oligomeric pore complex. Analysis of the structural models of the β-PFT pore structures (Song et al. 1996), including that of VCC (De and Olson 2011), suggests that the membrane-proximal rim-domain act as the structural scaffold for the β-PFT oligomeric pores. Rim-domain provides an interface that mediates interaction of the toxin with the lipid head-groups of the membrane lipid bilayer. It, therefore, appears that the cytolysin domain may also contribute toward interaction of the toxin with the membrane lipid head-groups. Indeed, a lipid-binding pocket in the cytolysin domain has been experimentally established within the membrane-proximal rim-domain of *S. aureus* α-hemolysin (Olson et al. 1999; Galdiero and Gouaux 2004). In case of VCC, however, no such lipid-binding pocket has been documented. Nevertheless, in one of our study, we have shown that a mutation within the membrane-proximal rim-domain of VCC confers compromised membrane-binding, and membrane pore-forming activity for the toxin (Paul and Chattopadhyay 2012). This observation, therefore, indicates the possibility of a potential lipid-binding motif in the cytolysin domain of VCC as well.

Pro-domain: As mentioned above, VCC is secreted by the bacteria in the form of an inactive precursor, Pro-VCC (Nagamune et al. 1996; Yamamoto et al. 1990). Structure of the Pro-VCC (Olson and Gouaux 2005) shows presence of ~15 kDa Pro-domain, which is attached to the N-terminus of the cytolysin domain via 29-residue long connecting flexible linker. This linker contains ~15 residue long sequence motif that acts as the cleavage site(s) for a battery of proteases (Nagamune et al. 1996). Consistent with such observation, proteolytic cleavage at this linker sequence results into the removal of the Pro-domain, and thus leads to the maturation of the VCC toxin.

Presence of the Pro-domain in the precursor form of VCC has been shown to be essential for efficient secretion and proper folding of the toxin molecule. One previous study has shown that the recombinant *V. cholerae* cells, harboring truncated variant of *hlyA* gene lacking the sequence for the Pro-region, fails to secrete the toxin outside the bacterial cells (Nagamune et al. 1997). In vitro denaturation/renaturation experiments have demonstrated that in absence of the Pro-domain VCC fails to refold back to its active form, while Pro-VCC can achieve proper refolding (Nagamune et al. 1997). In one of our recent study, we have shown that the presence of the Pro-domain increases the unfolding propensity of the precursor molecule in response to various denaturing conditions, while mature form of the toxin shows considerable resistance toward unfolding (Paul and Chattopadhyay 2011). Altogether, these studies suggest an intramolecular chaperone-like activity of the Pro-domain in terms of providing sufficient extent of structural plasticity in the VCC molecular structure, which might be required for efficient secretion of the toxin in its precursor form across the bacterial membrane. It is, however, still not properly elucidated how exactly the presence of the Pro-domain keeps the VCC toxin in its inactive precursor state.

β-Trefoil lectin-like domain: VCC contains a β-Trefoil lectin-like domain (~15 kDa) at the C-terminal boundary of the cytolysin domain (Olson and Gouaux 2005). This β-Trefoil lectin-like domain is also present in the closely related cytolysins from the *Vibrionaceae* bacteria, but it is absent in the archetypical β-PFT molecules like *S. aureus* α-hemolysin (Olson and Gouaux 2005). The β-Trefoil domain is connected to the cytolysin domain via a short linker sequence composed of Gly-Gly-Arg-Pro. VCC β-Trefoil domain shows sequence and structural similarity to the carbohydrate-binding domain of plant toxin ricin, and highlights presence of the QXW conserved carbohydrate-binding motif(s) observed in the archetypical β-trefoil lectin domains (Rutenber et al. 1987; Montfort et al. 1987; Loris 2002; Sharon and Lis 2004). However, carbohydrate-binding activity of the β-trefoil domain of VCC has not been explored yet experimentally. Also, the implication of the β-trefoil domain in the structure-function mechanism of VCC remains to be investigated.

β-Prism lectin-like domain: VCC contains another ~15 kDa domain that is connected to the C-terminus of the β-Trefoil domain via relatively long linker sequence (Olson and Gouaux 2005). This domain is not commonly documented in other β-PFTs, including closely related cytolysins from *V. vulnificus* and *Aeromonas hydrophila* (Olson and Gouaux 2005). This C-terminal domain of VCC shows structural similarity to several prototype β-Prism lectins, like jacalin (Sankaranarayanan et al. 1996) and *Maclura pomifera* agglutinin (MPA) (Lee et al. 1998). VCC β-Prism-like domain highlights presence of a binding pocket similar to the carbohydrate-binding site of the jacalin and MPA lectins. In one of our recent studies, we have established the role of the β-Prism domain in the lectin-like activity of VCC (Rai et al. 2013). In this study, we have shown that the truncation of the β-Prism domain completely abolishes the lectin activity of VCC toward β1-galactosyl-terminated glycoconjugates. Using structure-guided mutagenesis approach we have also mapped the critical residues within the β-Prism domain that are essential for the lectin activity. The study shows that an amino acid triad (composed of Asp617, Tyr654, and Tyr679) positioned within the putative carbohydrate-binding pocket constitute the essential element for the VCC lectin activity. Altogether, it has been conclusively established now that the β-Prism domain acts as the structural scaffold responsible for the lectin-like activity of VCC (Rai et al. 2013; Levan et al. 2013).

It is important to note that the β-Prism domain of VCC adopts two distinct positions with respect to the cytolysin domain, in the monomeric precursor form and in the oligomeric pore state (De and Olson 2011; Olson and Gouaux 2005) (Fig. 7.2). In the water-soluble precursor Pro-VCC, β-Prism domain is located on the opposite side of the Pro-domain on top of the pre-stem loop, while in the oligomeric pore structure it is repositioned in place of the Pro-domain (Fig. 7.2). Such rearrangement of the β-Prism domain is absolutely essential for the membrane insertion, and oligomeric pore formation process. In absence of the reorganization of the β-Prism domain, it would be positioned in such a way that would in turn block oligomerization of the toxin protomers, simply because of steric clash. Also, in absence of β-Prism domain's rearrangement, pre-stem loop would not be able to open up toward membrane insertion. Based on our recent study (Rai et al. 2013) it appears that the lectin-like activity of the β-Prism domain might act as a triggering mechanism so as to prompt its structural rearrangement against the cytolysin domain.

Presence of the β-Prism domain has been shown to be critical for efficient membrane pore-formation of VCC (Rai et al. 2013; Mazumdar et al. 2011; Olson and Gouaux 2005). Truncated variant of VCC lacking the C-terminal β-Prism domain shows abortive membrane pore formation. It has been observed that in absence of the β-Prism domain, the truncated toxin can form membrane-bound oligomers, but does not display functional pore-forming activity (Rai et al. 2013). Cryo electron microscopy-based analysis of the oligomers formed by the truncated variant suggests that in absence of the β-Prism domain VCC might form an abortive oligomeric pore having obstructed hole in the *cis*-side of pore lumen (Dutta et al. 2010; He and Olson 2010). It has been proposed that such pore occlusion happens, presumably due to collapse of the β-Trefoil domain in absence of the β-Prism domain.

Structural reorganizations during oligomeric pore formation: Comparison of the structures of the monomeric and oligomeric form of VCC suggests massive structural reorganization within the toxin monomers during the oligomeric membrane pore formation process (De and Olson 2011; Olson and Gouaux 2005) (Fig. 7.2). As mentioned above, the most critical structural rearrangement is the opening up of the 'pre-stem' loop from the cytolysin domain, and its subsequent membrane insertion in the form of 'stem' configuration. In the water-soluble monomeric Pro-VCC state, 'pre-stem' loop remains sandwiched between the β-Prism domain and the cytolysin domain. Thus, for 'pre-stem' to 'stem' conversion to occur, it requires movement of the β-Prism domain. Indeed, during the whole process β-Prism domain reorients with respect to the central cytolysin domain by almost 180° angle, and occupies the position where originally Pro-domain is located in the Pro-VCC structure. Rearrangement of the β-Prism domain represents the second most prominent structural change associated with the membrane pore formation process of VCC. Movement of the β-Prism domain makes way for the 'pre-stem' loop to undergo the conformational change required for the subsequent membrane insertion and oligomeric pore formation process.

Structural Features of the VCC β-Barrel Pore

Transmission electron microscopy (TEM)-based studies have initially characterized the transmembrane oligomeric pore complexes of VCC as membrane lesions of typical ring-like structures having inner diameter of approximately 1–2 nm (Ikigai et al. 1996). Studies showing inhibition of lytic activity by the osmoprotectants of defined molecular sizes have also suggested similar range of pore diameters for VCC (Ikigai et al. 1996). Single channel conductance measurement(s) using VCC pores generated in the supported lipid bilayer system have suggested that VCC forms anion-selective diffusion channels (Menzl et al. 1996; Ikigai et al. 1997). Single channel measurement studies have also suggested asymmetric lumen geometry for the VCC pores: larger opening in the *cis*-side than in the *trans*-side, with narrow constriction at the central part of the lumen (Yuldasheva et al. 2001). Crystal structure of the VCC oligomeric pore state confirms such 'cup-shaped' lumen geometry (De and Olson 2011). Crystal structure also elucidates that the lining of the VCC pore lumen is constituted by a combination of charged, as well as hydrophobic/aromatic amino acid residues. Analysis of the structural model also explains that the narrow constriction near the central region of the pore lumen is caused by the aromatic ring of a tryptophan residue contributed by each of the VCC protomer. Similar architecture has been documented in another β-PFT member, anthrax protective antigen (Krantz et al. 2005; Katayama et al. 2010; Sun et al. 2008). Implication of the lumen geometry of the VCC membrane pore has not been explored yet in the context of VCC mode of action.

Mechanism of Membrane Pore Formation

Membrane pore formation mechanism of β-PFT family of bacterial protein toxins, in general, has been proposed to involve three distinct steps: (i) binding of the water-soluble toxin monomers to the target cell membrane; (ii) assembly of the membrane-bound toxin monomers to generate a transient 'pre-pore' oligomeric intermediate on the membrane surface; (iii) conversion of the pre-pore intermediate into the functional transmembrane pore (Menestrina et al. 2001; Parker and Feil 2005). In the sequence of events, as mentioned above, stem-loops from each of the toxin protomers insert into the membrane lipid bilayer in a concerted manner, and generate the transmembrane β-barrel. Structural studies on a number of β-PFTs have validated overall generalization of such scheme (Song et al. 1996; Gilbert 2010; Dunstone and Tweten 2012; Hotze and Tweten 1818; Rossjohn et al. 2007). However, individual β-PFTs quite often deviate from such generalized scheme in terms of displaying significant variations in the details of the mechanism. For example, process of membrane interaction shows wide range of diversity in terms of receptor specificity, role of membrane lipid components, and so on. In many cases, discrete intermediate events are described only to a limited extent. More importantly, the mechanism(s) that regulate the discrete steps leading toward functional membrane pore-formation are not properly understood for most of the β-PFTs.

Membrane pore-formation process of VCC has been explored in large number of studies. Membrane pore-formation by VCC can be mimicked in the membrane lipid bilayer of synthetic lipid vesicles or liposomes suggesting that the membrane interaction process does not critically require any non-lipid components (Ikigai et al. 1997). However, membrane pore formation is found to be more efficient in biomembranes as compared to that in liposomes, suggesting accessory role of additional molecules present in the cellular membranes (Zitzer et al. 1999). For example, erythrocytes are significantly more susceptible compared to liposome. More interestingly, erythrocytes of different species show different extent of susceptibility against the lytic activity of VCC. Rabbit erythrocytes are found to be more sensitive as compared to human erythrocytes (Zitzer et al. 1997b). Previous studies have indicated role of different cell surface proteins (e.g., glycophorin B on human erythrocytes) as the potential receptor(s) for VCC (Zhang et al. 1999). Since, VCC contains specific lectin-like activity, it has also been speculated that cell surface glycoprotein/glycolipid molecules can act as probable receptor(s) for the toxin (Rai et al. 2013). However, exact identity of the specific cell surface receptor for VCC has not been elucidated yet.

Much of the mechanism(s) associated with the membrane pore formation process of VCC have been studied using synthetic liposome vesicles. As stated above, oligomeric β-barrel pore formation can be efficiently triggered in the membrane lipid bilayer of liposomes (Ikigai et al. 1996). In some earlier studies, association of VCC with the membrane lipid bilayer of liposomes has been suggested to be a non-specific process, driven mostly by global amphiphilicity of the water-soluble monomeric form of the toxin molecules (Chattopadhyay et al. 2002). However, oligomerization and membrane pore formation has been shown to be more specific events requiring assistance of distinct membrane components. Notably, presence of cholesterol has been shown to be absolutely essential for functionality of VCC (Ikigai et al. 1996). Cholesterol appears to regulate the mode of action of VCC in a stereospecific manner; enantiomeric form of cholesterol does not support the efficient activity of VCC in the liposome membrane (Zitzer et al. 2003). It therefore appears that cholesterol regulates VCC activity, not by regulating the physicochemical properties of the membrane environment, rather by physically interacting with the toxin molecule. In fact, requirement of specific structural elements present in cholesterol has been implicated for membrane-binding and oligomerization property of VCC (Ikigai et al. 2006). Nevertheless, specific cholesterol-binding structural motif present within the VCC structure has not been established yet. In one recent study, we have shown that a single point mutation within the potential membrane-binding rim domain of VCC critically abrogates cholesterol-dependent membrane pore-formation mechanism of VCC (Paul and Chattopadhyay 2012). More detail studies would be required, however, to conclusively elucidate the structural mechanism(s) of cholesterol-dependency in the context of VCC mode of action.

In addition to the requirement of cholesterol, efficiency of membrane pore formation by VCC varies depending on the lipid composition of the membrane. For example, membrane permeabilization activity of VCC is found to be more pronounced in presence of sphingolipids (Zitzer et al. 1999). In particular, ceramide moiety has been shown to be critically implicated for efficient membrane-damaging activity (Zitzer et al. 1999). It has also been demonstrated that the presence of specific cone-shaped lipids in conjugation with cholesterol favors efficient membrane pore-forming activity (Zitzer et al. 2001). It has been suggested that these specific lipids regulate VCC activity, not by directly interacting with the toxin molecule, rather by promoting favorable interaction of the membrane-bound toxins with cholesterol (Zitzer et al. 2001). Altogether, it appears that VCC employs complex lipid-dependent mechanism to exert its membrane pore-forming activity, the exact nature of which is still not fully understood. Once again, no information is available at present whether VCC utilizes any specific structural motif within its molecular structure to mediate interaction with the lipid components of the target membrane.

While the role(s) of membrane lipid components in regulating the VCC functionality have been studied extensively in the past, the dynamics of its membrane pore-formation events has been explored only in recent years. Existence of the pre-pore oligomeric intermediate in the process of pore formation for VCC has been documented only recently. It has been shown that the trapping of the membrane-spanning stem loop in its pre-stem configuration via engineered disulfide linkage could arrest the pre-oligomeric intermediate of VCC (Lohner et al. 2009). Also, a truncated variant of VCC lacking the pre-stem structure is found to remain trapped as the pre-pore oligomer on the membrane surface (Paul and Chattopadhyay 2014). These studies conclusively show that VCC follows the archetypical β-PFT mechanisms in terms of displaying involvement of the pre-pore oligomeric intermediates.

It has been suggested previously that the membrane-binding step precedes oligomerization of VCC (Zitzer et al. 2000). It has also been shown that the membrane oligomerization and functional pore formation for VCC require more stringent criteria as compared to the membrane binding step. For example, membrane binding can occur even at low temperature of 4 °C, whereas oligomerization and pore formation get critically arrested at temperature range less than 10 °C (Zitzer et al. 1997b, 2000). Such observation clearly suggests that the binding step is definitely distinct from the subsequent oligomerization and functional pore-formation events. However, membrane insertion event could not be delineated earlier from the binding and/or oligomerization steps of VCC. In a very recent study, we have now shown that the membrane insertion step depends critically on the prior oligomerization of the membrane-bound VCC monomers (Rai and Chattopadhyay 2014). Arresting of the VCC molecule in its membrane-bound monomeric state abrogates oligomerization, membrane insertion, as well as functional pore formation (Rai and Chattopadhyay 2014). These results provide valuable insights regarding the dynamics of the membrane pore-formation process employed by VCC as a prototype in the β-PFT family. Such issues are relevant not only in the context of the VCC mode of action, but are also crucial to understand and elaborate the generalized mechanisms of the β-PFT family of proteins. It is also important to note that such aspects are also relevant to understand the dynamics of the protein-membrane interactions in a broader canvas of membrane protein functionalities.

Conclusion

Existing information regarding the membrane pore formation process of VCC highlights an intriguing mechanism of enormous complexity. The mechanism of functional pore formation appears to be the outcome of a complex cross-talk between the structural elements of the protein's molecular structure and the components of the target membrane. It is really fascinating how exactly such mechanism is regulated in presence of the membrane lipid bilayer of the target cell systems.

Valuable insights have been obtained regarding the mechanism of actions of a large number of β-PFTs, including VCC. In particular, for VCC structures have been described for the monomeric and oligomeric pore form of the toxin. Information is also available regarding some of the intermediate steps involved in the process of membrane pore formation. However, the regulatory mechanisms that control the dynamics of the process have been elucidated only to a limited extent. Also, structural basis of the regulatory mechanisms imposed by the membrane lipid components needs to be elaborated in detail in future studies. Such information would be critical to enrich our insights regarding the membrane pore-formation mechanism of VCC in the context of its implications for the *V. cholerae* pathogenesis process. Also, such insights would enrich our knowledge regarding the generalized mechanism of β-PFT family of bacterial pore-forming toxins.

Acknowledgements We acknowledge Department of Biotechnology (DBT), India for financial support (DBT Grant No. BT/PR13350/BRB/10/751/2009). We also thank IISER Mohali for support.

References

Alouf JE (2001) Pore-forming bacterial toxins: an overview. In: van der Goot FG (ed) Pore-forming toxins. Springer, Heidelberg, pp 1–14

Alouf JE, Popoff MR (2006) The comprehensive sourcebook of bacterial protein toxins, 3rd edn. Academic, New York

Bhakdi S, Tranum-Jensen J (1978) Molecular nature of the complement lesion. Proc Natl Acad Sci U S A 75(11):5655–5659

Blumenthal R, Millard PJ, Henkart MP, Reynolds CW, Henkart PA (1984) Liposomes as targets for granule cytolysin from cytotoxic large granular lymphocyte tumors. Proc Natl Acad Sci U S A 81(17):5551–5555

Borsos T, Dourmashkin RR, Humphrey JH (1964) Lesions in erythrocyte membranes caused by immune haemolysis. Nature 202:251–252

Chakraborty DC, Mukherjee G, Banerjee P, Banerjee KK, Biswas T (2011) Hemolysin induces Toll-like receptor (TLR)-independent apoptosis and multiple TLR-associated parallel activation of macrophages. J Biol Chem 286(40):34542–34551. doi:10.1074/jbc.M111.241851. M111.241851 [pii]

Chattopadhyay K, Banerjee KK (2003) Unfolding of Vibrio cholerae hemolysin induces oligomerization of the toxin monomer. J Biol Chem 278(40):38470–38475. doi:10.1074/jbc.M305965200. M305965200 [pii]

Chattopadhyay K, Bhattacharyya D, Banerjee KK (2002) Vibrio cholerae hemolysin. Implication of amphiphilicity and lipid-induced conformational change for its pore-forming activity. Eur J Biochem 269(17):4351–4358. 3137 [pii]

Coelho A, Andrade JR, Vicente AC, Dirita VJ (2000) Cytotoxic cell vacuolating activity from Vibrio cholerae hemolysin. Infect Immun 68(3):1700–1705

De S, Olson R (2011) Crystal structure of the Vibrio cholerae cytolysin heptamer reveals common features among disparate pore-forming toxins. Proc Natl Acad Sci U S A 108(18):7385–7390. doi:10.1073/pnas.1017442108. 1017442108 [pii]

Dennert G, Podack ER (1983) Cytolysis by H-2-specific T killer cells. Assembly of tubular complexes on target membranes. J Exp Med 157(5):1483–1495

Dunstone MA, Tweten RK (2012) Packing a punch: the mechanism of pore formation by cholesterol dependent cytolysins and membrane attack complex/perforin-like proteins. Curr Opin Struct Biol 22(3):342–349

Dutta S, Mazumdar B, Banerjee KK, Ghosh AN (2010) Three-dimensional structure of different functional forms of the Vibrio cholerae hemolysin oligomer: a cryo-electron microscopic study. J Bacteriol 192(1):169–178. doi:10.1128/JB.00930-09. JB.00930-09 [pii]

Figueroa-Arredondo P, Heuser JE, Akopyants NS, Morisaki JH, Giono-Cerezo S, Enriquez-Rincon F, Berg DE (2001) Cell vacuolation caused by Vibrio cholerae hemolysin. Infect Immun 69(3):1613–1624

Galdiero S, Gouaux E (2004) High resolution crystallographic studies of alpha-hemolysin-phospholipid complexes define heptamer-lipid head group interactions: implication for understanding protein-lipid interactions. Protein Sci 13(6):1503–1511

Gilbert RJ (2010) Cholesterol-dependent cytolysins. Adv Exp Med Biol 677:56–66

Goldberg SL, Murphy JR (1984) Molecular cloning of the hemolysin determinant from Vibrio cholerae El Tor. J Bacteriol 160(1):239–244

Gonzalez MR, Bischofberger M, Pernot L, van der Goot FG, Freche B (2008) Bacterial pore-forming toxins: the (w)hole story? Cell Mol Life Sci 65(3):493–507

Hamon MA, Ribet D, Stavru F, Cossart P (2012) Listeriolysin O: the Swiss army knife of Listeria. Trends Microbiol 20(8):360–368

He Y, Olson R (2010) Three-dimensional structure of the detergent-solubilized Vibrio cholerae cytolysin (VCC) heptamer by electron cryomicroscopy. J Struct Biol 169(1):6–13

Heuck AP, Tweten RK, Johnson AE (2001) Beta-barrel pore-forming toxins: intriguing dimorphic proteins. Biochemistry 40(31):9065–9073

Honda T, Finkelstein RA (1979) Purification and characterization of a hemolysin produced by Vibrio cholerae biotype El Tor: another toxic substance produced by cholera vibrios. Infect Immun 26(3):1020–1027

Hotze EM, Tweten RK (1818) Membrane assembly of the cholesterol-dependent cytolysin pore complex. Biochim Biophys Acta 4:1028–1038

Humphrey JH, Dourmashkin RR (1969) The lesions in cell membranes caused by complement. Adv Immunol 11:75–115

Iacovache I, van der Goot FG, Pernot L (2008) Pore formation: an ancient yet complex form of attack. Biochim Biophys Acta 1778(7–8):1611–1623

Iacovache I, Bischofberger M, van der Goot FG (2010) Structure and assembly of pore-forming proteins. Curr Opin Struct Biol 20(2):241–246

Ichinose Y, Yamamoto K, Nakasone N, Tanabe MJ, Takeda T, Miwatani T, Iwanaga M (1987) Enterotoxicity of El Tor-like hemolysin of non-O1 Vibrio cholerae. Infect Immun 55(5):1090–1093

Ikigai H, Akatsuka A, Tsujiyama H, Nakae T, Shimamura T (1996) Mechanism of membrane damage by El Tor hemolysin of Vibrio cholerae O1. Infect Immun 64(8):2968–2973

Ikigai H, Ono T, Iwata M, Nakae T, Shimamura T (1997) El Tor hemolysin of Vibrio cholerae O1 forms channels in planar lipid bilayer membranes. FEMS Microbiol Lett 150(2):249–254

Ikigai H, Otsuru H, Yamamoto K, Shimamura T (2006) Structural requirements of cholesterol for binding to Vibrio cholerae hemolysin. Microbiol Immunol 50(10):751–757

Kaper JB, Morris JG Jr, Levine MM (1995) Cholera. Clin Microbiol Rev 8(1):48–86
Katayama H, Wang J, Tama F, Chollet L, Gogol EP, Collier RJ, Fisher MT (2010) Three-dimensional structure of the anthrax toxin pore inserted into lipid nanodiscs and lipid vesicles. Proc Natl Acad Sci U S A 107(8):3453–3457
Kondos SC, Hatfaludi T, Voskoboinik I, Trapani JA, Law RH, Whisstock JC, Dunstone MA (2010) The structure and function of mammalian membrane-attack complex/perforin-like proteins. Tissue Antigens 76(5):341–351
Krantz BA, Melnyk RA, Zhang S, Juris SJ, Lacy DB, Wu Z, Finkelstein A, Collier RJ (2005) A phenylalanine clamp catalyzes protein translocation through the anthrax toxin pore. Science 309(5735):777–781
Kristan KC, Viero G, Dalla Serra M, Macek P, Anderluh G (2009) Molecular mechanism of pore formation by actinoporins. Toxicon 54(8):1125–1134
Law RH, Lukoyanova N, Voskoboinik I, Caradoc-Davies TT, Baran K, Dunstone MA, D'Angelo ME, Orlova EV, Coulibaly F, Verschoor S, Browne KA, Ciccone A, Kuiper MJ, Bird PI, Trapani JA, Saibil HR, Whisstock JC (2010) The structural basis for membrane binding and pore formation by lymphocyte perforin. Nature 468(7322):447–451
Lee X, Thompson A, Zhang Z, Ton-that H, Biesterfeldt J, Ogata C, Xu L, Johnston RA, Young NM (1998) Structure of the complex of Maclura pomifera agglutinin and the T-antigen disaccharide, Galbeta1,3GalNAc. J Biol Chem 273(11):6312–6318
Levan S, De S, Olson R (2013) Vibrio cholerae cytolysin recognizes the heptasaccharide core of complex N-glycans with nanomolar affinity. J Mol Biol 425(5):944–957. doi:10.1016/j.jmb.2012.12.016. S0022-2836(12)00949-7 [pii]
Lohner S, Walev I, Boukhallouk F, Palmer M, Bhakdi S, Valeva A (2009) Pore formation by Vibrio cholerae cytolysin follows the same archetypical mode as beta-barrel toxins from gram-positive organisms. FASEB J 23(8):2521–2528. doi:10.1096/fj.08-127688. fj.08-127688 [pii]
Loris R (2002) Principles of structures of animal and plant lectins. Biochim Biophys Acta 1572(2–3):198–208
Los FC, Randis TM, Aroian RV, Ratner AJ (2013) Role of pore-forming toxins in bacterial infectious diseases. Microbiol Mol Biol Rev 77(2):173–207
Mayer MM (1972) Mechanism of cytolysis by complement. Proc Natl Acad Sci U S A 69(10):2954–2958
Mazumdar B, Ganguly S, Ghosh AN, Banerjee KK (2011) The role of C-terminus carbohydrate-binding domain of Vibrio cholerae haemolysin/cytolysin in the conversion of the pre-pore beta-barrel oligomer to a functional diffusion channel. Indian J Med Res 133:131–137. IndianJMedRes_2011_133_2_131_78112 [pii]
McCardell BA, Madden JM, Shah DB (1985) Isolation and characterization of a cytolysin produced by Vibrio cholerae serogroup non-O1. Can J Microbiol 31(8):711–720
McCormack R, de Armas L, Shiratsuchi M, Podack ER (2013) Killing machines: three pore-forming proteins of the immune system. Immunol Res 57(1–3):268–278
Menestrina G, Serra MD, Prevost G (2001) Mode of action of beta-barrel pore-forming toxins of the staphylococcal alpha-hemolysin family. Toxicon 39(11):1661–1672
Menzl K, Maier E, Chakraborty T, Benz R (1996) HlyA hemolysin of Vibrio cholerae O1 biotype El Tor. Identification of the hemolytic complex and evidence for the formation of anion-selective ion-permeable channels. Eur J Biochem 240(3):646–654
Mitra R, Figueroa P, Mukhopadhyay AK, Shimada T, Takeda Y, Berg DE, Nair GB (2000) Cell vacuolation, a manifestation of the El tor hemolysin of Vibrio cholerae. Infect Immun 68(4): 1928–1933
Montfort W, Villafranca JE, Monzingo AF, Ernst SR, Katzin B, Rutenber E, Xuong NH, Hamlin R, Robertus JD (1987) The three-dimensional structure of ricin at 2.8 A. J Biol Chem 262(11):5398–5403
Mueller M, Grauschopf U, Maier T, Glockshuber R, Ban N (2009) The structure of a cytolytic alpha-helical toxin pore reveals its assembly mechanism. Nature 459(7247):726–730
Mukherjee G, Biswas A, Banerjee KK, Biswas T (2008) Vibrio cholerae hemolysin is apoptogenic to peritoneal B-1a cells but its oligomer shepherd the cells for IgA response. Mol Immunol 45(1):266–270

Nagamune K, Yamamoto K, Naka A, Matsuyama J, Miwatani T, Honda T (1996) In vitro proteolytic processing and activation of the recombinant precursor of El Tor cytolysin/hemolysin (pro-HlyA) of Vibrio cholerae by soluble hemagglutinin/protease of V. cholerae, trypsin, and other proteases. Infect Immun 64(11):4655–4658
Nagamune K, Yamamoto K, Honda T (1997) Intramolecular chaperone activity of the pro-region of Vibrio cholerae El Tor cytolysin. J Biol Chem 272(2):1338–1343
Olson R, Gouaux E (2005) Crystal structure of the Vibrio cholerae cytolysin (VCC) pro-toxin and its assembly into a heptameric transmembrane pore. J Mol Biol 350(5):997–1016. doi:10.1016/j.jmb.2005.05.045. S0022-2836(05)00600-5 [pii]
Olson R, Nariya H, Yokota K, Kamio Y, Gouaux E (1999) Crystal structure of staphylococcal LukF delineates conformational changes accompanying formation of a transmembrane channel. Nat Struct Biol 6(2):134–140
Parker MW, Feil SC (2005) Pore-forming protein toxins: from structure to function. Prog Biophys Mol Biol 88(1):91–142
Paul K, Chattopadhyay K (2011) Unfolding distinguishes the Vibrio cholerae cytolysin precursor from the mature form of the toxin. Biochemistry 50(19):3936–3945. doi:10.1021/bi200332g
Paul K, Chattopadhyay K (2012) Single point mutation in Vibrio cholerae cytolysin compromises the membrane pore-formation mechanism of the toxin. FEBS J 279(21):4039–4051. doi:10.1111/j.1742-4658.2012.08809.x
Paul K, Chattopadhyay K (2014) Pre-pore oligomer formation by Vibrio cholerae cytolysin: insights from a truncated variant lacking the pore-forming pre-stem loop. Biochem Biophys Res Commun 443(1):189–193. doi:10.1016/j.bbrc.2013.11.078. S0006-291X(13)01979-7 [pii]
Podack ER, Dennert G (1983) Assembly of two types of tubules with putative cytolytic function by cloned natural killer cells. Nature 302(5907):442–445
Prevost G, Mourey L, Colin DA, Menestrina G (2001) Staphylococcal pore-forming toxins. Curr Top Microbiol Immunol 257:53–83
Rader AE, Murphy JR (1988) Nucleotide sequences and comparison of the hemolysin determinants of Vibrio cholerae El Tor RV79(Hly+) and RV79(Hly-) and classical 569B(Hly-). Infect Immun 56(6):1414–1419
Rai AK, Chattopadhyay K (2014) Trapping of Vibrio cholerae Cytolysin in the Membrane-bound Monomeric State Blocks Membrane Insertion and Functional Pore Formation by the Toxin. J Biol Chem 289(24):16978–16987. doi:10.1074/jbc.M114.567099
Rai AK, Paul K, Chattopadhyay K (2013) Functional mapping of the lectin activity site on the beta-prism domain of vibrio cholerae cytolysin: implications for the membrane pore-formation mechanism of the toxin. J Biol Chem 288(3):1665–1673. doi:10.1074/jbc.M112.430181. M112.430181 [pii]
Ray A, Chattopadhyay K, Banerjee KK, Biswas T (2003) Macrophage distinguishes Vibrio cholerae hemolysin from its protease insensitive oligomer by time dependent and selective expression of CD80-CD86. Immunol Lett 89(2–3):143–147. S0165247803001354 [pii]
Richardson K, Michalski J, Kaper JB (1986) Hemolysin production and cloning of two hemolysin determinants from classical Vibrio cholerae. Infect Immun 54(2):415–420
Rosado CJ, Kondos S, Bull TE, Kuiper MJ, Law RH, Buckle AM, Voskoboinik I, Bird PI, Trapani JA, Whisstock JC, Dunstone MA (2008) The MACPF/CDC family of pore-forming toxins. Cell Microbiol 10(9):1765–1774
Rossjohn J, Polekhina G, Feil SC, Morton CJ, Tweten RK, Parker MW (2007) Structures of perfringolysin O suggest a pathway for activation of cholesterol-dependent cytolysins. J Mol Biol 367(5):1227–1236
Rutenber E, Ready M, Robertus JD (1987) Structure and evolution of ricin B chain. Nature 326(6113):624–626
Saha N, Banerjee KK (1997) Carbohydrate-mediated regulation of interaction of Vibrio cholerae hemolysin with erythrocyte and phospholipid vesicle. J Biol Chem 272(1):162–167
Saka HA, Bidinost C, Sola C, Carranza P, Collino C, Ortiz S, Echenique JR, Bocco JL (2008) Vibrio cholerae cytolysin is essential for high enterotoxicity and apoptosis induction produced by a cholera toxin gene-negative V. cholerae non-O1, non-O139 strain. Microb Pathog 44(2): 118–128

Sankaranarayanan R, Sekar K, Banerjee R, Sharma V, Surolia A, Vijayan M (1996) A novel mode of carbohydrate recognition in jacalin, a Moraceae plant lectin with a beta-prism fold. Nat Struct Biol 3(7):596–603
Schnupf P, Portnoy DA (2007) Listeriolysin O: a phagosome-specific lysin. Microbes Infect 9(10):1176–1187
Sharon N, Lis H (2004) History of lectins: from hemagglutinins to biological recognition molecules. Glycobiology 14(11):53R–62R
Smith R, Separovic F, Milne TJ, Whittaker A, Bennett FM, Cornell BA, Makriyannis A (1994) Structure and orientation of the pore-forming peptide, melittin, in lipid bilayers. J Mol Biol 241(3):456–466
Song L, Hobaugh MR, Shustak C, Cheley S, Bayley H, Gouaux JE (1996) Structure of staphylococcal alpha-hemolysin, a heptameric transmembrane pore. Science 274(5294):1859–1866
Sun J, Lang AE, Aktories K, Collier RJ (2008) Phenylalanine-427 of anthrax protective antigen functions in both pore formation and protein translocation. Proc Natl Acad Sci U S A 105(11):4346–4351
Tejuca M, Anderluh G, Dalla Serra M (2009) Sea anemone cytolysins as toxic components of immunotoxins. Toxicon 54(8):1206–1214
Valeva A, Walev I, Weis S, Boukhallouk F, Wassenaar TM, Endres K, Fahrenholz F, Bhakdi S, Zitzer A (2004) A cellular metalloproteinase activates Vibrio cholerae pro-cytolysin. J Biol Chem 279(24):25143–25148
Valeva A, Walev I, Boukhallouk F, Wassenaar TM, Heinz N, Hedderich J, Lautwein S, Mocking M, Weis S, Zitzer A, Bhakdi S (2005) Identification of the membrane penetrating domain of Vibrio cholerae cytolysin as a beta-barrel structure. Mol Microbiol 57(1):124–131. doi:10.1111/j.1365-2958.2005.04684.x. MMI4684 [pii]
van der Goot FG (2003) Membrane-damaging toxins: pore-formation. In: Burns DL, Barbieri JT, Iglewski BH, Rappuoli R (eds) Bacterial protein toxins. ASM, Washington, pp 189–202
Voskoboinik I, Trapani JA (2006) Addressing the mysteries of perforin function. Immunol Cell Biol 84(1):66–71
Voskoboinik I, Smyth MJ, Trapani JA (2006) Perforin-mediated target-cell death and immune homeostasis. Nat Rev Immunol 6(12):940–952
Wiener M, Freymann D, Ghosh P, Stroud RM (1997) Crystal structure of colicin Ia. Nature 385(6615):461–464
Wu PL, Chiu CR, Huang WN, Wu WG (2012) The role of sulfatide lipid domains in the membrane pore-forming activity of cobra cardiotoxin. Biochim Biophys Acta 1818(5):1378–1385
Yamamoto K, Ichinose Y, Shinagawa H, Makino K, Nakata A, Iwanaga M, Honda T, Miwatani T (1990) Two-step processing for activation of the cytolysin/hemolysin of Vibrio cholerae O1 biotype El Tor: nucleotide sequence of the structural gene (hlyA) and characterization of the processed products. Infect Immun 58(12):4106–4116
Young JA, Collier RJ (2007) Anthrax toxin: receptor binding, internalization, pore formation, and translocation. Annu Rev Biochem 76:243–265
Yuldasheva LN, Merzlyak PG, Zitzer AO, Rodrigues CG, Bhakdi S, Krasilnikov OV (2001) Lumen geometry of ion channels formed by Vibrio cholerae EL Tor cytolysin elucidated by nonelectrolyte exclusion. Biochim Biophys Acta 1512(1):53–63
Zhang D, Takahashi J, Seno T, Tani Y, Honda T (1999) Analysis of receptor for Vibrio cholerae El tor hemolysin with a monoclonal antibody that recognizes glycophorin B of human erythrocyte membrane. Infect Immun 67(10):5332–5337
Zitzer A, Walev I, Palmer M, Bhakdi S (1995) Characterization of Vibrio cholerae El Tor cytolysin as an oligomerizing pore-forming toxin. Med Microbiol Immunol 184(1):37–44
Zitzer A, Wassenaar TM, Walev I, Bhakdi S (1997a) Potent membrane-permeabilizing and cytocidal action of Vibrio cholerae cytolysin on human intestinal cells. Infect Immun 65(4):1293–1298
Zitzer A, Palmer M, Weller U, Wassenaar T, Biermann C, Tranum-Jensen J, Bhakdi S (1997b) Mode of primary binding to target membranes and pore formation induced by Vibrio cholerae cytolysin (hemolysin). Eur J Biochem 247(1):209–216

Zitzer A, Zitzer O, Bhakdi S, Palmer M (1999) Oligomerization of Vibrio cholerae cytolysin yields a pentameric pore and has a dual specificity for cholesterol and sphingolipids in the target membrane. J Biol Chem 274(3):1375–1380

Zitzer A, Harris JR, Kemminer SE, Zitzer O, Bhakdi S, Muething J, Palmer M (2000) Vibrio cholerae cytolysin: assembly and membrane insertion of the oligomeric pore are tightly linked and are not detectably restricted by membrane fluidity. Biochim Biophys Acta 1509(1–2):264–274. S0005-2736(00)00303-5 [pii]

Zitzer A, Bittman R, Verbicky CA, Erukulla RK, Bhakdi S, Weis S, Valeva A, Palmer M (2001) Coupling of cholesterol and cone-shaped lipids in bilayers augments membrane permeabilization by the cholesterol-specific toxins streptolysin O and Vibrio cholerae cytolysin. J Biol Chem 276(18):14628–14633. doi:10.1074/jbc.M100241200. M100241200 [pii]

Zitzer A, Westover EJ, Covey DF, Palmer M (2003) Differential interaction of the two cholesterol-dependent, membrane-damaging toxins, streptolysin O and Vibrio cholerae cytolysin, with enantiomeric cholesterol. FEBS Lett 553(3):229–231

Chapter 8
New Vis-Tas in Lactosylceramide Research

Subroto Chatterjee, Sumita Mishra, and Sara Kimiko Suzuki

Introduction

Lactosylceramide (LacCer) is a member of a large family of compounds collectively called the glycosphingolipids (GSL). These molecules are present in all mammalian cells, some bacteria and fungus. GSLs are composed of an amino acid serine, fatty acids and sugars and are usually localized on the cell surface wherein they serve as receptors for diverse physiologically relevant molecules, bacteria and viruses. However, LacCer is predominantly stored within cytoplasmic vesicles located in the perinuclear area though some LacCer is present on the cell surface. The dynamics of these two pools of LacCer is not known. Nevertheless, recent efforts by several groups of investigators have opened up new vis-tas in LacCer research. The present article is to bring to forth these findings for further experimental validation and for use in translational research to develop better diagnostics and therapeutics for use humans and certain veterinary purposes. In particular, this article will focus on two areas: 1. Inflammation and the LacCer–phospholipase-A-2 (PLA2) connection and 2. Implications of LacCer modulation on cardiac hypertrophy.

Briefly, the biosynthesis of GSL begins upon the condensation of L-serine with palmitoyl-CoA to form sphingosine (Fig. 8.1). In mammalian cells, sphingosine is then metabolized in a sequential manner to synthesize several intermediates leading up to the synthesis of ceramide. Ceramide forms the non-polar tail of all GSL (presumed to be buried within the cell membrane) to which glucose and galactose (from respective nucleotide sugars) is added consecutively to yield glucosylceramide and LacCer, respectively.

S. Chatterjee (✉) • S. Mishra • S.K. Suzuki
Department of Pediatrics, Johns Hopkins University, School of Medicine,
Baltimore, MD 21044, USA
e-mail: Schatte2@jhmi.edu

A. Chakrabarti, A. Surolia (eds.), *Biochemical Roles of Eukaryotic Cell Surface Macromolecules*, Advances in Experimental Medicine and Biology 842,
DOI 10.1007/978-3-319-11280-0_8

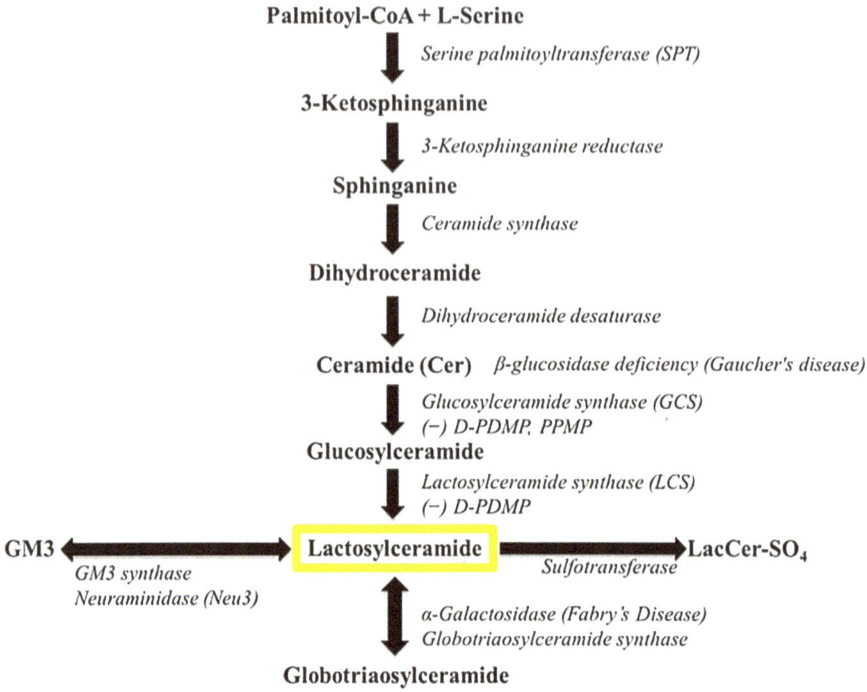

Fig. 8.1 Metabolic pathways involved in lactosylceramide biosynthesis and its role as a precursor to complex glycosphingolipids

In nature, at least two LacCer synthases (LCS) have been reported. According to the recent nomenclature, they are termed GalT-V and GalT-VI. Before the human genome was unraveled, only one LCS was known and was termed GalT-2, now referred to as GalT-VI. While GalT-V is a constitutively expressed in most tissues, GalT-VI is expressed in a tissue specific manner-in the brain (Lo et al. 1998). In this context, the readers are referred to another chapter in this series where a detailed description of the biosynthesis of complex GSL and nomenclature of these enzymes are described by Basu and co-workers (Ref). The important feature about GalT-V is that it is the major LCS in human endothelium (Chatterjee et al. 2008). Therefore, it plays a critical role in the biosynthesis of LacCer and LacCer-regulated phenotypes and diseases (Chatterjee and Alsaeedi 2012). For example, LacCer plays a critical role in vascular endothelial growth factor (VEGF)/fibroblast growth factor (FGF)-induced angiogenesis (Rajesh 2005; Kolmakova et al. 2009), a phenotype central to tumor metastasis, and tumor growth. Thus, the use of siRNA to ablate GalT-V gene in vitro and in vivo was found to mitigate angiogenesis and tumorigenic potential in B16-F10 mouse melanoma cells, respectively (Rajesh 2005; Wei et al. 2010; Furukawa et al. 2014). Also, the use of inhibitor's of LCS such as D-PDMP can reverse VEGF and FGF-induced angiogenesis. Further, the observation that the tumor necrosis factor (TNF) induced expression of intercellular cell adhesion molecule-1 (ICAM-1) requires the activation of the endothelial cell derived GalT-V which may be central to both inflammation and atherosclerosis. Also ICAM-1 serves as a receptor for Mac-1/CD11b present on the surface of monocytes and neutrophils. Thus, the adhesion of these blood cells to the endothelium and their subsequent intravasation is a first critical step in the initiation of inflammation and atherosclerosis seen below (Bhunia et al. 1997).

Another major source of LacCer production is due to the action of a sialidase termed Neu3 on a ganglioside GM3 (Miyagi and Yamaguchi 2012). This enzyme is highly enriched with the plasma membrane in cancer cells. This reaction seems to be utilized largely in human cancer cells and cancer tissue noted for its highly malignant properties as its contribution to an induction of phenotypes, e.g. cell migration and invasion. Studies using colon cancer cells have revealed that Neu3 activates Wnt receptor by phosphorylation of Ras/MAPK upon stimulation by EGF (Miyagi et al. 2012). Thus, Neu-3 induced LacCer production may well partake in signaling pathways leading to tumor metastasis. Neu3 activation was also shown to occur upon exposure of human dermal fibroblast to elastin that can activate this enzyme to generate LacCer (Rusciani et al. 2011). Furthermore, LacCer is shown to generate reactive oxygen species (ROS): a superoxide to activate the phosphorylation of mitogen activated protein kinase. In a cancer cell line, this helps increase in cell proliferation (Miyagi and Yamaguchi 2012), and in elastic tissue, the phosphorylation of MAPK facilitates its elasticity (Rusciani et al. 2011). These findings are summarized diagrammatically (Fig. 8.2).

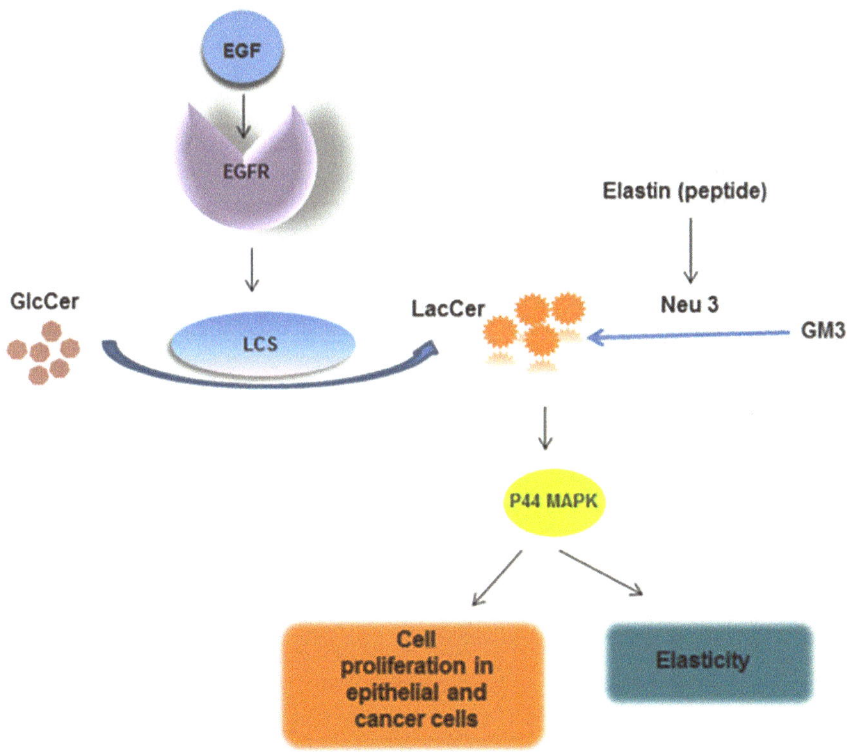

Fig. 8.2 Signaling pathways by which epidermal growth factor recruits LacCer to induce cell proliferation in cancer cells and elasticity

Inflammation and the LacCer: Phospholipase-A-2 (PLA2) Connection

Inflammation is probably the earliest process leading to the two major killers of mankind: heart disease and cancer, perhaps other inflammatory diseases as well. This involves the participation of cells in the circulation notably platelets, macrophages, neutrophils, leukocyte/monocytes, and the vascular cells such as arterial smooth muscle cells and the endothelium. Infections and other stressor's allow the release of various growth factors and pro-inflammatory cytokines from the various blood cells and smooth muscle cells. Since endothelial cell surface forms a barrier between blood cells and its components and the vascular wall, pro-inflammatory cytokines such as tumor necrosis factor (TNF-α, inflammatory cytokines, growth factors etc.) bind to their receptors on the surface of the these cells, thus activating them and producing signaling molecules such as LacCer (via activation of LacCer synthase) (Chatterjee and Alsaeedi 2012). In turn, LacCer specifically induces the expression of a cell adhesion molecule-intercellular cell adhesion molecule (ICAM-1) through an "oxygen-sensitive" signaling pathway. Studies in vitro and in vivo demonstrate that ICAM-1 serves as a receptor for another protein Mac-1/CD11b expressed on the surface of neutrophils and monocytes. This allows the capture of circulating neutrophils and monocytes and their intravasation into the sub-endothelial space. Herein, the monocytes undergo proliferation and differentiation into macrophages due to the action of several growth factors. Since macrophages express scavenger receptors e.g. SRB-1, CD-36 etc., it allows them to take up oxidized LDL. Studies show that oxidized LDL not only contribute to the deposition of cholesterol esters but also can inhibit their hydrolysis contributing to fatty streaks in the arterial wall, plaque development, and its subsequent pathological pathway. Additional studies show that LacCer can directly interact with neutrophils and monocytes to activate phospholipase-A-2 to increase the expression of Mac-1/CD-11b to facilitate adhesion to the endothelium (Fig. 8.3) (Arai et al. 1998).

The emerging view is that LacCer taken up by cells from lipoproteins, other cell membranes due to cell–cell interaction or simply by an exogenous supply which may form a LacCer membrane microdomain (Fig. 8.4). And such microdomains are involved in generating superoxides and/or activating phospholipase to bring about profound phenotypic changes in vitro. First, using human arterial smooth muscle cells it was shown that LacCer dose and time dependently raised the cellular levels of superoxides by activating NAD(P)H oxidase activity (Bhunia et al. 1997). Next, LacCer was shown to facilitate the migration of several components of the NAD(P)H complex such as c47phox and c67phox from the cytosol to the plasma membrane to bind with the other components of NAD(P)H oxidase, generating superoxides (Martin et al. 2006). Additional studies revealed that the LacCer microdomain together with a Src kinase; Lyn, expressed on the neutrophil plasma membrane, may well be implicated in innate immune response (Yoshizaki et al. 2008).

Another instance of direct LacCer protein interaction is the case with phospholipase-A-2. The roles of phospholipase-A-2 are many including initiation and propagation of inflammation, cellular damage, modulation of chemotaxis, phagocytosis

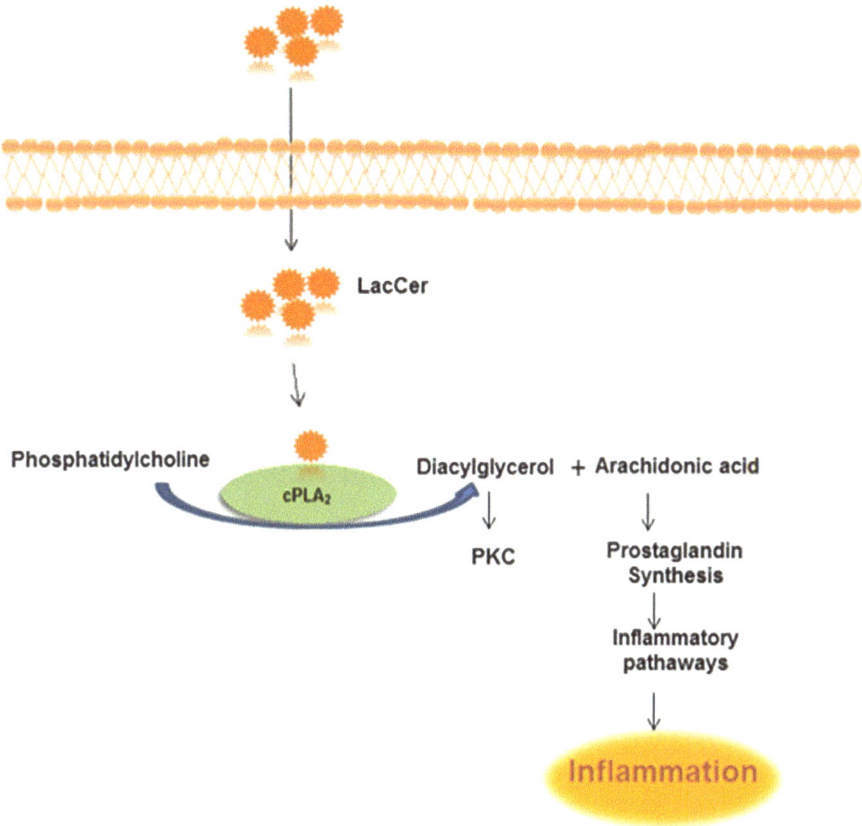

Fig. 8.3 Signaling pathways by which exogenous LacCer is involved in the activation of phospholipase-A-2, leading to inflammation

and superoxide generation. It also modulates vascular tone, enhances vascular permeability and may also impact T cell function. The substrate for PLA-2 is a phospholipid phosphatidylcholine wherein it cleaves the sn-2 fatty acid, arachidonic acid (Fig. 8.3). While LacCer can directly activate PLA-2, studies show that FcER1 cross linking may well activate PLA-2 via a receptor independent tyrosine kinase (src, lyn, yes, syk etc.). The IP3 generated interacts with the Ca2+ channel on the endoplasmic reticulum to increase the concentration of cytosolic Ca2+. In turn, this allows the translocation of cPLA2 from the cytosol to cell membrane compartment thus cleaving arachidonic acid. The phosphorylation of Ser 505 in PLA-2 increases intrinsic enzyme activity.

The proof of concept that indeed LacCer directly activated PLA2 arrived from several experimental observations. First, treatment of cells with PPMP an inhibitor of glucosylceramide synthase and LacCer synthase, reduced the level of LacCer and the release of arachidonic acid. This was bypassed by treating cells with just LacCer

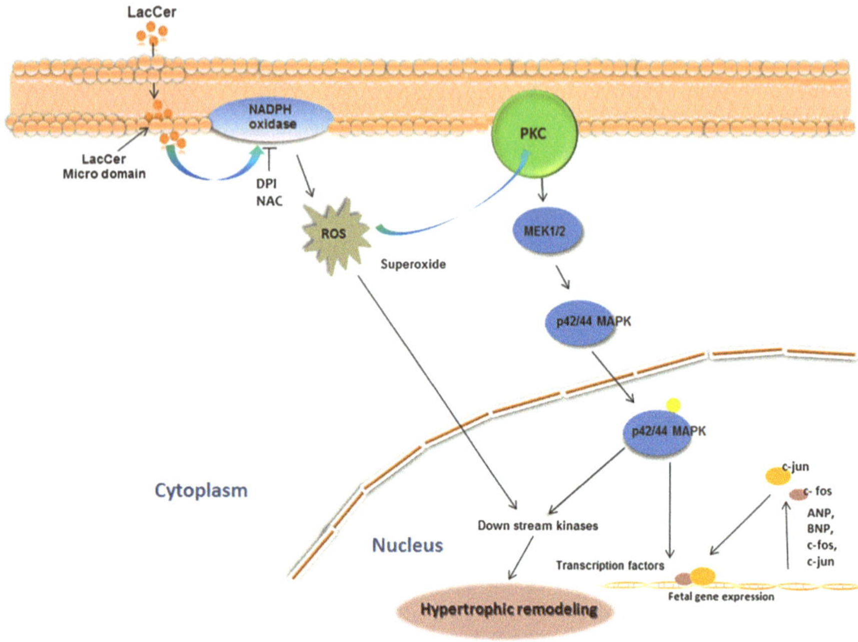

Fig. 8.4 Lactosylceramide induces hypertrophy in cardiomyocytes via ROS generation and activation of P44 MAP kinase (Mishra and Chatterjee 2014)

and no other glycosphingolipids (Nakamura et al. 2013). Second, silencing the enzyme cPLA2 or the use of an inhibitor of cPLA-2 also mitigated LacCer induced cPLA2 activation. These studies conducted using Chinese hamster ovary cells (CHO-W11A) showed that LacCer translocated D43N mutant of cPLA-2. Additional studies in a human monocytic cell line (U-937) have revealed that LacCer recruited PKC-α/ε to activate PLA-2 and the intrinsic expression of platelet endothelial cell adhesion molecule (PECAM-1). Since COX-2 inhibitors mitigated arachidonic acid-induced PECAM-1 expression, prostaglandins may mediate PECAM-1 expression in monocytes (Gong, NL 2004 PNAS). Previous studies show that PECAM-1 plays a critical role in the trans-endothelial migration of monocytes into the sub-endothelial space thus initiating atherogenensis and involving LacCer in the pathology in this disease (Chatterjee and Pandey 2007).

Implications of LacCer Modulation on Cardiac Hypertrophy

Lipids are required by all organs, including heart, for its function. They consist of fatty acids (FA) that supply calories required for numerous cellular activities and are also the important structural component of cells. The majority of plasma fatty acid constitutes triglycerides and phospholipids that exist in esterified form. Oxidation of various substrates like FA, glucose, lactate and ketone bodies generate ATP in normal adult hearts. Among these substrates, glucose and fatty acids are the most important for ATP production in the heart. Approximately 70 % of the ATP essential for regular cardiac function is provided by the FA. There are multiple pathways that modulate the attainment of FA by the cardiomyocytes, and any shift in these pathways affects cardiac metabolism and function (Taegtmeyer 1994). High blood cholesterol is a major risk factor for heart diseases, and hyperlipidemia for atherosclerosis and cardiovascular disease, respectively, including coronary heart disease. Epidemiologic studies have shown that hypercholesterolemia is associated with increased left ventricular mass and cardiac hypertrophy (Jung et al. 2010; Luo et al. 2010; Miguel-Carrasco et al. 2010; Planavila et al. 2011; Singh and Krishan 2010; Takayama et al. 2011; Wang et al. 2010; Wojciechowski et al. 2010). Hypertrophic cardiomyopathy is a pathological hypertrophy of the heart due to an increase in the size of myocytes in various heart diseases including long-term hypertension, myocardial infarction, chronic pressure overload, valvular defects and endocrine disorders (Frey et al. 2004; Grossman et al. 1975; Hood et al. 1968; Sandler and Dodge 1963). Myocardial hypertrophy is an adaptive response of the heart to increased workload. Cardiac hypertrophy is one of the main responses of cardiomyocytes to mechanical and neuro-hormonal stimuli. Although cardiac hypertrophy may initially represent an adaptive response of the myocardium, it often progresses to ventricular dilatation leading to heart failure, one of the leading causes of mortality in the world. Increased left ventricular mass (LVM) and decreased fractional shortening (FS) are risk factors in cardiac morbidity and mortality in the general population (Baumgartner et al. 2007; Lorell and Carabello 2000; Movahed and Saito 2009). Cardiac hypertrophy and fibrosis, which are the most common responses of the heart to all forms of injury, are the major determinants of morbidity and mortality from cardiovascular disease in both developing and developed countries.

The role of diet is crucial in the development and prevention of cardiovascular disease. It also impacts all other cardiovascular risk factors. Previous studies have demonstrated that dyslipidemia, hypercholesterolemia and cardiac lipotoxicity are associated with cardiac hypertrophy (Balakumar et al. 2011; Berger et al. 2005; Borradaile and Schaffer 2005; Lopaschuk et al. 2007; Poornima et al. 2006; Semeniuk et al. 2002; Smith and Yellon 2011; Unger and Orci 2001; Yang and Barouch 2007).

Hypertrophy induced by fat diet intake is steadily becoming one of the primary causes of myocardial infarction, morbidity, and stroke and is a major clinical concern in cardiovascular medicine. Increased levels of FAs from fatty diets can impact the heart harmfully due to the formation of toxic derivatives of glucose and lipid metabolism (Bayeva et al. 2013). Epidemiological studies showed that hypercholesterolemia is associated with higher left ventricular mass and that dyslipidemia is an independent determinant of increased left ventricular mass (Lee et al. 2005). In patients with Fabry's disease, glycosphingolipid deposition in heart causes progressive left ventricular hypertrophy that mimics the morphological and clinical picture of hypertrophic cardiomyopathy, with dyspnea on effort, palpitation and angina as the typical symptoms (Nakao et al. 1995). A close association between GSLs level and cardiac hypertrophy in vivo in apoE -/- mice fed a western diet was suggested by us recently (Chatterjee et al. 2013). We have observed that feeding a high fat and cholesterol diet to apoE-/- mice results in marked increase in the level of GSL e.g. glucosylceramide (GlcCer) and LacCer in heart tissue accompanied by an increase in the activity of glycosphingolipid glycosyltransferases (GTs) (Chatterjee et al. 2013). However, these in vivo studies did not elaborate whether one or more GSLs were implicated in cardiac hypertrophy. To address this issue, we used cultured neonatal rat cardiomyocytes and H9C2 cells and sought to determine whether GSL's affect cardiac hypertrophy. The cardiomyocytes were treated with different glycosphingolipids and their effect on hypertrophy was measured using multiple biochemical molecular and morphological parameters (Mishra and Chatterjee 2014). Among several glycosphingolipids examined, Lactosylceramide specifically stimulated hypertrophic parameters to a similar extent as PE (Phenylephrine) in these cells. Cardiac hypertrophy *in vivo* involves the enlargement of the heart caused by an overload of blood volume and increased blood pressure. Cardiac hypertrophy *in vitro* is induced by the use of agonists such as PE that binds to its cognate receptors and transduces downstream components of a ROS-mediated signal transduction pathway to eventually induce hypertrophy.

In this study, PE was used as a positive control. We demonstrated that at a similar concentration (100 μM), LacCer could serve as a bonafide agent to induce cardiac hypertrophy in H9c2 cells and freshly cultured primary rat cardiomyocytes. In contrast, the other classes of GSL, such as sulfatides, complex gangliosides and other neutral GSLs, failed to induce hypertrophy in cardiomyocytes. This shows that an intact molecule of LacCer is required to induce cardiac hypertrophy. Importantly, the catabolic or anabolic products of LacCer failed to induce this phenotype. At the cellular level, hypertrophy is characterized by an increase in the size of cells, protein synthesis, reactivation of fetal genes e.g. ANP (Atrial natriuretic peptide) and BNP (Brain natriuretic peptide), changes in the signal transduction pathways and reorganization of sarcomere structure. The increase in cell size is mainly accompanied by an increase in protein synthesis. Our studies employed multiple criteria to assess hypertrophy in these cardiomyocytes e.g. increased cell volume, increased protein synthesis using [^{3}H]-Leucine as a precursor, determination of cell size and the measurement of mRNA levels of ANP and BNP-established biomarkers of cardiac hypertrophy. These studies suggest that LacCer specifically induced hypertrophy in cardiomyocytes.

We observed that in cardiomyocytes LacCer induces the generation of superoxides in a time and concentration-dependent manner. This was mitigated by the use of antioxidants such as N-acetyl cysteine, a scavenger of free oxygen radicals and diphenylamine iodonium (DPI), an inhibitor of NAD(P)H oxidase (Hsieh et al. 2013; Yang et al. 2013a, b). Use of these inhibitors also mitigated LacCer induced cardiac hypertrophy biomarkers mRNA levels e.g. ANP and BNP. This observation suggests that, by activating NAD(P)H oxidase, LacCer generates superoxide radicals which in turn activates a downstream signaling cascade leading to cardiac hypertrophy (Fig. 8.4).

The immediate early genes activated during hypertrophic stimulus include c-jun, c-fos, c-myc etc. In our study we found that, LacCer induced hypertrophy also involved the upregulation of both c-fos and c-jun genes. We also demonstrated that the activation of these immediate early genes involves oxidative stress.

Subsequent studies have shown the effects of LacCer on Protein Kinase C (PKC) activation and cardiac hypertrophy. The involvement of PKC in cardiac hypertrophy has been reported previously (Bowman et al. 1997; Braz et al. 2002; Vijayan et al. 2004). We observed marked inhibition of LacCer-induced ANP and BNP mRNA levels in cardiomyocytes in the presence of PKC inhibitor, suggesting that PKC plays a central role in LacCer induced hypertrophy.

Previous studies have placed p44 MAPK activation as a central component in agonist induced cardiac hypertrophy (Araujo et al. 2010; Dai et al. 2011; Fahmi et al. 2013; Ferguson et al. 2013; Lopez-Contreras et al. 2013; Ruppert et al. 2013; Sbroggio et al. 2011). Also transforming growth factor-β 1 induces hypertrophy and fibrosis via activation of p44 MAPK (Bujak and Frangogiannis 2007). Therefore, we examined the effects of LacCer on p44 MAPK and cardiac hypertrophy. LacCer induced the rapid phosphorylation of p44 MAPK and this activation process was required to induce cardiac hypertrophy (Fig. 8.4). Our study suggests that LacCer alone can induce hypertrophy in cardiomyocytes and therefore exposes both LacCer and LacCer synthase as novel drug targets to mitigate this phenotype. This emphasizes the need for a better understanding of GSLs and GTs in cardiac hypertrophy and other cardiovascular diseases.

Perspectives

As of to date, it is still unclear whether or not PLA 2 is a bonafide inflammatory marker for cardiovascular risk due to the lipoprotein-PLA2 possibly playing a dual role as a pro-atherogenic and anti-atherogenic molecule. On one hand, Lp-PLA2 generates arachidonic acid, a precursor for prostaglandins and relevant to the inflammatory pathway contributing to atherosclerosis. On the other hand, LP-PLA 2 is implicated in the degradation of platelet activating factor (PAF), a potent mediator of inflammation. These characteristic of Lp-PLA 2 raises a burning question ,under what conditions does the PLA2 become atherogenic vs. anti atherogenic. Another aspect of PLA2 action is the generation of oxidized phospholipids and lysoPC. While the oxidized phospholipids could render LDL prone to oxidation and consequently contribute to atherosclerosis, the lysoPC is a potent fusogenic compound and may also serve in accelerating atherosclerosis.

Literature on the role of LacCer in vascular stiffness and in cardiac hypertrophy is just beginning to unravel. These studies must be validated using large mammals. Since GalT-V gene ablation is embryo - lethal alternative experimental models designed to deplete LacCer and or the use of highly specific inhibitors of LacCer synthesis are needed to explore this area of research. Since one in three people worldwide suffers from high blood pressure which can contribute to cardiac hypertrophy, research in this area could be lucrative for the industry and to the NIH effort to bolster their program of excellence in Glycosciences.

Acknowledgments This work was supported by grants from the NIH, PO-1-HL-107-153 and 3PO1HL 107153-03S1.

References

Arai T, Bhunia AK, Chatterjee S et al (1998) Lactosylceramide stimulates human neutrophils to upregulate Mac-1, adhere to endothelium, and generate reactive oxygen metaboties in vitro. Circ Res 82:54054–54057

Araujo AS, Fernandes T, Ribeiro MF, Khaper N, Bello-Klein A (2010) Redox regulation of myocardial ERK 1/2 phosphorylation in experimental hyperthyroidism: role of thioredoxin-peroxiredoxin system. J Cardiovasc Pharmacol 56:513–517

Balakumar P, Rohilla A, Mahadevan N (2011) Pleiotropic actions of fenofibrate on the heart. Pharmacol Res 63:8–12

Baumgartner D, Scholl-Burgi S, Sass JO, Sperl W, Schweigmann U, Stein JI, Karall D (2007) Prolonged QTc intervals and decreased left ventricular contractility in patients with propionic acidemia. J Pediatr 150:192–197

Bayeva M, Sawicki KT, Ardehali H (2013) Taking diabetes to heart—deregulation of myocardial lipid metabolism in diabetic cardiomyopathy. J Am Heart Assoc 2:e000433

Berger JP, Akiyama TE, Meinke PT (2005) PPARs: therapeutic targets for metabolic disease. Trends Pharmacol Sci 26:244–251

Bhunia AK, Han H, Snowden A, Chatterjee S (1997) Redox regulated signaling by lactosylceramide in the proliferation of aortic smooth muscle cells. J Biol Chem 272:15642–15649

Borradaile NM, Schaffer JE (2005) Lipotoxicity in the heart. Curr Hypertens Rep 7:412–417

Bowman JC, Steinberg SF, Jiang T, Geenen DL, Fishman GI, Buttrick PM (1997) Expression of protein kinase C beta in the heart causes hypertrophy in adult mice and sudden death in neonates. J Clin Invest 100:2189–2195

Braz JC, Bueno OF, De Windt LJ, Molkentin JD (2002) PKC alpha regulates the hypertrophic growth of cardiomyocytes through extracellular signal-regulated kinase1/2 (ERK1/2). J Cell Biol 156:905–919

Bujak M, Frangogiannis NG (2007) The role of TGF-beta signaling in myocardial infarction and cardiac remodeling. Cardiovasc Res 74:184–195

Chatterjee S, Alsaeedi N (2012) Lactosylceramide synthase as a therapeutic target to mitigate multiple human diseases in animal models. Adv Exp Med 749:153–169

Chatterjee S, Pandey A (2007) The Yin and Yang of lactosylceramide metabolism: implications in cell function. Biochim Biophys Acta 1780:370–382

Chatterjee S, Kolmakova A, Mohanraj R (2008) Regulation of lactosylceramide synthase; implications as a drug target. Curr Drug Targets 9:272–281

Chatterjee S, Bedja D, Mishra S, Kass D (2013) Inhibiting glycosphingolipid glycosyltransferase activity prevents cardiac hypertrophy in apoE-/-mice fed western diet and C57 Bl-6 mice subject to trans-aortic constriction. Glycobiology 23:1412

Dai HY, He T, Li XL, Xu WL, Ge ZM (2011) Urotensin-2 promotes collagen synthesis via ERK1/2-dependent and ERK1/2-independent TGF-beta1 in neonatal cardiac fibroblasts. Cell Biol Int 35:93–98

Fahmi A, Smart N, Punn A, Jabr R, Marber M, Heads R (2013) p42/p44-MAPK and PI3K are sufficient for IL-6 family cytokines/gp130 to signal to hypertrophy and survival in cardiomyocytes in the absence of JAK/STAT activation. Cell Signal 25:898–909

Ferguson BS, Harrison BC, Jeong MY, Reid BG, Wempe MF, Wagner FF, Holson EB, McKinsey TA (2013) Signal-dependent repression of DUSP5 by class I HDACs controls nuclear ERK activity and cardiomyocyte hypertrophy. Proc Natl Acad Sci U S A 110:9806–9811

Frey N, Katus HA, Olson EN, Hill JA (2004) Hypertrophy of the heart: a new therapeutic target? Circulation 109:1580–1589

Furukawa K, Shirane K, Kuji R, Tareyanage C, Sato T, Kobayashi Y, Furukawa S, Murata T, Kubota S, Ishikawa Y, Segawa K (2014) Gene expression levels of β4-galactosyltransferase 5 correlate with the tumorigenic potentials of B16-F10 mouse melanoma cells. Glycobiology 24(6):532–541

Grossman W, Jones D, McLaurin LP (1975) Wall stress and patterns of hypertrophy in the human left ventricle. J Clin Invest 56:56–64

Hood WP Jr, Rackley CE, Rolett EL (1968) Wall stress in the normal and hypertrophied human left ventricle. Am J Cardiol 22:550–558

Hsieh YC, Hsu SL, Gu SH (2013) Involvement of reactive oxygen species in PTTH-stimulated ecdysteroidogenesis in prothoracic glands of the silkworm, Bombyx mori. Insect Biochem Mol Biol 43:859–866

Jung HJ, Lee WY, Yoo YS, Chung BC, Choi MH (2010) Database-dependent metabolite profiling focused on steroid and fatty acid derivatives using high-temperature gas chromatography-mass spectrometry. Clin Chim Acta 411:818–824

Kolmakova A, Rajesh M, Zang D, Pili R, Chatterjee S (2009) VEGF recruits lactosylceramide to induce endothelial cell adhesion molecule expression and angiogenesis in vitro and in vivo. Glycoconj J 26:546–558

Lee TM, Lin MS, Chou TF, Chang NC (2005) Effect of simvastatin on left ventricular mass in hypercholesterolemic rabbits. Am J Physiol Heart Circ Physiol 288:H1352–H1358

Lo NW, Shaper JH, Pevsner J et al (1998) The expanding beta 4-galactosyltransferase gene family: messages from the databanks. Glycobiology 8(5):517–526

Lopaschuk GD, Folmes CD, Stanley WC (2007) Cardiac energy metabolism in obesity. Circ Res 101:335–347

Lopez-Contreras AJ, de la Morena ME, Ramos-Molina B, Lambertos A, Cremades A, Penafiel R (2013) The induction of cardiac ornithine decarboxylase by beta2 -adrenergic agents is associated with calcium channels and phosphorylation of ERK1/2. J Cell Biochem 114:1978–1986

Lorell BH, Carabello BA (2000) Left ventricular hypertrophy: pathogenesis, detection, and prognosis. Circulation 102:470–479

Luo J, Wu S, Liu J, Li Y, Yang H, Kim T, Zhelyabovska O, Ding G, Zhou Y, Yang Y et al (2010) Conditional PPARgamma knockout from cardiomyocytes of adult mice impairs myocardial fatty acid utilization and cardiac function. Am J Transl Res 3:61–72

Martin S, Williams N, Chatterjee S (2006) Lactosylceramide is required in apoptosis induced by neutral sphingomyelinase. Glycoconj J 23:147–157

Miguel-Carrasco JL, Monserrat MT, Mate A, Vazquez CM (2010) Comparative effects of captopril and l-carnitine on blood pressure and antioxidant enzyme gene expression in the heart of spontaneously hypertensive rats. Eur J Pharmacol 632:65–72

Mishra S, Chatterjee S (2014) Lactosylceramide promotes hypertrophy through ROS generation and activation of ERK1/2 in cardiomyocytes. Glycobiology 24(6):518–531, PMID:24658420

Miyagi T, Yamaguchi K (2012) Mammalian sialidases: physiological and pathological roles in cellular functions. Glycobiology 22(7):880–896

Miyagi T, Takahashi K, Hata K, Shiozaki K, Yamaguchi K (2012) Sialidase significance for cancer progression. Glycoconj J 29(8–9):567–577

Movahed MR, Saito Y (2009) Lack of association between obesity and left ventricular systolic dysfunction. Echocardiography 26:128–132

Nakamura H, Moriyama Y, Mkiyama T, Emori S et al (2013) Lactosylceramide interacts with and activates cytosolic phospholipase A2α. J Biol Chem 288(32):23264–23272. doi:10.1074/jbc.M113.491431

Nakao S, Takenaka T, Maeda M, Kodama C, Tanaka A, Tahara M, Yoshida A, Kuriyama M, Hayashibe H, Sakuraba H et al (1995) An atypical variant of Fabry's disease in men with left ventricular hypertrophy. N Engl J Med 333:288–293

Planavila A, Iglesias R, Giralt M, Villarroya F (2011) Sirt1 acts in association with PPARalpha to protect the heart from hypertrophy, metabolic dysregulation, and inflammation. Cardiovasc Res 90:276–284

Poornima IG, Parikh P, Shannon RP (2006) Diabetic cardiomyopathy: the search for a unifying hypothesis. Circ Res 98:596–605

Rajesh M, Kolmakova A, Chatterjee S (2005) Novel role of lactosylceramide in vascular endothelial growth factor mediated angiogenesis in human endothelial cells. Circ Res 97:796–804

Ruppert C, Deiss K, Herrmann S, Vidal M, Oezkur M, Gorski A, Weidemann F, Lohse MJ, Lorenz K (2013) Interference with ERK(Thr188) phosphorylation impairs pathological but not physiological cardiac hypertrophy. Proc Natl Acad Sci U S A 110:7440–7445

Rusciani A, Duca L, Startlet H et al (2011) Elastin peptides signaling relies on neuroaminidase-1-dependent lactosylceramide generation. PLoS One 5(11):e14010

Sandler H, Dodge HT (1963) Left ventricular tension and stress in man. Circ Res 13:91–104

Sbroggio M, Carnevale D, Bertero A, Cifelli G, De Blasio E, Mascio G, Hirsch E, Bahou WF, Turco E, Silengo L et al (2011) IQGAP1 regulates ERK1/2 and AKT signalling in the heart and sustains functional remodelling upon pressure overload. Cardiovasc Res 91:456–464

Semeniuk LM, Kryski AJ, Severson DL (2002) Echocardiographic assessment of cardiac function in diabetic db/db and transgenic db/db-hGLUT4 mice. Am J Physiol Heart Circ Physiol 283:H976–H982

Singh R, Krishan P (2010) Modulation of impact of high fat diet in pathological and physiological left ventricular cardiac hypertrophy by fluvastatin. Biomed Pharmacother 64:147–153

Smith CC, Yellon DM (2011) Adipocytokines, cardiovascular pathophysiology and myocardial protection. Pharmacol Ther 129:206–219

Taegtmeyer H (1994) Energy metabolism of the heart: from basic concepts to clinical applications. Curr Probl Cardiol 19:59–113

Takayama N, Kai H, Kudo H, Yasuoka S, Mori T, Anegawa T, Koga M, Kajimoto H, Hirooka Y, Imaizumi T (2011) Simvastatin prevents large blood pressure variability induced aggravation of cardiac hypertrophy in hypertensive rats by inhibiting RhoA/Ras-ERK pathways. Hypertens Res 34:341–347

Unger RH, Orci L (2001) Diseases of liporegulation: new perspective on obesity and related disorders. FASEB J 15:312–321

Vijayan K, Szotek EL, Martin JL, Samarel AM (2004) Protein kinase C-alpha-induced hypertro- p h y of neonatal rat ventricular myocytes. Am J Physiol Heart Circ Physiol 287:H2777–H2789

Wang P, Liu J, Li Y, Wu S, Luo J, Yang H, Subbiah R, Chatham J, Zhelyabovska O, Yang Q (2010) Peroxisome proliferator-activated receptor {delta} is an essential transcriptional regulator for mitochondrial protection and biogenesis in adult heart. Circ Res 106:911–919

Wei Y, Zhou F, Ge Y, Chen H, Cui C et al (2010) β1,4-galactoslytransferase V regulates self-renewal of glioma-initiating cell. Biochem Biophys Res Commun 396:602–607

Wojciechowski P, Juric D, Louis XL, Thandapilly SJ, Yu L, Taylor C, Netticadan T (2010) Resveratrol arrests and regresses the development of pressure overload- but not volume overload-induced cardiac hypertrophy in rats. J Nutr 140:962–968

Yang R, Barouch LA (2007) Leptin signaling and obesity: cardiovascular consequences. Circ Res 101:545–559

Yang L, Qu M, Wang Y, Duan H, Chen P, Shi W, Danielson P, Zhou Q (2013a) Trichostatin A inhibits transforming growth factor-beta-induced reactive oxygen species accumulation and myofibroblast differentiation via enhanced NF-E2-related factor 2-antioxidant response element signaling. Mol Pharmacol 83:671–680

Yang SJ, Chen CY, Chang GD, Wen HC, Chang SC, Liao JF, Chang CH (2013b) Activation of Akt by advanced glycation end products (AGEs): involvement of IGF-1 receptor and caveolin-1. PLoS One 8:e58100

Yoshizaki F et al (2008) Role of glycosphingolipid-enriched microdomains in innate immunity: microdomain-dependent phagocytic cell functions. Biochim Biophys Acta 1780:383–392

Chapter 9
Plasma Membrane-Associated Sialidase Confers Cancer Initiation, Promotion and Progression

Taeko Miyagi, Kohta Takahashi, Kazuhiro Shiozaki, Kazunori Yamaguchi, and Masahiro Hosono

Introduction

An increase in sialylation is often found in cell surface glycoproteins of malignant cells (Lau and Dennis 2008), and altered sialylation of glycolipids is also observed as a ubiquitous phenotype (Hakomori 2010). Despite the number of reports describing involvement of sialic acids in cancer, the molecular mechanism and significance are not fully understood. To understand further ganglioside neoplastic alterations, we have focused on a human ganglioside-specific sialidase NEU3, which cleaves sialic acids preferentially from gangliosides. Ganglioside sialidase activity levels fluctuate consistently with cell differentiation, cell growth, and malignant transformation. Alterations of the activity levels associated with malignant transformation were described in 3T3-transformed cells (Yogeeswaran and Hakomori 1975), in BHK-transformed cells (Schengrund et al. 1973) and in mouse epidermal JB6 cells exposed to phorbol esters (Miyagi et al. 1990). However, little was known about the

T. Miyagi (✉) • K. Takahashi
Cancer Glycosylation Research, Institute of Molecular Biomembrane and Glycobiology, Tohoku Pharmaceutical University, Sendai 981-8558, Japan
e-mail: tmiyagi@tohoku-pharm.ac.jp; ktakahashi@tohoku-pharm.ac.jp

K. Shiozaki
Marine Biochemistry, Kagoshima University, Kagoshima 890-0056, Japan
e-mail: shiozaki@fish.kagoshima-u.ac.jp

K. Yamaguchi
Miyagi Cancer Center Research Institute, Natori 981-1293, Japan
e-mail: yamaguchi-ka959@miyagi-pho.jp

M. Hosono
Cell Recognition Study, Institute of Molecular Biomembrane and Glycobiology, Tohoku Pharmaceutical University, Sendai 981-8558, Japan
e-mail: mhosono@tohoku-pharm.ac.jp

A. Chakrabarti, A. Surolia (eds.), *Biochemical Roles of Eukaryotic Cell Surface Macromolecules*, Advances in Experimental Medicine and Biology 842,
DOI 10.1007/978-3-319-11280-0_9

molecular mechanisms underlying such sialidase alterations until development of gene cloning studies. Mammalian sialidases so far identified (NEU1–NEU4) are encoded by different genes and differ in their major subcellular localization and enzymatic properties. Each sialidase has been found to play a unique role depending on its particular properties (Miyagi and Yamaguchi 2012) and behave in different manner in cancers (Miyagi et al. 2012). We have found that NEU3 is a key enzyme for ganglioside degradation because of its strict substrate preference to gangliosides, which co-localize with this sialidase in the surface membranes (Miyagi et al. 1999), and plays as a signal modulator (Miyagi et al. 2008). In various human cancers, NEU3 shows a remarkable upregulation. Although NEU4 can hydrolyze gangliosides as well as glycoproteins and oligosaccharides, our recent studies have exhibited its opposite behavior to NEU3 in carcinogenesis with a tendency of downregulation especially in colon cancer (Yamanami et al. 2007). We have now presented evidence of NEU3 for crucial involvement in cancer initiation and promotion in addition to progression in colon cancer through ganglioside modulation.

Upregulation of NEU3 Promotes Cancer Progression

Our previous studies demonstrated a significant upregulation of NEU3 in human various cancers including colon (Kakugawa et al. 2002), renal (Ueno et al. 2006), and prostate (Kawamura et al. 2012) cancers. The expression was increased in the surgical specimens of tumor tissues as compared to adjacent non-tumor tissues in the levels of mRNA and sialidase activity. Our attempt has been made to elucidate the significance and molecular mechanisms underlying increased NEU3 expression. In colon cancer cells, the increase caused suppression of cell differentiation and apoptosis, accompanied with increased Bcl-2 and decreased caspase3 expression. The endogenous sialidase level was downregulated in the process of differentiation and apoptosis of the cells induced by sodium butyrate. Compared to non-tumor mucosa, colon cancer tissues exhibited a marked accumulation of lactosylceramide (Lac-cer), a possible NEU3 product, and addition of Lac-cer to the culture confirmed to reduction of apoptotic cells during sodium butyrate treatment, indicating that high expression of NEU3 leads to protection against programmed cell death, probably through modulation of gangliosides (Kakugawa et al. 2002).

Further investigation of the mechanisms of NEU3-mediated cell survival revealed that its silencing caused apoptosis without specific stimuli, accompanied by decreased Bcl-xL and increased mda7 (melanoma differentiation associated gene-7, differentiation and apoptosis-inducing protein) and GM3 synthase mRNA levels in HeLa cells, whereas overexpression resulted in the opposite effects (Wada et al. 2007). Human colon and breast carcinoma cell lines, HT-29 and MCF-7 cells, appeared to be similarly affected by treatment with the NEU3 siRNA, but interestingly non-cancerous human WI-38 and NHDF fibroblasts and NHEK keratinocytes showed no significant changes. NEU3 silencing inhibited Ras activation and its overexpression to stimulate it with consequent influence on ERK and Akt. NEU3 promoted EGFR phosphorylation in response to EGF and co-immunoprecipitated with EGFR in the cells, suggesting that NEU3 suppresses apoptosis of cancer cells by promoting EGFR phosphorylation, probably through its association with EGF receptors and consequent activation of Ras cascades, especially via the Ras/ERK pathway.

NEU3-transfected colon cancer cells exhibited increased adhesion to laminins and consequent cell proliferation, but a decrease in cell adhesion to fibronectin, collagen I and IV, compared to control cells (Kato et al. 2006). On laminin-5, NEU3 clearly stimulated phosphorylation of FAK and ERK, and markedly enhanced tyrosine phosphorylation of integrin β4, with recruitment of Shc and Grb-2. These results indicate that NEU3 differentially regulate cell proliferation through integrin-mediated signaling depending on the extracellular matrix. This selective NEU3 effect may be favorable for cancer cell growth, because laminin-5 has been reported to increase and in contrast fibronectin to reduce progression of carcinoma.

NEU3 expression was found to be increased in renal cell carcinomas (RCCs) compared to adjacent non-tumor tissues, significantly correlating with elevation of interleukin (IL)-6, a pleiotropic cytokine, which has been implicated in immune responses and pathogenesis of several cancers (Ueno et al. 2006). Up-regulation of NEU3 in the tumor tissues was strongly linked to the IL-6 expression level and NEU3 in renal cancer ACHN cells was activated by IL-6 in a positive feed back manner on the cytokine function, mainly through the PI3K/Akt pathway, resulting in suppression of apoptosis and promotion of migration. Either NEU3 transfection or IL-6 treatment resulted in suppression of apoptosis and promotion of cell motility, and the combination resulted in synergistic effects. NEU3 hardly affected MAPK or IL-6-induced STAT3 activation but promoted the PI3K/Akt cascade in both IL-6 dependent and independent ways. Furthermore, IL-6 promoted Rho activation and the effect was potentiated by NEU3, leading to increased cell motility whereas NEU3 silencing resulted in decreased Akt phosphorylation and inhibition of Rho activation. Glycolipid analysis showed a decrease in ganglioside GM3 and an increase in Lac-cer after NEU3 transfection, and addition of these lipids to the culture apparently affected cell apoptosis and motility, consistent with to the observations in colon cancer cells. The results indicate that NEU3 activated by IL-6 exerts IL-6-mediated signaling largely via the PI3K/Akt cascade in a positive feedback manner and contributes to expression of a malignant phenotype in RCCs.

Up-regulation of NEU3 was also detected in prostate cancer, showing a significant correlation with malignancy as assessed by the Gleason score (Kawamura et al. 2012). In androgen-sensitive LNCaP cells, forced overexpression of NEU3 significantly induced progression-related transcription factor EGR-1, androgen receptors and prostate specific antigen, PSA, both with and without androgen, the cells becoming sensitive to low concentration of hormone. This NEU3-mediated induction was abrogated by inhibitors of PI3K and MAPK, in line with increased phosphorylation of Akt and ERK1/2 in NEU3-overexpressing cells. NEU3 silencing moreover caused a reduction in the cell growth of androgen-independent PC-3 cells in culture and of transplanted tumors in nude mice. These data strongly suggest that NEU3 regulates the progression of prostate cancer through androgen receptor signaling.

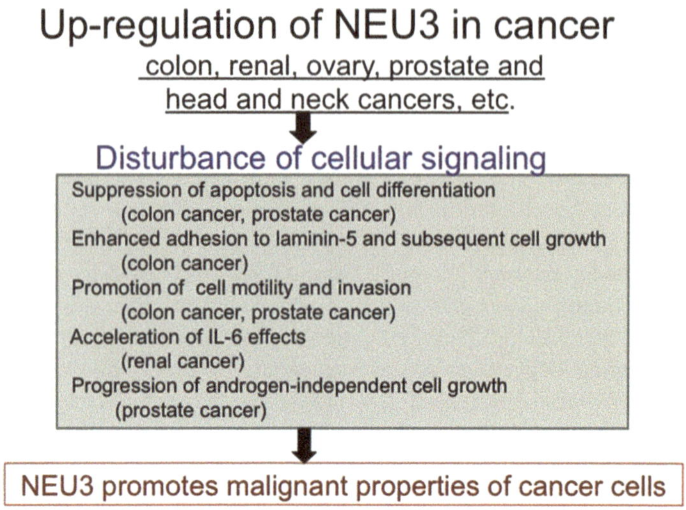

Fig. 9.1 Upregulation of NEU3 observed in various human carcinomas causes disturbance of cellular signaling, leading to augmentation of malignant properties of cancer cells

As illustrated in Fig. 9.1, all the results together display that NEU3 augments cancer progression through promoting cell survival, adhesion, growth and motility by potentiating signaling including MAPK/ERK, PI3K/AKT and FAK/ERK pathways, probably via Ras activation. To gain insights into regulation mechanisms of *NEU3* gene, we have recently determined the gene structure and assessed transcription factor involvement (Yamaguchi et al. 2010). NEU3 expression was found to be diversely regulated by Sp1/Sp3 transcription factors binding to alternative promoters. Such transcriptional control might also account for the upregulation of NEU3 in cancer, because Sp1 and Sp3 are now known to play critical roles in regulating the transcription of genes involved in cell growth control and tumorigenesis (Wierstra 2008). The expression of NEU3, in fact, exhibited good correlations with those of Sp1 or Sp3 in cancer, implying a promoting role in NEU3 gene transcription.

NEU3 Is Involved in Cancer Initiation and Promotion

To investigate whether NEU3 participates in cancer initiation and promotion as well as progression, a possible role of NEU3 in promoting tumorigenesis in vivo has been demonstrated in human NEU3 transgenic mice treated with a carcinogen, azoxymethane (AOM), for induction of precancerous colonic aberrant crypt foci (ACF) (Shiozaki et al. 2009). ACF were induced in the mice significantly more frequently than in their control wild-type counterparts. Enhanced phosphorylation of EGF receptor, Akt and ERK and up-regulation of Bcl-xL protein were observed

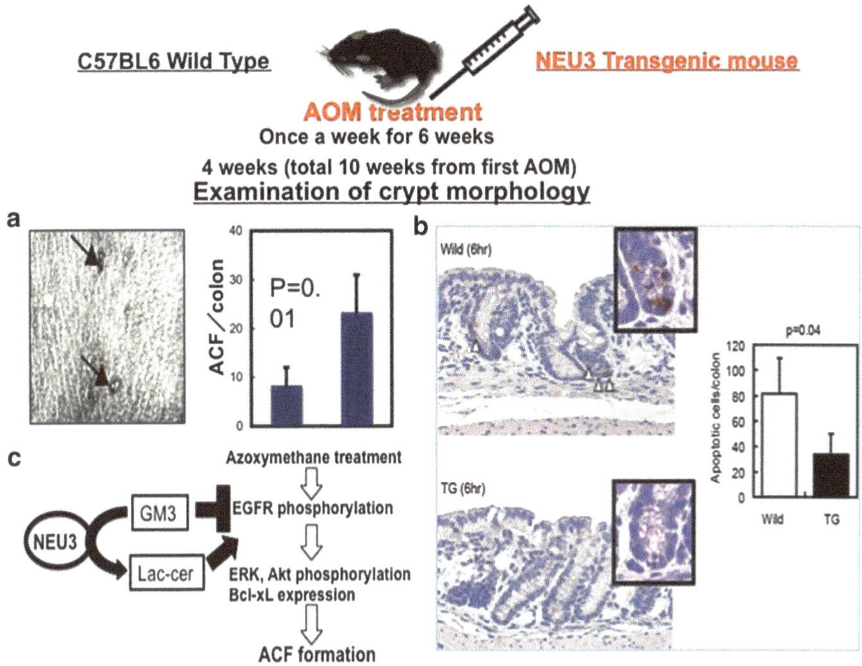

Fig. 9.2 Promotion of aberrant cript foci (ACF) formation in NEU3 transgenic mice. The susceptibility of NEU3 transgenic mice to induction of ACF was examined by treatment with azoxymethane (AOM) (Shiozaki et al. 2009). (**a**) Mice were injected with AOM (i.p., 15 mg/kg/week) for 6 weeks, and 4 weeks later ACF had formed in the NEU3 transgenic mice significantly more than in the control wild-type mice. (**b**) Numbers of apoptotic cells were stained in immuno-histochemical sections with anti-cleaved caspase 3 antibody. (**c**) Enhanced phosphorylation of EGFR, Akt and ERK and up-regulation of Bcl-xL protein were observed in the transgenic colon mucosa

in the transgenic colon mucosa, but no changes were found in cell proliferation, suggesting that the increased ACF formation was due to suppression of apoptosis, as shown in Fig. 9.2. Thus, NEU3 up-regulation may be important to the promotion stage of colorectal carcinogenesis in vivo. When *Neu3*-deficient mice were exposed to dimethylhydrazine, there were no differences in the incidence or growth of tumors from wild-type mice. On the other hand, the *Neu3*-deficient mice were less susceptible to colitis-associated colon carcinogenesis induced by AOM and dextran sodium sulfate, indicating an involvement of NEU3 in inflammation-dependent tumor development (Yamaguchi et al. 2012). In addition to the observations in genetically engineered mouse models, we have provided evidence of a close link between NEU3 expression and Wnt/β-catenin signaling in colon cancer cells by analyzing cancer stem-like characteristics and tumor initiating capability (Takahashi et al. in submission). NEU3-silencing in colon cancer cells resulted in a significant decrease in clonogenicity on soft agar and in vivo tumor growth, along with down-regulation of stemness genes. Under sphere-forming conditions, endogenous

NEU3 expression was significantly increased. Null-activity mutants of NEU3 failed to activate relevant signaling, indicating that the activation is dependent on ganglioside changes.

The available data strongly suggest participation of NEU3 in tumor initiation and promotion, since constitutive activation of Wnt/β-catenin signaling is implicated in the maintenance of cancer stem cells and initiation of the process of colon carcinogenesis (Clevers 2006). However, it is thought to be insufficient for progression without additional Ras activation (Phelps et al. 2009). In this context, NEU3 may be a pivotal molecule involved in both cancer initiation and progression, by regulating Wnt/β-catenin and Ras/MAPK signaling pathways.

Conclusion

In this review, we describe possible roles of the plasma membrane-associated sialidase NEU3 in cancer. We have so far demonstrated that NEU3 is markedly upregulated in various human cancers and the aberrant increase causes augmentation of malignant properties of cancer cells, including increased cell survival, adhesion, migration and invasion. Here, we present evidence of an involvement of NEU3 in cancer initiation and promotion as well as progression using genetically engineered mouse models and NEU3- silenced cancer cells. Treatment with NEU3 siRNA brings about reversal of some malignant properties of cancer cells in culture and tumor regression *in vivo*. Thus, NEU3 may be a prominent determinant in cancer initiation, promotion and progression, and represent an attractive target for treatment.

References

Clevers H (2006) Wnt/beta-catenin signaling in development and disease. Cell 127:469–480

Hakomori SI (2010) Glycosynaptic, microdomains controlling tumor cell phenotype through alteration of cell growth, adhesion, and motility. FEBS Lett 584:1901–1906

Kakugawa Y, Wada T, Yamaguchi K, Yamanami H, Ouchi K, Sato I, Miyagi T (2002) Up-regulation of plasma membrane-associated gangliosidesialidase (Neu3) in human colon cancer and its involvement in apoptosis suppression. Proc Natl Acad Sci U S A 99:10718–10723

Kato K, Shiga K, Yamaguchi K, Hata K, Kobayashi T, Miyazaki K, Saijo S, Miyagi T (2006) Plasma-membrane-associated sialidase (NEU3) differentially regulates integrin-mediated cell proliferation through laminin- and fibronectin-derived signalling. Biochem J 394:647–656

Kawamura S, Sato I, Wada T, Yamaguchi K, Li Y, Li D, Zhao X, Ueno S, Aoki H, Tochigi T, Kuwahara M, Kitamura T, Takahashi K, Moriya S, Miyagi T (2012) Plasma membrane-associated sialidase (NEU3) regulates progression of prostate cancer to androgen-independent growth through modulation of androgen receptor signaling. Cell Death Differ 19:170–179

Lau KS, Dennis JW (2008) N-Glycans in cancer progression. Glycobiology 18:750–760

Miyagi T, Yamaguchi K (2012) Mammalian sialidases: physiological and pathological roles in cellular functions. Glycobiology 22:880–896

Miyagi T, Sagawa J, Kuroki T, Matsuya Y, Tsuiki S (1990) Tumor-promoting phorbol ester induces alterations of sialidase and sialyltransferase activities of JB6 cells. Jpn J Cancer Res 81: 1286–1292

Miyagi T, Wada T, Iwamatsu A, Hata K, Yoshikawa Y, Tokuyama S, Sawada M (1999) Molecular cloning and characterization of a plasma membrane-associated sialidase specific for gangliosides. J Biol Chem 274:5004–5011

Miyagi T, Wada T, Yamaguchi K, Hata K, Shiozaki K (2008) Plasma membrane-associated sialidase as a crucial regulator of transmembrane signaling. J Biochem 144:279–285

Miyagi T, Takahashi K, Hata K, Shiozaki K, Yamaguchi K (2012) Sialidase significance for cancer progression. Glycoconj J 29:567–577

Phelps RA, Chidester S, Dehghanizadeh S, Phelps J, Sandoval IT, Rai K, Broadbent T, Sarkar S, Burt RW, Jones DA (2009) A two step model for colon adenoma initiation and progression caused by APC loss. Cell 137:623–634

Schengrund CL, Lausch RN, Rosenberg A (1973) Sialidase activity in transformed cells. J Biol Chem 248:4424–4428

Shiozaki K, Yamaguchi K, Sato I, Miyagi T (2009) Plasma membrane-associated sialidase (NEU3) promotes formation of colonic aberrant crypt foci in azoxymethane-treated transgenic mice. Cancer Sci 100:588–594

Ueno S, Saito S, Wada T, Yamaguchi K, Satoh M, Arai Y, Miyagi T (2006) Plasma membrane-associated sialidase is up-regulated in renal cell carcinoma and promotes interleukin-6-induced apoptosis suppression and cell motility. J Biol Chem 281:7756–7764

Wada T, Hata K, Yamaguchi K, Shiozaki K, Koseki K, Moriya S, Miyagi T (2007) A crucial role of plasma membrane-associated sialidase in the survival of human cancer cells. Oncogene 26:2483–2490

Wierstra I (2008) Sp1: emerging roles–beyond constitutive activation of TATA-less housekeeping genes. Biochem Biophys Res Commun 372:1–13

Yamaguchi K, Koseki K, Shiozaki M, Shimada Y, Wada T, Miyagi T (2010) Regulation of plasma-membrane-associated sialidase NEU3 gene by Sp1/ Sp3 transcription factors. Biochem J 430:107–117

Yamaguchi K, Shiozaki K, Moriya S, Koseki K, Wada T, Tateno H, Sato I, Asano M, Iwakura Y, Miyagi T (2012) Reduced susceptibility to colitis-associated colon carcinogenesis in mice lacking plasma membrane-associated sialidase. PLoS One 7:e41132

Yamanami H, Shiozaki K, Wada T, Yamaguchi K, Uemura T, Kakugawa Y, Hujiya T, Miyagi T (2007) Down-regulation of sialidase NEU4 may contribute to invasive properties of human colon cancers. Cancer Sci 98:299–307

Yogeeswaran G, Hakomori S (1975) Cell contact-dependent ganglioside changes in mouse 3T3 fibroblasts and a suppressed sialidase activity on cell contact. Biochemistry 14:2151–2156

Chapter 10
A Signal with a Difference: The Role of GPI Anchor Signal Sequence in Dictating Conformation and Function of the Als5 Adhesin in *Candida albicans*

Mohammad Faiz Ahmad *, Pareeta Gajraj Mann *, and Sneha Sudha Komath

Introduction

GPI Anchors in C. albicans

GPI anchors are unique flexible glycolipid anchors that are used by eukaryotes to anchor proteins in the extracellular leaflet of the plasma membrane or to the cell wall. Elsewhere in this volume, Prof. Taroh Kinosita provides a succinct review of the GPI anchor biosynthetic pathway and its importance for mammalian cells. Although there are some species-specific differences, the core structure of the GPI anchor precursor is conserved in eukarya. As with higher eukaryotes, in lower eukaryotes too GPI anchored proteins perform a variety of extremely important functions (McConville and Ferguson 1993; Paulick and Bertozzi 2008). In lower eukaryotes, the pathway appears to be essential for growth and viability (McConville and Ferguson 1993). In *S. cerevisiae* and its closely related pathogenic fungal cousin, *C. albicans*, inhibiting GPI anchor biosynthesis is lethal (Leidich et al. 1994; Grimme et al. 2004; Pittet and Conzelmann 2007; Yadav et al. 2014). In *C. albicans*, a number of GPI anchored proteins are specifically required for adhesion and virulence and several of these are regulated with the morphogenetic switch that converts the yeast form of the organism into the invasive hyphal form (Martinez-Lopez et al. 2004). Prominent among these are the Als family of eight adhesins, the hyphal wall protein 1 (Hwp1) and Elf4E-associated protein 1 (Eap1) (Staab et al. 1999; Cormack et al. 1999; Hoyer and Hecht 2001; Sundstrom 2002).

*Author contributed equally with all other contributors.

M.F. Ahmad • P.G. Mann • S.S. Komath (✉)
School of Life Sciences, Jawaharlal Nehru University, New Delhi, India
e-mail: sskomath@mail.jnu.ac.in; sskomath@yahoo.com

A. Chakrabarti, A. Surolia (eds.), *Biochemical Roles of Eukaryotic Cell Surface Macromolecules*, Advances in Experimental Medicine and Biology 842,
DOI 10.1007/978-3-319-11280-0_10

Not surprisingly, defects in GPI biosynthesis, therefore, also hamper the ability of the organism to successfully establish colonies on host tissues and inhibit its infection and virulence (Martinez-Lopez et al. 2004).

The Role of C-Terminal Signal Sequences in the Attachment of the Anchor

As interesting as the GPI anchors themselves are the GPI anchor attachment signal sequences (SSs) at the C-terminal ends of the proteins. These SSs dictate the attachment of pre-formed GPI anchors to the proteins within the lumen of the endoplasmic reticulum (ER).

Starting with their discovery in the late 1980s, a great deal of debate has centered upon the nature and role of the SSs in determining which proteins will receive the anchor and how they will function. We now know that the amino acid sequence of a SS varies with the protein. We also know that identities of the residues *per se* are not as important as conservation of the 'nature' of the residues in a sequence identified as a GPI-attachment SS. A general rule appears to be that the site of cleavage of SS, and its replacement with the GPI anchor precursor, is always an amino acid residue (ω-site) close to the C-terminal end of the protein with a very small side chain. This must be flanked on either side by residues with small side chains as well ($\omega-1$ to $\omega+2$). A polar linker segment of 8–12 residues beyond the $\omega+2$ site and a C-terminal hydrophobic stretch of 10–20 residues then completes the SS (Eisenhaber et al. 1998). Although present generally at the C-terminal end of proteins, the introduction of SS in the middle of a protein sequence can also apparently signal for cleavage and addition of the preformed glycolipid anchor at that position (Caras 1991). Subsequent to the transfer of the GPI anchor 'precursor', the anchor as well as the protein undergoes several remodeling steps/modifications within the ER as well as in the Golgi before it is finally targeted to the cell surface (as discussed in detail by Prof. Kinoshita in this book).

Do the SSs have any additional roles to play other than the obvious one of serving as flags for transfer of the GPI anchors? SSs also appears to work as markers for localization, dictating whether specific proteins should go to the cell wall or the plasma membrane in *S. cerevisiae* and other organisms that have a cell wall (Hamada et al. 1999; Frieman and Cormack 2004; Ouyang et al. 2013). There are also studies that indicate that the SS dictates sorting to the apical versus basolateral surfaces in mammalian cells (Ali et al. 1996; Paladino et al. 2008). Likewise, there are suggestions that the type of GPI anchor attached may itself be dependent on the SS (Nicholson and Stanners 2007). Since the GPI anchor undergoes several steps of remodeling within the endoplasmic reticulum (ER) as well as in the Golgi after it is attached to the protein, and before it is finally targeted to the cell surface, it is unclear whether the ER exit sites for the proteins are dictated by the SS or by the protein. Several studies suggest that the SS also dictates the fate of the protein under GPI deficiency conditions or when it enters the degradation pathway (Garg et al. 1997;

Ashok and Hegde 2008). Residues at and around the ω-site, appear to not only be important for GPI attachment, they also seem to influence the glycosylation pattern on the GPI-anchored proteins in the Golgi and thereby alter both oligomerization status as well as localization in mammalian cells (Miyagawa-Yamaguchi et al. 2014). There are also studies that suggest that the SS after its detachment from the protein could continue to dictate its fate (Guizzunti and Zurzolo 2014). However, it is not very clear how these functions are mediated by a peptide stretch that is removed from the protein at the time of the GPI anchor attachment.

Mutations in SS

What happens in the case of mutations in the SS? For proteins like the mammalian prion protein (PrP), for example, such a question also assumes clinical significance since mutations in the SS have been associated with neurodegenerative disorders (Poggiolini et al. 2013). One obvious outcome, for mutations that occur at or around the ω-site, would be that GPI anchors do not get attached to the proteins and these must then be routed either towards secretion as anchorless variants or towards the protein degradation pathways (Böhme and Cross 2002; Kiachopoulos et al. 2005; Hizume et al. 2010). The absence of the anchor could certainly affect the functionality of the protein in such cases, both because it alters its localization and because the anchor itself could have a say in the final conformation of some proteins (Frieman and Cormack 2003; Kiachopoulos et al. 2005; Nisbet et al. 2010). Other mutations in the SS that do not necessarily have any effect on anchor attachment can also nevertheless affect the functioning of mature proteins on the cell surface (Hoque et al. 1996; Windl et al. 1999). How and why this happens is not very clear. One hypothesis, based on the idea that the SS continues to have an independent existence even after its detachment from the protein, has been that these mutations alter the degradation patterns of the detached SS and thereby modulate the functionality of the mature GPI-anchored protein (Guizzunti and Zurzolo 2014). The mechanism for this however remains unclear. Another hypothesis, based on the hypothesis that the SS carries a functional specificity signal within itself, suggests that mutations could alter the kind of anchor attached to a protein and therefore alter its functionality/fate (Nicholson and Stanners 2007). Although differences are known to occur in processing of the GPI anchored proteins during their transport through the secretory pathway, including differences due to mutations in the SS (Miyagawa-Yamaguchi et al. 2014), no studies have so far indicated that the GPI precursor attached by the transamidase to nascently translocated proteins is different for different proteins.

Using the Als5 adhesin from *C. albicans* we propose a third possible hypothesis where SS has a role in controlling the conformation of the protein and thereby influencing GPI anchor attachment and functionality. The Als family of adhesins have three distinct features that mark them out for the unique tasks that they are required to perform on the surface of *C. albicans* (Sundstrom 2002). Besides binding to a wide variety of specific peptide ligands, they also possess the ability to adhere to several basal lamina proteins and to undergo oligomerization via amyloid formation/β-aggregation that assist in the establishment of fungal colonies. We had shown previously that the GPI anchor attachment SS could significantly alter the conformation of Als5 (Ahmad et al. 2012). The presence of SS could not only keep the protein in a relatively well folded and predominantly α-helical conformation, it could also prevent it from adhering to collagen type IV or aggregating, both important functions of the adhesin. Deletion of SS from the protein resulted in a protein with predominantly pre-molten globule characteristics and complete functionality. We had hypothesized that this difference between the two proteins could be because the SS at the C-terminal end of Als5-SS folds back and interacts with the amyloid patch within the adhesin, forcing the protein into adopting an α-helical conformation and suppressing its functionality (Fig. 10.1). Als5 lacking the SS would of

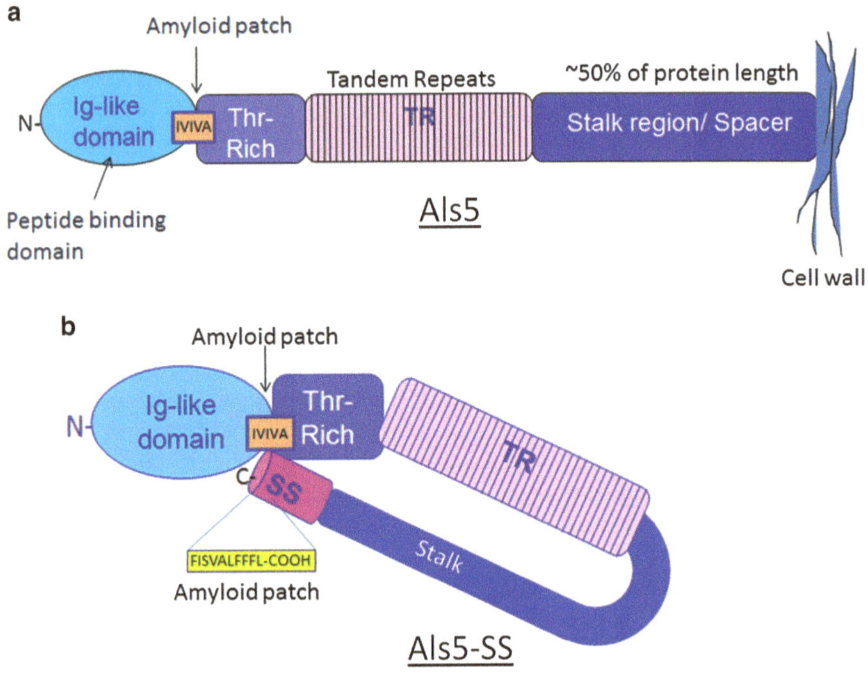

Fig. 10.1 Domain organization of Als5 andAls5-SS. Mature Als proteins on the cell surface are organized into several domains as shown in (**a**). Based on our previous studies on Als5, we had proposed that before the cleavage of the C-terminal SS and attachment of a preformed GPI anchor, SS folds back and interact with the N-terminal half of Als5 (Ahmad et al. 2012), as shown in (**b**). This folding back of SS influences the secondary structure of the protein leading to a more structured secondary conformation than predicted. We also proposed that the interaction takes place via the amyloidogenic regions within Als5-SS, one adjacent to the Ig-like peptide binding domain (shown in *orange*) and the other within its SS (shown in *yellow*), and thus inhibits the aggregation of Als5. Thus, presence of SS can influence the structure and function of Als5 protein. A mutation within the amyloidogenic regions could possibly disrupt or weaken this interaction, and thus affect the structure and function of the mutant protein

course be free from such a constraint and capable of adopting a more extended conformation as is expected for the functional protein on the cell surface. Likewise, mutations in Als5-SS that abrogate interaction between the two domains, as per our hypothesis, should result in a protein whose amyloid patch is exposed. In this manuscript we put this hypothesis to test. We show that mutant variants Als5-SS V309N (mutation in the amyloid patch) or Alss5-SS L1326R and Als5-SS F1327R (both mutations in the SS) are proteins which show better adhesion as well as aggregation, more like Als5 rather than Als5-SS.

Effect of Mutations on the Conformation and Function of Als5-SS

Identifying Mutations in Als5-SS That Would Abrogate β-Aggregation Potential of Putative Interacting Domains

In a previous manuscript, we showed by TANGO analysis that Als5-SS has two patches which have a significant amount of β-aggregation potential as predicted (Ahmad et al. 2012). One of these lies within the mature functional protein sequence, at the C-terminus of the Ig-like N-terminal domain of the protein that has been previously shown to bind specific peptide ligands (Klotz et al. 2004; Rauceo et al. 2006), while the other lies within the SS (Fig. 10.1). The internal site with the motif IVIVA was also shown by Peter Lipke and co-workers to be highly conserved in a variety of closely related adhesins, including the Als-family, and protein fragments of Als5 containing this motif were all capable of exhibiting aggregation (Ramsook et al. 2010). Using TANGO (http://tango.crg.es/), we also predicted what mutations within these sites would turn the amyloid patch into one with insignificant β-aggregation potential. We discovered that a V309N mutation within the protein reduced the β-aggregation potential of the first patch drastically (Fig. 10.2a). Likewise, L1326R or F1327R mutations reduced the β-aggregation potential of the SS to nearly zero (Fig. 10.2b, c). Thus, we hypothesized that if we introduced any of these mutations in Als5-SS, it should reduce the interactions between these two amyloidogenic regions of Als5-SS and result in better adhesion and aggregation.

In silico Predictions of GPI Anchor Attachment for Als5-SS and Its Mutant Variants

While introduction of the V309N mutation was not expected to affect GPI anchor attachment, it was pertinent to also ask whether mutations in the SS would affect GPI anchor addition. We used the GPI-SOM (http://gpi.unibe.ch/; Fankhauser and Mäser 2005) and Big-PI (http://mendel.imp.univie.ac.at/set/gpi/gpi_server.html;

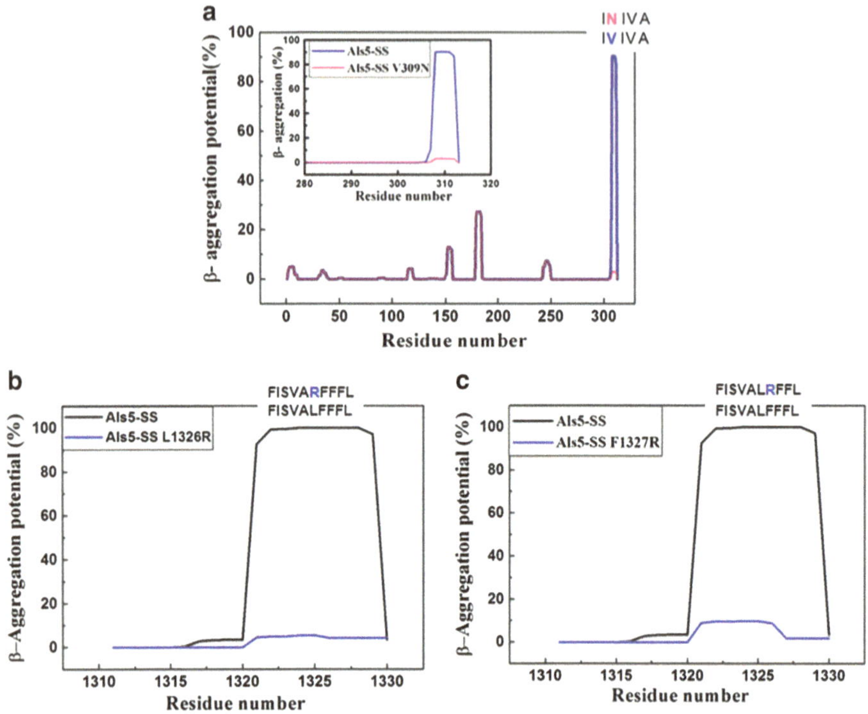

Fig. 10.2 TANGO prediction for **β**-aggregation potential. Introduction of V309N (**a**), or L1326R (**b**), or F1327R (**c**) mutations leads to reduced **β**-aggregation potential for Als5-SS. The sequences within which the mutations have been made are also shown. In each figure the sequence below represents the wild type and the one above shows the mutated sequence. The inset in (**a**) corresponds to the region containing the IVIVA domain

Table 10.1 Prediction of GPI anchor attachment by BIG-PI predictor and GPI-SOM

Proteins	BIG PI-predictor	GPI-SOM
Als5-SS	GPI anchored	GPI anchored
Als5-SS L1326R	Non GPI anchored	GPI anchored
Als5-SS F1327R	GPI anchored	GPI anchored
Als5-SS L1326R F1327R	Non GPI anchored	Non GPI anchored

Eisenhaber et al. 1999) software to predict whether the mutations affected GPI anchor attachment. We discovered that the L1326R mutation was drastic enough to convert the SS into a stretch that was no longer likely to be recognized as a putative SS (Table 10.1). On the other hand, mutating F1327R appeared to have no significant effect on its potential to act as a signal for GPI anchor attachment (Table 10.1). Needless to say, the double mutation L1326R F1327R resulted in a sequence with no potential to serve as a successful SS (Table 10.1).

Table 10.2 Prediction of ΔG_{app} for TM helix insertion of SS from Als5 by ΔG predictor software (http://dgpred.cbr.su.se)

	Predicted ΔG_{app} (kcal/mol)
SS	+2.955
SS L1326R	+3.675
SS F1327R	+3.585
SS L1326R F1327R	+4.313

An accompanying question we needed to address was about the nature of SS itself. For long it was believed that during translocation into the ER, a newly translated proprotein destined to be GPI anchored would have the hydrophobic segment of its SS inserted into the ER membrane where it would then be recognized by the GPI-transamidase (Berger et al. 1988; Caras and Weddell 1989). However, there is now growing experimental evidence to suggest that membrane-inserted SS domains are not absolutely essential (Wang et al. 1999; Dalley and Bulleid 2003). Several proproteins that get GPI anchored are perhaps soluble proteins. A thermodynamic assessment of the free energies (ΔG_{app}) for membrane insertion of a variety of SS suggested that SS in general have sequences with marginal hydrophobicities and $\Delta G_{app} \sim 0.0$ (Galian et al. 2012). Using ΔG predictor software (http://dgpred.cbr.su.se) developed by von Heijne's group (Hessa et al. 2007) the SS fragment of Als5-SS was predicted to have a helix insertion ΔG_{app} of +2.6 kcal/mol (Table 10.2). In other words, Als5-SS could perhaps be a soluble ER luminal protein. This also implies that the SS is likely to be free to interact with the rest of the protein, as hypothesized in our model, and may not be constrained to be inserted in the membrane. Mutation of V309N, as expected, does not in any way alter the value of ΔG_{app} for SS (+2.955 kcal/mol). Mutations within SS, both L1326R and F1327R, make the ΔG_{app} value more positive (Table 10.2), suggesting that these mutations too by themselves are unlikely to promote membrane insertion of the SS.

Generating ALS5-SS Variants Carrying the Different Mutations

Next we set about introducing these mutations, both L1326R and F1327R, within Als5-SS for wet lab experiments. Although the mutation L1326R was predicted to convert SS into a sequence that may not be recognized by the transamidase, as discussed above, we persisted in generating the mutant variant of Als5-SS possessing this mutation for two major reasons: 1) Theoretical predictions only give a general picture. They are neither fool-proof nor 100 % accurate and are based on the experimental data set available at this point in time. It is possible that predictions would change when more data or information become available. 2) The mutation, as per our hypothesis, should still be able to successfully disrupt the interaction between the internal amyloid patch and the SS. Thus, as a proof of concept, this is still a variant worth generating. The mutant, as a candidate for the transamidase, is currently being evaluated in wet lab experiments in the lab.

Table 10.3 List of primers used

	Primer sequence
Als5-SS FP	5′GCG*GGATCC*ATGATTCAACAATTTACATTGTTATTC3′
Als5-SS RP	5′GCG*CTCGAG*TCATAGAAAGAAGAATAATGCAACG3′
Als5-SS V309N FP	5′GATGCCGGATCTAACGGTATTAATATTGT 3′
Als5-SS V309N RP	5′TGTTCTAGTTGTAGCAACAATATTAATACC3′
Als5-SS SS L1326R FP	5′CTTAAATTTATTAGCGTTGCACGGTTCTTC3′
Als5-SS SS L1326R RP	5′CTCGAGTCATAGAAAGAAGAACCGTGCAAC3′
Als5-SS F1327R FP	5′AAATTTATTAGCGTTGCATTACGGTTCTTT 3′
Als5-SS F1327R RP	5′CCGCTCGAGTCATAGAAAGAACCGTAATGC3′

Since we required Als5-SS as a proprotein with its SS intact for these studies, we expressed the protein in bacterial cells that lack the GPI biosynthetic machinery. We previously generated the construct *Als5-SS-pGEX-6p-2* for expression of the protein in bacterial cells (Ahmad et al. 2012). We used the same plasmid to amplify these mutations in *GST-Als5-SS* using mutagenic primers (sequences of the primers are given in Table 10.3). The template plasmid, without any mutation, used for PCR was digested using Dpn1 enzyme and then transformed in DH5α strain of *E. coli*. The colonies obtained after transformation were screened by colony PCR using gene specific primers, FP and RP of *ALS5-SS*, and the colonies confirmed positive were further confirmed by complete sequencing of the gene of interest. We successfully generated three mutants (*GST-ALS5-SS V309N, GST-ALS5-SS L1326R* and *GST-ALS5-SS F1327R*) in the *GST-ALS5-SS* gene (data not shown).

The proteins were expressed and purified as already described previously (Ahmad et al. 2012). Briefly, plasmids without (for GST-Als5-SS) or with mutations (for GST-Als5-SS V309N, GST-Als5-SS L1326R and GST-Als5-SS F1327R) were transformed into BL21 strain of *E. coli*. Primary cultures were obtained by incubation of these transformed strains at 37 °C at 220 rpm overnight. Secondary cultures were generated using 1 % inoculum of the primary culture in each case after incubation at 37 °C at 220 rpm for 3 h. Protein expression was induced for 4 h using 0.1 mM IPTG at 16 °C when OD_{600nm} of the cultures reached ~0.6. The cells were pelleted down at 8,500 rpm at 4 °C for 10 min and the pellet resuspended in lysis buffer (50 mM sodium phosphate buffer, pH 8.0, 150 mM NaCl, 10 μM PMSF, 0.1 mg/mL lysozyme, and 5 % glycerol). After incubation at 4 °C on a rocker for 1 h, the cells were lysed using digital probe sonicator (30 s ON and 30 s OFF for 7 min). The cell lysate was centrifuged at 8,500 rpm for 45 min at 4 °C, the supernatant mixed with glutathione-agarose beads and incubated for 4 h on a rocker at 4 °C. The beads were washed using wash buffer (50 mM sodium phosphate buffer, pH 8.0, 3 M NaCl), and proteins eluted using 10 mM glutathione in elution buffer (50 mM sodium phosphate buffer, pH 8.0, 150 mM NaCl, 20 % glycerol). The homogeneity of the protein samples were >90 % as assessed from runs on 6 % SDS polyacrylamide gels stained with Coomassie brilliant blue R250 (data not shown).

The proteins were dialyzed to remove glutathione before they were characterized. As explained in our previous paper as well, the GST-tag is attached to the protein via a PreScission™ protease site, however, this site did not appear to be accessible to the protease and the tag could not be removed (Ahmad et al. 2012). Thus, all the wet lab experiments included purified GST as a control. The tag did not appear to significantly affect our results.

The Als5-SS Mutant Variants Had a Conformation That Was Different from That of Als5-SS

The mutant proteins were assessed for secondary structure formation by CD spectroscopy by a Chirascan CD spectrometer (Applied Photophysics). Dialyzed proteins at ~0.09 mg/mL were used for obtaining the spectra between 200 and 260 nm in a 1 mm path length cuvette with 1 nm step length. In all cases the buffer background was subtracted. The final spectra were averages of five repeat scans. The spectrum of GST was then subtracted from that of each of the fusion proteins to get the spectrum of Als5-SS V309N, Als5-SS L1326R and Als5-SS F1327R using Pro-Data software that came with the instrument.

We had previously shown that Als5 and Als5-SS had very different secondary structures as assessed by CD spectroscopy (Ahmad et al. 2012). While Als5-SS was a relatively well folded protein, Als5 itself was predominantly a pre-molten globular protein. This was one of the reasons why we had hypothesized that the SS at the C-terminal end probably folded back and interacted with the amyloid patch on the Als5 and thereby induced a significant amount of structure into a protein that would otherwise be largely a natively unfolded protein.

We found that each of the single mutations that we introduced during the course of this study perturbed the secondary structure of Als5-SS, although none of them had CD spectra that completely resembled that of Als5 (Fig. 10.3). In other words,

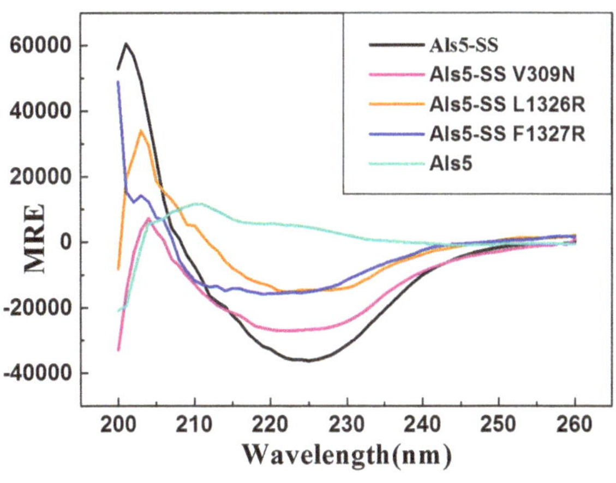

Fig. 10.3 Secondary structure prediction by CD. The secondary structures of Als5-SS V309N, Als5-SS L1326R and F1327R mutant proteins differed from Als5-SS, suggesting these mutations alter the conformation of the protein possibly by disruption of interaction between the N-terminus amyloidogenic patch and SS. The spectrum of Als5 lacking the SS is also shown for comparison

as per our model, the mutations appeared to alter the interactions between the domains without completely abrogating it.

We were now ready to examine adhesion by and aggregation of the mutant proteins.

The Als5-SS Mutant Variants Adhere Better to Collagen Type IV than Als5-SS Itself

We first examined the ability of the mutant proteins to adhere to collagen type IV. We had previously shown that GST-Als5 had approximately twofold better adherence to the basal lamina protein in comparison to GST-Als5-SS while GST by itself showed little or no adhesion to collagen type IV under similar conditions.

Using a similar protocol as before, we coated a 96 well plate with 0.5 mg/mL of collagen type-IV and incubated it at 4 °C overnight. Next day the wells were washed thrice with PBS and twice with elution buffer before adding 200 μL of ~0.07 mg/mL of each protein (GST, GST-Als5-SS, GST-Als5-SS V309N, GST-Als5-SS L1327R and GST-Als5-SS F1327R) to the respective wells. The plate was incubated at 37 °C for 1 h. Two controls, one using GST and the other without primary antibody were taken. 200 μL of 5 % skimmed milk in PBS were added to block the wells and incubated at 37 °C for another hour and washed 5 times with antibody buffer (1 % skimmed milk in PBS + 0.025 % Tween-20). Anti-GST primary antibody (1:1,000 diluted) was added to each well and incubated at 37 °C for 1 h. After incubation all the wells were washed 5 times with antibody buffer and 200 μL of HRP-conjugated secondary antibody (1:20,000 diluted) was added and incubated at 37 °C for 1 h. The wells were washed 5 times with antibody buffer and 100 μL of Tetramethylbenzidine (TMB) solution (1 mg/mL TMB in DMSO + 9 mL of 0.05 M phosphate citrate buffer, pH 5 and 2 μL H_2O_2) was added to each well and incubated at 37 °C for 1 h. OD_{650nm} was recorded using a plate reader (Thermo Scientific Multiscan GO).

We observed that the GST-Als5-SS mutants showed significantly better adhesion in comparison to GST-Als5-SS (Fig. 10.4).

The Als5-SS Mutant Variants Show Much Better Aggregation That Wild Type Als5-SS

We next explored the aggregation ability of the mutant variants of Als5 using transmission electron microscopy (TEM) with the help of a JEOL 2100F system. The proteins (0.09 mg/mL) were incubated at 37 °C for different time points (12, 24 h,

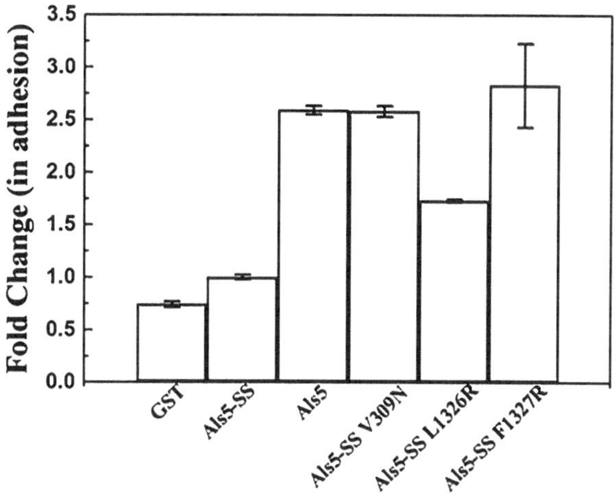

Fig. 10.4 GST-Als5-SS mutants show better adherence to collagen type IV than GST-Als5-SS. GST-Als5-SS V309N, GST-Als5-SS L1326R and GST-Als5-SS F1327R showed significantly higher adherence to collagen type IV (basal lamina protein) as compared to GST-Als5-SS. GST alone was used as control

1 week and 2 weeks) before analysis. Carbon coated copper grids were used for spotting the proteins, which were then stained with 2 % uranyl acetate and dried, before TEM images were recorded at 200 kV accelerating voltage.

We had previously shown that GST-Als5-SS could not form significant aggregates even if incubated for 2 weeks at 37 °C. On the other hand, GST-Als5 could form extensive aggregates within 24 h of incubation under similar conditions (Ahmad et al. 2012). As before, we observed that neither GST nor GST-Als5-SS formed aggregates even after 2 weeks of aggregation, while GST-Als5 formed significant aggregates within 12 h of incubation under similar conditions (Fig. 10.5). In contrast to GST-Als5-SS, GST-Als5-SS V309N, GST-Als5-SS L1326R and GST-Als5-SS F1327R could all show a significant amount of aggregation when incubated at 37 °C for 12 h (Fig. 10.5). The aggregation, as expected, was better when the mutant variants were incubated for longer periods of time.

Mutations in a Synthetic SS Peptide Prevent It from Inhibiting Self-Aggregation of Als5

In order to confirm the above results, we used a synthetic SS peptide (synthesized by Custom Peptide Synthesis) composed of just the last 20 residues at the C-terminus of Als5-SS as a competitor in our TEM studies. We previously showed that the synthetic SS peptide at 1:1 M concentrations could significantly inhibit the self-aggregation of GST-Als5 (Ahmad et al. 2012). If our hypothesis was correct, we

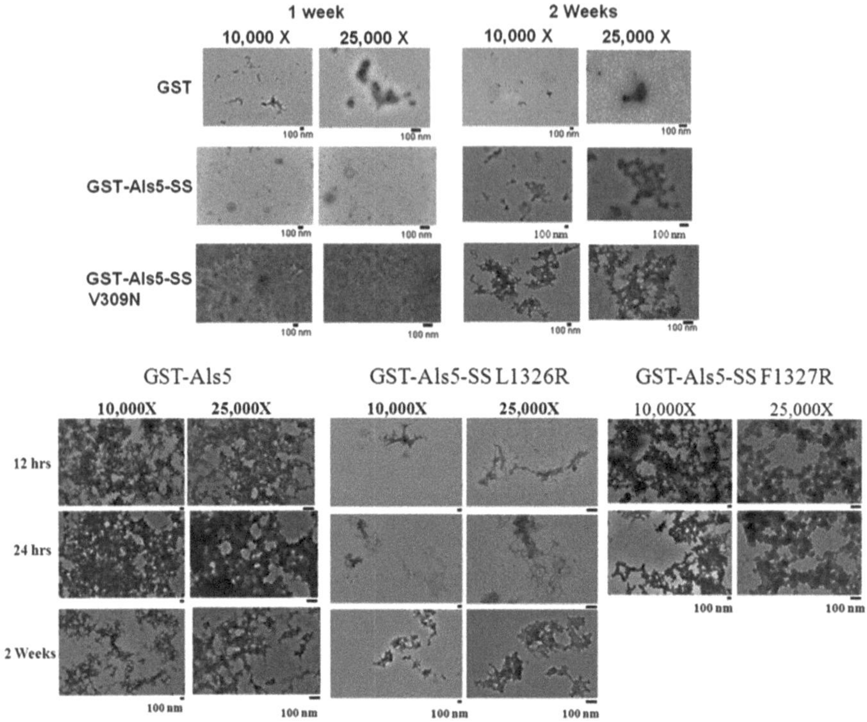

Fig. 10.5 Effects of mutations in amyloidogenic region and SS on the aggregation of GST-Als5-SS. GST-Als5-SS V309N, GST-Als5-SS L1326R and GST-Als5-SS F1327R showed much better aggregation than GST-Als5-SS. Thus, these mutations probably alter the interactions between SS and amyloidogenic domain near the Ig-like domain in GST-Als5-SS, thereby enhancing the aggregation of the mutant proteins. The aggregation of GST alone is also shown as control to demonstrate that the aggregation is not due to the N-terminal GST tag on the protein variants

expected to see reduced effect of the peptide on the self-aggregation of GST-Als5 if we introduced the L1326R or the F1327R mutation in it. Hence, we tested whether a synthetic peptide that included either the L1326R or the F1327R mutation would be able to inhibit GST-Als5 aggregation. Indeed, while SS continued to inhibit the aggregation of GST-Als5, neither SS L1326R nor SS F1327R could inhibit the self-aggregation of GST-Als5 to any significant degree (Fig. 10.6). As expected, the peptide with the double mutation also did not inhibit the aggregation of GST-Als5 (Fig. 10.6).

Thus, all in all, the data presented here confirms the hypothesis that we began with at the start of this work.

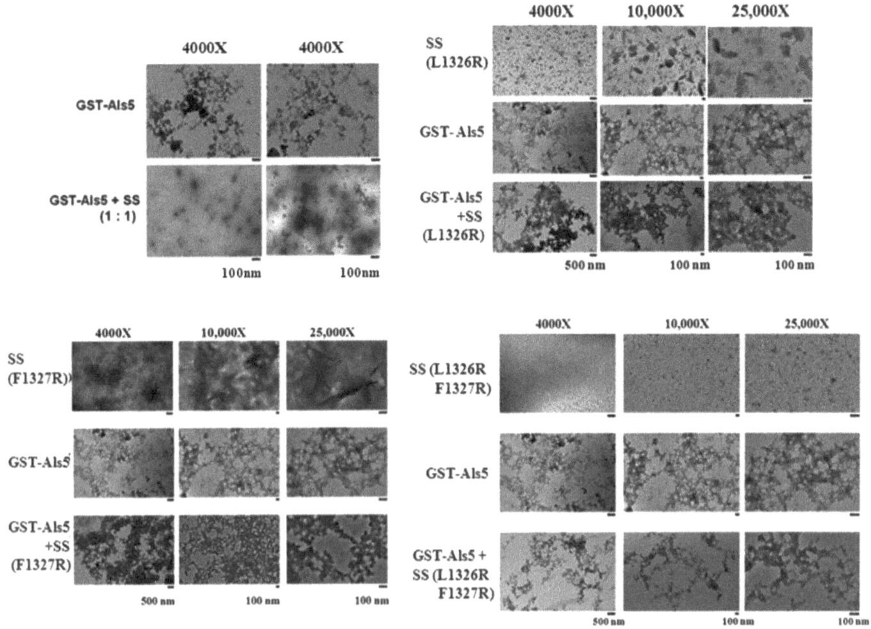

Fig. 10.6 Inhibition of aggregation of GST-Als5 by synthetic SS peptide variants. Incubation of GST-Als5 with equimolar concentration of synthetic SS peptide inhibits the aggregation property of GST-Als5, as also shown previously (Ahmad et al. 2012). However, SS peptide with L1326R and F1327R, or double mutations of L1326R F1327R failed to inhibit the aggregation of GST-Als5, suggesting the inability of these mutagenic SS peptides to interact with the amyloidogenic region of GST-Als5

Concluding Remarks

There is growing evidence to suggest that the SS plays multiple roles in GPI anchor attachment to proteins and these could vary depending both on the nature of the proteins and the SS themselves. But given the fact that the SS gets removed from the proprotein before GPI anchor attachment and the fact that the mature protein does not carry it, direct experimental proof on the exact mechanisms by which these additional roles may be performed by SS has remained sketchy.

The results described in this manuscript support a model wherein the SS domain of Als5-SS interacts with the rest of the protein and alters both its conformation and function. What could be the implications of these results? As we show from *in silico* analysis, Als5-SS is likely to be completely translocated into the lumen of the ER before the GPI-transamidase works on it to cleave off the SS and attach the pre-formed GPI anchor at its ω-site. As long as the protein remains in the ER and retains its SS, it is likely to be a reasonably well-folded protein with its amyloid domain protected from extraneous interactions by the SS. Thus, this proprotein has no

Table 10.4 Prediction of ΔG_{app} for TM helix insertion of SS from all Als adhesins by ΔG predictor software (http://dgpred.cbr.su.se)

SS of the following adhesins	Predicted ΔG_{app} (kcal/mol)
Als1	+2.347
Als2	+6.641
Als3	+1.689
Als4	+2.530
Als5	+2.955
Als6	+2.955
Als7	+2.045
Als9	+1.838

a

```
Als1    -SIIQHSTWLYGLITLLSLFI--
Als3    -SVIQHSTWLCGLITLLSLFI--
Als4    -SIVQHSGWLYVLLTAISIFF--
Als7    GSVSKYSLWMMAFYMLFGLF-
--
Als9    -SVIKLHSWFYGLVTIFFLFI--
Als5    --ATKHPSWLLKFISVALFFFL-
Als6    --ATKHPSWLLKFISVALFFFL-
Als2    ---NIPATGTSILLENQLSFHTQ
             :    *
```

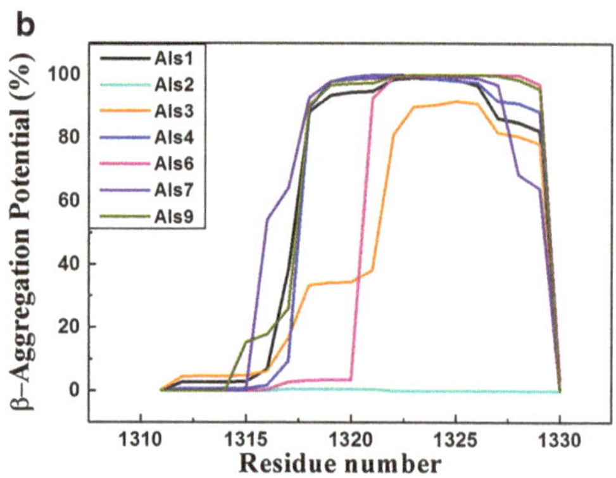

Fig. 10.7 Analysis of sequence similarity and **β**-aggregation potential for SS from all Als family adhesins: (**a**) multiple sequence alignment using Clustal Omega shows poor sequence conservation amongst the SS of the different Als adhesins from *C. albicans*. (**b**) TANGO analysis shows the presence of a **β**-aggregation patch with significant aggregation potential in all Als adhesins with the exception of Als2

tendency to self-aggregate, a property so critical for the functioning of the mature protein at the cell surface. Upon removal of the SS, however, the amyloid patch becomes available for interactions and makes the protein functional. The presence or absence of the SS not only alters the folding of the protein and its functionality, but it could possibly also determine its interactions with other proteins of the GPI biosynthetic machinery, the ER exit sites and those of the secretory pathway. Mutations in the SS, which do not affect attachment of the GPI anchor, could nevertheless alter the conformation of the proprotein sufficiently to modulate these interactions and thereby alter the final fates of the mutant variants.

Is there a bigger picture too? *In silico* analyses suggest that all other members of the Als family are also likely to be have signal sequences that are unlikely to be membrane-inserted by themselves (Table 10.4). In addition, despite poor conservation (Fig. 10.7a), the SSs of all the Als adhesins, with the exception of Als2, have C-terminal ends with >90 % β-aggregation potential (Fig. 10.7b) and a highly conserved amyloid domain (Ramsook et al. 2010). Thus, it is possible these proteins too could possibly fit our model. In other words, for the Als family of adhesins from *C. albicans*, the SS may not only provide the appropriate signal for the transamidase to recognize the site for transfer of the GPI anchor it could also probably provide a mechanism for keeping the proprotein variants from prematurely aggregating in the ER. The absence of conservation in the sequences, therefore, may not only serve to provide differential kinetics of GPI transfer to the Als proteins by the GPI-transamidase it may also be necessitated by the conformational requirements of the proproteins.

Whether this model could be extended beyond the Als family of proteins, remains to be tested. While a general set of rules appear to govern the overall arrangement of residues within the anchors, the significant variations between the sequences and their thermodynamic characteristics would argue against a one-model-fits-all hypothesis. Nevertheless, neither the evidence presented by Nicholson and Stanners (2007) nor that by Guizzunti and Zurzolo (2014) (which provide alternative models to explain how mutations in SS affect localization or functionality of the mature protein), would preclude such an interpretation. The issue in both cases may just as well be one of altered conformation of the proprotein itself resulting in alterations in its downstream interacting partners and therefore alterations in transportation and functionality of the mature protein.

Acknowledgements This work in SSK's lab is supported in parts by funds from DST-PURSE grant to Jawaharlal Nehru University as well as UGC-Resource network grant to the School of Life Sciences. PGM receives a Senior Research Fellowship from UGC.

References

Ahmad MF, Yadav B, Kumar P et al (2012) The GPI anchor signal sequence dictates the folding and functionality of the Als5 adhesin from Candida albicans. PLoS One 7:e35305. doi:10.1371/journal.pone.0035305

Ali S, Hall J, Hazlewood GP et al (1996) A protein targeting signal that functions in polarized epithelial cells in vivo. Biochem J 315(Pt 3):857–862

Ashok A, Hegde RS (2008) Retrotranslocation of prion proteins from the endoplasmic reticulum by preventing GPI signal transamidation. Mol Biol Cell 19:3463–3476. doi:10.1091/mbc.E08-01-0087

Berger J, Howard AD, Brink L et al (1988) COOH-terminal requirements for the correct processing of a phosphatidylinositol-glycan anchored membrane protein. J Biol Chem 263:10016–10021

Böhme U, Cross GAM (2002) Mutational analysis of the variant surface glycoprotein GPI-anchor signal sequence in Trypanosoma brucei. J Cell Sci 115:805–816

Caras IW (1991) An internally positioned signal can direct attachment of a glycophospholipid membrane anchor. J Cell Biol 113:77–85

Caras IW, Weddell GN (1989) Signal peptide for protein secretion directing glycophospholipid membrane anchor attachment. Science 243:1196–1198

Cormack BP, Ghori N, Falkow S (1999) An adhesin of the yeast pathogen Candida glabrata mediating adherence to human epithelial cells. Science 285:578–582

Dalley JA, Bulleid NJ (2003) How does the translocon differentiate between hydrophobic sequences that form part of either a GPI (glycosylphosphatidylinositol)-anchor signal or a stop transfer sequence? Biochem Soc Trans 31:1257–1259

Eisenhaber B, Bork P, Eisenhaber F (1998) Sequence properties of GPI-anchored proteins near the omega-site: constraints for the polypeptide binding site of the putative transamidase. Protein Eng 11:1155–1161

Eisenhaber B, Bork P, Eisenhaber F (1999) Prediction of potential GPI-modification sites in proprotein sequences. J Mol Biol 292:741–758. doi:10.1006/jmbi.1999.3069

Fankhauser N, Mäser P (2005) Identification of GPI anchor attachment signals by a Kohonen self-organizing map. Bioinformatics 21:1846–1852. doi:10.1093/bioinformatics/bti299

Frieman MB, Cormack BP (2004) Multiple sequence signals determine the distribution of glycosylphosphatidylinositol proteins between the plasma membrane and cell wall in Saccharomyces cerevisiae. Microbiology 150:3105–3114. doi:10.1099/mic.0.27420-0

Frieman MB, Cormack BP (2003) The omega-site sequence of glycosylphosphatidylinositol-anchored proteins in Saccharomyces cerevisiae can determine distribution between the membrane and the cell wall. Mol Microbiol 50:883–896

Galian C, Björkholm P, Bulleid N, von Heijne G (2012) Efficient glycosylphosphatidylinositol (GPI) modification of membrane proteins requires a C-terminal anchoring signal of marginal hydrophobicity. J Biol Chem 287:16399–16409. doi:10.1074/jbc.M112.350009

Garg N, Tarleton RL, Mensa-Wilmot K (1997) Proteins with glycosylphosphatidylinositol (GPI) signal sequences have divergent fates during a GPI deficiency. GPIs are essential for nuclear division in Trypanosoma cruzi. J Biol Chem 272:12482–12491

Grimme SJ, Colussi PA, Taron CH, Orlean P (2004) Deficiencies in the essential Smp3 mannosyltransferase block glycosylphosphatidylinositol assembly and lead to defects in growth and cell wall biogenesis in Candida albicans. Microbiology 150:3115–3128. doi:10.1099/mic.0.27254-0

Guizzunti G, Zurzolo C (2014) The fate of PrP GPI-anchor signal peptide is modulated by P238S pathogenic mutation. Traffic 15:78–93. doi:10.1111/tra.12126

Hamada K, Terashima H, Arisawa M et al (1999) Amino acid residues in the omega-minus region participate in cellular localization of yeast glycosylphosphatidylinositol-attached proteins. J Bacteriol 181:3886–3889

Hessa T, Meindl-Beinker NM, Bernsel A et al (2007) Molecular code for transmembrane-helix recognition by the Sec61 translocon. Nature 450:1026–1030. doi:10.1038/nature06387

Hizume M, Kobayashi A, Mizusawa H, Kitamoto T (2010) Amino acid conditions near the GPI anchor attachment site of prion protein for the conversion and the GPI anchoring. Biochem Biophys Res Commun 391:1681–1686. doi:10.1016/j.bbrc.2009.12.128

Hoque MZ, Kitamoto T, Furukawa H et al (1996) Mutation in the prion protein gene at codon 232 in Japanese patients with Creutzfeldt-Jakob disease: a clinicopathological, immunohistochemical and transmission study. Acta Neuropathol 92:441–446

Hoyer LL, Hecht JE (2001) The ALS5 gene of Candida albicans and analysis of the Als5p N-terminal domain. Yeast 18:49–60

Kiachopoulos S, Bracher A, Winklhofer KF, Tatzelt J (2005) Pathogenic mutations located in the hydrophobic core of the prion protein interfere with folding and attachment of the glycosylphosphatidylinositol anchor. J Biol Chem 280:9320–9329. doi:10.1074/jbc.M412525200

Klotz SA, Gaur NK, Lake DF, Chan V, Rauceo J, et al. (2004) Degenerate peptide recognition by Candida albicans adhesins Als5p and Als1p. Infect. Immun. 72, 2029–2034.

Leidich SD, Drapp DA, Orlean P (1994) A conditionally lethal yeast mutant blocked at the first step in glycosyl phosphatidylinositol anchor synthesis. J Biol Chem 269:10193–10196

Martinez-Lopez R, Monteoliva L, Diez-Orejas R et al (2004) The GPI-anchored protein CaEcm33p is required for cell wall integrity, morphogenesis and virulence in Candida albicans. Microbiology 150:3341–3354. doi:10.1099/mic.0.27320-0

McConville MJ, Ferguson MA (1993) The structure, biosynthesis and function of glycosylated phosphatidylinositols in the parasitic protozoa and higher eukaryotes. Biochem J 294(Pt 2):305–324

Miyagawa-Yamaguchi A, Kotani N, Honke K (2014) Expressed glycosylphosphatidylinositol-anchored horseradish peroxidase identifies co-clustering molecules in individual lipid raft domains. PLoS One 9:e93054. doi:10.1371/journal.pone.0093054

Nicholson TB, Stanners CP (2007) Identification of a novel functional specificity signal within the GPI anchor signal sequence of carcinoembryonic antigen. J Cell Biol 177:211–218. doi:10.1083/jcb.200701158

Nisbet RM, Harrison CF, Lawson VA et al (2010) Residues surrounding the glycosylphosphatidylinositol anchor attachment site of PrP modulate prion infection: insight from the resistance of rabbits to prion disease. J Virol 84:6678–6686. doi:10.1128/JVI.02709-09

Ouyang H, Chen X, Lü Y et al (2013) One single basic amino acid at the ω-1 or ω-2 site is a signal that retains glycosylphosphatidylinositol-anchored protein in the plasma membrane of Aspergillus fumigatus. Eukaryot Cell 12:889–899. doi:10.1128/EC.00351-12

Paladino S, Lebreton S, Tivodar S et al (2008) Different GPI-attachment signals affect the oligomerisation of GPI-anchored proteins and their apical sorting. J Cell Sci 121:4001–4007. doi:10.1242/jcs.036038

Paulick MG, Bertozzi CR (2008) The glycosylphosphatidylinositol anchor: a complex membrane-anchoring structure for proteins. Biochemistry (Mosc) 47:6991–7000. doi:10.1021/bi8006324

Pittet M, Conzelmann A (2007) Biosynthesis and function of GPI proteins in the yeast Saccharomyces cerevisiae. Biochim Biophys Acta 1771:405–420. doi:10.1016/j.bbalip.2006.05.015

Poggiolini I, Saverioni D, Parchi P (2013) Prion protein misfolding, strains, and neurotoxicity: an update from studies on mammalian prions. Int J Cell Biol 2013:910314. doi:10.1155/2013/910314

Ramsook CB, Tan C, Garcia MC et al (2010) Yeast cell adhesion molecules have functional amyloid-forming sequences. Eukaryot Cell 9:393–404. doi:10.1128/EC.00068-09

Rauceo JM, De Armond R, Otoo H, Kahn PC, Klotz SA, et al. (2006) Threonine-rich repeats increase fibronectin binding in the Candida albicans adhesin Als5p. Eukaryotic Cell 5, 1664–1673

Staab JF, Bradway SD, Fidel PL, Sundstrom P (1999) Adhesive and mammalian transglutaminase substrate properties of Candida albicans Hwp1. Science 283:1535–1538

Sundstrom P (2002) Adhesion in Candida spp. Cell Microbiol 4:461–469

Wang J, Maziarz K, Ratnam M (1999) Recognition of the carboxyl-terminal signal for GPI modification requires translocation of its hydrophobic domain across the ER membrane. J Mol Biol 286:1303–1310. doi:10.1006/jmbi.1999.2584

Windl O, Giese A, Schulz-Schaeffer W et al (1999) Molecular genetics of human prion diseases in Germany. Hum Genet 105:244–252

Yadav B, Bhatnagar S, Ahmad MF et al (2014) First step of glycosylphosphatidylinositol (GPI) biosynthesis cross-talks with ergosterol biosynthesis and Ras signaling in Candida albicans. J Biol Chem 289:3365–3382. doi:10.1074/jbc.M113.528802

Chapter 11
Novel Chondroitin Sulfate Oligosaccharide Motifs as Biomarkers: Insights into Their Involvement in Brain Development

Kazuyuki Sugahara

Abbreviations

2AB	2-Aminobenzamide
C4ST-1	Chondroitin 4-*O*-sulfotransferase 1
C4ST-2	Chondroitin 4-*O*-sulfotransferase 2
CS	Chondroitin sulfate
D4ST-1	Dermatan 4-*O*-sulfotransferase
DS	Dermatan sulfate
GAG	Glycosaminoglycan
GalNAc	*N*-Acetyl-D-galactosamine
GalNAc4S-6ST	GalNAc-4-sulfate 6-*O*-sulfotransferase
GlcUA	D-Glucuronic acid
IdoUA	L-Iduronic acid
mAb	Monoclonal antibody
PG	Proteoglycan
UST	Uronyl 2-*O*-sulfotransferase

K. Sugahara, Ph.D. (✉)
Faculty of Advanced Life Science Frontier Research Center for Post-Genomic Science and Technology 5F, North-21 West-11 Kita-ku, Sapporo 001-0021, Japan
e-mail: k-sugar@sci.hokudai.ac.jp

A. Chakrabarti, A. Surolia (eds.), *Biochemical Roles of Eukaryotic Cell Surface Macromolecules*, Advances in Experimental Medicine and Biology 842,
DOI 10.1007/978-3-319-11280-0_11

Introduction

Evidence is accumulating to suggest the various biological functions of the glycosaminoglycan (GAG) side chains of proteoglycans (PGs), which have largely been classified into chondroitin sulfate (CS), dermatan sulfate (DS) and heparan sulfate. These are known to be involved in cell division, cell growth, proliferation, differentiation, migration, neoplastic transformation, invasion, cell-cell adhesion, tissue formation, morphogenesis, development, neural network formation, and viral infection. These events mainly occur through interactions between the GAG side chains and a wide variety of functional proteins such as growth factors, cytokines, and neurotrophic factors. These effector proteins have been shown to specifically interact with the functional domain oligosaccharide sequences embedded in long linear GAG chains (Kjellén and Lindahl 1991; Perrimon and Bernfield 2001; Casu and Lindahl 2001; Sugahara et al. 2003; Sugahara and Mikami 2007; Bishop et al. 2007; Lindahl 2014), which regulates their functions. This concept of domain structures is well established for heparin and heparan sulfate chains including prototypical antithrombin-binding and basic fibroblast factor-binding pentasaccharides (Thunberg et al. 1982; Atha et al. 1987; Petitou and van Boeckel 2004). Although a large number of ligand proteins have been identified for heparin and heparan sulfate, studies on the functional domain sequences of CS or DS chains and their protein ligands are limited. However, similar situations are expected for CS, DS and CS-DS hybrid chains based on recent findings. CS-DS hybrid chains are composed of more than ten building units (Fig. 11.1a (Purushothaman et al. 2011)), and these can generate numerous polymer sequences by various arrangements. The sequences constructed by these units can be as diverse as peptide sequences generated by a combination of 20 amino acid residues. The developmentally regulated and cell type-specific expression of distinct sulfated GAG structures on cell surface PGs may provide a large amount of information relevant to cell-cell interactions and differentiation in developing organisms and various tissues including the brain. However, unlike nucleic acids and proteins, an automated, and widely applicable, sequencing method has not yet been established for GAGs due to the complexity of sugar sequences. Nonetheless, some functional fragments have been isolated and sequenced using elaborate techniques. In this chapter, the relationship between the structure and function of brain CS-DS has mainly been discussed based on our recent findings.

Developmental Changes in the Structures of CS Chains in the Brain

Disaccharide composition analysis of a fraction containing CS and DS prepared from chick whole brains revealed that composition of disaccharides changed during embryonic development (Fig. 11.2) (Kitagawa et al. 1997). The findings revealed

a

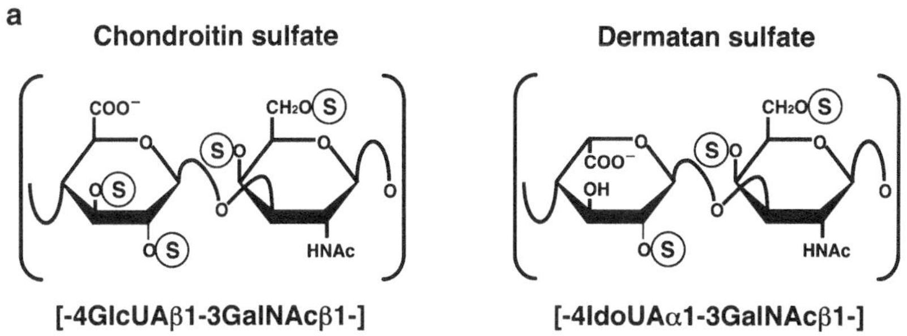

Symbol	Sequence
O unit	GlcUA-GalNAc
A unit	GlcUA-GalNAc(4S)
C unit	GlcUA-GalNAc(6S)
B unit	GlcUA(2S)-GalNAc(4S)
D unit	GlcUA(2S)-GalNAc(6S)
E unit	GlcUA-GalNAc(4S,6S)

Symbol	Sequence
iO unit	IdoUA-GalNAc
iA unit	IdoUA-GalNAc(4S)
iC unit	IdoUA-GalNAc(6S)
iB unit	IdoUA(2S)-GalNAc(4S)
iD unit	IdoUA(2S)-GalNAc(6S)
iE unit	IdoUA-GalNAc(4S,6S)

b

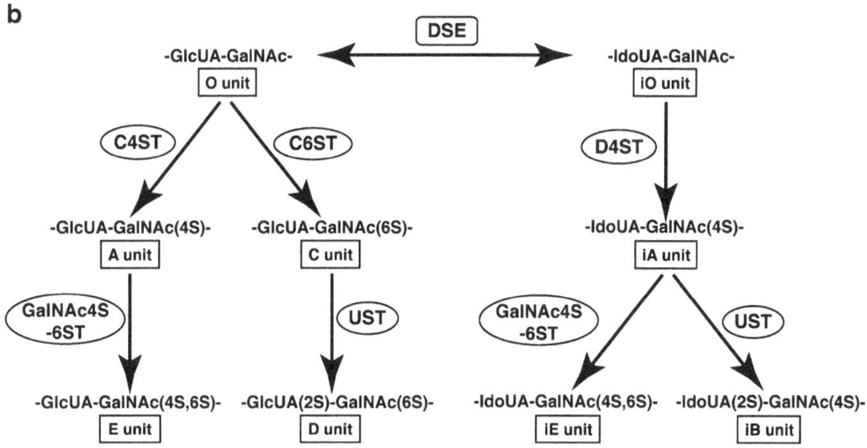

Fig. 11.1 Structures of CS and DS disaccharide units and their biosynthetic pathways. (**a**) The repeating disaccharide units of CS are composed of GlcUA and GalNAc residues (*left*). GlcUA often undergoes epimerization to form IdoUA and generates DS disaccharide units (*right*) along the CS chains that form CS-DS hybrid chains. CS units are traditionally named, and the corresponding DS units are indicated by "i", which stands for IdoUA. Sulfation of the sugar residues, indicated by "S," was shown to occur at the second, fourth, and sixth carbon positions in the ring (Perrimon and Bernfield 2001). Namely, 2S, 4S, and 6S represent the 2-*O*-, 4-*O*-, and 6-*O*-sulfate groups, respectively. Note that the 'B' unit, which is rich in CS-B (DS), conventionally represents IdoUA(2S)α1-3GalNAc(4S) (Sugahara et al. 2003), but also stands for GlcUA(2S)β1-3GalN Anc(4S) in this coding system (Sugahara and Mikami 2007). Depending on the position of sulfation, there are six main disaccharide units for both CS and DS, respectively, which together generate numerous CS-DS polymer sequences based on their different combinations. The large diversity thus generated by the various disaccharide units can result in the formation of various sets of "wobble motifs" along the CS-DS hybrid chains. Several chondroitin lyases, which were designated as chondroitinases, depolymerized CS, DS, or CS-DS polymers to produce unsaturated disaccharides containing unsaturated hexuronic acid, abbreviated as ΔHexUA, from the respective parent disaccharides (for example, ΔA from A and iA units). This figure was taken from (Purushothaman et al. 2011). (**b**) Biosynthetic pathways of CS/DS chains. *C4ST* chondroitin 4-*O*-sulfotransferase, *C6ST* chondroitin 6-*O*-sulfotransferase, *D4ST* dermatan 4-*O*-sulfotransferase, *epimerase* glucuronyl C5-epimerase, *GalNAc4S-6ST* GalNAc 4-sulfate 6-*O*-sulfotransferase, *UST* uronyl 2-*O*-sulfotransferase, *epimerase* glucuronyl C5-epimerase, *UST* uronyl 2-*O*-sulfotransferase

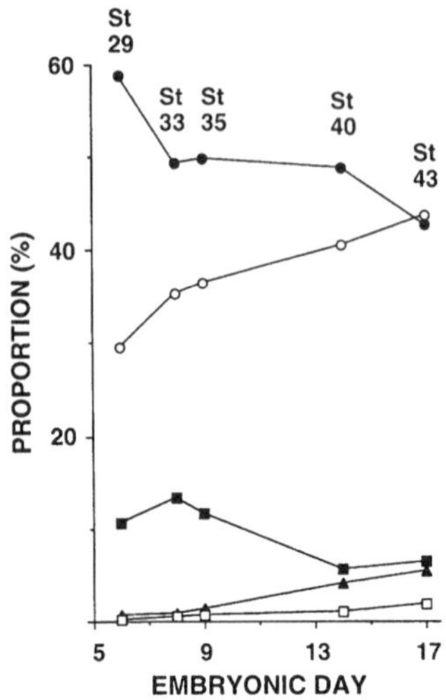

Fig. 11.2 Developmental changes in the proportion of disaccharide units in the CS-DS chains of embryonic chick brains. The PG fractions prepared from chick embryo brains at various developmental stages (stages 29–43) were digested with chondroitinase ABC. The digests were derivatized with the fluorophore 2AB to increase the sensitivity for detection, and then analyzed by HPLC to determine the amount of disaccharide units. ΔDi-6S (*filled circle*), ΔDi-4S (*open circle*), ΔDi-0S (*filled square*), ΔDi-diS$_D$ (*filled triangle*), and ΔDi-diS$_E$ (*open square*). The sum of the disaccharide units in each sample was taken as 100 %. This figure was taken from Kitagawa et al. (1997)

that although the main disaccharides detected were 4-*O*-sulfated and 6-*O*-sulfated units, appreciable proportions of the non-sulfated and disulfated disaccharide units D and E were also present, and also that the proportions of different units changed during embryonic development. Immunohistochemical studies using sagittal sections from postnatal mouse brains also indicated spatiotemporal changes in the structures of CS chains when sections were stained with various CS monoclonal antibodies (mAbs) (Fig. 11.3) (Maeda et al. 2003).

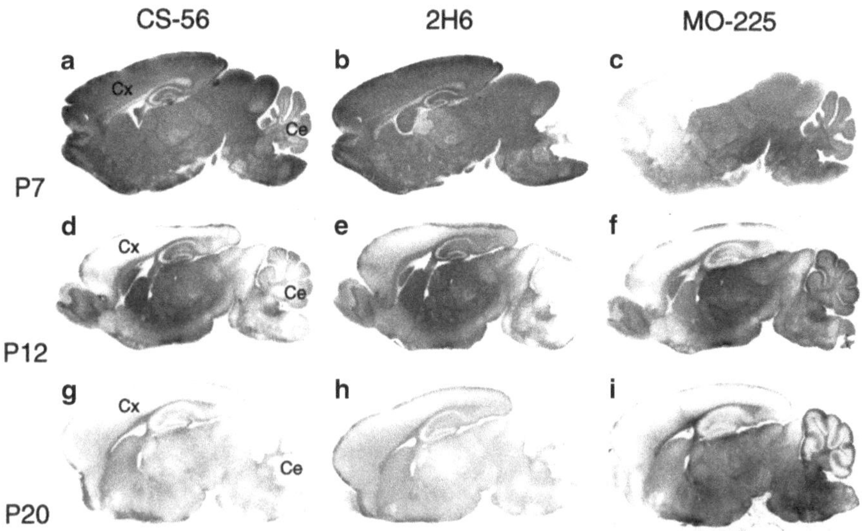

Fig. 11.3 Immunohistochemical localization of CS/DS epitopes in the postnatal mouse brain. Sagittal brain sections from postnatal day (P) 7 (**a**–**c**), P12 (**d**–**f**), and P20 (**g**–**i**) mice were immunostained with mAbs CS-56 (**a**, **d**, and **g**), 2H6 (**b**, **e**, and **h**), and MO-225 (**c**, **f**, and **i**) (Maeda et al. 2003). The respective staining patterns in the cerebral cortex (Cx) and cerebellum (Ce) changed during development. This figure was taken from Maeda et al. (2003)

Epitope Oligosaccharide Sequences of mAbs That Recognize Brain CS Chains

The mAbs used in the study described above included CS56, 2H6, and MO-225, the epitopes of which have been characterized as a set of multiple octasaccharide sequences (Table 11.1) (Sugahara and Mikami 2007). The epitopes of CS56 and MO-225 have the A-D tetrasaccharide sequence in common, whereas 2H6 does not always require the D unit and recognizes less sulfated octasaccharides composed of only four C units, as well as three C units plus one A or D unit at different positions in the sequences (Sugahara and Mikami 2007). Staining of the cerebellum in particular was strong and revealed dynamic changes during postnatal development.

CS-DS hybrid chains, which were derived from embryonic pig brains, exhibited neuritogenic and growth factor-binding activities, which was in contrast to those from adult pig brains and depended on their IdoUA content defining the DS-like structure, as described below (Fig. 11.4) (Bao et al. 2004). In biosynthesis, CS/DS chains are initiated by the transfer of a xylose (Xyl) residue from UDP-Xyl to a specific Ser residue in the GAG-attachment site of core proteins. It serves as acceptor of two Gal residues and a GlcUA residue to result in the so-called GAG-protein linkage region tetrasaccharide (GlcUA-Gal-Gal-Xyl-*O*-). Chain polymerization of

Table 11.1 The oligosaccharide sequences recognized by various mAbs that are used to stain mouse brain sections

Monoclonal antibodies				
	473HD	CS56	MO-225	2H1
Hexasaccharides	D-A-D[a]		D-A-D	
	D-A-A			
	C-A-D			
	(D-C-A)[b]			
Octasaccharides	[c]ΔC-A-D-C	ΔC-A-D-C	ΔC-A-D-C	C-C-A-C
	ΔC-A-D-A	ΔC-A-D-A	ΔC-A-D-A	C-C-C-C
	ΔA-A-D-C	ΔA-A-D-C	ΔA-A-D-C	C-C-C-A
	ΔA-A-D-A	ΔA-A-D-A	ΔA-A-D-A	D-C-C-C
		C-A-D-C	C-A-D-C	C-C-C-D
		C-C-A-D		(C-E-C-C)[b]
		A-D-C-C		(D-C-C-A)[b]
Decasaccharides			E-E-E-E-C	

Oligosaccharides with or without D were prepared after digestion with chondroitin lyases or testicular hyaluronidase, which is a hydrolase. The former contained a 4,5-unsaturated hexuronic acid residue (ΔHexUA) at their non-reducing termini, whereas the latter did not

[a]The underlined A-D or D-A tetrasaccharide sequences were common to the sequences recognized by mAbs 473HD, CS-56, and MO-225

[b]Reactivity toward the hexa- and octasaccharides in parentheses remains to be confirmed

[c]**Δ** represents the 4,5-unsaturated bond of unsaturated hexuronic acid (ΔHexUA) formed by the action of bacterial chondroitin lyases

the repeating disaccharide region in CS and DS chains is initiated by the transfer of the first GalNAc from UDP-GalNAc to the GlcUA residue in the linkage region tetrasaccharide by β1,4-GalNAc transferase-I (Sugahara et al. 2003), whereas the transfer of a GlcNAc residue from UDP-GlcNAc to the linkage region tetrasaccharide by α1,4-GlcNAc transferase-I is known to result in the initiation of the repeating disaccharide region of heparan sulfate and heparin chains (Sugahara et al. 2003). Six chondroitin synthase family members have been identified including chondroitin synthase, chondroitin-polymerizing factor, and CS GalNAc transferases. Chondroitin synthase is a bifunctional glycosyltransferase that exhibits CS GlcUA transferase-II and CS GalNAc transferase-II activities, which are required for the biosynthesis of the repeating disaccharide region, –4GlcUAβ1–3GalNAcβ1. However, chondroitin synthase itself is unable to construct the backbone of CS by the activity of polymerase, whereas the enzyme complex of chondroitin synthase with chondroitin-polymerizing factor can form the repeating disaccharide region, exhibiting chondroitin polymerizing activity. A precursor of CS, the chondroitin backbone, is then maturated by sulfation modified by various sulfotransferases such as uronosyl 2-*O*-sulfotransferase, chondroitin 4-*O*-sulfotransferases, chondroitin 6-*O*-sulfotransferase, and GalNAc 4-sulfate 6-*O*-sulfotransferase (Fig. 11.1b) (Sugahara et al. 2003). Epimerization of the C-5 position of GlcUA residues in a chondroitin polymer occurs as a precursor backbone is constructed, which results in

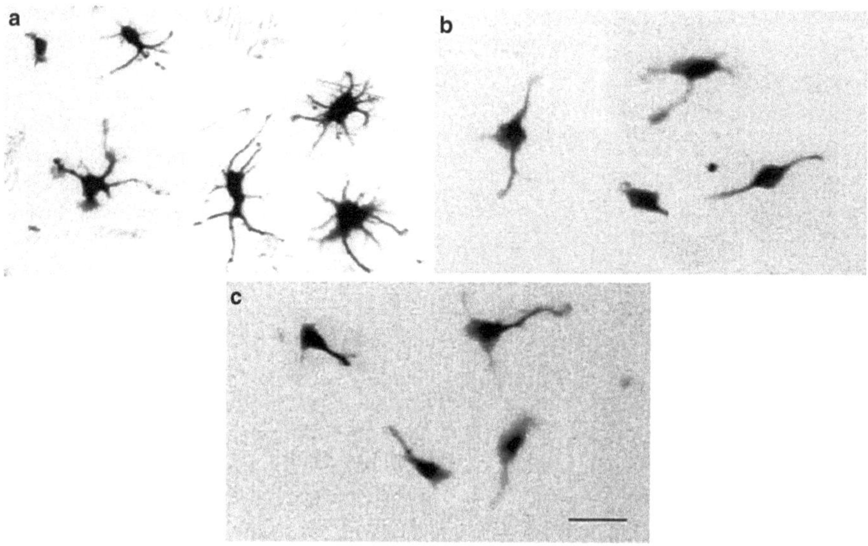

Fig. 11.4 Differential neurite outgrowth-promoting activities of the CS/DS chains from embryonic and adult pig brains toward hippocampal neurons. E16 mouse hippocampal neurons were cultured for 24 h on plastic coverslips coated with poly-D,L-ornithin and then with CS-DS derived from embryonic pig brains (**a**), or with CS/DS derived from adult pig brains (**b**). These cells were fixed and stained for microtubule-associated protein 2 and neurofilament. In the control experiments, neurons were cultured on coverslips coated with poly-D,L-ornithin alone (**c**). An increase was observed in the number of neurites for the neurons cultured on the substrate prepared with CS/DS from embryonic pig brains, but was not for the substrate prepared with CS/DS from adult pig brains. Scale bar, 20 μm. This figure was taken from Bao et al. (2004)

the formation of the repeating disaccharide region, –4IdoUAα1–3GalNAcβ1-, of DS chains. Depending on the degree of epimerization CS-DS hybrid chains with various proportions of GlcUA and IdoUA are formed. The dermatan chains fully develop through sulfation catalyzed by dermatan 4-*O*-sulfotransferases and uronosyl 2-*O*-sulfotransferase.

The expression of the sulfotransferases responsible for the synthesis of different disaccharide units was also examined to elucidate the distribution of the synthetic machinery of such functional sugar chains during brain development (Mitsunaga et al. 2006). *In situ* hybridization was performed to examine the expression of three CS and DS GalNAc 4-*O*-sulfotransferases; dermatan 4-*O*-sulfotransferase (D4ST-1), chondroitin 4-*O*-sulfotransferase 1 (C4ST-1), and chondroitin 4-*O*-sulfotransferase 2 (C4ST-2), as well as the uronyl 2-*O*-sulfotransferase (UST), which was involved in the biosynthesis of DS in addition to CS intermediates (Fig. 11.1b). C4ST-1 and C4ST-2 were ubiquitously expressed in the postnatal mouse brain, whereas the expression of D4ST-1 and UST was restricted in the developing cerebellum and peaked on postnatal day 14, as shown by reverse transcriptase-PCR analysis (data not shown). *In situ* analysis of the disaccharides of CS-DS in brain sections revealed that the concentration of CS-DS increased twofold during development (postnatal day 7–7 weeks). The proportions of DS-specific, principal disaccharides, IdoUA-GalNAc(4-*O*-sulfate) (iA), and IdoUA(2-*O*-sulfate)-GalNAc(4-*O*-sulfate) (iB), which were produced by the sequential actions of D4ST-1 and UST, were higher in CS-DS chains from the cerebellum than in those from whole brain sections. The marked dramatic increase (tenfold) observed in the proportion of iB during development was important. In contrast, GlcUA/IdoUA(2-*O*-sulfate)-GalNAc(6-*O*-sulfate) (D/iD) and GlcUA/IdoUA-GalNAc(4,6-*O*-disulfate) (E/iE) decreased to 50 and 30 %, respectively, in the developing cerebellum. These findings suggested that the IdoUA-containing iA and iB units along with D/iD and E/iE units in the CS-DS hybrid may play important roles in the formation of the cerebellar neural network during postnatal brain development.

The critical importance of the disulfate disaccharide units E and/or iE has also been suggested based on the findings of immunohistochemical and *in situ* hybridization studies (Purushothaman et al. 2007). An evaluation of the specificity of GD3G7 toward various GAG preparations revealed that this antibody specifically reacted with squid cartilage-derived CS-E rich in E-disaccharide units [GlcUAβ1–3GalNAc (4,6-*O*-sulfate)], hagfish notochord-derived CS-H (or alternatively designated DS-E) rich in iE units [IdoUAα1–3GalNAc(4,6-*O*-sulfate)], and shark skin DS rich in both E and iE units. The immunohistochemical localization of the GD3G7 epitope in the mouse brain demonstrated that the epitope was relatively abundant in the cerebellum, hippocampus, and olfactory bulb (data not shown). GD3G7 suppressed the growth of neurites in embryonic hippocampal neurons mediated by CS-E when added to embryonic mouse hippocampal cell cultures (see below), which suggested that the epitope may be embedded in the neurite outgrowth-promoting motif of CS-E.

In situ hybridization for the expression of GalNAc-4-sulfate 6-*O*-sulfotransferase (GalNAc4S-6ST) in the postnatal mouse brain, which is responsible for the biosynthesis of CS/DS-E by transferring sulfate to the C6 position of GalNAc-4-*O*-sulfate of A and iA units (Fig. 11.1b), showed the widespread expression of the transcript in the developing brain, except on postnatal day 7 when the transcript was strongly expressed in the external granule cell layer of the cerebellum (Fig. 11.5) (Purushothaman et al. 2007). The expression of GalNAc4S-6ST switched from the external to internal granule cell layer with development. In addition, a CS-E decasaccharide fraction was found to be the critical minimal structure required for recognition by GD3G7. Four to five discrete decasaccharide epitopic sequences were identified, whereas none of the octasaccharides containing E-units were recognized by GD3G7 (Table 11.2) (Purushothaman et al. 2007; Deepa et al. 2007). Taken together, these findings demonstrated the dynamic changes and structure–function relationship of CS-DS during brain development.

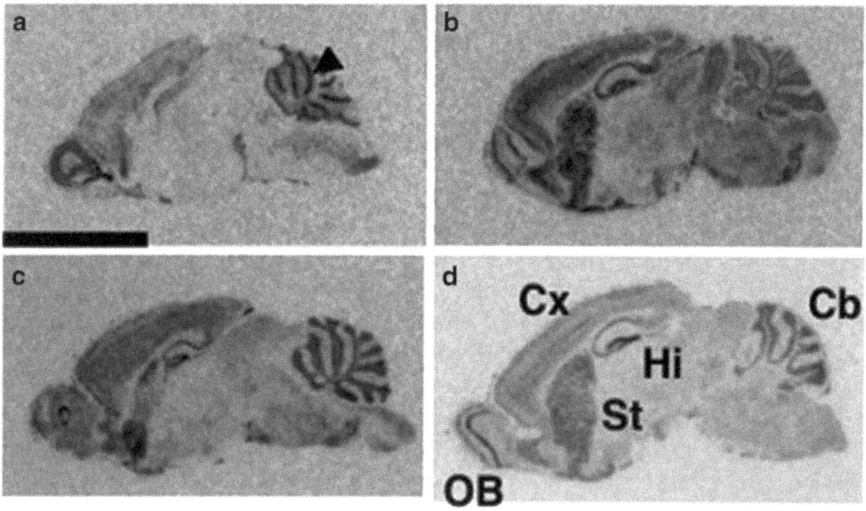

Fig. 11.5 *In situ* hybridization for GalNAc4S-6ST in the mouse brain during postnatal development. Consecutive sagittal sections of the brain from P7 (**a**), P14 (**b**), P21 (**c**), and 7-week-old (**d**) mice were hybridized with ^{35}S-labeled antisense (**a**–**d**) cRNA probes for GalNAc4S-6ST. The expression of the GalNAc4S-6ST transcript was widespread in the developing brain, except at P7 when its expression was strong in the external granule cell layer (*arrowhead*) of the cerebellum. Its expression in the granule cell layer switched to the internal region and appeared to decrease slightly with development. The strong expression of GalNAc4S-6ST was also observed in the olfactory bulb, hippocampus, cerebral cortex, and striatum of the developing brain. *OB* olfactory bulb, *Cx* cerebral cortex, *Hi* hippocampus, *Cb* cerebellum, *St* striatum. Scale bar, 5 mm. This figure was taken from Purushothaman et al. (2007)

Table 11.2 Oligosaccharide sequences that were or were not recognized by the single chain phage display antibody, GD3G7, that is used to stain mouse brain sections

Decasaccharides that were recognized by GD3G7	Octasaccharides that were not recognized by GD3G7
E-E-E-E-E	E-E-E-E
E-E-E-E-A	E-E-E-A
E-E-E-E-C	E-C-E-E
E-E-E-A-A and/or C-E-E-E-A	A-E-E-E

These oligosaccharide fractions were isolated from CS chains derived from squid cartilage, sequenced (Purushothaman et al. 2007), and were then utilized to determine the binding specificity of the antibody GD3G7 (Mitsunaga et al. 2006)

Neurite Outgrowth-Promoting Activity of Oversulfated CS Chains

In 1994, Faissner et al. purified CS-PG from postnatal days 7–14 mouse brain extracts, which was designated DSD-1-PG, carry CS-DS hybrid chains, and are expressed in the central nervous system (Faissner et al. 1994). A mAb against DSD-1-PG, which was specific to the CS-DS side chains, was also developed and designated 473HD. DSD-1-PG exhibited neurite outgrowth-promoting properties towards embryonic day 14 mesencephalic and embryonic day 18 hippocampal neurons from the rat. These properties were attributed to the CS-DS side chains of DSD-1-PG, because they were lost when the CS-DS chains were removed following a treatment with chondroitinases, which remove CS and/or DS chains, and could be blocked by 473HD. These findings demonstrated that the hybrid GAG structure DSD-1-PG supported the morphological differentiation of neurons in the central nervous system. The epitopes of 473HD was subsequently characterized as a series of octasaccharides containing the A-D tetrasaccharide sequence (Table 11.1) (Sugahara and Mikami 2007; Ito et al. 2005).

Oversulfated CS-D and CS-E were subsequently reported to promote the neurite outgrowth of embryonic hippocampal neurons (Nadanaka et al. 1988; Clement et al. 1998; 1999). CS-D and CS-E stimulated the outgrowth of multiple dendritic neurites and a single axon-like neurite, respectively, which suggested distinct molecular mechanisms underlying these phenomena. However, since CS-D and CS-E are derived from the cartilages of shark and squid, respectively, CS and DS chains in mammalian brains were also characterized as described below.

Isolation of Functional Domain Oligosaccharides from Embryonic Pig Brains

The CS-DS hybrid chains were purified from embryonic pig brains, and exhibited prominent neuritogenic (Fig. 11.4a) and growth factor-binding activities toward various growth factors including fibroblast growth factor 2, pleiotrophin (PTN), and midkine, all of which have neuroregulatory activities in the brain. In contrast, the CS-DS preparation derived from adult pig brains exhibited markedly less binding activity to these growth factors and neuritogenic activity was absent (Fig. 11.4b) (Bao et al. 2004). Structural analysis indicated that the average sizes of the CS-DS chains were similar (40 kDa) in these two preparations. However, the compositions of the disaccharides were markedly different, with a significant proportion of IdoUA-containing disaccharides (8–9 %) being observed in the CS/DS chains from embryos but not in those from adults (<1 %). Both the neurite outgrowth-promoting and growth factor-binding activities of the CS-DS chains from embryos were abolished by digestion not only with chondroitinase ABC, but also chondroitinase B, which is specific to the DS structure, suggesting that the IdoUA-containing motifs may be essential for these activities (Bao et al. 2004). These findings implied that the temporal expression of CS-DS hybrid structures containing both GlcUA and IdoUA and binding activities toward various growth factors play important roles in neurogenesis in the early stages of brain development.

Mechanism Underlying the Neurite Outgrowth-Promoting Activities of CS Chains

The molecular mechanism underlying the neuritogenic activities of CS chains have not yet been investigated in detail. An affinity column was prepared by immobilizing purified CS-DS hybrid chains purified from embryonic pig brains to identify the main growth factor involved in this neurite outgrowth-promoting activity. PTN, a heparin-binding growth factor that is mainly produced by glia cells, was identified as the predominant binding partner in the membrane fraction of the neonatal rat brain for the CS-DS chains derived from embryonic pig brains (Bao et al. 2005a). CS-DS chains from embryonic pig brains were then separated on a PTN-immobilized affinity column into unbound, low affinity, and high affinity fractions by stepwise elution using salts. The latter two fractions promoted the outgrowth of dendrite- and axon-like neurites, respectively, whereas the unbound fraction did not. The activity of the low affinity fraction was abolished by an anti-PTN antibody or when glia cells were removed from the culture. In contrast, the high affinity fraction displayed activity under both these conditions. Hence, PTN mainly from glia cells mediated activity in the low affinity, but not high affinity fraction. The anti-CS-DS antibody 473HD neutralized the neuritogenic activities of both fractions. Interaction analysis indicated that the 473HD epitope and PTN-binding domains in the CS-DS chains from embryonic pig brains largely overlapped (Bao et al. 2005a). The composition of disaccharides and distribution of IdoUA-containing disaccharides along the chains differed among the three affinity subfractions. Oversulfated disaccharides and nonconsecutive IdoUA-containing units were required for the CS-DS hybrid chains to bind PTN and exhibit neuritogenic activities. Thus, CS subpopulations with distinct structures in the mammalian brain play different roles in neuritogenesis through distinct molecular mechanisms, at least in part, by regulating the functions of growth factors.

Isolation of Functional Domain Oligosaccharides from the CS-DS Hybrid Chains Purified from Embryonic Pig Brains

The CS-DS hybrid chain fractions purified from embryonic pig brains were digested with chondroitinase ABC, AC-I, or B, which degrade both CS and DS, CS alone, or DS alone, respectively. The PTN-binding activity of each digest was then evaluated. The results obtained showed that the former two enzymes almost completely lost the PTN-binding activity of the CS-DS chains, whereas chondroitinase B, which cleaved IdoUA-GalNAc, but not GlcUA-GalNAc linkages, only partially (by approximately 35 %) reduced the PTN-binding activity of the CS-DS chains. Hence, the CS-DS fraction was exhaustively digested with chondroitinase B to isolate the PTN-binding domain, and the digest was then fractionated by gel filtration. The minimum size fraction that showed inhibitory activity against the binding of PTN to the parent CS-DS chain fraction was the octasaccharide fraction, which suggested that PTN bound to octasaccharides that were resistant to chondroitinase B. This fraction was labeled with 2-aminobenzamide (2AB) to increase the sensitivity for detection. It was then fractionated by affinity chromatography using a PTN-immobilized column, which bound a small portion of the CS-DS fraction, and was eluted with 0.2 M NaCl to separate PTN-bound from PTN-unbound oligosaccharides. The PTN-bound and -unbound fractions were further purified individually by strong anion-exchange chromatography, which yielded a series of 2AB-labeled oligosaccharide peaks.

Each peak was structurally characterized first by MALDITOF-MS analysis, which was performed using a basic peptide $(\text{Arg-Gly})_{15}$ to prepare a noncovalent complex with octasaccharides (2–4 pmol of individual octasaccharides, or 10–20 pmol of mixed octasaccharide fractions) (Bao et al. 2005b). The molecular mass of each octasaccharide component was calculated by subtracting the measured m/z value of the protonated peptide from that of the protonated peptide/octasaccharide complex, which unequivocally revealed the amino sugar and uronic acid composition as well as the number of sulfate groups of each octasaccharide component. Secondly, enzymatic digestions in conjunction with HPLC, which revealed the disaccharide composition (molar ratio of different disaccharide units) of the unsaturated octasaccharides in the main component of each octasaccharide fraction.

Each 2AB-labeled octasaccharide fraction was then digested with either chondroitinase ABC or AC-II to yield 2AB-labeled tetra- or disaccharides, respectively. The fragments produced were directly analyzed by strong anion-exchange chromatography to identify the structure of each fragment according to authentic standards. Each enzyme digest was re-labeled with 2AB and subjected to anion-exchange HPLC in order to identify the released di- or tetrasaccharide fragments. The sequences thus determined are summarized in Table 11.3 (Bao et al. 2005b). All the PTN-bound octasaccharides contained at least one D-unit, whereas none of the PTN-unbound octasaccharides contained a D-unit. These findings suggested the critical importance of the D-disaccharide unit for the binding of PTN to CS-DS hybrid chains. Since chondroitinase B was used in the digestion to prepared the octasaccharides, the unsaturated uronic acid (ΔHexUA) at the non-reducing end of each octasaccharide had to be derived from IdoUA, and the degraded disaccharide unit, which was located immediately next to the reducing end of each octasaccharide, had to be bound to an IdoUA-containing disaccharide in the CS-DS hybrid chains. Another important structural feature was that although the content of D-units in embryonic pig brains was less than 2 %, as reported previously (Bao et al. 2005b), they were concentrated in the PTN-binding region. A previous study proposed that the PTN-binding sequences and epitope sequences of mAb 473HD most likely overlapped in the CS-DS hybrid chains in the embryonic brain (Bao et al. 2005a).

Table 11.3 PTN-bound and PTN-unbound octasaccharide sequences isolated from embryonic pig brains

PTN-bound octasaccharides	PTN-unbound octasaccharides
ΔC-C-D-C	ΔC-O-C-C
ΔA-C-D-C or ΔC-A-D-C	ΔA-O-A-C
ΔD-C-D-C	ΔA-O-A-A
ΔC-D-D-C or ΔC-D-iD-C	ΔC-C-C-C
ΔE-D-A-D or ΔE-D-iA-D	ΔA-C-C-C
	ΔC-C-A-C
	ΔA-C-A-C
	ΔA-C-C-A
	ΔC-A-A-A

The CS-DS fraction was prepared from embryonic pig brains after extraction with a EDTD-containing buffer, followed by digestion with protease (actinase E) and the removal of heparan sulfate and hyaluronan by enzymatic digestions. The CS-DS preparation was fractionated into PTN-bound and -unbound fractions using a PTN-immobilized affinity column. The PTN-bound fraction was exhaustively digested with chondroitinase B, which specifically cleaved galactosaminidic linkages bound to IdoUA residues, but not those bound to GlcUA residues, and the digest was then fractionated by gel-filtration into various oligosaccharides based on size (Bao et al. 2005b). The octa- and larger oligosaccharide fractions exhibited PTN-binding activities. The octasaccharide fraction was further fractionated by the PTN-column into PTN-bound and unbound fractions by strong anion-exchange chromatography. Each separated fraction was labeled with the fluorophore 2AB and then, subjected to sequencing by MALDI-TOF-MS and enzymatic fragmentation in conjunction with anion-exchange chromatography, as reported previously (Bao et al. 2005b). These octasaccharides were assumed to have been flanked by two IdoUA in the parent CS-DS hybrid chains because they were prepared after digestion with chondroitinase B

Novel Highly Sulfated Hexasaccharide Sequences Isolated from CS Derived from Shark Fin Cartilage

The findings of the above-described and other studies (Tanaka et al. 2003) strongly suggested the importance of the roles of D-disaccharide units in brain development, and also that CS chains containing D-units were involved in the development of cerebellar Purkinje cells and neurite outgrowth-promoting activity by interacting with the neurotrophic factor, PTN, which regulates signaling. Hence, to obtain further structural information on the CS chains containing D-disaccharide units, oligosaccharides containing D-units were isolated from shark fin cartilage, which is known to be rich in D-units. Seven novel hexasaccharide sequences: ΔO-D-D, ΔA-D-D, ΔC-D-D, ΔE-A-D, ΔD-D-C, ΔE-D-D and ΔA-B-D, in addition to three previously reported sequences: ΔC-A-D, ΔC-D-C and ΔA-D-A, were isolated from a CS preparation of shark fin cartilage following exhaustive digestion with

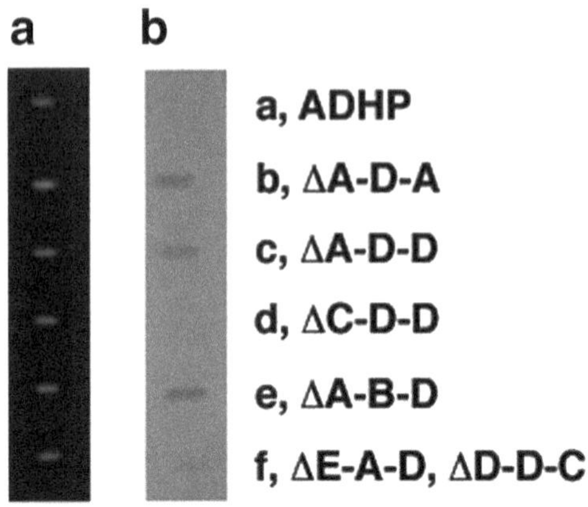

Fig. 11.6 Reactivities of isolated hexasaccharide fractions with mAb MO-225. The isolated hexasaccharide fractions a–f were individually labeled with the fluorescent lipid, ADHP. ADHP-labeled hexasaccharide fractions (2.5 pmol/spot) were applied onto a nitrocellulose membrane using a sample applicator and were then photographed under ultraviolet light (254 nm) to confirm the immobilization of similar amounts of neoglycolipids. (**a**) The nitrocellulose membrane carrying the ADHP-derivatized hexasaccharides was then probed with the mAb MO-225, followed by detection with mouse anti-IgM conjugated with horseradish peroxidase and 3,3′-diaminobenzidine. (**b**) Free ADHP (2.5 pmol) was used as a negative control. This figure was taken from Mizumoto et al. (2012)

chondroitinase AC-I, which cannot act on galactosaminidic linkages adjacent to the D-units (Table 11.5) (Mizumoto et al. 2012).

Of these, relatively abundant hexasaccharides could be evaluated for their binding activity to mAb MO-225, which had been used to stain mouse brain sections. The results obtained revealed significant binding capacities of ΔA-D-D, ΔA-D-D, and ΔA-B-D, as well as the weak activity of a mixture fraction containing ΔE-A-D and ΔD-D-C, whereas ΔC-D-D showed no binding activity even though it contained two D-units (Fig. 11.6) (Mizumoto et al. 2012). These results confirmed the importance of the underlined trisaccharide moiety of the A-D tetrasaccharide sequence GlcUA-GalNAc(4-*O*-sulfate)-GlcUA(2-*O*-sulfate)-GalNAc(6-*O*-sulfate) for recognition by mAb MO-225, as described above (Mizumoto et al. 2012).

Importance of the Three-Dimensional Structure and Electrostatic Potential Distribution at the Molecular Surfaces of Functional Oligosaccharides

mAbs, which are assumed to be specific to a single epitope structure, recognized multiple similar, yet distinct oligosaccharide sequences, it was hypothesized that these oligosaccharides may have similar conformations. Therefore, conformational

Table. 11.4 Major hexasaccharde sequences resistant to chondroitinase AC-I

Major hexasaccharde sequences in chondroitinase	AC-I-resistant fractions
ΔC-A-D[a]	ΔO-D-D
ΔC-D-C[a]	ΔA-D-A[a]
ΔA-D-D	ΔC-D-D
ΔE-A-D	ΔD-D-C
ΔA-B-D	ΔE-D-D

These unsaturated hexasaccharides were isolated and sequenced after the exhaustive digestion of a CS preparation derived from shark fin cartilage with chondroitinase AC-I, which cannot cleave galactosaminidic linkages adjacent to D-units, and were then sequenced, as reported previously (Mizumoto et al. 2012). Of these sequences, seven have not yet been reported (Mizumoto et al. 2012)

[a]Three were previously reported (see Mizumoto et al. (2012))

Table 11.5 CS octasaccharide sequences that were or were not recognized by the mAb WF6

Octasaccharides recognized by WF6	Octasaccharides not recognized by WF6
ΔD-C-C-C	ΔC-A-D-C
ΔC-C-A-D	ΔC-C-C-C

These octasaccharides were isolated from CS chains from shark cartilage after partial digestion with chondroitinase ABC, followed by gel filtration and anion-exchange chromatography. The sequential arrangement of the disaccharide units in the octasaccharides were determined in conjunction of chondroitinase digestion with anion-exchange HPLC after derivatization of the oligosaccharides with the fluorophore 2AB (Pothacharoen et al. 2007). Binding studies were performed on nitrocellulose membranes using immobilizing neoglycolipids, as described in the legend to Fig. 11.7 in Mizumoto et al. (2012)

studies were performed using NMR spectroscopy and the molecular modeling for four octasaccharide sequences, including ΔC-A-D-C, ΔC-AD-A, ΔA-A-D-C, and ΔA-A-D-A (Table 11.1), which were recognized by the mAbs 473HD, CS56, and MO-225 that had been used to stain embryonic brain sections. As a negative control, ΔC-C-C-C, which was not recognized by these mAbs, was also studied. The most and second most stable conformations of the CS octasaccharides recognized by CS-56, MO-225, and 473HD were very similar with subtle energy differences and were consistent with the structures determined by NMR spectroscopy. This combination of experimental and theoretical approaches revealed that the sulfate group at the C2 position of GlcUA in disaccharide D and the presence of an exocyclic negative tail in disaccharides C, [GlcUA(β1-3)GalNAc-6-*O*-sulfate], and ΔC, [ΔHexUA(α1-3)GalNAc-6-*O*-sulfate], were important for antibody recognition (Blanchard et al. 2007).

The mAb designated WF6, which recognized the CS epitopes expressed in the sera of patients with ovarian cancer, osteoarthritis, and rheumatoid arthritis, was previously raised against embryonic shark cartilage (Pothacharoen et al. 2006). Although it remains unknown whether this Ab can stain brain sections, two distinct

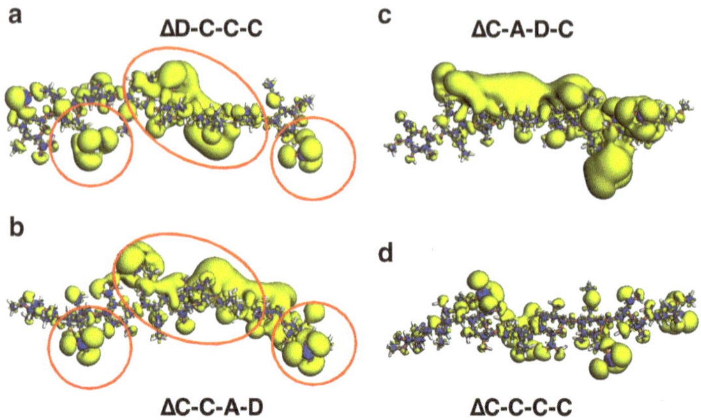

Fig. 11.7 Calculated ESP distribution of the isolated CS octasaccharides. ESP distribution was calculated for each isolated octasaccharide, and the electronegative zone (*yellow*) and electropositive zone (*blue*) are shown. (**a**) **Δ**D-C-C-C (preferred); (**b**) **Δ**C-C-A-D (preferred); (**c**) **Δ**C-A-D-C (non-preferred); (**d**) **Δ**C-C-C-C (non-preferred). The predicted binding sites of the two octasaccharide sequences, **Δ**D-C-C-C and **Δ**C-C-A-D, which were recognized by the mAb WF6, were indicated by *red circles* in (**a**) and (**b**). This figure was taken from Pothacharoen et al. (2007)

octasaccharide sequences, ΔD-C-C-C and ΔC-C-A-D, that were recognized by mAb WF6 were isolated from shark cartilage following partial digestion with chondroitinase ABC, whereas two other related octasaccharides, ΔC-A-D-C and ΔC-C-C-C, were not (Table 11.4) (Pothacharoen et al. 2007). The structures and sequences of both the binding and nonbinding octasaccharides were compared by computer modeling performed at a quantum mechanics level, which gave a more accurate information than geometry optimization at a classical mechanics level, and the results obtained revealed common and marked structural similarities between the shape and distribution of the electrostatic potential in the two different octasaccharide sequences that bound to WF6, and also differences from the nonbinding octasaccharides (Fig. 11.7). The strong similarities in the structures predicted for the two binding CS octasaccharides (ΔD-C-C-C and ΔC-C-A-D) provided a possible explanation for their similar affinities for WF6, even though their sequences differed and, thus, formed two specific epitopes for the antibody. A previous study proposed that such flexible functional oligosaccharide motifs should be termed "wobble oligosaccharide motifs" (Purushothaman et al. 2011).

Perspectives

The potential 'neuronal plasticity' of the brain declines during postnatal development, and this is accompanied by the emergence of perineuronal nets, which are specialized, reticular extracellular matrices rich in CS-PGs that encapsulate neuronal cell bodies and proximal dendrites and also inhibit axonal growth in the adult brain as well as spinal cord injury. Cortical plasticity is most evident during a critical period in early life. A recent study unequivocally demonstrated that a developmental increase in the 4-*O*-sulfation/6-*O*-sulfation ratio of CS-PGs in the brain extracellular matrix terminated the critical period for ocular dominance plasticity in the mouse visual cortex (Miyata et al. 2012). These findings indicated that the critical period for cortical plasticity may be regulated by the 4-*O*-sulfation/6-*O*-sulfation ratio of CS-PGs, which determines the maturation of interneurons. In another study, CS GalNAc transferase 1 knockout mice, in which CS chains were systemically lacking, were found to recover more completely from spinal cord injuries than wild-type mice (Takeuchi et al. 2013). This enzyme transfers the very first GalNAc residue to the core protein-GAG linkage region tetrasaccharide (GlcA-Gal-Gal-Xylose) (Kjellén and Lindahl 1991), which is covalently attached to the Ser residues in the GAG attachment sites of the CS chains of PGs. Thus, the regulation of a single gene of this particular enzyme mediates excellent recovery from spinal cord injuries by optimizing the counteracting effector molecules of axon regeneration. However, it remains to be investigated whether defined sugar sequences and some effector binding proteins are also involved in the mechanism underlying the cortical plasticity and the recovery of spinal cord injuries. However, a recent study reported that CS chains in perineuronal nets specifically interact with CS containing E-disaccharide units, and that the combination of CS chains in the perineuronal nets and semaphrin3A was shown to be a potent inhibitor of axonal growth (Dick et al. 2013).

A set of overlapping oligosaccharides, which are recognized by a single functional protein, may be the entity of the "wobble CS-DS oligosaccharide motifs" described above (Purushothaman et al. 2011). In contrast, a hypothesis has recently been proposed based on sequencing data obtained using mass spectroscopy of the structurally simple CS chain of bikunin CS-PG, the complex roles of GAGs, including CS, DS, and even heparan sulfate may reflect combinations of single-defined sequence-dependent or sequence-specific mechanisms (Ly et al. 2011). Emerging glycomic technologies including mass spectrometry (Ly et al. 2011), GAG microarrays (Shipp and Hsieh-Wilson 2007), and microsequencing (Deepa et al. 2007; Bao et al. 2005b) will accelerate the identification of protein ligands and functional GAG domain sequences. Thus, the development of a high-throughput universal sequencing method is essential for GAG-directed drug discovery. Once functional primary sequences become available, computer simulations can construct models of molecular shapes with an electrostatic potential surface in order to identify crucial structural elements for chemical synthesis.

Acknowledgments This work was supported in part by a Grant-in-Aid for Challenging Exploratory Research 25670018 and the Japan–Thailand Research Cooperative Program from the Japan Society for the Promotion of Science (JSPS), and a Grant in-Aid for Scientific Research on Innovative Areas 24110501 from the Ministry of Education, Culture, Sports, Science, and Technology of Japan (MEXT).

References

Atha DH, Lormeau JC, Petitou M, Rosenberg RD, Choay J (1987) Contribution of 3-*O*- and 6-*O*-sulfated glucosamine residues in the heparin-induced conformational change in antithrombin III. Biochemistry 26:6454–6461

Bao X, Nishimura S, Mikami T, Yamada S, Itoh N, Sugahara K (2004) Chondroitin sulfate/dermatan sulfate hybrid chains from embryonic pig brain, which contain a higher proportion of L-iduronic acid than those from the adult pig brain, exhibit neuritogenic and growth factor-binding activities. J Biol Chem 279(11):9765–9776

Bao X, Mikami T, Yamada S, Faissner F, Muramatsu T, Sugahara K (2005a) Heparin-binding growth factor, pleiotrophin, mediates neuritogenic activity of embryonic pig brain-derived chondroitin sulfate/dermatan sulfate hybrid chains. J Biol Chem 280(10):9180–9191

Bao X, Muramatsu T, Sugahara K (2005b) Demonstration of the pleiotrophin-binding oligosaccharide sequences isolated from chondroitin sulfate/dermatan sulfate hybrid chains of embryonic pig brains. J Biol Chem 280(42):35318–35328

Bishop JR, Schuksz M, Esko JD (2007) Heparan sulfate proteoglycans fine-tune mammalian physiology. Nature 446:1030–1037

Blanchard V, Chevalier F, Imberty A, Leeflang BR, Basappa, Sugahara K, Kamerling JP (2007) Conformational studies on five octasaccharides isolated from chondroitin sulfate using NMR spectroscopy and molecular modeling. Biochemistry 46(5):1167–1175

Casu B, Lindahl U (2001) Structure and biological interactions of heparin and heparan sulfate. Adv Carbohydr Chem Biochem 57:159–206

Clement AM, Nadanaka S, Masayama K, Mandl C, Sugahara K, Faissner A (1998) The DSD-1-epitope depends on sulfation, correlates with chondroitin sulfate D motifs, and is sufficient to promote neurite outgrowth. J Biol Chem 273(43):28444–28453

Clement A, Sugahara K, Faissner A (1999) Chondroitin sulfatae E promotes neurite outgrowth of embryonic day 18 hippocampal neurons. Neurosci Lett 269(3):125–128

Deepa SS, Kalayanamitra K, Ito Y, Kongtaweelert P, Fukui S, Yamada S, Mikami T, Sugahara K (2007) Novel sulfated octa- and decasaccharides from squid cartilage chondroitin sulfate-E: sequencing and application for determination of the epitope structure of monoclonal antibody MO-225. Biochemistry 46(9):2453–2465

Dick G, Tan CL, Alves JN, Ehlert EM, Miller GM, Hsieh-Wilson LC, Sugahara K, Oosterhof A, van Kuppevelt TH, Verhaagen J, Fawcett JW, Kwok JC (2013) Semaphorin 3A binds to the perineuronal nets via chondroitin sulfate type E motifs in rodent brains. J Biol Chem 288:27384–27395

Faissner A, Clement A, Lochter A, Streit A, Mandl C, Schachner M (1994) Isolation of a neural chondroitin sulfate proteoglycan with neurite outgrowth promoting properties. J Cell Biol 126(3):783–799

Ito Y, Hikino M, Yajima Y, Mikami T, Sirko L, von Holst A, Faissner A, Fukui S, Sugahara K (2005) Structural characterization of the epitopes of the monoclonal antibodies 473HD, CS-56, and MO-225 specific for chondroitin sulfate D-type using the oligosaccharide library. Glycobiology 15(6):593–603

Kitagawa H, Tsutsumi K, Tone Y, Sugahara K (1997) Developmental regulation of the sulfation profile of chondroitin sulfate chains in the chicken embryo brain. J Biol Chem 272(50): 31377–31381

Kjellén L, Lindahl U (1991) Proteoglycans: structures and interactions. Annu Rev Biochem 60:443–475

Lindahl U (2014) A personal voyage through the proteoglycan field. Matrix Biol 35:3–7. doi:10.1016/j.matbio.2014.01.001

Ly M, Leach FE 3rd, Laremore TN, Toida T, Amster IJ, Linhardt RJ (2011) The proteoglycan bikunin has a defined sequence. Nat Chem Biol 7(11):827–833

Maeda N, He J, Yajima Y, Mikami T, Sugahara K, Yabe T (2003) Heterogeneity of the chondroitin sulfate portion of phosphacan/6B4 proteoglycan regulates its binding affinity for pleiotrophin/HB-GAM. J Biol Chem 278(37):35805–35811

Mitsunaga C, Mikami T, Mizumoto S, Fukuda J, Sugahara K (2006) Chondroitin sulfate/dermatan sulfate hybrid chains in the development of cerebellum: spatiotemporal regulation of the expression of critical disulfated disaccharides by specific sulfotransferases. J Biol Chem 281(28):18942–18952

Miyata S, Komatsu Y, Yoshimura Y, Taya C, Kitagawa H (2012) Persistent cortical plasticity by upregulation of chondroitin 6-sulfation. Nat Neurosci 15(3):414–422

Mizumoto S, Murakoshi S, Kalayanamitra K, Deepa SS, Fukui S, Kongtawelert P, Yamada S, Sugahara K (2012) Highly sulfated hexasaccharide sequences isolated from chondroitin sulfate of shark fin cartilage: insights into the sugar sequences with bioctivities. Glycobiology 23(2):155–168

Nadanaka S, Clement A, Masayama K, Faissner A, Sugahara K (1988) Characteristic hexasaccharide sequences in octasaccharides derived from shark cartilage chondroitin sulfate D with a neurite outgrowth promoting activity. J Biol Chem 273(6):3296–3307

Perrimon N, Bernfield M (2001) Cellular functions of proteoglycans—an overview Semin. Semin Cell Dev Biol 12:65–67

Petitou M, van Boeckel CA (2004) A synthetic antithrombin III binding pentasaccharide is now a drug! What comes next? Angew Chem Int Ed Engl 43:3118–3133

Pothacharoen P, Ong-Chai S, Supaphun J, Kumja P, Wanaphirak C, Sugahara K, Hardingham T, Kongtawelert P (2006) Raised serum chondroitin sulfate epitope level in ovarian epithelial cancer. J Biochem 140(4):517–524

Pothacharoen P, Kalayanamitra K, Deepa SS, Fukui S, Hattori T, Fukushima N, Hardingham T, Kongtawelert P, Sugahara K (2007) Two related but distinct chondroitin sulfate mimetope octasaccharide sequences recognized by monoclonal antibody WF6. J Biol Chem 282(48): 35232–35246

Purushothaman A, Fukuda J, Mizumoto S, ten Dam GB, van Kuppevelt TH, Kitagawa H, Mikami T, Sugahara K (2007) Functions of chondroitin sulfate/dermatan sulfate chains in the brain development: critical roles of E and iE disaccharide units recognized by a single chain antibody GD3G7. J Biol Chem 282(27):19442–19452

Purushothaman A, Sugahara K, Faissner A (2011) Chondroitin sulfate "wobble motifs" modulates the maintenance and differentiation of neural stem cells and their progeny. J Biol Chem 287(5): 2935–2942

Shipp EL, Hsieh-Wilson LC (2007) Profiling the sulfation specificities of glycosaminoglycan interactions with growth factors and chemotactic proteins using microarrays. Chem Biol 14(2):195–208

Sugahara K, Mikami T (2007) Chondroitin/dermatan sulfate in the central nervous system. Curr Opin Struct Biol 17:536–545

Sugahara K, Mikami T, Uyama U, Mizuguchi S, Nomura K, Kitagawa H (2003) Recent advances in the structural biology of chondroitin sulfate and dermatan sulfate. Curr Opin Struct Biol 13:612–620

Takeuchi K, Yoshioka N, Higa Onaga S, Watanabe Y, Miyata S, Wada Y, Kudo C, Okada M, Ohko K, Oda K, Sato T, Yokoyama M, Matsushita N, Nakamura M, Okano H, Sakimura K, Kawano H, Kitagawa H, Igarashi M (2013) Chondroitin sulphate N-acetylgalactosaminyl-transferase-1 inhibits recovery from neural injury. Nat Commun 4:2740. doi:10.1038/ncomms3740

Tanaka M, Maeda N, Noda M, Marunouchi T (2003) A chondroitin sulfate proteoglycan PTPzeta /RPTPbeta regulates the morphogenesis of Purkinje cell dendrites in the developing cerebellum. J Neurosci 23(7):2804–2814

Thunberg L, Bäckström G, Lindahl U (1982) Further characterization of the antithrombin-binding sequence in heparin. Carbohydr Res 100:393–410

Chapter 12
Role of Hyaluronidases in the Catabolism of Chondroitin Sulfate

Shuhei Yamada

Abbreviations

2AB	2-Aminobenzamide
Chn	Chondroitin
CS	Chondroitin sulfate
GalNAc	*N*-Acetyl-D-galactosamine
GlcUA	D-Glucuronic acid
GPI	Glycosylphosphatidylinositol
HA	Hyaluronan
HPLC	High performance liquid chromatography

Introduction

Chondroitin sulfate (CS) proteoglycans are ubiquitous components in the extracellular matrix as well as at the surface of various cell types (Rodén 1980; Iozzo 1998). They are known to be involved in various biological processes including cell proliferation, cell differentiation, cell migration, cell–cell recognition, extracellular matrix deposition, and tissue morphogenesis (Fig. 12.1) (Esko and Selleck 2002; Sugahara et al. 2003; Rauch and Kappler 2006; Uyama et al. 2007). CS chains are composed of repeating disaccharide units alternatively consisting of D-glucuronic acid (GlcUA) and *N*-acetyl-D-galactosamine (GalNAc), which are sulfated at

S. Yamada (✉)
Department of Pathobiochemistry, Faculty of Pharmacy, Meijo University,
150 Yagotoyama, Tempaku-ku, Nagoya 468-8503, Japan
e-mail: shuheiy@meijo-u.ac.jp

A. Chakrabarti, A. Surolia (eds.), *Biochemical Roles of Eukaryotic Cell Surface Macromolecules*, Advances in Experimental Medicine and Biology 842,
DOI 10.1007/978-3-319-11280-0_12

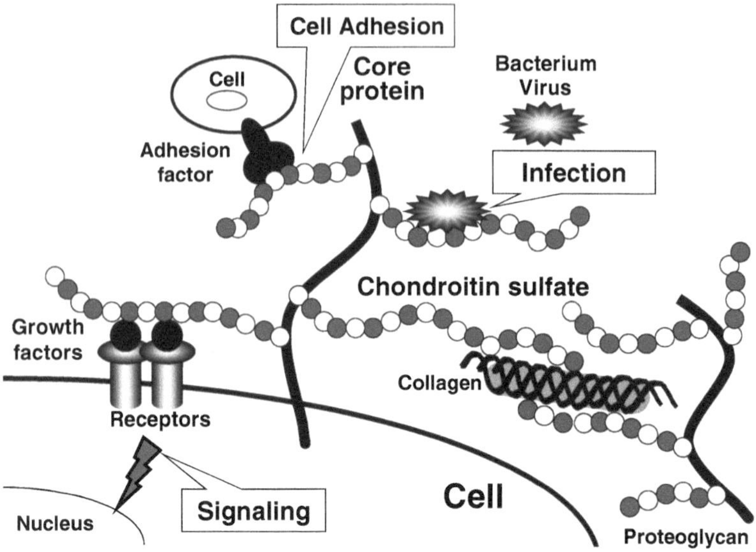

Fig. 12.1 Biological functions of CS on the cell surface or in the extracellular matrix. CS chains play indispensable roles on the cell surface or in the extracellular matrix, including cell adhesion by interacting with adhesion factors, the regulation of growth factor signaling, and involvement in infections by microorganisms. CS chains are represented as alternating *open* and *closed circles*

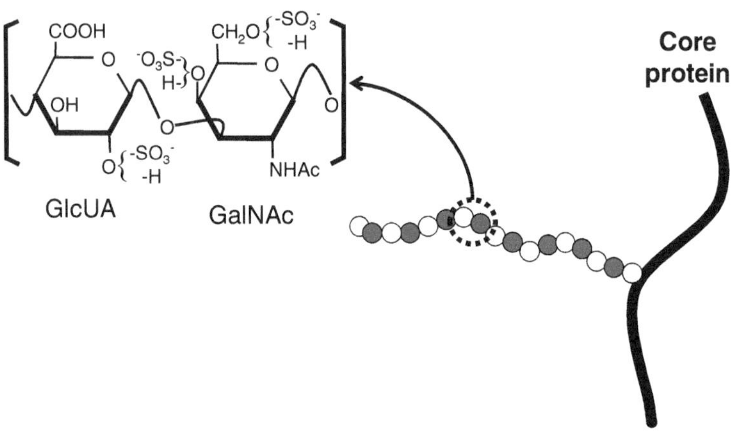

Fig. 12.2 Structure of CS. CS is a linear polymer composed of alternating disaccharide units of GlcUA (*white circles*) and GalNAc (*gray circles*). CS chains are extensively modified by sulfation at the C-2 of GlcA, and C-4 and/or C-6 of GalNAc residues

different hydroxyl groups in various combinations to give rise to structural variations (Fig. 12.2) (Rodén 1980; Sugahara and Yamada 2000; Sugahara et al. 2003). Such structural variety is the basis for the multiple functions of CS.

Chn/CS-Specific Hydrolase

CS chains are predominantly degraded in lysosomes (Fig. 12.3) (Prabhakar and Sasisekharan 2006). In the first step of degradation, CS polysaccharides are fragmented by an endo-type hydrolase into oligosaccharides. The products are then cleaved sequentially from the nonreducing end by exo-type glycosidases as well as sulfatases to liberate monosaccharide moieties. The hyaluronan (HA)-degrading enzymes, hyaluronidases, have been considered to play the role of the endo-type enzymes responsible for the fragmentation of CS at the initial stage of degradation. The structure of HA is similar to that of the nonsulfated CS, chondroitin (Chn). The stereoconfiguration of sugars, substitution pattern of the backbone of hydroxy groups, and glycosidic linkages are identical between Chn and HA, and only the configuration at the C-4 position of the hexosamine residues differs. Most human hyaluronidases can degrade both HA and CS, whereas their preferred substrate was previously reported to be HA, not CS (Csoka et al. 2001; Jedrzejas and Stern 2005).

Since endoglycosidases specific to CS had not been identified, we used *Caenorhabditis elegans* as a model to investigate the initial stage of the degradation process of CS because it contains Chn, but not HA (Yamada et al. 1999, 2011). Thus, it was an ideal system for studying the hyaluronidase-independent catabolic mechanism of CS. A homolog of human hyaluronidase has been detected in the *C. elegans* genome even though *C. elegans* does not synthesize HA.

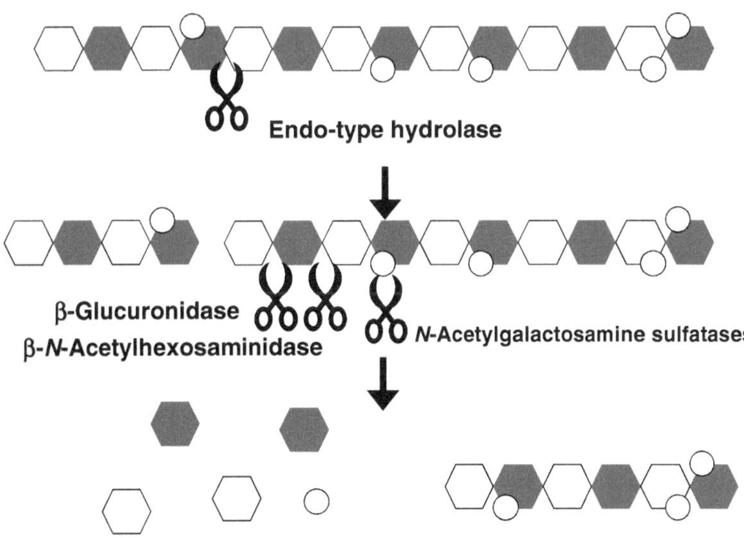

Fig. 12.3 Cellular degradation of CS. The in vivo degradation of CS polymers proceeds sequentially via the actions of a combination of endoglycosidases, sulfatases, and exoglycosidases. Degradation is initiated by the endolytic cleavage of the long polysaccharide chain into smaller fragments. The actions of sulfatases and exoglycosidases (β-glucuronidase and β-*N*-acetylhexosaminidase) then mediate further degradation from the nonreducing ends of oligosaccharides. *White hexagon* GlcUA, *gray hexagon* GalNAc, *open circle* sulfate group

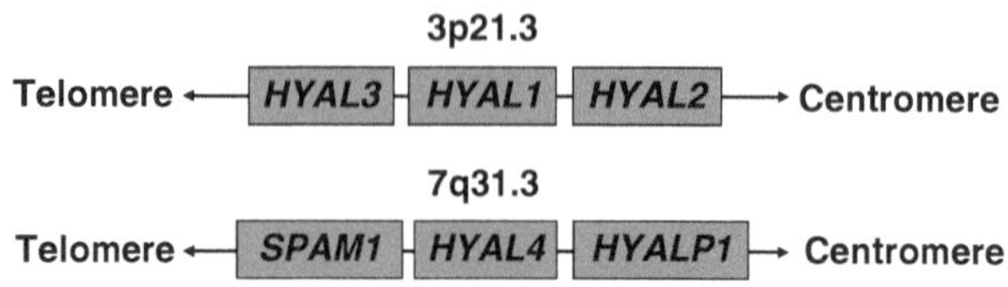

Fig. 12.4 Hyaluronidase or hyaluronidase-like genes. Six genes encode a hyaluronidase or a hyaluronidase-like enzyme in the human genome

Table 12.1 Hyaluronidase genes and their products

Gene	Protein	Localization
HYAL1	HYAL1	Almost ubiquitous
HYAL2	HYAL2	Almost ubiquitous
HYAL3	HYAL3	Including the bone marrow, brain, and testis
HYAL4	HYAL4	Placenta, skeletal muscle, testis
SPAM1	PH-20	Testis
HYALP1	None[a]	–

[a]*HYALP1* is a pseudogene that is translated as a truncated and inactive form

The recombinant protein of the gene product was prepared and its substrate specificity was characterized. We demonstrated that the enzyme specifically acted on Chn/CS (Kaneiwa et al. 2008; Yamada et al. 2009a). Based on this finding, we hypothesized that uncharacterized members of the human hyaluronidase family may have the activity of a CS-specific hydrolase.

Six genes are encoded as hyaluronidases or hyaluronidase-like enzymes in the human genome; HYAL1, HYAL2, HYAL3, HYAL4, SPAM1, and HYALP1 (Csoka et al. 2001) (Fig. 12.4, Table 12.1). Of these, the enzymatic properties, including catalytic activities, of HYAL3 and HYAL4 had not yet been characterized. Therefore, they were investigated as candidates for CS-specific hydrolases. The enzymatic activity of HYAL3 has not yet been elucidated. However, we demonstrated the unique activity of HYAL4 toward CS (Kaneiwa et al. 2010). HYAL4 preferentially acts on the galactosaminidic bond in the GlcUA(2-*O*-sulfate)-GalNAc(6-*O*-sulfate)-GlcUA-GalNAc(4-*O*-sulfate or 6-*O*-sulfate) sequence in CS, and was shown to be a CS-specific endo-beta-*N*-acetylgalactosaminidase. We also identified the amino acid residues essential for this enzymatic activity as well as the substrate specificity of HYAL4 in site-directed mutagenesis studies (Kaneiwa et al. 2012).

Cellular Localization of Hyal4

Although we demonstrated that HYAL4 exhibited hydrolytic activity toward CS chains and degraded them into oligosaccharides, but hardly depolymerizes HA, it does not appear to be the hydrolase that is responsible for the systemic catabolism of CS because it is not ubiquitously expressed. In contrast, CS is widely distributed in

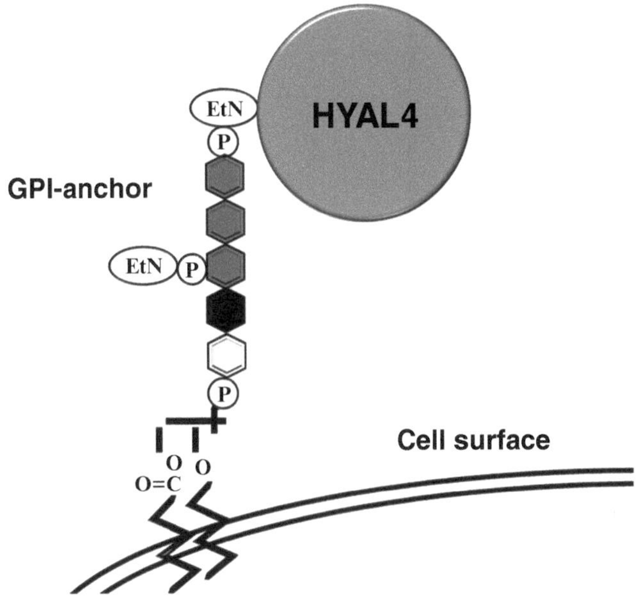

Fig. 12.5 Schematic drawing of the HYAL4 protein on the cell surface. HYAL4 possesses a putative GPI-anchored domain in the C-terminal region and appears to be a GPI-anchored protein. HYAL4 may function on specific cell surfaces. *EtN* ethanolamine, *P* phosphate group, *gray hexagon* D-mannose, *black hexagon* D-glucosamine, *white hexagon* inositol

various tissues and organs. The results obtained by RT-PCR analysis confirmed that the expression of human HYAL4 (hHYAL4) mRNA was restricted to the placenta, skeletal muscle, and testis (Kaneiwa et al. 2012). We also characterized the mouse counterpart, mHyal4, and showed that it was a CS-specific hydrolase (Kaneiwa et al. 2012). mHyal4 mRNA was also not expressed ubiquitously, and we found that it was restricted to the testis and 17-day-old embryos (Kaneiwa et al. 2012).

Both hHYAL4 and mHyal4 possess a putative glycosylphosphatidylinositol (GPI)-anchored domain in the C-terminal region and appear to be GPI-anchored proteins (Fig. 12.5). They may function on the cell surface even though their optimum pH is 4.5–5.0 (Kaneiwa et al. 2010, 2012). The cellular localization of mHyal4 was examined after the transfection of COS-7 cells with a vector containing *mHyal4*. The periphery of the cells was stained by anti-mHyal4 polyclonal antibodies under non-permeabilized conditions, which implied the cell surface expression of mHyal4 (Kaneiwa et al. 2012). The natural localization of the mHyal4 protein in sperm was also investigated in an immunofluorescent study. We demonstrated that the surface of sperm heads, presumably the acrosomal membrane, was stained (Kaneiwa et al. 2012). mHyal4 appears to secure itself to the acrosomal membranes of mouse sperm using its GPI anchor, similar to testicular hyaluronidase PH-20.

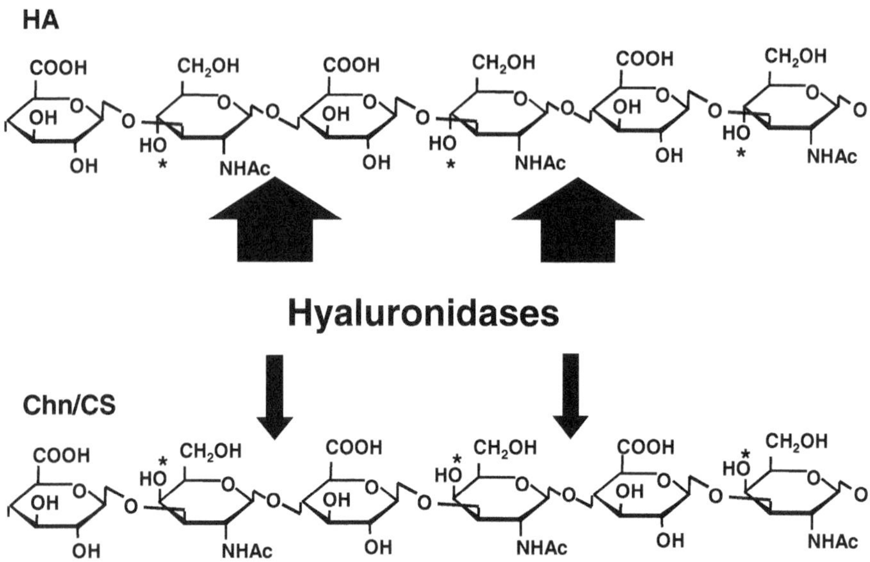

Fig. 12.6 Action of hyaluronidases toward HA and Chn/CS. Although both HA and Chn/CS are the substrates of the hyaluronidases, HYAL1 and testicular hyaluronidase PH-20, their preferred substrate had been considered to be HA, not CS because they were previously reported to degrade CS chains to a limited extent. The structural difference between them is the configuration at the C-4 position of the hexosamine residues, as indicated by *asterisks*. *Arrows* indicate the hexosaminidic bonds cleaved by hyaluronidases

Systemic Catabolism of CS

Since we confirmed that HYAL4 was not involved in the systemic catabolism of CS, we next attempted to identify a candidate enzyme. HYAL1 and HYAL2 are the only members among the hyaluronidase family that are ubiquitously expressed. Although HYAL1 was previously shown to digest CS more slowly than HA, with its preferred substrate being HA, not CS (Csoka et al. 2001; Jedrzejas and Stern 2005) (Fig. 12.6), the activities of HYAL1 and HYAL2 toward CS remained unclear and their activities toward HA and CS had yet to be kinetically compared. Furthermore, CS has structural heterogeneity due to its sulfation (Fig. 12.7); therefore, it is intrinsically more difficult to determine hydrolytic activity toward CS variants. CS chains contain multiple disaccharide units; A, C, D, and E (Table 12.2), in various proportions, and have been designated as CS-A, CS-B, CS-C, CS-D, and CS-E depending on the ratio of the building blocks (Sugahara and Yamada 2000; Yamada and Sugahara 2008). Although the degradation velocity of CS-A by a hyaluronidase may be distinct from that of CS-C, hydrolytic activity toward these CS variants has not been quantitatively compared.

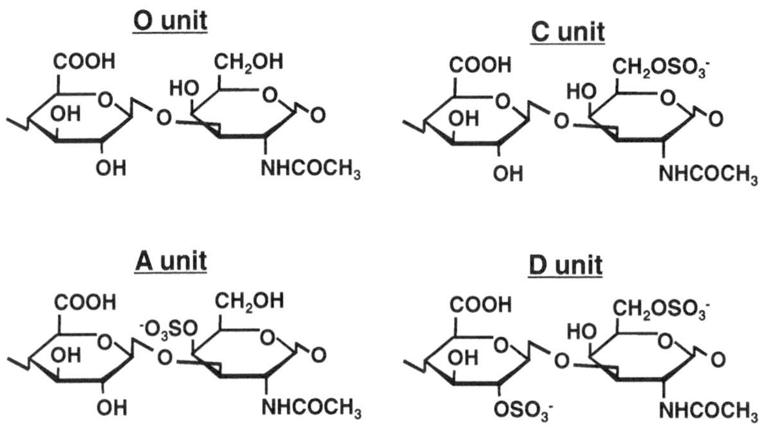

Fig. 12.7 Typical disaccharide unit in each CS variant. The repeating disaccharide unit of CS is sulfated to various extents at different positions. CS polysaccharide chains are formed by the combination of these various disaccharide units. CS variants differ in their disaccharide compositions, as shown in Table 12.2

Table 12.2 Disaccharide compositions of the CS/Chn preparations used in the present study

	% composition			
CS variant	O unit	A unit	C unit	D unit
Chn[a]	100	N.D.[b]	N.D.	N.D.
CS-A	N.D.	76	24	N.D.
CS-C	N.D.	13	79	8

[a]Chn, a chemically desulfated derivative of CS-A, CS-A from whale cartilage, CS-C from shark cartilage. Regarding the nomenclature of the disaccharide units, see Fig. 12.7

[b]*N.D.* not detected

Measurement of HYAL1 Activity Toward CS Variants

HYAL1 was expressed in COS-7 cells as a recombinant protein fused with the FLAG tag, and its hydrolytic activity toward HA, Chn, CS-A, and CS-C was determined. CS-A contains the A unit (Fig. 12.7) as its major disaccharide unit, approximately 76 %, but is also composed of the C unit (Fig. 12.7) (24 %). CS-C also contains not only the C unit (79 %), but also other disaccharide units (Table 12.2).

The strategy for the quantitative method to detect the hydrolytic activity of HYAL1 is shown in Fig. 12.8. Recombinant HYAL1 was incubated with Chn, CS-A, CS-C, or HA in 50 mM formate buffer containing 150 mM NaCl under various pH conditions. Each digestion product was derivatized with the fluorophore, 2-aminobenzamide (2AB) to label the newly formed reducing ends, and was then analyzed using anion-exchange HPLC after being digested with bacterial chondroitinases (Honda et al. 2012). The amount of 2AB-labeled oligosaccharides corresponded to

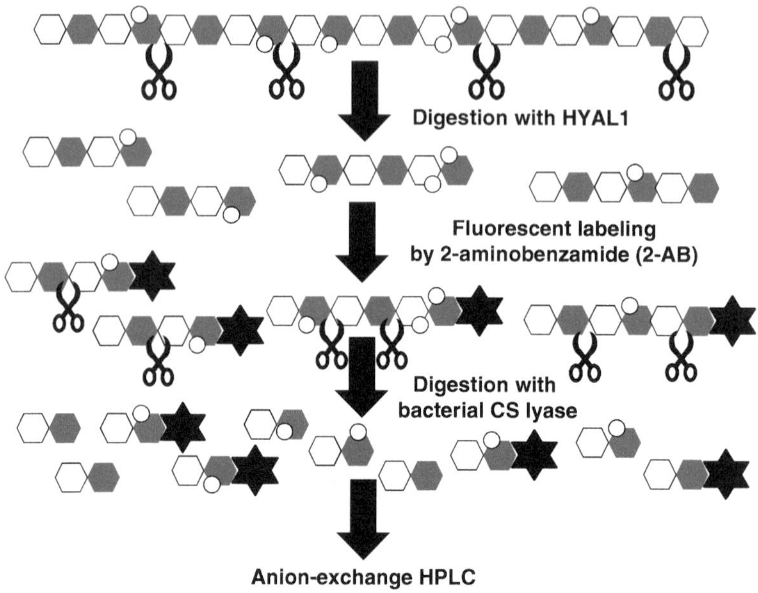

Fig. 12.8 Strategy for quantifying the hydrolytic activity of HYAL1. Non-labeled HA or CS variant chains were digested with HYAL1, and the newly formed reducing ends of the digestion products were derivatized with 2AB. The amount of 2AB-derivatives was determined by anion-exchange HPLC after the digestion of 2AB-oligosaccharides by CS lyases into 2AB-disaccharides. *White hexagon* GlcUA, *gray hexagon* *N*-acetylhexosamine, *open circle* sulfate group, *star* 2AB

that of the newly formed reducing ends. The fluorescent intensity of 2AB was used to quantify the number of cleaved sites. The more HYAL1 that acted on a substrate, the more 2AB-derivatives that formed. The initial velocities of HYAL1 toward various glycosaminoglycan substrates were determined and assessed. The hydrolytic activity of HYAL1 toward CS-A was higher than that toward HA at pH 4.0–4.5. The relative rate of degradation of HA: CS-A: CS-C: Chn by HYAL1 at pH 4.0 was calculated to be 1.0: 1.3: 0.6: 0.3, respectively (Honda et al. 2012).

The kinetic analysis of HYAL1 towards different substrates was performed using the initial reaction rates and substrate concentrations (Fig. 12.9). The apparent Michaelis-Menten constants as well as V_{max} values for Chn, CS-A, CS-C, and HA were determined and are shown in Table 12.3 to compare the ratio V_{max}/K_m. The value towards CS-A was higher than that towards HA, which indicated that CS-A was more effectively hydrolyzed by HYAL1 than HA at pH 4.0. Therefore, HYAL1 preferentially depolymerized CS-A over HA even though the genuine substrate of hyaluronidases had been considered to be HA. Since HYAL1 is ubiquitously expressed and hydrolyzes CS-A more rapidly than HA, it is a candidate for the systemic catabolism of CS, as suggested by earlier investigators.

Based on the X-ray crystallography studies of HYAL1 (Chao et al. 2007) as well as a bee venome hyaluronidase bound to a HA tetrasaccharide substrate (Marković-Housley et al. 2000), the architecture of the HA-binding site of HYAL1

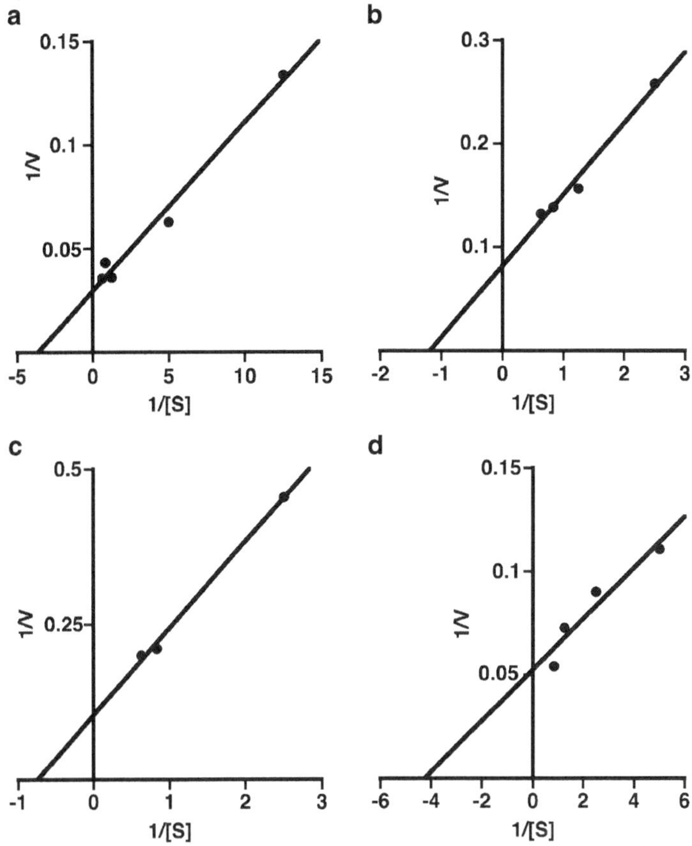

Fig. 12.9 Lineweaver–Burk plots of the initial velocities obtained by enzymatic hydrolysis at various concentrations of CS isoforms and HA with HYAL1. HYAL1 was incubated in 50 mM formate buffer, pH 4.0, containing 150 mM NaCl with different concentrations of CS-A (**a**), CS-C (**b**), Chn (**c**), or HA (**d**). The plots show linearity and the reaction velocity was used in the kinetic analysis to determine apparent K_m, V_{max}, and V_{max}/K_m values for HYAL1

Table 12.3 Kinetic parameters of recombinant HYAL1

Substrate	Apparent K_m (mM) as disaccharides	Apparent V_{max} (pmol/min)	Apparent V_{max}/K_m
HA	0.24	19.0	79.2
CS-A	0.28	33.9	121
CS-C	0.85	12.3	14.5
Chn	1.34	9.6	7.2

has been inspected. A long substrate-binding groove was observed to extend perpendicularly to the barrel structure of HYAL1. Several point mutants of HYAL1 have been generated and some amino acid residues have been identified to be important for HA-binding (Zhang et al. 2009). However, it remains to be investigated whether they can also contribute to the binding of HYAL1 to CS. Since 4-*O*-sulfate group of GalNAc residue in CS-A play an indispensable role for interaction with HYAL1 (Table 12.3) (Honda et al. 2012), the amino acid residues in HYAL1 responsible for the CS-A recognition seem to be different from those for HA. Analysis of the three-dimensional structure of HYAL1 cocrystallized with a CS-A oligosaccharide may be required.

Distribution of Glycosaminoglycans in the Animal Kingdom

We previously characterized the structures of glycosaminoglycans in various model organisms including a nematode (Yamada et al. 1999), hydra (Yamada et al. 2007), fruit fly (Yamada et al. 2002), planarian (Yamada et al. 2011), and African clawed frog (Yamada et al. 2009b). Based on our results as well as the findings reported in other studies, the distribution of glycosaminoglycans in the animal kingdom has been summarized in Fig. 12.10. Although CS and heparan sulfate were previously detected in the nematode, hydrozoan, and planarian, HA was not. HA appears to be common among vertebrates. CS most likely occurred in evolution prior to HA, which emerged at a relatively late stage of evolution. As described in "Chn/CS-specific hydrolase", a homolog of the human hyaluronidase gene exists in *C. elegans* genome even though *C. elegans* itself does not contain HA. Since the hydrolytic activity of HYAL1 toward CS-A is known to be higher than that toward HA, CS chains may be the genuine substrate for HYAL1. Therefore, hyaluronidases appear to have originally been CS hydrolases that later acquired hydrolytic activity toward HA.

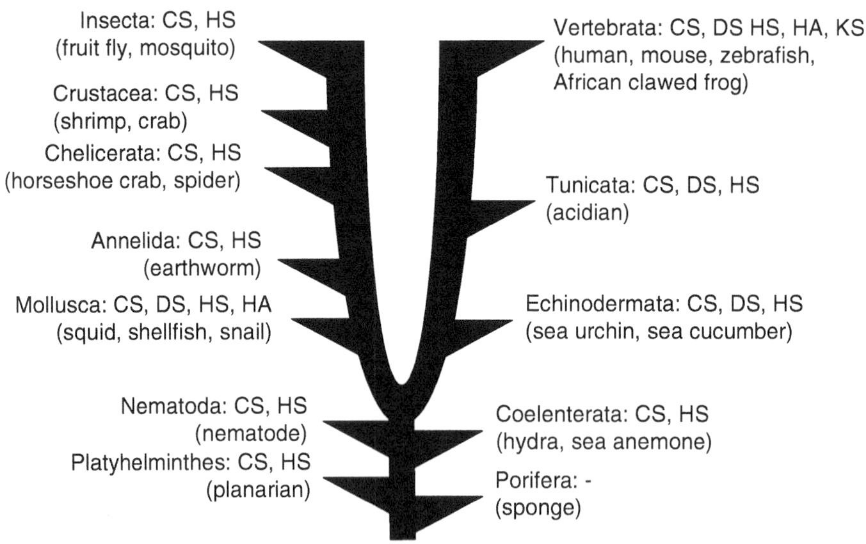

Fig. 12.10 Distribution of glycosaminoglycans in the animal kingdom. Representative animals in which glycosaminoglycan structure has been characterized are shown in *parentheses*. *DS* dermatan sulfate, *HS* heparan sulfate, *KS* keratan sulfate

Evidence for the Involvement of HYAL1 in the Systemic Degradation of CS

The *in vivo* functions of HYAL1 in the systemic degradation of CS have recently been reported. Gushulak et al. (2012) described the accumulation of CS in mice deficient in Hyal1. Mikami et al. (2012) showed that the level of CS chains was markedly diminished at the stage of extensive syncytial myotube formation and that HYAL1 appeared to be involved in this reduction. These findings were consistent with our observation that HYAL1 exhibits strong hydrolytic activity toward CS-A. Furthermore, HYAL1 does not appear to play a major role in the catabolism of HA. Yoshida et al. (2013) identified a new HA-degrading enzyme, KIAA1199, and demonstrated that KIAA1199, not hyaluronidases plays a central role in the depolymerization of HA, at least in the skin. Based on these findings as well as our results, we concluded that the primary substrate of HYAL1 was CS, not HA.

Perspectives

Hyaluronidases appear to play a role in various biological processes by degrading CS as well as HA chains. Their functions may be exhibited by not only eliminating polysaccharides, but also by generating bioactive oligosaccharides. A previous study reported that HA oligosaccharides were involved in inflammation and tumor migration (Stern 2008) and that CS-E oligosaccharides enhanced the cleavage of CD44 and tumor cell motility (Sugahara et al. 2008). Therefore, elucidating the function and influence of the degradation of CS by hyaluronidases *in vivo* under physiological as well as pathological conditions is important.

HYAL4 does not appear to be present in lysosomes, but exists on cell surfaces through a GPI anchor. HYAL4 may not be involved in the systemic catabolism of CS, but appears to have specific temporal functions in particular organs or tissues. The physiological functions of HYAL4 have yet to be clarified. Although the expression of hHYAL4 is known to be restricted to the placenta, skeletal muscle, and testis under normal conditions, it may be augmented under pathological conditions such as inflammation and cancer, similar to HYAL1 (Stern 2008). The enhanced expression of HYAL4 mRNA has been detected in some cultured tumor cell lines (Kaneiwa, Mizumoto, Sugahara, and Yamada, unpublished results). An analysis of the expression of HYAL4 in some disease conditions may provide a deeper insight into the biological functions of HYAL4.

Acknowledgments This work was supported in part by Grants-in-aid for Scientific Research (C) (24590071) and for Scientific Research on Innovative Areas (26110719) from the Ministry of Education, Culture, Sports, Science, and Technology of Japan (MEXT), and the Research Institute of Meijo University, Japan. I thank Professor Kazuyuki Sugahara (Hokkaido University) for his helpful suggestions as well as Tomoko Honda, Tomoyuki Kaneiwa, and Shuji Mizumoto (Hokkaido University) for their substantial contribution.

References

Chao KL, Muthukumar L, Herzberg O (2007) Structure of human hyaluronidase-1, a hyaluronan hydrolyzing enzyme involved in tumor growth and angiogenesis. Biochemistry 46:6911–6920

Csoka AB, Frost GI, Stern R (2001) The six hyaluronidase-like genes in the human and mouse genomes. Matrix Biol 20:499–508

Esko JD, Selleck SB (2002) Order out of chaos: assembly of ligand binding sites in heparan sulfate. Annu Rev Biochem 71:435–471

Gushulak L, Hemming R, Martin D, Seyrantepe V, Pshezhetsky A, Triggs-Raine B (2012) Hyaluronidase 1 and β-exosaminidase have redundant functions in hyaluronan and chondroitin sulfate degradation. J Biol Chem 287:16689–16697

Honda T, Kaneiwa T, Mizumoto S, Sugahara K, Yamada S (2012) Hyaluronidases have strong hydrolytic activity toward chondroitin 4-sulfate comparable to that for hyaluronan. Biomolecules 2:549–563

Iozzo RV (1998) Matrix proteoglycans: from molecular design to cellular function. Annu Rev Biochem 67:609–652

Jedrzejas MJ, Stern R (2005) Structures of vertebrate hyaluronidases and their unique enzymatic mechanism of hydrolysis. Proteins 61:227–238

Kaneiwa T, Yamada S, Mizumoto S, Montaño AM, Mitani S, Sugahara K (2008) Identification of a novel chondroitin hydrolase in *Caenorhabditis elegans*. J Biol Chem 283:14971–14979

Kaneiwa T, Mizumoto S, Sugahara K, Yamada S (2010) Identification of human hyaluronidase-4 as a novel chondroitin sulfate hydrolase that preferentially cleaves the galactosaminidic linkage in the trisulfated tetrasaccharide sequence. Glycobiology 20:300–309

Kaneiwa T, Miyazaki A, Kogawa R, Mizumoto S, Sugahara K, Yamada S (2012) Identification of amino acid residues required for the substrate specificity of human and mouse chondroitin sulfate hydrolase (conventional hyaluronidase-4). J Biol Chem 287:42119–42128

Marković-Housley Z, Miglierini G, Soldatova L, Rizkallah PJ, Müller U, Schirmer T (2000) Crystal structure of hyaluronidase, a major allergen of bee venom. Structure 8:1025–1035

Mikami T, Koyama S, Yabuta Y, Kitagawa H (2012) Chondroitin sulfate is a crucial determinant for skeletal muscle development/regeneration and improvement of muscular dystrophies. J Biol Chem 287:38531–38542

Prabhakar V, Sasisekharan R (2006) The biosynthesis and catabolism of galactosaminoglycans. Adv Pharmacol 53:69–115

Rauch U, Kappler L (2006) Chondroitin/dermatan sulfates in the central nervous system: their structures and functions in health and disease. Adv Pharmacol 53:337–356

Rodén L (1980) Structure and metabolism of connective tissue proteoglycans. In: Lennarz WJ (ed) The biochemistry of glycoproteins and proteoglycans. Plenum Publishing Corp, New York, pp 267–371

Stern R (2008) Hyaluronidases in cancer biology. Semin Cancer Biol 18:275–280

Sugahara K, Yamada S (2000) Structure and function of oversulfated chondroitin sulfate variants: unique sulfation patterns and neuroregulatory activities. Trends Glycosci Glycotechnol 12:321–349

Sugahara K, Mikami T, Uyama T, Mizuguchi S, Nomura K, Kitagawa H (2003) Recent advances in the structural biology of chondroitin sulfate and dermatan sulfate. Curr Opin Struct Biol 13:612–620

Sugahara KN, Hirata T, Tanaka T, Ogino S, Takeda M, Shimada I, Tamura J, ten Dam GB, van Kuppevelt TH, Miyasaka M (2008) Chondroitin sulfate E fragments enhance CD44 cleavage and CD44-dependent motility in tumor cells. Cancer Res 68:7191–7199

Uyama T, Kitagawa H, Sugahara K (2007) Biosynthesis of glycosaminoglycans and proteoglycans. In: Kamerling JP (ed) Comprehensive glycoscience, vol 3. Elsevier, Amsterdam, pp 79–104

Yamada S, Sugahara K (2008) Potential therapeutic application of chondroitin sulfate/dermatan sulfate. Curr Drug Discov Technol 5:289–301

Yamada S, Van Die I, Van den Eijnden DH, Yokota A, Kitagawa H, Sugahara K (1999) Demonstration of glycosaminoglycans in *Caenorhabditis elegans*. FEBS Lett 459:327–331

Yamada S, Okada Y, Ueno M, Iwata S, Deepa SS, Nishimura S, Fujita M, Van Die I, Hirabayashi Y, Sugahara K (2002) Determination of the glycosaminoglycan-protein linkage region oligosaccharide structures of proteoglycans from *Drosophila melanogaster* and *Caenorhabditis elegans*. J Biol Chem 277:31877–31886

Yamada S, Morimoto H, Fujisawa T, Sugahara K (2007) Glycosaminoglycans in *Hydra magnipapillata* (Hydrozoa, Cnidaria): demonstration of chondroitin in the developing nematocyst, sting organelle, and structural characterization of glycosaminoglycans. Glycobiology 17:886–894

Yamada S, Mizumoto S, Sugahara K (2009a) Chondroitin hydrolase in *Caenorhabditis elegans*. Trends Glycosci Glycotechnol 21:149–162

Yamada S, Onishi M, Fujinawa R, Tadokoro Y, Okabayashi K, Asashima M, Sugahara K (2009b) Structural and functional changes of sulfated glycosaminoglycans in *Xenopus laevis* during embryogenesis. Glycobiology 19:488–498

Yamada S, Sugahara K, Özbek S (2011) Evolution of glycosaminoglycans: comparative biochemical study. Commun Integr Biol 4:150–158

Yoshida H, Nagaoka A, Kusaka-Kikushima A, Tobiishi M, Kawabata K, Sayo T, Sakai S, Sugiyama Y, Enomoto H, Okada Y, Inoue S (2013) KIAA1199, a deafness gene of unknown function, is a new hyaluronan binding protein involved in hyaluronan depolymerization. Proc Natl Acad Sci U S A 110:5612–5617

Zhang L, Bharadwaj AG, Casper A, Barkley J, Barycki JJ, Simpson MA (2009) Hyaluronidase activity of human Hyal1 requires active site acidic and tyrosine residues. J Biol Chem 284:9433–9442

Chapter 13
Pattern Recognition in Legume Lectins to Extrapolate Amino Acid Variability to Sugar Specificity

Nisha Jayaprakash Grandhi, Ashalatha Sreshty Mamidi, and Avadhesha Surolia

Introduction

Lectins are proteins of non immunologic origin that bind to carbohydrates with high fidelity. Lectins form a large class of multivalent recognition molecules that specifically interact with their cognate sugar moieties for decoding the information underlying the structural heterogeneity (Sharon and Lis 2004). Although, their occurrence in nature was known only during the early nineteenth century, by 1960s, tremendous research in this field was carried out to explore their functional importance in a range of biologic processes, both across plant and animal kingdom. However, the plant lectins were the most extensively studied (da Silva and Correia 2014), among which those from legume in particular were foremost to be investigated and were found to be a rich source of lectins and are most widely studied.

Legume lectins have been pivotal to the study of the molecular basis of protein carbohydrate interactions (Sharon and Lis 1995). They are a large family of homologues proteins possessing great overall similarities in terms of their physical, chemical and biological properties, despite their origin from different taxonomically distant species. They display remarkable divergence in their carbohydrate specificities ranging from monosaccharides to oligosaccharides. Some legume lectins are synthesized as prolectins in the endoplasmic reticulum and undergo post translational modifications in the Golgi apparatus to function as secretory proteins (Moreira et al. 2013).

Since the advent of recombinant techniques in 1970s, intensified studies were performed for determining the physico-chemical and physiological properties of lectins, amino acid sequences and elucidating their 3D structures. The 3D structure

N.J. Grandhi • A.S. Mamidi • A. Surolia (✉)
Molecular Biophysics Unit, Indian Institute of Science, Bangalore 560 012, India
e-mail: surolia@mbu.iisc.ernet.in

A. Chakrabarti, A. Surolia (eds.), *Biochemical Roles of Eukaryotic Cell Surface Macromolecules*, Advances in Experimental Medicine and Biology 842,
DOI 10.1007/978-3-319-11280-0_13

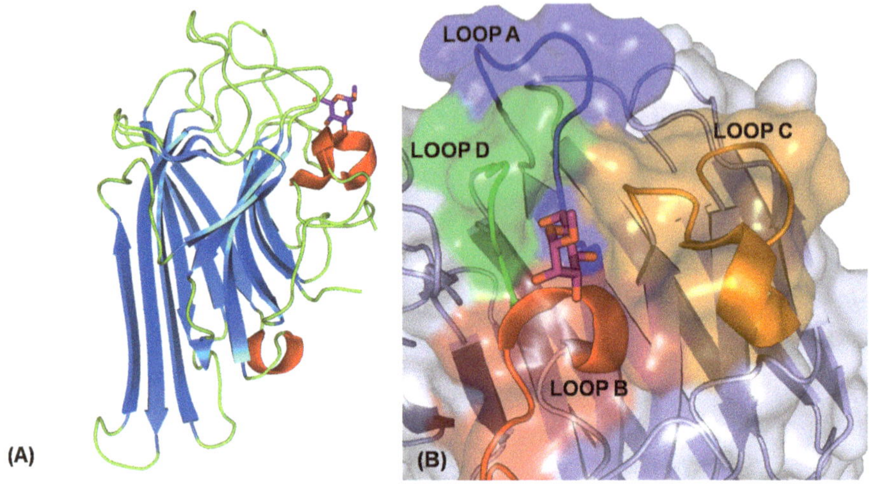

Fig. 13.1 (**a**) Structure of *Canavalia* A as a model for legume lectin fold represented as cartoon. (**b**) Binding site loops A, B, C and D of *Canavalia* with mannose in its carbohydrate recognition domain

of concanavalin A amongst legume lectins was the first lectin for which a high resolution X-ray crystallographic structure became available (Edelman et al. 1972). Soon thereafter 3D structures for a diverse group of lectins were elucidated.

The basic architecture of the protomer is the "lectin fold", which is related to the "jelly-roll fold" comprising three anti-parallel sandwiched β sheets which are connected by α turns, β turns and bends along with short loops. The three anti-parallel β sheets constitute a flat six stranded "back" sheet, a concave seven stranded "front" sheet and a short "top" sheet which holds the two sheets together (Fig. 13.1). They are usually devoid of α helices with the exception of occasional 3_{10} helices. Each protomer is dome shaped with dimensions of 42×40×39 Å and molecular weight of 25–30 kDa. The carbohydrate recognition domain (CRD) is a shallow depression on the surface located at the apex of the dome like structure, accessible to both monosaccharides and oligosaccharides for binding (Sharon and Lis 2002). The basic architecture of CRD in legume lectins constitute four binding site loops A, B, C, D, which are adjacent to each other in the pocket in the 3D structure but are not close together in the sequence. The residues in the binding pocket are known to show the greatest variability and are inferred to be involved in specificity determination (Young and Oomen 1992; Benevides et al. 2012). The floor of the binding site consists of few conserved key amino acids residues in the loops including Asp in Loop A, Gly or Arg (in *Concanavalia* and *Dioclea* lectins) in Loop B, Asn and an aromatic residue in Loop C, which contribute to hydrogen bonds and vander Waals interactions with the sugar. The variation in loop C and D is possibly a primary determinant of the monosaccharide specificity (Sharma and Surolia 1997; Rao et al. 1998). The CRD in these lectins lies in close proximity with the metal binding sites and require Ca^{2+} and transition metal ion Mn^{2+} for their binding activity (Etzler et al. 2009).

Despite emulating a common β-sandwich fold, variability among the legume lectins occurs both at the level of the quaternary fold, with a variety of dimeric and tetrameric arrangement (Srinivas et al. 2001; Manoj and Suguna 2001) and at the level of the binding site. Other modes of interaction that contribute to the variability in specificity are interaction with water, post translational modification, carbohydrate-aromatic interactions, etc. Thus, classification of lectins into distinct groups based on their monosaccharide specificity that is the best hapten inhibitor of the lectin and its extrapolation it to amino acid sequence variations will shed light on the features of the design of their combining sites.

So far, the relationship between the variation of the amino acid composition of legume lectins in the context of their diverse specificities has been examined only to a limited extent in the past (Swamy et al. 1985). In this piece of work, we identify broad features that allow generation of a spectrum of specificities in them without a fundamental alteration of their 3D structural fold. For this, we employ a new approach to simultaneously visualize and analyse the amino acid variations in 46 legume lectins categorized under five different sugar specific groups through pattern recognition method using heatmaps.

Methodology

1. Generation of dataset of 3D structures

   Nearly 1,094 plant lectins belonging to leguminosae family were deposited (with 159 unique source entries) in the comprehensive database of UNIPROT (http://www.uniprot.org/) with amino acid sequence and functional information. Of which, 235 PDB structures have been deposited in the Protein Data Bank (http://www.rcsb.org/pdb/home/home.do), where each lectin has been complexed with one or more ligands (Berman et al. 2000). For this study, a set of 46 legume lectins were short-listed from the large dataset based on "unique source" as the criteria and whose 3D structures were elucidated.

   These legume lectins were categorize into five groups according to their monosaccharide specificity, i.e. (1) Mannose/Glucose (MG), (2) Galactose (GA), (3) N-acetyl-Glucosamine (GLN), (4) N-Acetyl-Galactosamine (GAN) and (5) Fucose (FU), based on the literature. Table 13.1 provides the complete details of 46 lectins along with their source and monosaccharide specificity and PDB IDs. The final dataset constitutes 24 MG, 1 GLN, 8 GA, 10 GAN and 3 FU lectins.
2. Obtaining amino acid sequences

   Complete canonical sequences were only selected for these entries and retrieved in "FASTA format" from RCSB-PDB. Chain A was only chosen to maintain consistency in the data, except for the lectins with PDB IDs: 1LEN, 1LOB, 2B7Y and 2LTN, we have considered both chains A and B as they were fragments of the same protomer which had been truncated. As the lectins belonging to the genus *Canavalia*, *Dioclea*, *Cratylia* and *Cymbosema* of MG group

Table 13.1 Dataset of 46 legume lectins considered in this study

Sl No.	PDB ID	Source	Specificity	Uniprot ID	Reference
1	3JU9	*Canavalia brasiliensis*	MG	P55915	Bezerra et al.(2011)
2	3QLQ	*Canavalia cathartica*	MG	P81461	Sundberg et al. (2012)
3	1I3H	*Canavalia ensiformis*	MG	P02866	Sanders et al. (2001)
4	2OVU	*Canavalia gladiata*	MG	P14894	Moreno et al. (2007)
5	2OW4	*Canavalia maritima*	MG	P81364	Moreno et al. (2008)
6	2A7A	*Canavalia virosa*	MG	P81461	Mueller-Dieckmann et al. (2005)
7	4I30	*Canavalia lineata*	MG	P81460	–
8	2JEC	*Dioclea grandiflora*	MG	P08902	Nagano et al. (2008)
9	2ZBJ	*Dioclea rostrata*	MG	P58908	de Oliveira et al. (2008)
10	2GDF	*Dioclea violacea*	MG	P08902	–
11	3RS6	*Dioclea virgata*	MG	P58907	Nóbrega et al. (2012)
12	3SH3	*Dioclea wilsonii*	MG	P86624	Rangel et al. (2012)
13	2JE7	*Dioclea guianensis*	MG	P81637	Nagano et al. (2008)
14	3A0K	*Cymbosema roseum*	MG	D5MNX4	Rocha et al. (2011)
15	2D3P	*Cratylia argentea*	MG	P81517	Del Sol et al. (2007)
16	1MVQ	*Cratylia mollis*	MG	P83721	de Souza et al. (2003)
17	3U4X	*Camptosema pedicellatum*	MG	J9PBR3	Rocha et al. (2012)
18	2FMD	*Bowringia mildbraedii*	MG	P42088	Buts et al. (2006)
19	3ZYR	*Platypodium elegans*	MG	G1EUI6	Benevides et al. (2012)
20	1UKG	*Pterocarpus angolensis*	MG	Q8GSD2	Loris et al. (2004)
21	1LEN	*Lens culinaris*	MG	P02870	Loris et al. (1994)
22	2LTN	*Pisum sativum*	MG	P02867	Suddath et al. (1989)
23	2B7Y	*Vicia faba*	MG	P02871	Reeke and Becker (1986)
24	1LOB	*Lathyrus ochrus I*	MG	P04122	Bourne et al. (1990)
25	1QNW	*Ulex europaeus II*	GLN	P22973	Loris et al. (2000)
26	1AXZ	*Erythrina corallodendron*	GA	P16404	Shaanan and Elgavish (1998)
27	1DBN	*Maackia amurensis*	GA	P93248	Imberty et al. (2000)
28	1GZC	*Erythrina crista-galli*	GA	P83410	Svensson et al. (2002)
29	1HQL	*Griffonia simplicifolia 1-B4*	GA	Q8W1R6	Tempel et al. (2002)
30	1V6I	*Arachis hypogaea*	GA	P02872	Kundhavai Natchiar et al. (2004)
31	3IPV	*Spatholobus parviflorus*	GA	P86352	Geethanandan et al. (2011)
32	3UJO	*Dolichos lablab*	GA	B3EWQ9	Shetty et al. (2013)
33	3USU	*Butea monosperma*	GA	H2L2M6	–
34	1AVB	*Phaseolus vulgaris—Arcelin 1*	GAN	P19329	Mourey et al. (1998)

(continued)

Table 13.1 (continued)

Sl No.	PDB ID	Source	Specificity	Uniprot ID	Reference
35	1F9K	*Psophocarpus tetragonolobus (Acidic)*	GAN	Q9SM56	Manoj et al. (2000)
36	1FAT	*Phaseolus vulgaris-L*	GAN	P05087	Hamelryck et al. (1996a)
37	1FNY	*Robinia pseudoacacia*	GAN	Q41159	Rabijns et al. (2001)
38	1IOA	*Phaseolus vulgaris—Arcelin 5A*	GAN	Q42460	Hamelryck et al. (1996b)
39	1G7Y	*Vigna unguiculata* subsp. *(DB58)*	GAN	P19588	Buts et al. (2001)
40	1 LU1	*Vigna unguiculata* subsp. *(DBL)*	GAN	P05045	Hamelryck et al. (1999)
41	1 N47	*Vicia villosa*	GAN	P56625	Babino et al. (2003)
42	1SBF	*Glycine max*	GAN	P05046	Olsen et al. (1997)
43	1WBF	*Psophocarpus tetragonolobus (Basic)*	GAN	024313	Manoj et al. (1999)
44	1GSL	*Griffonia simplicifolia IV*	FU	P24146	Delbaere et al. (1993)
45	2EIG	*Lotus tetragonolobus*	FU	D0VWW1	Moreno et al. (2008)
46	1FX5	*Ulex europaeus I*	FU	P22972	Audette et al. (2000)

exhibit circular homology, their sequences were manually re-transposed to align them with other sequences of legume lectins.

3. Protein secondary structure prediction using PSSPRED

   For the secondary structure prediction, PSSPRED (Protein Secondary Structure PREDiction server), a webserver (http://zhanglab.ccmb.med.umich.edu/PSSpred/) was employed based on the Rumelhart error back-propagation method (Xu and Zhang 2013) using amino acid sequence. This tool uses a simple neural network training algorithm for accurate prediction (Zhang 2012). Based on these calculations, the amino acid sequence of the four binding site loops were determined for the lectin dataset.

4. Multiple sequence alignment and analysis

   Multiple sequence alignment was performed using ClustalW2 (http://www.ebi.ac.uk/Tools/msa/clustalw2/) (Larkin et al. 2007), using all default parameters. BLOSUM protein weight matrix was employed along with penalties for GAP opening as 10, GAP extension as 0.20 and a GAP distance penalty as 5.

5. Phylogenetic analysis of legume lectins

   Phylogenetic analysis was based on amino acid sequence alignment. Multiple sequence alignments were performed for the entire set of 46 lectins considering the complete sequences as well as only the amino acid sequences of binding site loops using ClustalW2. For this, alignment was generated using PAM matrix with all other default settings. For phylogenetic analysis based on sequence alignment, a software tool MEGA6 (Molecular Evolutionary Genetics Analysis) (Tamura et al. 2013) was used, in which the output sequence alignment file was provided as input for inferring phylogenetic trees.

6. Calculation of percentage identity matrix

   Pairwise percentage identity scores for all the 46 lectins and their respective binding sites were computed based on the sequence alignment in ClustalW2. The alignment scores were rearranged as a matrix to indicate the pairwise identity scores calculated between every pair of sequences among the legume lectin dataset. These indicate the number of identities between the two sequences, divided by the length of the alignment, and represented as a percentage.
7. Computation of amino acid composition

   Amino acid composition of the complete protein and only the binding site loops were computed separately. ProtParam, a webserver (http://web.expasy.org/protparam/) was employed to obtain the percentage compostion of each amino acid in a given protein sequence (Gasteiger et al. 2005). Similarly, this procedure was repeated for the four binding site loops. The values were tabulated into a 20×46 matrix to generate a clustergram for the same.
8. Pattern recognition and clustering

   In order to demonstrate characteristic features among sugar specific lectin groups, we computed heatmaps to display specific patterns in the entire lectin structure and particularly binding site, based on two aspects: (1) Percentage identities to demonstrate (dis)similarities and (2) Percentage amino acid compositions to study the significance of amino acid variation. Heatmaps were generated using MATLAB v7.5 (MathWorks 2007) (Distance measure: Euclidean). Clustergrams based on amino acids were also generated by employing Kmeans clustering algorithm (MacQueen 1967; Weisstein 1995) using a function module CVAP 3.7 (Cluster Validity Analysis Platform) in MATLAB v7.5.

Results and Discussion

In the present study, we have employed pattern recognition for demonstrating the influence of amino acid variability on legume lectin specificity. Pattern recognition allows making inferences from observations using a statistical approach. Pattern recognition enables discrimination between seemingly similar entities based on their quantitative features (Duin and Pekalska 2007). Accordingly, we have used heatmaps and clustergrams to highlight the characteristic features of each of the five lectin groups classified based on their monosaccharide binding abilities.

(Dis)similarites in Legume Lectins Based on Percentage Identities

Figure 13.2 shows the percentage identity matrix as a heatmap, the top diagonal half computed based on binding site loops and the lower second half represents the full lectin sequences. From the heatmap, it is evident that there is a clear demarcation

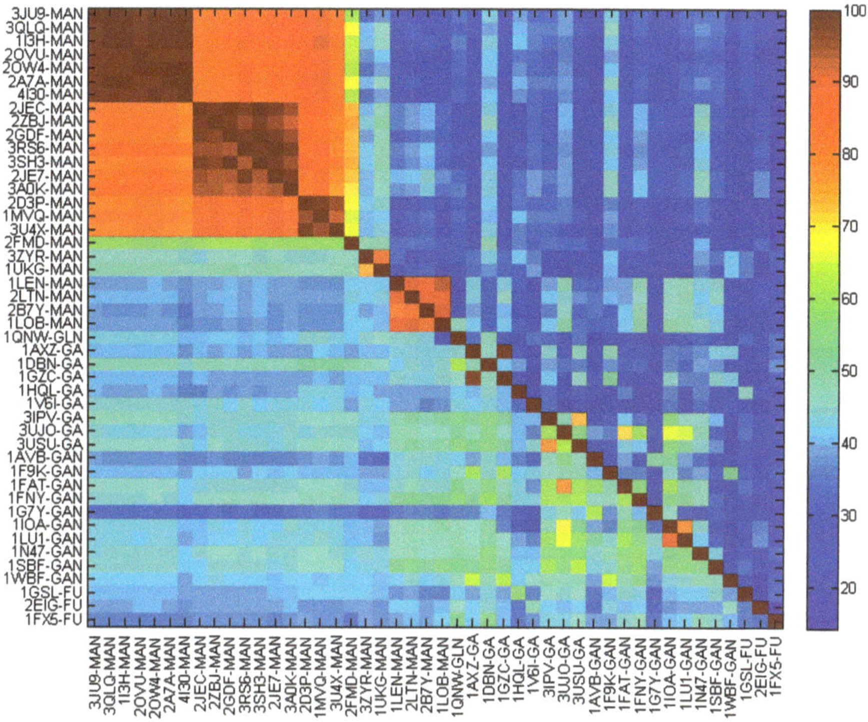

Fig. 13.2 Heatmap generated for 46 legume lectins using percentage identity matrix. The *upper diagonal half* represents the identities computed for the four binding site loops and the *lower diagonal half* is based on the full protein sequence

between the five groups of legume lectins, which were differentiated based on their pair wise comparisons. We observed that the overall percentage identity for the entire protein across 46 lectins was in the range of 28.24–100 %, while it was only 14.29–100 % for the binding site loops. This clearly represents the variability in the carbohydrate binding site residues relative to the whole protein sequence, with the highest identities shared among the same species in the same sugar specific group. The intra-group percentage identities for the MG specific proteins was found to be between 35.68 and 100 %; 33.78 and 96.65 % for GA, 37.78 and 61.61 % for GAN and 35.71 and 36.89 % for FU lectins. Similarly, the identities between the amino acids of binding site loops fall in the range of 21.05–100 % for MG; 17.07–98.04 % for GA; 20.83–56.6 % for GAN and 14.49–34 % for FU binding legume proteins. Table 13.2 illustrates the inter-group percentage identities across the five different groups of legume lectins.

1. MG lectin group: This set includes lectins from *Canavalia* sp., *Dioclea* sp., *Cratylia* sp., *Cymbosema* sp., *Camptosema* sp., *Bowringia*, *Platypodium*, *Pterocarpus*, *Lens culinaris*, *Pisum sativum*, *Vicia faba* and *Lathyrus ochrus I.*

Table 13.2 Intra and Inter group percentage identities calculated from full protein sequences and binding site loops

	MG	GLN	GA	GAN	FU
Full protein					
	MG	GLN	GA	GAN	FU
MG	35.68–100	40–89.7	36.65–51.52	30.6–54.19	35.5–42.27
GLN	36.96–89.7	–	38.01–62.29	35.45–56.52	36.73–52.97
GA	36.65–51.52	38.01–62.29	33.78–96.65	28.24–86.96	35.5–57.92
GAN	30.6–54.19	35.45–56.52	28.24–86.96	37.78–61.61	30.53–46.78
FU	35.5–42.27	36.73–52.97	35.5–57.92	30.53–46.78	35.71–36.89
Four loops					
	MG	GLN	GA	GAN	FU
MG	21.05–100	21.57–46.67	19.23–50	17.24–50	16.22–37.14
GLN	21.57–46.67	–	26–49.09	17.39–32.08	23.08–35.19
GA	19.23–50	26–49.09	17.07–98.04	19.15–71.7	19.05–46
GAN	17.24–50	17.39–32.08	19.15–71.7	20.83–60.96	15.91–38.3
FU	16.22–37.14	23.08–35.19	19.05–46	15.91–38.3	14. 49–34

There are seven *Canavalia* sp. in the dataset, which shared more than 97 % intra-species identity and in particular, three lectins (PDB ID: 1I3H, 3QLQ and 2A7A) possessed 100 % identity for the full protein sequence while, the binding site loops of the six *Canavalia* lectins, except 2OVU, exhibited 100 % identity in their carbohydrate binding residues indicating high conservation in the binding site architecture. Phylogenetic trees based on the sequences of entire protein and its binding site shows that all the lectins of *Canavalia* sp. are closely clustered (Figs. 13.3 and 13.4). Similarly, there are six lectins in *Dioclea* sp. with a percent identity greater than 95 %. However, we have noticed that the three proteins 2JEC, 2GDF, 3SH3 showed 100 % sequence identity in their binding site loops, which also corroborated with the formation of a single clade in the cladogram obtained based on the binding site. We have also noted that *Cymbosema* lectin (3A0K) also shared an high identity (>93 %) with the *Dioclea* lectins and hence was grouped together (Loris et al. 1998). The two *Cratylia* (2D3P and 1MVQ) proteins along with *Camptosema* (3U4X) formed a third clade closer to the origin of *Canavalia* sp. in both the cladograms. It was interesting to note that all these lectins of above mentioned species are known to have an unusual type of homology called the circular homology. Initially, these are synthesized as glycosylated precursors having nearly 290 amino acids and are known to undergo transposition by domain swapping followed by transpeptidation (Sharon and Lis 1990). On the other hand, *Platypodium* (3ZYR) and *Pterocarpus* (1UKG) proteins of this MG group are found to cluster together, while lectin from *Bowringia* (2FMD) stands as an individual clade. The four lectins—*Lens culinaris* (1LEN), *Pisum sativum* (2LTN), *Vicia faba* (2B7Y) and *Lathyrus ochrus I* (1LOB) shared high percentage identity (>80 %) and were grouped together in the heatmaps as well as the cladograms as these four lectins have identical B-chain (Kolberg et al. 1980; Debraya and Rougé 1984).

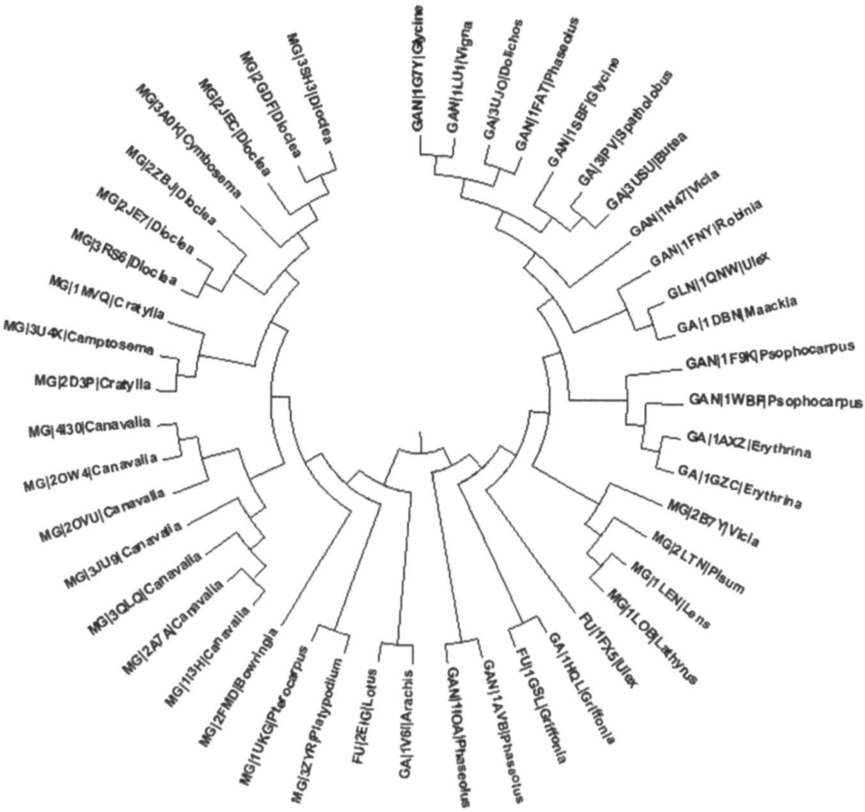

Fig. 13.3 Cladogram for the 46 lectins obtained from full protein sequences

2. GLN lectin group: Among the 46 lectin dataset, only single legume lectin belonging to *Ulex europaeus* II (1QNW) was found to be GLN specific. The whole lectin differed with MG, GA, GAN, FU by a range of 40–89.7 %, 38.01–62.29 %, 35.45–56.52 % and 36.73–52.97 %, respectively. Similarly, the binding site differed with the above sugar specific groups by 21.57–46.67 %, 26–49.09 %, 17.39–32.08 % and 23.08–35.19 %, respectively.
3. GA and GAN lectin group: Until recently, the GA and GAN specific lectins were grouped together (early 1990s), but due to their amino acid variability in the binding regions attributing differences to their biologic recognition process, they were considered as separate entities (Sharma et al. 1998). The findings of our present study reinforces these observations as the heatmap computed based on the percentage identities showed higher identities among these two groups while considering entire lectin (28.24–86.96 %) whereas the same was not observed for the binding site residues (19.15–71.7 %). This demonstrates the difference in specificity between the two sets of lectins. The intra-group identity of GA was 33.78–96.65 % and 37.78–61.61 % for GAN binding proteins for the full sequence,

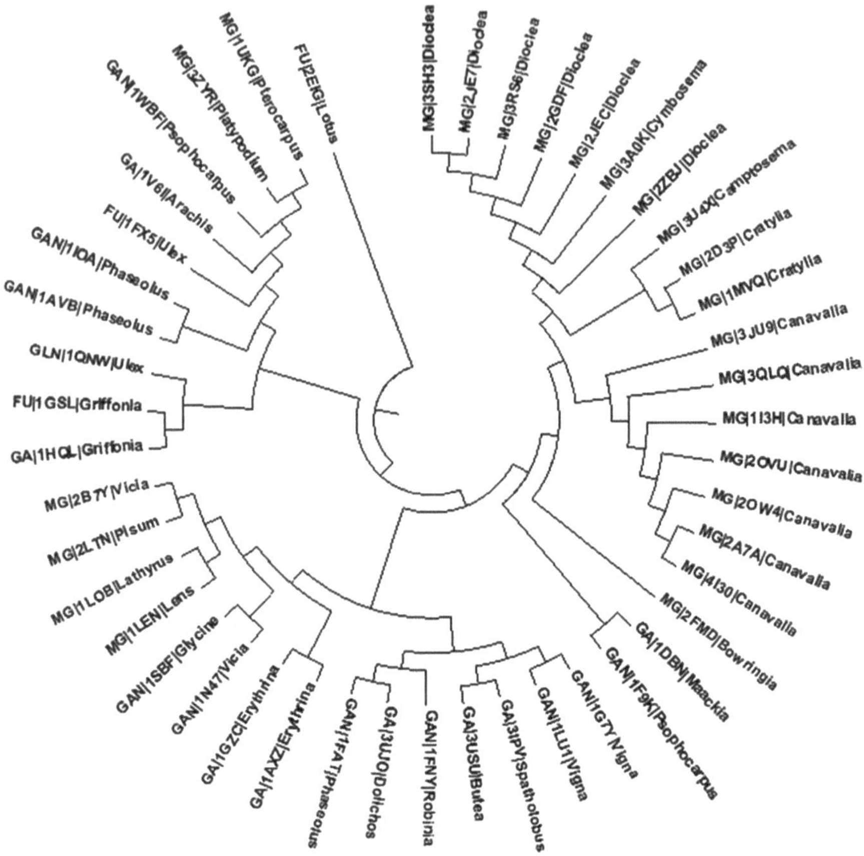

Fig. 13.4 Cladogram showing the relationship between 46 legume lectins based on the four loops of their binding site

while for the binding site loops, 17.07–98.04 % and 20.83–60.96 % identities were observed for GA and GAN proteins, respectively. Analysis of the cladogram demonstrated a similar trend, wherein the lectins of GA and GAN were distributed in common clades, which in turn supported the fact that the structural characteristics of GA and GAN lectins are dependent on their phylogeny rather than their differences in sugar specificity (Liener et al 1986).

4. FU lectin group: This set included three proteins—*Griffonia simplicifolia* IV (1GSL), *Lotus tetragonolobus* (2EIG) and *Ulex europaeus* I (1FX5). These lectins had an intra-group identity range of 35.7–36.89 % and 14.49–34 % for the full protein sequence and binding site loops, respectively. Despite their specificities, these FU specific lectins are clustered with their respective genus clades in both the cladograms (Thomas and Surolia 2000).

Amino Acid Variability in Legume Lectins Based on Percentage Composition

The basic differences in the binding site architecture of legume lectins can be attributed to their amino acid variability and thereby their biologic function. The amino acids in the clustergrams were grouped based on their abundance (Figs. 13.5 and 13.6). Table 13.3 details the relative abundance of 20 amino acids in all 46 lectins classified as high, moderate and low.

In the clustergram of full protein, it was evident that Ser and Thr were highly present across the 46 lectins and Ser was prevalent in MG, GLN and GA binding proteins with 12.39 %, 12.10 % and 11.7 %, respectively. Unlike other MG lectins, Thr was found in relatively high percentage in *Lens culinaris* (1LEN), *Pisum sativum*

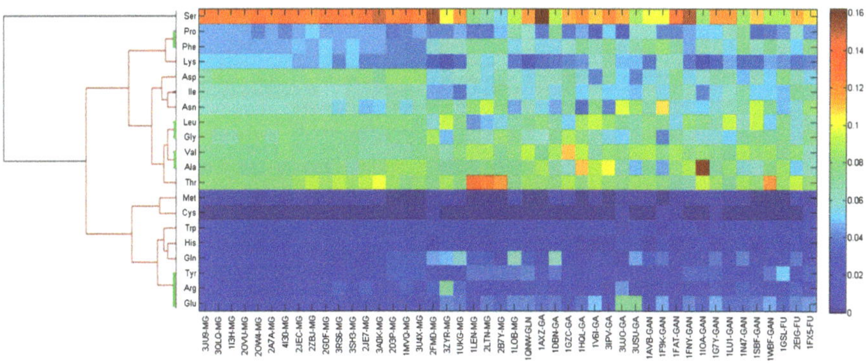

Fig. 13.5 Heatmaps with the dendrogram constructed using percentage composition of amino acids for the 46 lectins on full protein sequences

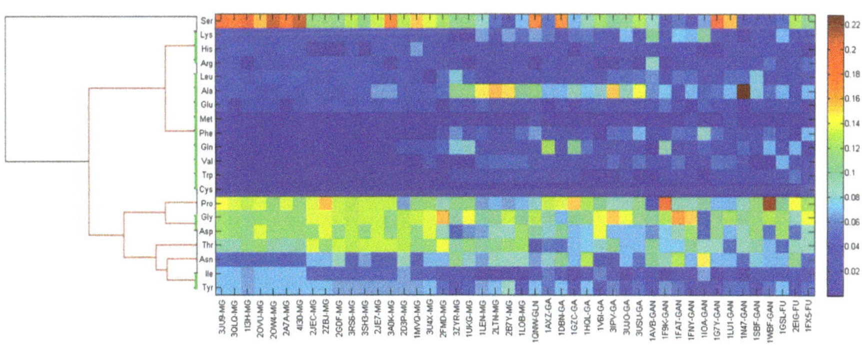

Fig. 13.6 Heatmaps with the dendrogram constructed using percentage composition of amino acids in the binding site loops for the 46 lectins

Table 13.3 Categorisation of amino acids based on their abundance in the full protein and the binding site loops

	Percentage range	MG	GLN	GA	GAN	FU
Full protein						
High	11–15	Ser	Ser	Ser	–	–
Moderate	6–10	Thr, Ala, Val, Leu, Asp, Gly, Asn, Ile	Val, Thr, Ala, Leu, Asn, Gly	Val, Thr, Ala, Leu, Gly, Asn, Phe	Ser, Thr, Val, Ala, Leu, Asn, Phe, Gly, Asp	Ser, Thr, Val, Gly, Leu, Ala, Asp, Ile, Asn, Phe
Low	0–5	Phe, Lys, Pro, Tyr, Arg, Gln, Glu, His, Trp, Met	Pro, Phe, Asp, Ile, Tyr, Glu, Gln, Lys, Trp, Arg, His	Asp, Ile, Lys, Pro, Glu, Tyr, Gln, Arg, Trp, His, Met, Cys	Ile, Pro, Lys, Glu, Gln, Tyr, Arg, Trp, His, Met, Cys	Pro, Lys, Glu, Tyr, Arg, Gln, Trp, His, Met, Cys
Binding site loops						
High	11–15	Ser, Gly, Pro, Thr	Ser, Gly, Ala	Pro, Gly	Gly	Pro, Gly
Moderate	6–10	Asp, Tyr, Asn, Ala	Asp, Pro, Lys, Phe, Asn	Ser, Ala, Asp, Asn, Thr	Pro, Asn, Thr, Asp, Ala, Ser	Ser, Asp, Thr, Asn
Low	0–5	Ile, Leu, Lys, Glu, Arg, His, Gln, Phe, Val, Trp	Gln, Val, His, Leu, Glu, Met, Trp, Thr, Ile, Thr, Arg, Cys	Tyr, Gln, Leu, Lys, Phe, Val, Ile, Glu, His, Arg, Trp	Lys, Phe, Leu, Tyr, Gln, Arg, Val, Glu, Ile, His, Trp, Met	Tyr, Ala, Val, Lys, Arg, Trp, Leu, Phe, Gln, Ile, Glu, His, Cys, Met

(2LTN) and *Vicia faba* (2B7Y). MG proteins had a high percentage of acidic amino acids in comparison with the others. Residues Val and Thr were present in moderate percentages in the full protein but their presence was found to be low at the binding site indicating their importance for the protein structure stability than binding site specificity.

Binding site loop composition clearly differentiates the high presence of Gly over other residues with the maximum in MG constituting 12.37 and 11.39 % in GA binding proteins. GA, GAN and FU are found to be Pro rich at the carbohydrate binding site. The other major residue Ser has a high percentage in MG, GLN and GA specific lectins. Tyr in particular has a moderate percentage of 6.47 and is reported to be involved in CH-pi interaction in MG binding proteins. GAN and FU binding lectins have basic residues Lys and Arg in considerably higher percentage in the loops.

The residues Asp, Asn and Ala have an intermediate percentage in the binding site loops in comparison to the full protein, which is in accordance to their role in non-covalent interactions with the monosaccharide. Similarly Phe, His, Tyr and Trp found in low percentages have been reported to be necessary for stacking interactions with the sugar unit.

Cys and Met were either found in very low percentages or absent and thus were grouped together in the dendrogram.

Conclusion

Pattern recognition through heatmaps assists in reducing data complexity and enhances data interpretation by visualisation. Hence, we have exploited it in this study to analyse the data generated from amino acid variability in a set of 46 legume lectins. Our findings on sequence based variability and phylogenetic analysis are complementary to the previous studies, revealing that legume lectins arose from divergent evolution while retaining a common beta sandwich fold. There is a clear distinction in the sequence identity among these proteins specific to a particular monosaccharide. The results from percentage composition justify the plausible role of certain amino acid residues in the carbohydrate binding site for non-covalent interactions with the sugar.

Acknowledgements This work has been funded by Council of Scientific and Industrial Research (CSIR), India. A.S. is a Bhatnagar fellow, N.G.J. thanks Department of Science and Technology (DST), Govt of India for INSPIRE Fellowship. M.A.S. is a D.S. Kothari fellow supported by University Grants Commission (UGC), India.

References

Audette GF, Vandonselaar M, Delbaere LT (2000) The 2.2 Å resolution structure of the O(H) blood-group-specific lectin I from *Ulex europaeus*. J Mol Biol 304(3):423–433

Babino A, Tello D, Rojas A, Bay S, Osinaga E, Alzari PM (2003) The crystal structure of a plant lectin in complex with the Tn antigen. FEBS Lett 536(1–3):106–110

Benevides RG, Ganne G, Simoes Rda C, Schubert V, Niemietz M, Unverzagt C, Chazalet V, Breton C, Varrot A, Cavada BS, Imberty A (2012) A lectin from *Platypodium elegans* with unusual specificity and affinity for asymmeteric complex N glycans. J Biol Chem 287(31):26352–26364

Berman HM, Westbrook J, Feng Z, Gilliland G, Bhat TN, Weissig H, Shindyalov IN, Bourne PE (2000) The Protein Data Bank. Nucleic Acids Res 28:235–242

Bezerra EH, Rocha BA, Nagano CS, Bezerra Gde A, Moura TR, Bezerra MJ, Benevides RG, Sampaio AH, Assreuy AM, Delatorre P, Cavada BS (2011) Structural analysis of ConBr reveals molecular correlation between the carbohydrate recognition domain and endothelial NO synthase activation. Biochem Biophys Res Commun 408(4):566–570

Bourne Y, Roussel A, Frey M, Rougé P, Fontecilla-Camps JC, Cambillau C (1990) Three-dimensional structures of complexes of *Lathyrus ochrus* isolectin I with glucose and mannose: fine specificity of the monosaccharide-binding site. Proteins 8(4):365–376

Buts L, Dao-Thi MH, Loris R, Wyns L, Etzler M, Hamelryck T (2001) Weak protein-protein interactions in lectins: the crystal structure of a vegetative lectin from the legume *Dolichos biflorus*. J Mol Biol 309(1):193–201

Buts L, Garcia-Pino A, Wyns L, Loris R (2006) Structural basis of carbohydrate recognition by a Man(alpha1-2)Man-specific lectin from *Bowringia milbraedii*. Glycobiology 16(7):635–640

da Silva LC, Correia MT (2014) Plant lectins and Toll-like receptors: implications for therapy of microbial infections. Front Microbiol 5:20

de Oliveira TM, Delatorre P, da Rocha BA, de Souza EP, Nascimento KS, Bezerra GA, Moura TR, Benevides RG, Bezerra EH, Moreno FB, Freire VN, de Azevedo WF Jr, Cavada BS (2008) Crystal structure of *Dioclea rostrata* lectin: insights into understanding the pH-dependent dimer-tetramer equilibrium and the structural basis for carbohydrate recognition in Diocleinae lectins. J Struct Biol 164(2):177–182

de Souza GA, Oliveira PS, Trapani S, Santos AC, Rosa JC, Laure HJ, Faça VM, Correia MT, Tavares GA, Oliva G, Coelho LC, Greene LJ (2003) Amino acid sequence and tertiary structure of *Cratylia mollis* seed lectin. Glycobiology 13(12):961–972

Debraya H, Rougé P (1984) The fine sugar specificity of the *Lathyrus ochrus* seed lectin and isolectins. FEBS Lett 176(1):120–124

Del Sol FG, Cavada BS, Calvete JJ (2007) Crystal structures of *Cratylia floribunda* seed lectin at acidic and basic pHs. Insights into the structural basis of the pH-dependent dimer-tetramer transition. J Struct Biol 158(1):1–9

Delbaere LT, Vandonselaar M, Prasad L, Quail JW, Wilson KS, Dauter Z (1993) Structures of the lectin IV of *Griffonia simplicifolia* and its complex with the Lewis b human blood group determinant at 2.0 Å resolution. J Mol Biol 230(3):950–965

Duin RPW, Pekalska E (2007) The science of pattern recognition; achievements and perspectives. In: Duch W, Mandziuk J (eds) Challenges for computational intelligence, studies in computational intelligence, vol 63. Springer, Heidelberg, pp 221–259

Edelman GM, Cunningham BA, Reeke GN Jr, Becker JW, Waxdal MJ, Wang JL (1972) The covalent and three dimensional structure of concanavalin A. Proc Natl Acad Sci 69:2580–2584

Etzler ME, Surolia A, Cummings RD (2009) L-type lectins (Chap. 29). In: Varki A, Cummings RD, Esko JD et al (eds) Essentials of glycobiology, 2nd edn. Cold Spring Harbor Laboratory Press, Cold Spring Harbor

Gasteiger E, Hoogland C, Gattiker A, Duvaud S, Wilkins MR, Appel RD, Bairoch A (2005) Protein identification and analysis tools on the ExPASy server. In: Walker JM (ed) The proteomics protocols handbook. Humana, Totowa, pp 571–607

Geethanandan K, Abhilash J, Bharath SR, Sadasivan C, Haridas M (2011) X-ray structure of a galactose-specific lectin from *Spatholobous parviflorous*. Int J Biol Macromol 49(5): 992–998

Hamelryck TW, Dao-Thi MH, Poortmans F, Chrispeels MJ, Wyns L, Loris R (1996a) The crystallographic structure of phytohemagglutinin-L. J Biol Chem 271(34):20479–20485

Hamelryck TW, Poortmans F, Goossens A, Angenon G, Van Montagu M, Wyns L, Loris R (1996b) Crystal structure of arcelin-5, a lectin-like defense protein from *Phaseolus vulgaris*. J Biol Chem 271(51):32796–32802

Hamelryck TW, Loris R, Bouckaert J, Dao-Thi MH, Strecker G, Imberty A, Fernandez E, Wyns L, Etzler ME (1999) Carbohydrate binding, quaternary structure and a novel hydrophobic binding site in two legume lectin oligomers from *Dolichos biflorus*. J Mol Biol 286(4):1161–1177

Imberty A, Gautier C, Lescar J, Pérez S, Wyns L, Loris R (2000) An unusual carbohydrate binding site revealed by the structures of two *Maackia amurensis* lectins complexed with sialic acid-containing oligosaccharides. J Biol Chem 275(23):17541–17548

Kolberg J, Michaelsen TE, Sletten K (1980) Subunit structure and N terminal sequences of the *Lathyrus odoratus* lectin. FEBS Lett 117:281–283

Kundhavai Natchiar S, Arockia Jeyaprakash A, Ramya TN, Thomas CJ, Suguna K, Surolia A, Vijayan M (2004) Structural plasticity of peanut lectin: an X-ray analysis involving variation in pH, ligand binding and crystal structure. Acta Crystallogr D Biol Crystallogr 60(2):211–219

Larkin MA, Blackshields G, Brown NP, Chenna R, McGettigan PA, McWilliam H, Valentin F, Wallace IM, Wilm A, Lopez R, Thompson JD, Gibson TJ, Higgins DG (2007) ClustalW and ClustalX version 2. Bioinformatics 23(21):2947–2948

Liener IE, Sharon N, Goldstein IJ (1986) The lectins: properties, functions, and applications in biology and medicine. Academic Press, Orlando

Loris R, Van Overberge D, Dao-Thi MH, Poortmans F, Maene N, Wyns L (1994) Structural analysis of two crystal forms of lentil lectin at 1.8 Å resolution. Proteins 20(4):330–346

Loris R, Hamelryck T, Bouckaert J, Wyns L (1998) Legume lectin structure. Biochim Biophys Acta 1383:9–36

Loris R, De Greve H, Dao-Thi MH, Messens J, Imberty A, Wyns L (2000) Structural basis of carbohydrate recognition by lectin II from *Ulex europaeus*, a protein with a promiscuous carbohydrate-binding site. J Mol Biol 301(4):987–1002

Loris R, Van Walle I, De Greve H, Beeckmans S, Deboeck F, Wyns L, Bouckaert J (2004) Structural basis of oligomannose recognition by the *Pterocarpus angolensis* seed lectin. J Mol Biol 335(5):1227–1240

MacQueen JB (1967) Some methods for classification and analysis of multivariate observations. In: Proceedings of 5th Berkeley symposium on mathematical statistics and probability, vol 1, University of California Press, Berkeley, pp 281–297

Manoj N, Suguna K (2001) Signature of quaternary structure in the sequences of legume lectins. Protein Eng 14(10):735–745

Manoj N, Srinivas VR, Suguna K (1999) Structure of basic winged-bean lectin and a comparison with its saccharide-bound form. Acta Crystallogr D Biol Crystallogr 55(4):794–800

Manoj N, Srinivas VR, Surolia A, Vijayan M, Suguna K (2000) Carbohydrate specificity and salt-bridge mediated conformational change in acidic winged bean agglutinin. J Mol Biol 302(5):1129–1137

MATLAB v7.5, MathWorks (2007) Bioinformatics toolbox: user's guide (R2007b)

Moreira GM, Conceição FR, McBride AJ, Pinto Lda S (2013) Structure predictions of two *Bauhinia variegata* lectins reveal patterns of C-terminal properties in single chain legume lectins. PLoS One 8(11)

Moreno FB, Bezerra GA, Oliveira TM, de Souza EP, da Rocha BA, Benevides RG, Delatorre P, de Azevedo WF, Jr CBS (2007) Structural analysis of *Canavalia maritima* and *Canavalia gladiata* lectins complexed with different dimannosides: new insights into the understanding of the structure-biological activity relationship in legume lectins. J Struct Biol 160(2): 168–176

Moreno FB, de Oliveira TM, Martil DE, Viçoti MM, Bezerra GA, Abrego JR, Cavada BS, Filgueira de Azevedo W Jr (2008) Identification of a new quaternary association for legume lectins. J Struct Biol 161(2):133–143

Mourey L, Pédelacq JD, Birck C, Fabre C, Rougé P, Samama JP (1998) Crystal structure of the arcelin-1 dimer from *Phaseolus vulgaris* at 1.9 Å resolution. J Biol Chem 273(21): 12914–12922

Mueller-Dieckmann C, Panjikar S, Tucker PA, Weiss MS (2005) On the routine use of soft X-rays in macromolecular crystallography. Part III The optimal data-collection wavelength. Acta Crystallogr D Biol Crystallogr 61(9):1263–1272

Nagano CS, Calvete JJ, Barettino D, Pérez A, Cavada BS, Sanz L (2008) Insights into the structural basis of the pH-dependent dimer-tetramer equilibrium through crystallographic analysis of recombinant *Diocleinae* lectins. Biochem J 409(2):417–428

Nóbrega RB, Rocha BA, Gadelha CA, Santi-Gadelha T, Pires AF, Assreuy AM, Nascimento KS, Nagano CS, Sampaio AH, Cavada BS, Delatorre P (2012) Structure of *Dioclea virgata* lectin: Relations between carbohydrate binding site and nitric oxide production. Biochimie 94(3):900–906

Olsen LR, Dessen A, Gupta D, Sabesan S, Sacchettini JC, Brewer CF (1997) X-ray crystallographic studies of unique cross-linked lattices between four isomeric biantennary oligosaccharides and soybean agglutinin. Biochemistry 36(49):15073–15080

Rabijns A, Verboven C, Rougé P, Barre A, Van Damme EJ, Peumans WJ, De Ranter CJ (2001) Structure of a legume lectin from the bark of *Robinia pseudoacacia* and its complex with N-acetylgalactosamine. Proteins 44(4):470–478

Rangel TB, Rocha BA, Bezerra GA, Assreuy AM, Pires Ade F, do Nascimento AS, Bezerra MJ, do Nascimento KS, Nagano CS, Sampaio AH, Gruber K, Delatorre P, Fernandes PM, Cavada BS (2012) Crystal structure of a pro-inflammatory lectin from the seeds of Dioclea wilsonii Standl. Biochimie 94(2):525–532

Rao VSR, Lam K, Qasba PK (1998) Architecture of the sugar binding sites in carbohydrate binding proteins—a computer modeling study. Int J Biol Macromol 23(4):295–307

Reeke GN Jr, Becker JW (1986) Three-dimensional structure of favin: saccharide binding-cyclic permutation in leguminous lectins. Science 234(4780):1108–1111

Rocha BA, Delatorre P, Oliveira TM, Benevides RG, Pires AF, Sousa AA, Souza LA, Assreuy AM, Debray H, de Azevedo WF, Jr SAH, Cavada BS (2011) Structural basis for both pro- and anti-inflammatory response induced by mannose-specific legume lectin from *Cymbosema roseum*. Biochimie 93(5):806–816

Rocha BA, Souza Teixeira C, da Silva HC, de Moura TR, Pereira-Júnior FN, do Nascimento KS, Nagano CS, Sampaio AH, Delatorre P, Cavada BS (2012) Crystal structure of the lectin of Camptosema pedicellatum: implications of a conservative substitution at the hydrophobic subsite. J Biochem 152(1):87–98

Sanders DA, Moothoo DN, Raftery J, Howard AJ, Helliwell JR, Naismith JH (2001) The 1.2 Å resolution structure of the Con A-dimannose complex. J Mol Biol 310(4):875–884

Shaanan B, Elgavish S (1998) Structures of the *Erythrina corallodendron* lectin and of its complexes with mono- and disaccharides. J Mol Biol 277(4):917–932

Sharma V, Surolia A (1997) Analyses of carbohydrate recognition by legume lectins: size of the combining site loops and their primary specificity. J Mol Biol 267:433–445

Sharma V, Srinivas VR, Adhikari P, Vijayan M, Surolia A (1998) Molecular basis of recognition by Gal/GalNAc specific legume lectins: influence of Glu 129 on the specificity of peanut agglutinin (PNA) towards C2-substituents of galactose. Glycobiology 8(10):1007–1012

Sharon N, Lis H (1990) Legume lectins: a large family of homologous proteins. FASEB J 4:3198–3208

Sharon N, Lis H (1995) Lectins-proteins with a sweet tooth: functions in cell recognition. Essays Biochem 30:59–75

Sharon N, Lis H (2002) How proteins bind carbohydrates: lessons from legume lectins. J Agric Food Chem 50:6586–6591

Sharon N, Lis H (2004) History of lectins: from hemagglutinins to biological recognition molecules. Glycobiology 14(11):53–62

Shetty KN, Latha VL, Rao RN, Nadimpalli SK, Suguna K (2013) Affinity of a galactose-specific legume lectin from *Dolichos lablab* to adenine revealed by X-ray crystallography. IUBMB Life 65(7):633–644

Srinivas VR, Reddy GB, Ahmad N, Swaminathan CP, Mitra N, Surolia A (2001) Legume lectin family, the 'natural mutants of the quaternary state', provide insights into the relationship between protein stability and oligomerization. Biochim Biophys Acta 1527:102–111

Suddath FL, Prasthofer T, Phillips SR, Engler JA (1989) Design, expression, and crystallization of recombinant lectin from the garden pea (*Pisum sativum*). J Biol Chem 264(12):6793–6796

Sundberg EJ, Trastoy B, Bonsor DA, Perez-Ojeda ME, Jimeno ML, Garcia-Fernandez JM, Chiara JL (2012) Synthesis and biophysical study of disassembling nano hybrid bioconjugates with a cubic octasilsesquioxane core. Adv Funct Mater 22:3191–3201

Svensson C, Teneberg S, Nilsson CL, Kjellberg A, Schwarz FP, Sharon N, Krengel U (2002) High-resolution crystal structures of *Erythrina cristagalli* lectin in complex with lactose and 2′-alpha-L-fucosyllactose and correlation with thermodynamic binding data. J Mol Biol 321(1):69–83

Swamy MJ, Sastry MVK, Surolia A (1985) Prediction and comparison of the secondary structure of legume lectins. J Biosci 9(3–4):203–212

Tamura K, Stecher G, Peterson D, Filipski A, Kumar S (2013) MEGA6: molecular evolutionary genetics analysis version 6.0. Mol Biol Evol 30:2725–2729

Tempel W, Tschampel S, Woods RJ (2002) The xenograft antigen bound to *Griffonia simplicifolia* lectin 1-B(4). X-ray crystal structure of the complex and molecular dynamics characterization of the binding site. J Biol Chem 277(8):6615–6621

Thomas CJ, Surolia A (2000) Mode of molecular recognition of l-fucose by fucose-binding legume lectins. Biochem Biophys Res Commun 268:262–267

Weisstein E (1995) K-Means clustering algorithm. From MathWorld—A Wolfram Web Resource. http://mathworld.wolfram.com/K-MeansClusteringAlgorithm.html

Xu D, Zhang Y (2013) Toward optimal fragment generations for ab initio protein structure assembly. Proteins 81(2):229–239

Young NM, Oomen RP (1992) Analysis of sequence variation among legume lectins. J Mol Biol 228:924–934

Zhang Y (2012) http://zhanglab.ccmb.med.umich.edu/PSSpred

Chapter 14
Conformational Dynamics of Oligosaccharides Characterized by Paramagnetism-Assisted NMR Spectroscopy in Conjunction with Molecular Dynamics Simulation

Ying Zhang, Takumi Yamaguchi, Tadashi Satoh, Maho Yagi-Utsumi, Yukiko Kamiya, Yoshitake Sakae, Yuko Okamoto, and Koichi Kato

Abbreviations

Aβ	Amyloid β
HSQC	Heteronuclear single-quantum coherence
MD	Molecular dynamics
NMR	Nuclear magnetic resonance
NOE	Nuclear Overhauser effect

Y. Zhang • T. Yamaguchi
Okazaki Institute for Integrative Bioscience and Institute for Molecular Science, National Institutes of Natural Sciences, 5-1 Higashiyama, Myodaiji, Okazaki 444-8787, Japan

School of Physical Sciences, The Graduate University for Advanced Studies, 5-1 Higashiyama, Myodaiji, Okazaki 444-8787, Japan

Graduate School of Pharmaceutical Sciences, Nagoya City University, 3-1 Tanabe-dori, Mizuho-ku, Nagoya 467-8603, Japan

T. Satoh
JST, PRESTO, 3-1 Tanabe-dori, Mizuho-ku, Nagoya 467-8603, Japan

Graduate School of Pharmaceutical Sciences, Nagoya City University, 3-1 Tanabe-dori, Mizuho-ku, Nagoya 467-8603, Japan

M. Yagi-Utsumi
Present address: Department of Chemistry, University of Cambridge, Lensfield Road, Cambridge CB2 1EW, UK

Okazaki Institute for Integrative Bioscience and Institute for Molecular Science, National Institutes of Natural Sciences, 5-1 Higashiyama, Myodaiji, Okazaki 444-8787, Japan

Graduate School of Pharmaceutical Sciences, Nagoya City University, 3-1 Tanabe-dori, Mizuho-ku, Nagoya 467-8603, Japan

A. Chakrabarti, A. Surolia (eds.), *Biochemical Roles of Eukaryotic Cell Surface Macromolecules*, Advances in Experimental Medicine and Biology 842,
DOI 10.1007/978-3-319-11280-0_14

PCS	Pseudocontact shift
REMD	Replica-exchange molecular dynamics
αSN	α-Synuclein

Introduction: Static and Dynamic Views of Carbohydrate–Protein Interactions

The biological functions of oligosaccharides predominantly occur through their interactions with proteins (Drickamer and Taylor 1993; Sharon and Lis 2003). In cells, a series of high-mannose-type oligosaccharides present on secretory proteins serve as quality tags, which are decoded by intracellular lectins involved in the folding, transport, and degradation of glycoproteins (Kamiya et al. 2012). On the cell surface, oligosaccharides that modify membrane proteins and lipids act as acceptors for extracellular proteins, thereby mediating a variety of physiological and pathological events involved in cellular communication and development, cancer metastasis, viral infections, and the promotion of neurodegenerative disorders (Ernst and Magnani 2009). In order to elucidate the mechanisms underlying these oligosaccharide functions, an in-depth description of carbohydrate–protein interactions is essential, which will provide important clues for future therapeutic approaches (Kamiya et al. 2011a).

One of the most powerful means of characterizing carbohydrate–protein interactions is X-ray crystallography, which can provide the atomic coordinates of biomacromolecules and their complexes. For example, we recently determined the crystal

Y. Kamiya
EcoTopia Science Institute and Graduate School of Engineering, Nagoya University, Furo-cho, Chikusa-ku, Nagoya 464-8603, Japan

Okazaki Institute for Integrative Bioscience and Institute for Molecular Science, National Institutes of Natural Sciences, 5-1 Higashiyama, Myodaiji, Okazaki 444-8787, Japan

Y. Sakae • Y. Okamoto
School of Science, Nagoya University, Furo-cho, Chikusa-ku, Nagoya 464-8602, Japan

K. Kato (✉)
Okazaki Institute for Integrative Bioscience and Institute for Molecular Science, National Institutes of Natural Sciences, 5-1 Higashiyama, Myodaiji, Okazaki 444-8787, Japan

School of Physical Sciences, The Graduate University for Advanced Studies, 5-1 Higashiyama, Myodaiji, Okazaki 444-8787, Japan

Graduate School of Pharmaceutical Sciences, Nagoya City University, 3-1 Tanabe-dori, Mizuho-ku, Nagoya 467-8603, Japan

The Glycoscience Institute, Ochanomizu University, 2-1-1 Ohtsuka, Bunkyo-ku, Tokyo 112-8610, Japan

GLYENCE Co., Ltd., 2-22-8 Chikusa, Chikusa-ku, Nagoya 464-0858, Japan
e-mail: kkatonmr@ims.ac.jp

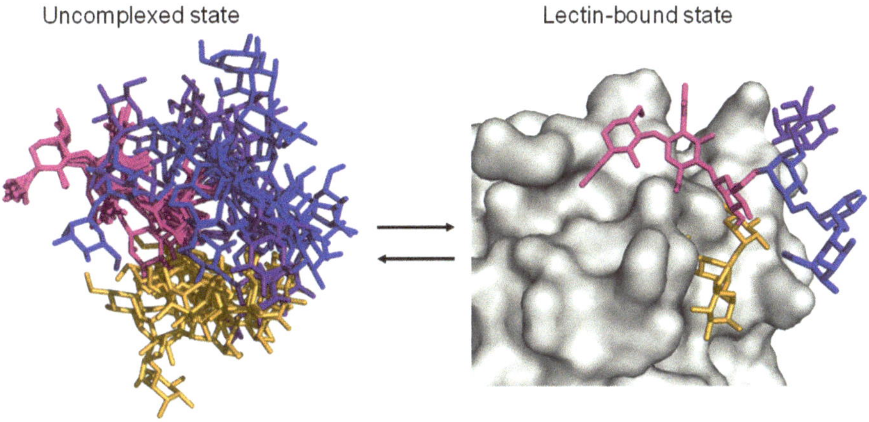

Fig. 14.1 Static and dynamic views of carbohydrate–protein interactions. 3D-structure models of the high-mannose-type oligosaccharide $Man_9GlcNAc_2$ exhibiting conformational dynamics in solution (*left*) and accommodated in the sugar-binding pocket of ERGIC-53 (*right*)

structures of ternary complexes composed of the carbohydrate-recognition domain of ERGIC-53, its binding partner MCFD2, and α1,2 mannotriose (Satoh et al. 2014). This allowed us to build a 3D-structural model of the complex formed between this lectin and a high-mannose-type oligosaccharide $Man_9GlcNAc_2$, which visualizes atomic contacts mediating the carbohydrate–protein interaction (Fig. 14.1). This model demonstrates how the carbohydrate-binding pocket of ERGIC-53 accommodates the non-reducing terminal residues of the cognate ligand, thereby explaining the structural basis of the sugar-binding specificities of this lectin in comparison to its homologs. However, it should be noted that the sugar chains are highly flexible and mobile, at least in their uncomplexed forms (Fig. 14.1). This property endows the oligosaccharides with conformational adaptabilities upon interaction with proteins resulting in a loss of conformational entropy. For quantitative evaluation of the energetics of the carbohydrate–protein interactions, it is crucial to understand such oligosaccharide conformational dynamics. Nuclear magnetic resonance (NMR) spectroscopy has a powerful potential for characterizing the conformational dynamics of oligosaccharides in solution. However, several significant issues need to be addressed when applying NMR spectroscopy for detailed analyses of the dynamic conformation of oligosaccharides, as described below.

How to Prepare NMR Samples

First, it is necessary to obtain sufficient quantities of homogeneous oligosaccharides. Here we primarily used the pentasaccharide moiety of gagnglioside GM1, Galβ1-3GalNAcβ1-4(Neu5Acα2-3)Galβ1-4Glc, as a model molecule because this glycosphingolipid is abundant in animal brains and commercially available as a source of the homogeneous oligosaccharide. In general, however, glycoprotein glycans are structurally heterogeneous and much less abundant. Sophisticated synthetic approaches have proved useful in providing large quantities of oligosaccharides with complicated branched structures (Ando et al. 2010; Lepenies et al. 2010; Takeda et al. 2009; Wang et al. 2013). Another promising approach has been the use of eukaryotic glycoprotein expression systems along with genetic engineering (Kamiya et al. 2014). We used *Saccharomyces cerevisiae* strains with deletion of genes associated with *N*-glycan processing in the secretory pathway, for producing glycoproteins with homogeneous glycoforms (Fig. 14.2a) (Kamiya et al. 2011b, 2013; Nakanishi-Shindo et al. 1993). From the engineered yeast cells, high yields of specific high-mannose-type oligosaccharides could be isolated. Key advantages of this technique include the efficient production of stable-isotope-labeled oligosaccharides using yeast metabolic labeling, which is extremely useful for detailed NMR analyses (Fig. 14.2b, c).

How to Obtain Atomic Long-Distance Information

NMR determination of biomolecular conformations are typically performed on the basis of the nuclear Overhauser effect (NOE) and scalar coupling (or *J*-coupling) (Peters and Pinto 1996; Wormald et al. 2002; Zhao et al. 2007). However, the major limitation of NMR analyses of oligosaccharides is the insufficiency of conformational restraints provided by these local conformational parameters because of the low number of protons that contribute to defining inter-glycosidic linkage conformation. To overcome this problem, we applied paramagnetic effects that can provide the long-distance information of oligosaccharides (Fig. 14.3) (Zhang 2014; Zhang et al. 2013). In this approach, lanthanide ions with unpaired electrons are attached to the reducing terminal of the oligosaccharide using a metal chelator (Yamamoto et al. 2011, 2012; Zhang et al. 2012). The magnetic dipole–dipole interaction of unpaired electrons with their spatially proximal nuclear spins perturb the chemical shifts and relaxation observed in NMR spectra of the tagged oligosaccharides, thus reflecting the geometrical arrangements of the individual ^{1}H and ^{13}C with respect to the paramagnetic probe (Luchinat and Parigi 2007). Figure 14.4 shows the method of introduction of a lanthanide ion at the reducing end of an oligosaccharide: The pentasaccharide moiety of GM1 was enzymatically cleaved from the ceramide part and chemically connected with a phenylenediamine-derived lanthanide-chelating group. Upon coordination of paramagnetic lanthanide ions, such as Tm^{3+}, spectral changes were induced due to pseudocontact shifts (PCSs), in comparison with the reference spectrum observed with a diamagnetic La^{3+} ion as the source of the atomic long-distance information (Fig. 14.5). We also used reducing-terminal spin labeling with a nitroxide radical, which accelerates NMR relaxation of nuclei in its spatial proximity, for characterizing overall conformation of high-mannose-type oligosaccharides (Yamaguchi et al. 2013a).

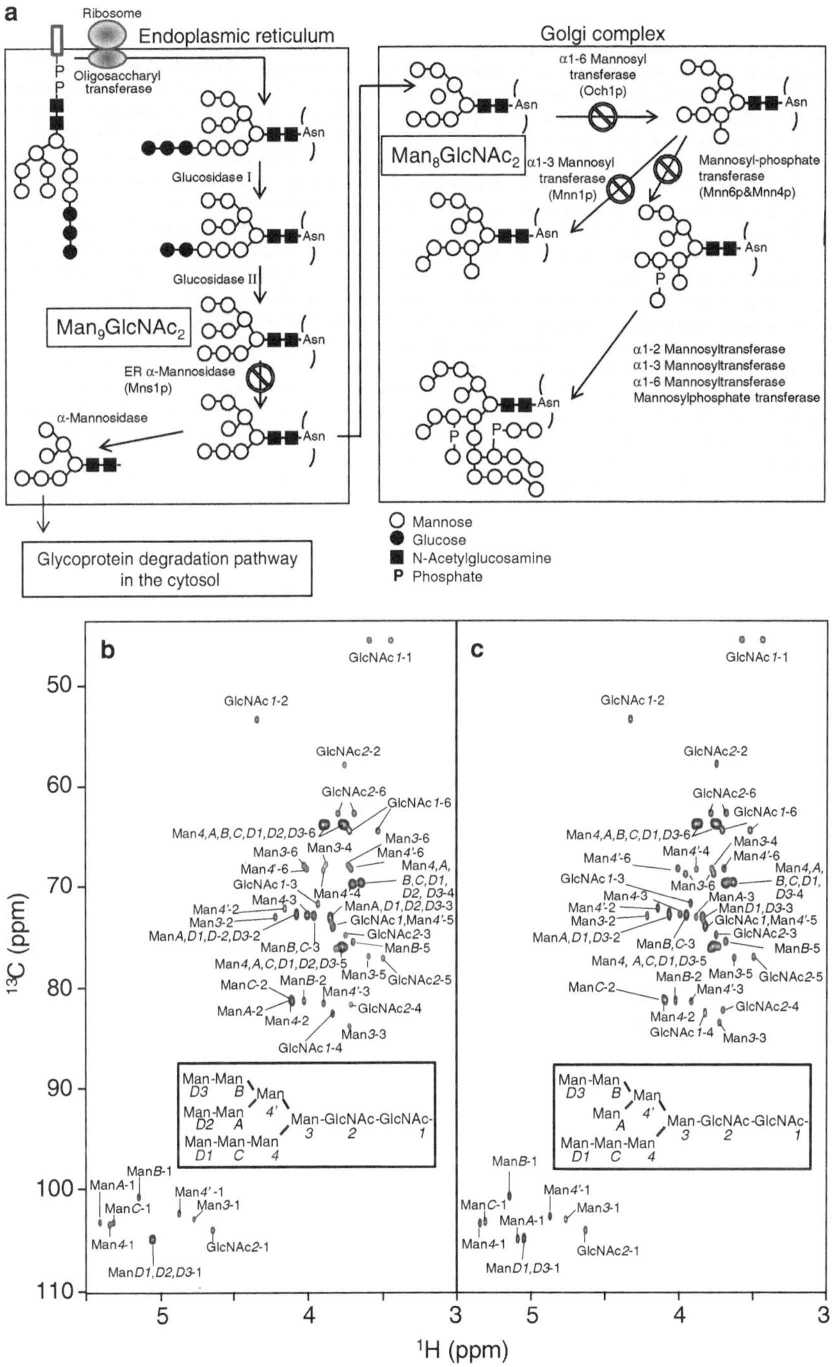

Fig. 14.2 (**a**) Genetically engineered yeast cells for the overexpression of homogeneous high-mannose-type oligosaccharides $Man_9GlcNAc_2$ and $Man_8GlcNAc_2$. ^{1}H-^{13}C HSQC spectra of uniformly ^{13}C-labeled (**b**) $Man_9GlcNAc_2$ and (**c**) $Man_8GlcNAc_2$, respectively, produced in the yeast cells. These figures were adapted from Kamiya et al. (2011b, 2013) with the permissions of Springer and MDPI

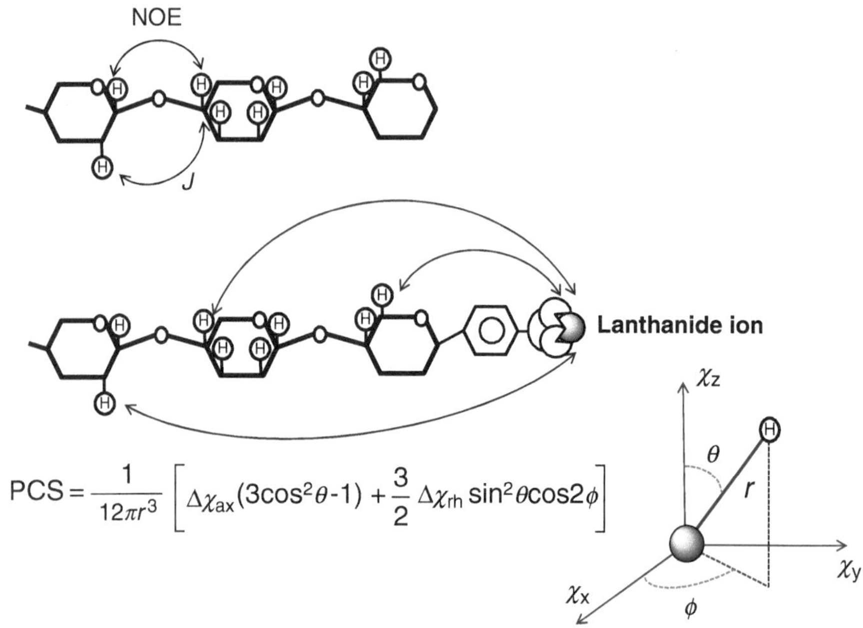

Fig. 14.3 Conformational restraints provided by NOE, *J*, and PCS for characterizing glycosidic linkage conformation. In the equation for PCS, $\Delta\chi_{ax}$ and $\Delta\chi_{rh}$ are the axial and rhombic components, respectively, of the anisotropic magnetic susceptibility ($\boldsymbol{\Delta\chi}$) tensor, and r, θ, and φ are the spherical coordinates of the nucleus defined with respect to the paramagnetic center and the principal axis of the $\boldsymbol{\Delta\chi}$ tensor

How to Deal with Dynamic Conformational Ensembles

Structural information provided by NMR is averaged over dynamic conformational ensembles because conformational transitions of the free oligosaccharides occur very rapidly on the NMR time scale. By contrast, computational approaches including molecular dynamics (MD) simulation can describe the molecular behavior of an oligosaccharide in solution at the atomic level. However, the simulation results heavily depend on calculation protocols such as force field, initial state, and simulation time, and therefore have to be validated on the basis of experimental observations. We combined the paramagnetism-assisted NMR methods with MD simulation to explore the conformational space occupied by a flexible oligosaccharide in solution (Fig. 14.6) (Zhang 2014; Zhang et al. 2013; Yamaguchi et al. 2014). Here, we show the results obtained through this methodology using a series of ganglioside oligosaccharides, the GM3 trisaccharide, GM2 tetrasaccharide, and GM1 pentasaccharide (Yamamoto et al. 2012; Zhang 2014; Zhang et al. 2012).

To explore the conformational spaces of the carbohydarate moieties of GM3 and GM2, ten MD runs were performed for each oligosaccharide in explicit water with GLYCAM_06 force field for 12 ns at 300 K and combined after excluding the first

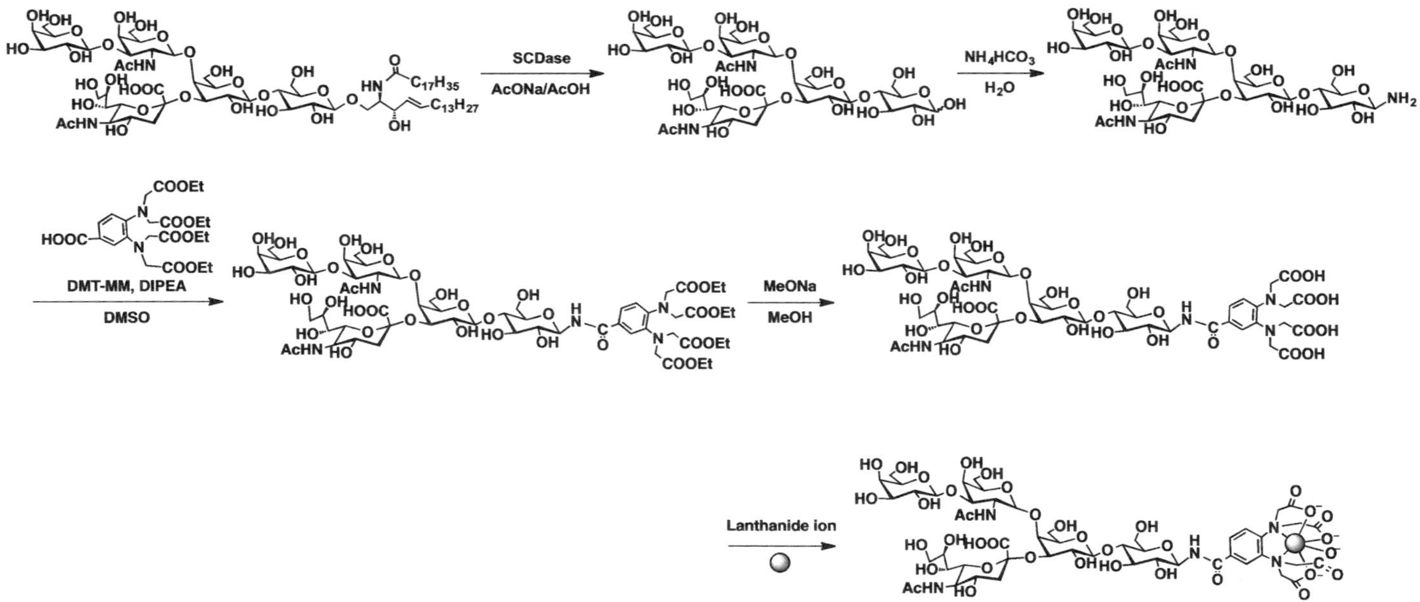

Fig. 14.4 Scheme of lanthanide-tagging of the GM1 pentasaccharide. SCDase = sphingolipid ceramide *N*-deacylase, DIPEA = *N*,*N*-diisopropylethylamine, DMT-MM = 4-(4,6-dimethoxy-1,3,5-triazin-2-yl)-4-methylmorpholinium chloride

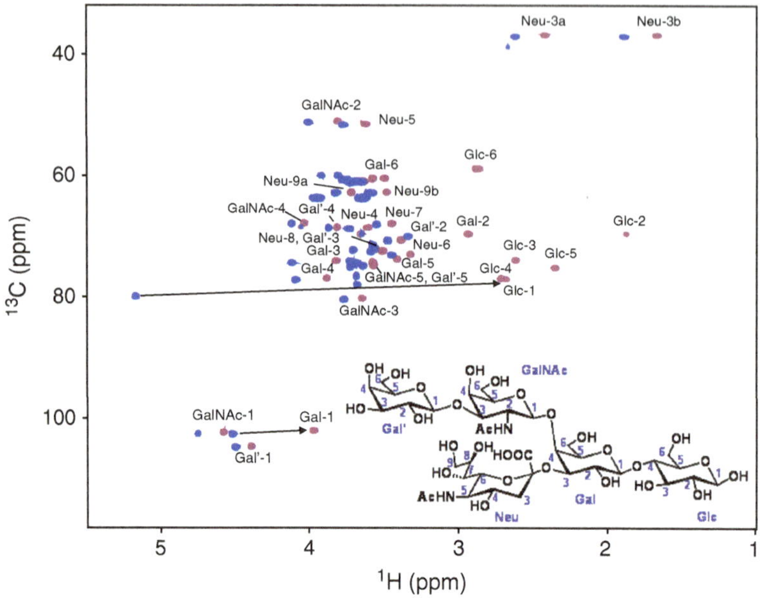

Fig. 14.5 ^{1}H-^{13}C HSQC spectral changes of the tagged GM1 pentasaccharide resulting from paramagnetic lanthanide-induced PCSs. The spectra of 1:1 complexes (0.1 mM each) of the pentasaccharide with paramagnetic Tm^{3+} (*magenta*) and diamagnetic La^{3+} (*blue*) in D_2O were recorded at 300 K on a Bruker AVANCE 800 spectrometer equipped with a cryogenic probe (at Instrument Center, IMS)

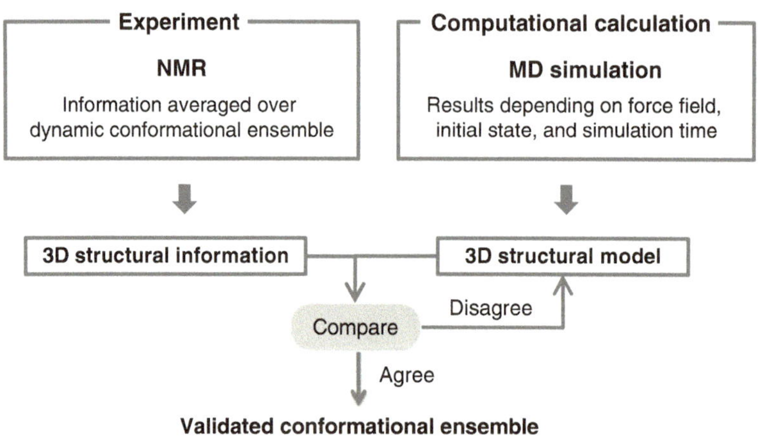

Fig. 14.6 A coalition between NMR spectroscopy and MD simulation for characterizing dynamic conformational ensembles of oligosaccharides

2 ns trajectories. Theoretical PCS values were computed from ensemble models made of 2,000 conformers that were extracted from the MD trajectories at equal intervals. By confirming the close agreement between the experimentally observed and computationally calculated PCS values, the MD-derived conformational spaces were validated for the GM3 trisaccharide and GM2 tetrasaccharide.

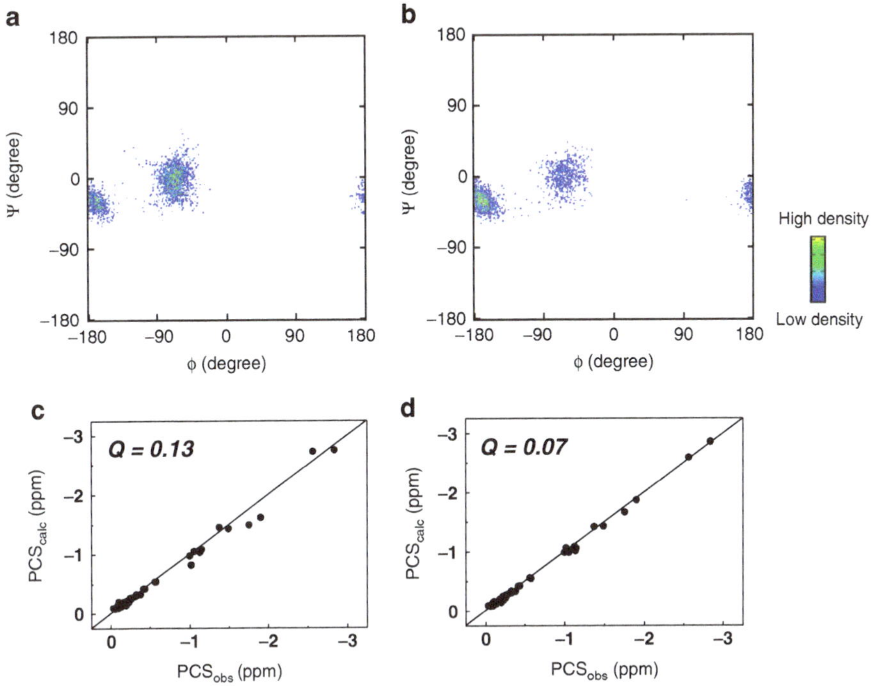

Fig. 14.7 Torsion angle density maps of the Neu5Ac–Gal linkage of the GM1 pentasaccharides obtained by (**a**) the conventional MD and (**b**) the REMD simulation. The torsion angles **Φ** and **ψ** were defined as C1–C2–O′3–C′3 and C2–O′3–C′3–H′3, respectively. (**c**, **d**) Correlations between experimentally observed PCS values with Tm^{3+} and computationally calculated PCS data for the GM1 pentasaccharide. The PCS values were back calculated from ensemble models derived from (**c**) the conventional MD and (**d**) the REMD simulation using $\Delta\chi_{ax}$ and $\Delta\chi_{rh}$ determined based on the protocol described in the literature (Yamamoto et al. 2012). The REMD simulations of the carbohydrate moiety in explicit water were performed by using the AMBER12 program package. The simulations were carried out in *NVT* ensemble for 60 ns with 32 replicas with an exponential temperature distribution between 300 and 500 K. *Q* values defined by the equation $Q = \text{rms}(PCS_{calc} - PCS_{obs})/\text{rms}(PCS_{obs})$ are criteria of agreement between experimental and calculated values

Using the same simulation protocol, an ensemble model of the GM1 pentasaccharide was created and subjected to the validation analysis. In contrast to the GM3 and GM2 cases, there was a significant disagreement between the experimental and theoretical PCS data, even though the total simulation time used for creation of the ensemble model was extended up to 240 ns (Fig. 14.7a). The discordance is most probably explained by insufficient sampling due to higher energy barriers between the multiple minima in the conformational energy-landscape of the large, branched oligosaccharides. To efficiently explore such a rough energy landscape dealing with the multiple-minima issue, we used replica-exchange MD (REMD) simulations, in which replicated simulations were run at different temperatures and exchanged during simulations to avoid being trapped into a local-energy-minimum state (Sugita and Okamoto 1999). We performed a REMD simulation using

GLYCAM_06 force field with a total simulation time of 1.9 μs and thereby obtained torsion angle density maps of the GM1 pentasaccharide. Significantly different conformational spaces were computed from the conventional MD and REMD simulations as exemplified by the Neu5Ac–Gal glycosidic linkage conformation (Fig. 14.7). The experimentally observed PCS data of this oligosaccharide are well represented in the ensemble model composed of 2,000 conformers that were extracted from the REMD simulation, demonstrating the utility of the method for the exploration of conformational spaces of large, branched oligosaccharides (Fig. 14.7).

Figure 14.8 compares the PCS-validated conformational spaces of the three ganglioside oligosaccharides, indicating similarities between the GM1 and GM2 oligosaccharides in terms of the conformational space of their common parts, i.e., the Gal-Glc, Neu5Ac-Gal, and GalNAc-Gal glycosidic linkage conformations. By contrast, the GM3 trisaccharide is distinct from the others regarding the Neu5Ac-Gal conformation. In the GM3 trisaccharide, the conformation of this linkage is most

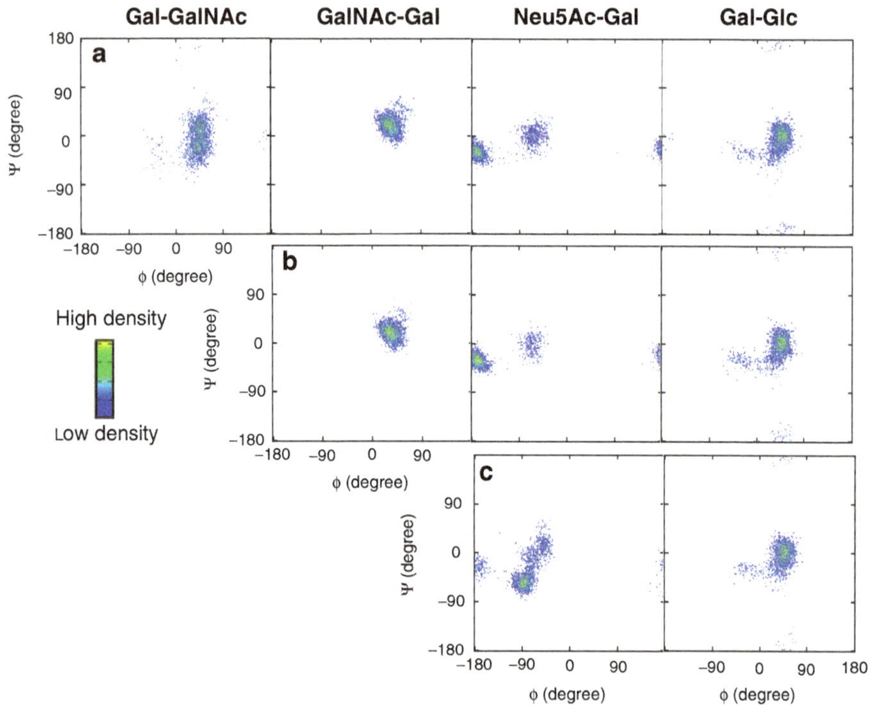

Fig. 14.8 Torsion angle density maps of the experimentally validated MD trajectory of the (**a**) GM1, (**b**) GM2, and (**c**) GM3 oligosaccharides. The definitions of **Φ** and **ψ** were used for the Gal–GalNAc linkage, **Φ**=H1–C1–O′3–C′3 and **ψ**=C1–O′3–C′3–H′3, for the GalNAc–Gal and Gal–Glc linkages, **Φ**=H1–C1–O′4–C′4 and **ψ**=C1–O′4–C′4–H′4, and for the Neu5Ac–Gal linkage, **Φ**=C1–C2–O′3–C′3 and **ψ**=C2–O′3–C′3–H′3. This figure was from Zhang (2014) and parts of this figure were originally reproduced from Yamamoto et al. (2012) and Zhang et al. (2013) with the permissions of The Royal Society of Chemistry and MDPI

populated in the cluster ($(\phi, \psi)=(-90° \pm 11°, -57° \pm 11°)$), while the corresponding cluster is missing in the sialyl linkage conformations of GM1 and GM2. These data show that the GalNAc branch restricts the conformational freedom of the Neu5Ac-Gal glycosidic linkage, while the outermost Gal residue has no significant impact on the conformation of the remaining parts of the carbohydrate moiety.

Concluding Remarks and Perspectives: Glycolipid clusters as a Platform for Protein Interactions

We developed the paramagnetism-assisted NMR approach in association with MD simulations, and this has been successfully applied to the characterization of the conformational dynamics of oligosaccharides derived from gangliosides. This methodology is applicable to the conformational NMR analyses of oligosaccharides in complex with proteins and would provide a new avenue toward atomic descriptions of dynamic oligosaccharide behaviors involved in interactions with proteins, which would in turn enable quantitative understanding of the energetics of carbohydrate recognition events. However, it should be noted that the gangliosides form clusters on cell surfaces and thereby promote sophisticated biomolecular functions.

Recently, growing evidence has indicated that ganglioside clusters on neuronal cell surfaces act as unique platforms for binding coupled with conformational transition of intrinsically disordered proteins involved in neurodegenerative diseases, e.g., amyloid β (Aβ) in Alzheimer's disease and α-synuclein (αSN) in Parkinson's disease (Fantini and Yahi 2010; Matsuzaki et al. 2010; Piccinini et al. 2010). These proteins interact with specific ganglioside clusters, and thereby undergo conformational changes resulting in the formation of their toxic, aggregated forms. To gain deeper insights into the molecular mechanisms underlying these cell-surface events, it is of vital importance to elucidate the dynamic conformation and interactions of the proteins and glycolipids involved therein by employing appropriate membrane models. In this context, it would be interesting to compare conformational dynamics between the liberated oligosaccharides and their assembled states.

At the hydrophilic/hydrophobic interfaces in membrane-like environments, Aβ and αSN have been reported to assume α-helical structures (Bodner et al. 2010; Ulmer et al. 2005; Utsumi et al. 2009; Yagi-Utsumi et al. 2010). Recently developed ganglioside-embedding small bicelles serve as nanoscale standardized membrane mimics for detailed NMR characterization of the carbohydrate–protein interactions on the glycolipid clusters (Fig. 14.9) (Gayen and Mukhopadhyay 2008; Khatun and Mukhopadhyay 2013; Yamaguchi et al. 2013b). By using these systems, we have successfully reveled that αSN interacts with GM1 and GM2 but not GM3, through its most ganglioside-philic site located in the N-terminal segment because of the limited sizes of the bicelles (Fig. 14.9). This success opens up new possibilities for probing the initial encounter complex transiently formed between proteins and glycolipids on membranes depending on their outer carbohydrate structure. Further developments of stable-isotope- and paramagnetism-assisted

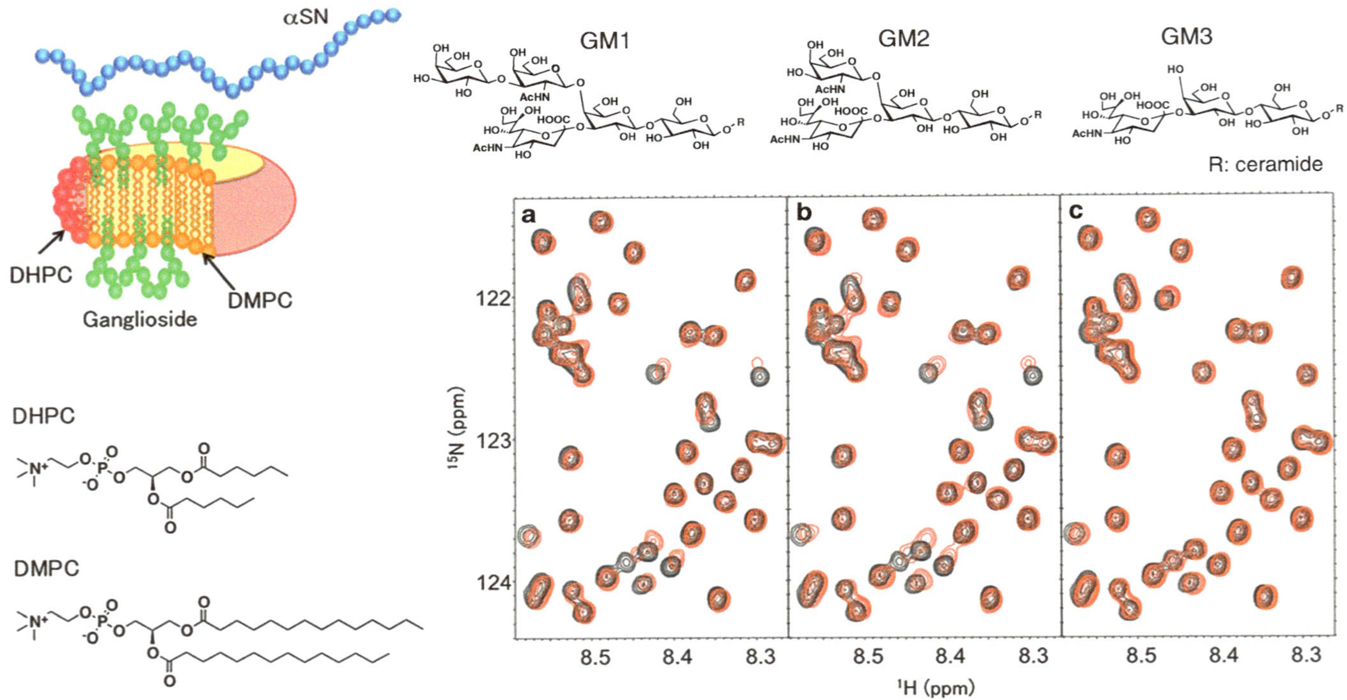

Fig. 14.9 NMR characterization of interactions of **αSN** with the ganglioside-embedding small bicelles as nanoscale standardized membrane mimics for NMR analyses. ^{1}H-^{15}N HSQC spectral changes of **αSN** induced by interactions with gangliosides (**a**) GM1, (**b**) GM2, and (**c**) GM3 embedded in the bicelles. This figure was from Zhang et al. (2013) and parts of this figure were originally reproduced from Yamaguchi et al. (2013b) with the permission of The Royal Society of Chemistry. DMPC = 1,2-dimyristoyl-*sn*-glycero-3-phosphocholine, DHPC = 1,2-diheptanoyl-*sn*-glycero-3-phosphocholine

NMR techniques using neo-glycolipids will facilitate detailed characterization of the biomolecular systems on cell surfaces with dynamical ordering for the creation of integrated functions.

Acknowledgements We wish to acknowledge our former colleagues, Dr. Yoshinori Uekusa, Dr. Kotaro Yanagi, Mr. Tsuyoshi Uno, and Ms. Sayoko Yamamoto, who contributed to the studies reported in this paper. We also thank Dr. Tomoshi Kameda (The National Institute of Advanced Industrial Science and Technology) for useful discussions. Finally, we are grateful to Drs. Yasunori Chiba, Toshihiko Kitajima, and Yoshifumi Jigami (The National Institute of Advanced Industrial Science and Technology) for providing the engineered yeast cells. This work was partly supported by the Okazaki ORION project, the Nanotechnology Platform Program (Molecule and Material Synthesis) of MEXT, Japan, and the JSPS/MEXT Grants in Aid for Scientific Research on Innovation Areas (25102008, 25102009, and 26102518), Scientific Research (A) (24249002), Challenging Exploratory Research (26560451), and Young Scientists (B) (24750170).

References

Ando H, Ishida H, Kiso M (2010) Renewed synthetic approach to gangliosides exploiting versatile and powerful synthetic units. Methods Enzymol 478;521–540

Bodner CR, Maltsev AS, Dobson CM, Bax A (2010) Differential phospholipid binding of α-synuclein variants implicated in Parkinson's disease revealed by solution NMR spectroscopy. Biochemistry 49:862–871

Drickamer K, Taylor ME (1993) Biology of animal lectins. Annu Rev Cell Biol 9:237–264

Ernst B, Magnani JL (2009) From carbohydrate leads to glycomimetic drugs. Nat Rev Drug Discov 8:661–677

Fantini J, Yahi N (2010) Molecular insights into amyloid regulation by membrane cholesterol and sphingolipids: common mechanisms in neurodegenerative diseases. Expert Rev Mol Med 12:e27

Gayen A, Mukhopadhyay C (2008) Evidence for effect of GM1 on opioid peptide conformation: NMR study on leucine enkephalin in ganglioside-containing isotropic phospholipid bicelles. Langmuir 24:5422–5432

Kamiya Y, Yagi-Utsumi M, Yagi H, Kato K (2011a) Structural and molecular basis of carbohydrate-protein interaction systems as potential therapeutic targets. Curr Pharm Des 17:1672–1684

Kamiya Y, Yamamoto S, Chiba Y, Jigami Y, Kato K (2011b) Overexpression of a homogeneous oligosaccharide with ^{13}C labeling by genetically engineered yeast strain. J Biomol NMR 50:397–401

Kamiya Y, Satoh T, Kato K (2012) Molecular and structural basis for *N*-glycan-dependent determination of glycoprotein fates in cells. Biochim Biophys Acta 1820:1327–1337

Kamiya Y, Yanagi K, Kitajima T, Yamaguchi T, Chiba Y, Kato K (2013) Application of metabolic ^{13}C labeling in conjunction with high-field nuclear magnetic resonance spectroscopy for comparative conformational analysis of high mannose-type oligosaccharides. Biomolecules 3:108–123

Kamiya Y, Satoh T, Kato K (2014) Recent advances in glycoprotein production for structural biology: toward tailored design of glycoforms. Curr Opin Struct Biol 26:44–53

Khatun UL, Mukhopadhyay C (2013) Interaction of bee venom toxin melittin with ganglioside GM1 bicelle. Biophys Chem 180–181:66–75

Lepenies B, Yin J, Seeberger PH (2010) Applications of synthetic carbohydrates to chemical biology. Curr Opin Chem Biol 14:404–411

Luchinat C, Parigi G (2007) Paramagnetic systems in biochemistry: solution NMR studies eMagRes. Wiley, New York

Matsuzaki K, Kato K, Yanagisawa K (2010) Abeta polymerization through interaction with membrane gangliosides. Biochim Biophys Acta 1801:868–877

Nakanishi-Shindo Y, Nakayama K, Tanaka A, Toda Y, Jigami Y (1993) Structure of the *N*-linked oligosaccharides that show the complete loss of α-1,6-polymannose outer chain from *och1*,

och1 mnn1, and *och1 mnn1 alg3* mutants of *Saccharomyces cerevisiae*. J Biol Chem 268:26338–26345
Peters T, Pinto BM (1996) Structure and dynamics of oligosaccharides: NMR and modeling studies. Curr Opin Struct Biol 6:710–720
Piccinini M, Scandroglio F, Prioni S, Buccinnà B, Loberto N, Aureli M, Chigorno V, Lupino E, Demarco G, Lomartire A, Rinaudo MT, Sonnino S, Prinetti A (2010) Deregulated sphingolipid metabolism and membrane organization in neurodegenerative disorders. Mol Neurobiol 41:314–340
Satoh T, Suzuki K, Yamaguchi T, Kato K (2014) Structural basis for disparate sugar-binding specificities in the homologous cargo receptors ERGIC-53 and VIP36. PLoS One 9:e87963
Sharon N, Lis H (2003) Lectins, 2nd edn. Springer, Dordrecht
Sugita Y, Okamoto Y (1999) Replica-exchange molecular dynamics method for protein folding. Chem Phys Lett 314:141–151
Takeda Y, Totani K, Matsuo I, Ito Y (2009) Chemical approaches toward understanding glycan-mediated protein quality control. Curr Opin Chem Biol 13:582–591
Ulmer TS, Bax A, Cole NB, Nussbaum RL (2005) Structure and dynamics of micelle-bound human α-synuclein. J Biol Chem 280:9595–9603
Utsumi M, Yamaguchi Y, Sasakawa H, Yamamoto N, Yanagisawa K, Kato K (2009) Up-and-down topological mode of amyloid beta-peptide lying on hydrophilic/hydrophobic interface of ganglioside clusters. Glycoconj J 26:999–1006
Wang Z, Chinoy ZS, Ambre SG, Peng W, Mcbride R, De Vries RP, Glushka J, Paulson JC, Boons GJ (2013) A general strategy for the chemoenzymatic synthesis of asymmetrically branched *N*-glycans. Science 341:379–383
Wormald MR, Petrescu AJ, Pao YL, Glithero A, Elliott T, Dwek RA (2002) Conformational studies of oligosaccharides and glycopeptides: complementarity of NMR, X-ray crystallography, and molecular modelling. Chem Rev 102:371–386
Yagi-Utsumi M, Kameda T, Yamaguchi Y, Kato K (2010) NMR characterization of the interactions between lyso-GM1 aqueous micelles and amyloid beta. FEBS Lett 584:831–836
Yamaguchi T, Kamiya Y, Choo YM, Yamamoto S, Kato K (2013a) Terminal spin labeling of a high-mannose-type oligosaccharide for quantitative NMR analysis of its dynamic conformation. Chem Lett 42:544–546
Yamaguchi T, Uno T, Uekusa Y, Yagi-Utsumi M, Kato K (2013b) Ganglioside-embedding small bicelles for probing membrane-landing processes of intrinsically disordered proteins. Chem Commun (Camb) 49:1235–1237
Yamaguchi T, Sakae Y, Zhang Y, Yamamoto S, Okamoto Y, Kato K (2014) Exploration of conformational spaces of high-mannose-type oligosaccharides by an NMR-validated simulation. Angew Chem Int Ed Engl in press
Yamamoto S, Yamaguchi T, Erdélyi M, Griesinger C, Kato K (2011) Paramagnetic lanthanide tagging for NMR conformational analyses of *N*-linked oligosaccharides. Chem Eur J 17:9280–9282
Yamamoto S, Zhang Y, Yamaguchi T, Kameda T, Kato K (2012) Lanthanide-assisted NMR evaluation of a dynamic ensemble of oligosaccharide conformations. Chem Commun (Camb) 48:4752–4754
Zhang Y (2014) Paramagnetism–assisted NMR analyses of conformational dynamics of ganglioside glycans. The Graduate University for Advanced Studies
Zhang Y, Yamamoto S, Yamaguchi T, Kato K (2012) Application of paramagnetic NMR-validated molecular dynamics simulation to the analysis of a conformational ensemble of a branched oligosaccharide. Molecules 17:6658–6671
Zhang Y, Yamaguchi T, Kato K (2013) New NMR tools for characterizing the dynamic conformations and interactions of oligosaccharides. Chem Lett 42:1455–1462
Zhao H, Pan Q, Zhang W, Carmichael I, Serianni AS (2007) DFT and NMR studies of $^{2}J_{COH}$, $^{3}J_{HCOH}$, and $^{3}J_{CCOH}$ spin-couplings in saccharides: C-O torsional bias and H-bonding in aqueous solution. J Org Chem 72:7071–7082

Chapter 15
Characterization of Cholesterol Crystalline Domains in Model and Biological Membranes Using X-Ray Diffraction

R. Preston Mason and Robert F. Jacob

Introduction

Cholesterol is an important component of most biological membranes where it regulates structural and dynamic properties of the lipid bilayer through its direct interactions with membrane phospholipids (Chen et al. 1995; Leonard and Dufourc 1991; McIntosh 1978; Yeagle 1985). Free or unesterified cholesterol is amphipathic in nature and consists of a planar, alkyl-substituted, tetracyclic steroid nucleus, modified at carbon three by a polar hydroxyl substituent in the β-position. The polar hydroxyl group anchors cholesterol at the membrane surface, causing the molecule to orient in the membrane with its long-axis parallel to the surrounding phospholipid acyl chains (Yeagle 1985; Schroeder and Wood 1995). This orientation increases order in the upper acyl chain region of the membrane while decreasing packing constraints among the terminal methyl segments located in the hydrocarbon core, effectively *condensing* the spatial arrangement of phospholipids within the membrane bilayer (Yeagle 1985; Schroeder and Wood 1995; Shinitzky and Inbar 1976). The effects of cholesterol on the conformation and rotational dynamics of neighboring molecules are highly dependent on the acyl chain composition and structural integrity of membrane phospholipids (Tulenko et al. 1998). Oxidative modification of membrane lipid acyl chains, for example, can affect the behavior of cholesterol and even its tendency to associate with other sterol molecules.

R.P. Mason, Ph.D. (✉)
Cardiovascular Division, Department of Medicine, Brigham and Women's Hospital, Harvard Medical School, Boston, MA 02115, USA

Elucida Research LLC, 100 Cummings Center, Suite 135 L, Beverly, MA 01915, USA
e-mail: rpmason@elucidaresearch.com

R.F. Jacob, Ph.D.
Elucida Research LLC, 100 Cummings Center, Suite 135 L, Beverly, MA 01915, USA

A. Chakrabarti, A. Surolia (eds.), *Biochemical Roles of Eukaryotic Cell Surface Macromolecules*, Advances in Experimental Medicine and Biology 842,
DOI 10.1007/978-3-319-11280-0_15

Membrane Effects of Cholesterol Enrichment

The amount of cholesterol present in a biological membrane influences its biophysical properties, including the activity of membrane-restricted proteins. Changes in membrane cholesterol content can alter the conformation and activity of various channel proteins, including calcium channels (Bialecki and Tulenko 1989) and potassium channels (Bolotina et al. 1991). Cholesterol enrichment of the cell membrane has also been shown to inhibit Na^+/K^+ ATPase activity in erythrocytes (Broderick et al. 1989), endothelial cells (Lau 1994), and renal cells (Yeagle et al. 1988). In vascular smooth muscle cells derived from an animal model of dietary atherosclerosis, calcium transport mechanisms and basal intracellular calcium levels were observed to change as a function of increased membrane cholesterol content (Gleason et al. 1991). In addition, cholesterol enrichment has been shown to alter the conformation of calcium-activated potassium channels, forcing the ion channel pore to favor the closed state under otherwise normal stimulatory conditions (Chang et al. 1995). These functional effects of cholesterol enrichment correlated directly with changes in structural stress and lateral elastic stress energy (Chang et al. 1995). Changes in cholesterol content have also been shown to influence G-protein coupled receptors, including the serotonin receptor (Shrivastava et al. 2010). Collectively, these observations provide compelling evidence for the hypothesis that membrane cholesterol levels must be maintained within certain physiologic limits in order to ensure proper cell and cell membrane function.

Lipid Rafts

The cell plasma membrane is a complex structure consisting of numerous microdomains assembled from specific lipid and protein constituents. These membrane domains compartmentalize cellular processes by serving as organizing centers for the assembly of signaling molecules while also modulating membrane fluid dynamics and regulating protein trafficking, receptor function, and other cellular activity such as neurotransmission.

One type of domain that has been the subject of intensive investigation is the *lipid raft*, which is more highly-ordered as compared to the surrounding membrane bilayer (Simons and Toomre 2000). Lipid rafts contain 3–5 times the amount of cholesterol as compared to the surrounding bilayer and are also enriched in sphingolipids, particularly sphingomyelin, which interacts favorably with cholesterol due to its accommodating headgroup structure and the highly-saturated nature of its hydrocarbon chains. Although not all phospholipids associated with lipid rafts are fully saturated, the acyl chains present in this domain are typically more saturated and more tightly-packed than those in the surrounding membrane bilayer. Cholesterol, by virtue of its inherent structural properties as well as its affinity for lipids with more rigid acyl chains, plays an essential role in stabilizing lipid rafts.

Lipid rafts possess properties consistent with the gel state, including extended acyl chains and a relatively high melting temperature, but also properties associated with the liquid crystalline state, such as rapid lateral molecular mobility (Brown and London 2000; Ostermeyer et al. 1999). These membrane domains also host specific cellular proteins and mediate a variety of biologic processes, including signal transduction, adhesion, and sorting of membrane components. The insulin receptor, for example, is known to form functional dimers in lipid rafts but not in other regions of the membrane. T cell antigen receptor activation on the surface of T lymphocytes is regulated by their association with lipids rafts. Viruses, as obligate intracellular parasites, bind to cellular receptors expressed in lipid rafts in order to gain access to target cells. Many vertebrate cell types also contain specialized lipid rafts known as caveolae, which appear (by microscopic analysis) as small, flask-shaped invaginations of the plasma membrane. These rafts are enriched with cholesterol, sphingomyelin, and unique proteins, such as caveolin, and engage in various cell functions, including endocytosis and signal transduction (Edidin 1997).

Cholesterol Domains in Model and Biological Membranes

The systematic addition of cholesterol to biological membranes eventually results in lateral phase separation and the formation of membrane-restricted cholesterol domains (Tulenko et al. 1998; Bach et al. 1998; Engelman and Rothman 1972; Houslay and Stanley 1982; Rice and McConnell 1989; Ruocco and Shipley 1984; Slotte 1995a, b). In model membranes prepared largely from lecithin, cholesterol was shown to aggregate into clusters at cholesterol-to-phospholipid mole ratios greater than 0.3:1 (Engelman and Rothman 1972) and to form separate domains at ratios greater than 1:1 (Houslay and Stanley 1982). Similar effects have been observed in well-defined lipid monolayer systems using various microscopy approaches (Rice and McConnell 1989; Slotte 1995a, b). Cholesterol domains have also been characterized in membrane bilayers using small angle X-ray diffraction and other biophysical techniques. Ruocco and Shipley showed that increasing the cholesterol content of model membrane bilayers to levels greater than 50 mol % resulted in the formation of an immiscible cholesterol monohydrate phase, with a characteristic unit cell periodicity of 34 Å, that was coexistent with a bulk, liquid-crystalline lipid bilayer phase (Ruocco and Shipley 1984). The repeat unit associated with the cholesterol phase corresponds to a tail-to-tail arrangement of cholesterol molecules, as the long axis of cholesterol monohydrate is 17 Å in the crystalline state (Craven 1976). This interpretation has been confirmed in other model membrane systems, as well as select native membrane preparations such as myelin membranes, using a variety of techniques (Schroeder and Wood 1995; Bloom and Thewalt 1995; Harris et al. 1995; Hui 1995; Tocanne 1992; Kirschner and Caspar 1972).

Lipid rafts isolated from neuronal cell membranes, and identified as detergent-insoluble membrane fractions, were shown to contain relatively low amounts of sphingomyelin but very high amounts of cholesterol (Maekawa et al. 1999). Epand and coworkers extended this work and showed that the formation of cholesterol-rich domains could be induced in model membranes by introducing a neuronal protein—namely NAP-22, a myristoylated, calcium-dependent , calmodulin-binding protein found largely in the synapse and shown to be a major component of neuron-associated, detergent-insoluble, low-density membrane fractions. Differential scanning calorimetry analysis demonstrated that NAP-22 changed the shape and enthalpy of the phase transition of phosphatidylcholine and induced the appearance of cholesterol "crystalline" domains in membranes composed of phosphatidylcholine with either saturated or unsaturated acyl chains. Using atomic force microscopy, NAP-22 was further shown to cause a marked change in the surface morphology of dioleoylphosphatidylcholine bilayers containing cholesterol at 40 mol %. In the absence of protein, the membrane bilayer appeared as a smooth structure of uniform thickness; the addition of NAP-22 resulted in the formation of a more convoluted surface consisting of raised bilayer domain structures measuring approximately 1.5 nm in height (Epand et al. 2001).

Role of Cholesterol Domains in Membrane Function

Cholesterol is typically associated with separate kinetic domains (or pools) and is thus considered to be distributed non-randomly within the plasma membrane (Yeagle 1985; Liscum and Underwood 1995; Phillips et al. 1997; Schroeder et al. 1991, 1995). Regulation of the size and physico-chemical properties of these kinetic domains may influence extracellular and intracellular cholesterol transport pathways (Schroeder et al. 1991; Bretscher and Munro 1993). Investigators have proposed that cholesterol domains may modulate the activity of membrane proteins that localize specifically to cholesterol-rich domains (*e.g.*, nicotinic acetylcholine receptor, human erythrocyte band 3 protein, glycophorin, as well as Na^+/K^+-ATPase) or cholesterol-poor domains (*e.g.*, Ca^{2+}ATPase) (see Mukherjee and Chattopadhyay for review (Mukherjee and Chattopadhyay 1996)). Sterol-rich regions have also been hypothesized to play a crucial role in other cellular functions, including signal transduction, cell adhesion, cell motility, and the sorting and trafficking of membrane components (Janes et al. 2000; Langlet et al. 2000; Simons and Ikonen 1997, 2000).

Membrane Structural Analysis Using Small Angle X-ray Diffraction

The use of X-ray diffraction approaches to study the structural properties of biological membranes has been well established. Membrane diffraction studies were first reported in the 1930s; however, this area of inquiry remained somewhat esoteric until the 1960s, when the field experienced rapid growth (Franks and Levine 1981). Since that time, small angle X-ray diffraction has been used extensively to study various model and native membrane preparations.

In order to appreciate the use of X-ray diffraction approaches in analyzing membrane structure, it is important to consider the *lipid bilayer theory*. According to this theory, lipids that comprise a membrane are arranged in a bilayer structure as a result of their amphipathic properties. All typical lipids have a polar, hydrophilic headgroup region and a nonpolar, hydrophobic fatty acyl chain region. In order to avoid energetically unfavorable interactions with water, lipids will associate with one another such that their headgroups form two surfaces in contact with the surrounding aqueous environment, with their acyl chains oriented into the space between the two surfaces. The acyl chain region of a bilayer formed in this manner is called the membrane or hydrocarbon core, while the hydrophilic surfaces are known as the membrane headgroup layers (Blaurock 1982). If cholesterol is present (which is true of almost all naturally occurring membranes), this molecule is positioned almost entirely within the acyl chain region of the bilayer.

This specific arrangement of membrane lipids is important in that it serves as the basis for the structural continuity of a membrane repeat unit. If membranes are "stacked" into multiple layers, this basic bilayer structure becomes a periodic function that yields coherent scattering in diffraction analyses. The unit cell of such a system is represented by the membrane lipid bilayer that is repeated in these preparations. Numerous X-ray diffraction experiments have been conducted using membrane multibilayers, including myelin membranes (Blaurock 1971; Moody 1963), disk membranes from the outer segments of retinal rod cells (Blaurock and Wilkins 1972; Corless 1972), erythrocyte ghosts (Knutton et al. 1970), and artificial multilayers derived from the sarcoplasmic reticulum (Dupont and Hasselbach 1973; Worthington and Liu 1973). Some membrane systems, such as nerve myelin membranes and rod outer segment membranes, occur naturally as repeating, multibilayer structures. Fiber cells of the ocular lens, as discussed below, also appear to be organized into regular, repeating membrane units, making them particularly well-suited to X-ray diffraction analysis.

X-ray diffraction analysis of a multibilayer membrane sample results in the production of discrete diffraction peaks also known as Bragg reflections. These reflections result from the coherent (constructive) scattering of secondary X-rays produced by atoms comprising a sample. Coherent scattering from membranes follows the same rules as required for the diffraction of crystals: (1) the spacing between the scattering planes must be roughly equal to the wavelength of the incident X-rays, (2) the scattering centers (membrane layers) must be spatially distributed in a highly regular manner, and (3) the repeating membrane units must be oriented so that the diffraction angle (θ) satisfies Bragg's law, $h\lambda = 2d\ sin\theta$, where h is the diffraction order, λ is the wavelength of the X-ray radiation, d is the membrane lipid bilayer unit cell periodicity, and θ is the Bragg angle equal to one-half the angle between the incident beam and scattered beam. The relationship of Bragg's law to the diffraction analysis of a membrane multibilayer sample is illustrated in Fig. 15.1. In this case, the individual

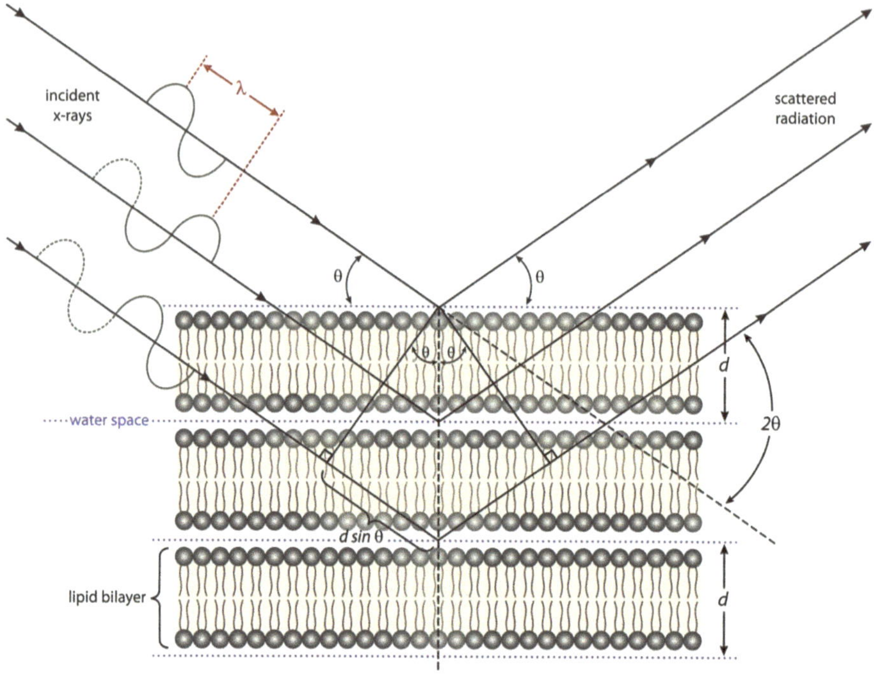

Fig. 15.1 Schematic representation of membrane bilayers as an X-ray diffraction lattice. The unit cell periodicity, *d*, represents the distance spanning a single bilayer plus half the water space on each side of the bilayer. θ is the Bragg angle and is equal to one-half the angle between incident and scattered radiation

lipid bilayer represents the minimum volume of information that is being repeated in the sample (*i.e.*, unit cell periodicity). The unit cell periodicity, *d*, is often referred to as the *d*-space, and represents the distance from the center of one water space to the next across the lipid bilayer.

X-Ray Diffraction Method

A typical X-ray diffraction experiment consists of placing a multilammelar membrane specimen into a monochromatic, collimated beam of X-rays and measuring the intensity of the scattered radiation (Fig. 15.2). Diffraction occurs only when the plane of each sample bilayer is oriented around an axis perpendicular to the incident X-ray beam. After orienting a membrane sample using relatively gentle centrifugation approaches (Franks and Levine 1981), the membrane sample is positioned relative to the incident X-ray beam to allow for the parallel alignment of the repeating membrane planes with the imaginary Bragg planes, thus achieving the specific angles required for diffraction as described by Bragg's law.

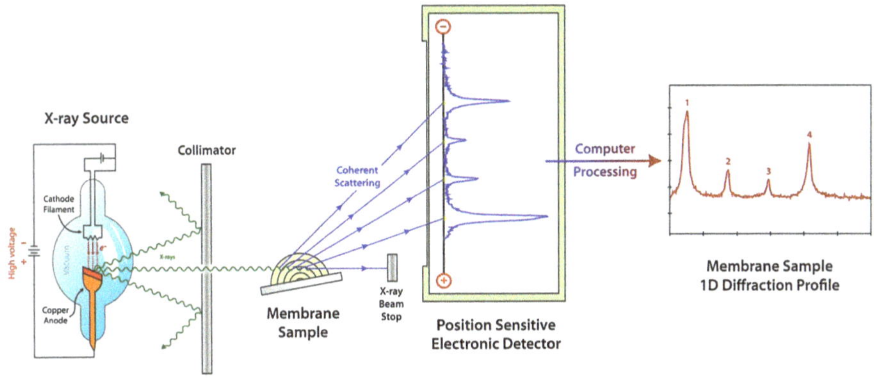

Fig. 15.2 Schematic representation of the small angle X-ray scattering method. Monochromatic radiation ($\lambda = 1.54$ Å) is produced by a high-brilliance, microfocus generator. An oriented sample is placed on a curved mount at near-grazing incidence with respect to the focused beam. Coherent scattering data are then collected on a one-dimensional, position-sensitive electronic detector

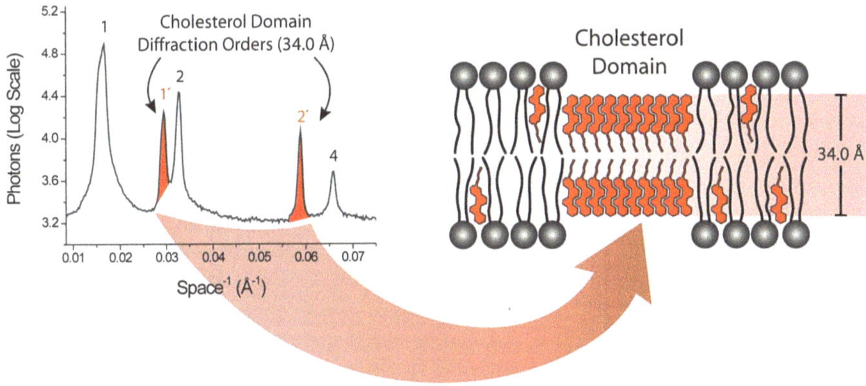

Fig. 15.3 Identification of membrane cholesterol domains using small angle X-ray diffraction approaches. Cholesterol domains yield characteristic diffraction peaks (1′ and 2′) that correspond to a unit cell periodicity (*d*-space) of 34 Å

In our laboratory, membrane diffraction is accomplished by aligning the sample at grazing incidence with respect to a collimated, monochromatic X-ray beam produced by a Rigaku Rotaflex RU-200, high-brilliance microfocus generator (Rigaku Americas, The Woodlands, TX). The fixed geometry beamline utilizes a single Franks mirror providing nickel-filtered radiation ($K\alpha_1$ and $K\alpha_2$ unresolved) at the detection plane. Diffraction data are collected on a one-dimensional, position-sensitive electronic detector (Hecus X-ray Systems, Graz, Austria) at a sample-to-detector distance of 150 mm and calibrated using cholesterol monohydrate crystals.

The presence of cholesterol domains in a given membrane sample results in the production of a distinct set of Bragg peaks having a singular unit cell periodicity of 34 Å (Fig. 15.3). Under typical temperature and relative humidity conditions, the second-order cholesterol domain peak is well-delineated from other, neighboring cholesterol or phospholipid diffraction peaks and can be used to quantitate relative cholesterol domain peak intensity (calculated as the fraction of total peak area).

Cholesterol Domains in Vascular Smooth Muscle Cell Membranes

Atherosclerosis is the product of endothelial dysfunction, inflammation, and excessive lipid accumulation in the arterial wall (Libby 2002). Given its principal role in the structural and functional properties of low-density lipoproteins (LDL), cholesterol has been the primary focus of much of the research conducted in this field. Cholesterol exists in free form, as previously described, or as cholesteryl esters, which are formed by the action of acyl-coenzyme A (CoA)-cholesterol acyl transferase (ACAT). This enzyme catalyzes the covalent attachment of a fatty acid moiety to the hydroxyl group on cholesterol, converting the molecule into a more hydrophobic form for improved storage and transport.

Cholesterol uptake into cells is regulated by the expression of LDL receptors. Through various feedback mechanisms, LDL expression is reduced when cholesterol biosynthesis occurs at adequate levels in the cell. If ACAT is inhibited or rendered ineffective by some perturbation, free cholesterol levels become elevated in the cell. Under such conditions, cholesterol also accumulates in the plasma membrane where it can aggregate into discrete, crystalline domains. These domains are believed to precede the development of microscopic cholesterol crystals that are typically observed in the extracellular space associated with the atherosclerotic plaque (Kellner-Weibel et al. 1999; Small 1988). These cholesterol crystals are toxic and contribute to the instability of the atherosclerotic lesion by increasing local inflammation and plaque mass (Small 1988).

Using X-ray diffraction approaches, we directly examined the effects of cholesterol enrichment on membrane lipid structural organization in cultured smooth muscle cells and cells obtained *ex vivo* from an animal model of dietary atherosclerosis (Tulenko et al. 1998). The comparative effects of cholesterol enrichment in these separate systems were remarkably consistent. Following 9 weeks of feeding with a cholesterol-enriched diet, the cholesterol-to-phospholipid mole ratio measured in aortic smooth muscle cell membranes increased from 0.4:1 to approximately 1:1. Under such atherosclerotic-like conditions, prominent cholesterol domains (identified by their characteristic periodicity of 34 Å) could be observed in the smooth muscle cell plasma membranes (Tulenko et al. 1998). The formation of cholesterol domains was also reproduced in this study using membranes reconstituted as binary mixtures of bovine cardiac phosphatidylcholine and cholesterol at a cholesterol-to-phospholipid mole ratio of 1:1 (Tulenko et al. 1998).

In another study, cultured mouse peritoneal macrophage foam cells were treated with an ACAT inhibitor, which induced the formation of free cholesterol crystals that extended away the cell membrane with various morphologies, including plates, needles and helices (Kellner-Weibel et al. 1999). With the use of X-ray diffraction approaches, the early stages of crystal formation could be identified in whole cell and isolated membranes obtained from either diseased tissue *ex vivo* or cultured cells *in vitro* following ACAT inhibition (Tulenko et al. 1998; Kellner-Weibel et al. 1999). Preventing crystal formation is an important goal as cholesterol in this state is pro-inflammatory and does not respond well to pharmacologic interventions that promote lesion regression due to its high stability (Small 1988; Katz et al. 1982).

Cholesterol Domains in Model Membranes Exposed to Atherogenic Conditions

The formation of cholesterol crystalline domains in the membrane can also occur in the absence of sterol enrichment. In particular, such domains can form following oxidative modification to membrane lipids in a manner that can be inhibited with certain lipophilic antioxidants or stimulated with pro-oxidant agents (Jacob and Mason 2005; Mason et al. 2006; Jacob et al. 2013). A similar increase in cholesterol domains with oxidative stress was also observed under conditions of hyperglycemia (Self-Medlin et al. 2009). In these models of disease, the formation of well-defined cholesterol domains did not require a change in the overall membrane cholesterol content but were directly related to levels of lipid hydroperoxides generated during oxidative stress. These findings indicate that the interactions of cholesterol with surrounding phospholipids are influenced by their physico-chemical properties, including chemical modifications resulting from their interactions with reactive oxygen species. We also observed that changes in membrane width and even cholesterol domain formation are highly dependent on the length and degree of phospholipid acyl chain composition (Tulenko et al. 1998). Finally, cholesterol itself can undergo oxidative modification during various disease processes. At high levels, these oxidized sterols also form domains and extracellular crystals with dimensions that differ from that of non-oxidized forms but that still cause apoptosis (Phillips et al. 2001; Geng et al. 2003).

Cholesterol Domains in Ocular Lens Fiber Cell Membranes

The human ocular lens has been a particularly interesting tissue for analysis of cholesterol domains in the plasma membrane. Unlike cells associated with atherosclerosis, it appears that the presence of cholesterol crystalline domains is essential for normal ocular function and light transparency. Through the controlled regulation of its shape, the ocular crystalline lens allows for light to be efficiently transmitted through the eye and focused onto the retina. The lens is an encapsulated structure consisting almost entirely of a large number of rigid, elongated cells known as lens fibers or fiber cells, which are produced by the differentiation of a single layer of epithelial cells located just beneath the anterior surface of the lens capsule. These cells are deposited in successive layers through a process that begins in early embryogenesis and continues throughout life. Existing fiber cells are displaced toward the center of the lens as new layers are formed. In the *lens nucleus*, mature fiber cells are compacted into the center of the lens; cells peripheral to this region, including new and mitotically active cells of the adult lens, are collectively referred to as the *lens cortex*. Fibers cells eventually lose all subcellular organelles during their progressive displacement toward the lens nucleus. As a result, the plasma membrane becomes the only substantive organelle of the adult lens (Rafferty 1985). A unique biochemical characteristic of the fiber cell plasma membrane is its relatively high level of free cholesterol. The cholesterol-to-phospholipid mole ratio of the fiber cell membrane ranges from 1 to 2 in the cortex to as high as 3–4 in the lens nucleus (Li et al. 1985, 1987). This stands in sharp contrast to the levels of cholesterol found in other mammalian plasma membranes, which range between 0.5 and 1.0. The fiber cell plasma membrane is also distinct from other biologic membranes in that it contains only trace amounts of polyunsaturated fatty acid (Broekhuyse and Soeting 1976) and, in the human lens, a phospholipid composition of more than 50 % sphingomyelin and sphingomyelin derivatives (Byrdwell and Borchman 1997; Byrdwell et al. 1994).

The unusual lipid composition of fiber cell plasma membrane makes it an intriguing biologic system for conducting structural studies. Moreover, fiber cells can be efficiently removed from the lens and plasma membranes isolated for X-ray diffraction analysis. It was predicted, based on previous studies in model membrane systems, that these biologic membranes would be organized as a "mosaic of phospholipid and cholesterol patches" (Li et al. 1985).

Using small angle X-ray diffraction approaches, we observed that cholesterol domains are clearly present in both reconstituted and intact (protein-containing) fiber cell plasma membrane preparations (Fig. 15.4). These domains were identified by their characteristic, 34 Å repeat orders, which remained stable over a broad range of temperature and relative humidity conditions (Jacob et al. 1999). By contrast, the dimensions of the surrounding liquid crystalline phase increased by as much as 30.9 Å (60 %) in reconstituted lens plasma membrane. Interestingly, the dimension of the sterol-poor region of the membrane was less affected by experimental conditions in the *intact* fiber cell plasma membranes. Thus, while the presence of protein is not necessary for the formation of immiscible cholesterol domains, it does appear to significantly influence both the size of the cholesterol domains and the dimensions of the surrounding sterol-poor region. In these membrane preparations, the ratio of cholesterol to phospholipid exceeded 2:1 under normal conditions.

The functional role for discrete cholesterol regions in ocular lens fiber cell plasma membrane is an intriguing question. The essential activity of the lens fiber cell is to facilitate the efficient transmission of visible light through the eye. By ordering membrane lipid constituents, higher cholesterol levels may provide such transparency. Infrared spectroscopy approaches have demonstrated that the highest membrane cholesterol content is associated with the center or more visually-significant region of the ocular lens (Borchman et al. 1996). Another role for cholesterol domains may be to interfere with the membrane association of the protein

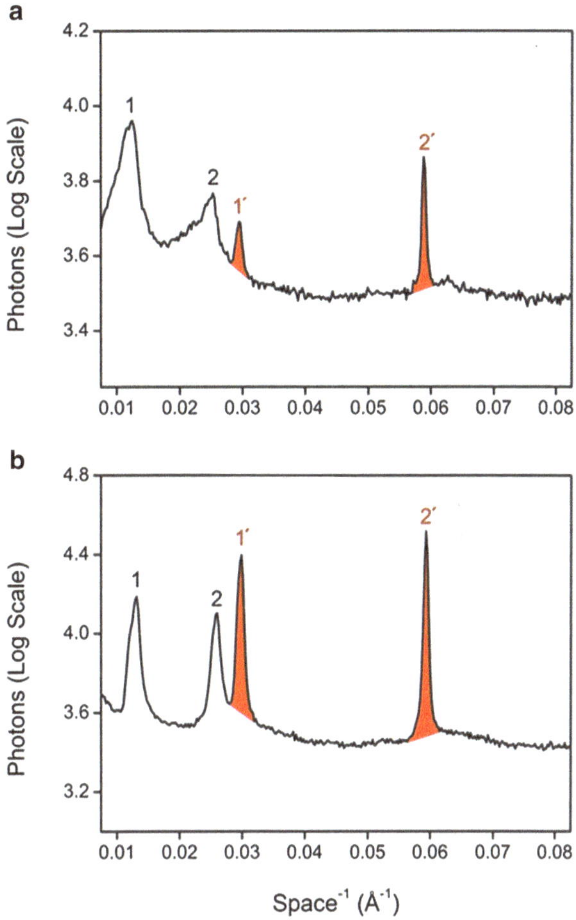

Fig. 15.4 Representative X-ray diffraction patterns obtained from intact (A) and reconstituted (B) human ocular lens fiber cell plasma membranes. Data were collected on a one-dimensional, position-sensitive electronic detector at 20 °C, 92 % RH. Diffraction peaks labeled as 1′ and 2′ correspond to immiscible cholesterol domains; other peaks correspond to the surrounding membrane lipid bilayer. This research was originally published in the *Journal of Biological Chemistry*. R.F. Jacob, R.J. Cenedella, and R.P. Mason. Evidence for distinct cholesterol domains in fiber cell membranes from cataractous human lenses. *J Biol Chem*. 2001; **276**:13573-13578. © the American Society for Biochemistry and Molecular Biology

crystallin (especially α-crystallin), an important feature of human and experimental animal cataracts (Chandrasekher and Cenedella 1995). In fact, cataractogenesis can be accelerated in an animal model by reducing membrane cholesterol content in the lens with specific biosynthesis inhibitors (Cenedella and Bierkamper 1979). The results of these membrane structure studies suggest that the coexistence of distinct sterol-rich and -poor regions may interfere with the ability of extrinsic proteins to aggregate at the membrane surface (Jacob et al. 1999).

Conclusion

Small-angle X-ray diffraction approaches have been used to successfully evaluate the organization of lipids in plasma membranes derived from distinct mammalian cell types, including arterial smooth muscle cells and ocular lens fiber cells.

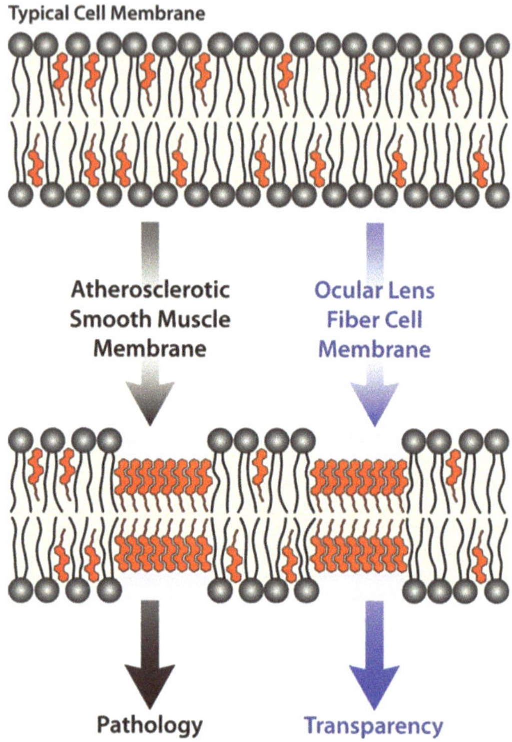

Fig. 15.5 A schematic model describing the formation of immiscible cholesterol domains in atherosclerotic-like smooth muscle cell plasma membranes following cholesterol enrichment and leading to disease. The cholesterol microdomain has a unit cell periodicity of 34 Å and coexists in the same plane as the surrounding sterol-poor region. At the same time, these same cholesterol domains are essential for normal ocular lens function and transparency. These studies show cholesterol can self-associate into immiscible domains within the plasma membrane, a phenomenon that contributes to both physiologic and pathologic cellular processes

These studies show that at elevated cholesterol concentrations or under conditions of oxidative stress, cholesterol can self-associate into immiscible domains within the plasma membrane, a phenomenon that contributes to both physiologic and pathologic cellular processes (Fig. 15.5). In fiber cell plasma membranes isolated from the ocular lens, by contrast, cholesterol domains appear to be essential to normal physiology. The unique structural heterogeneity of the lens fiber cell plasma membrane appears to facilitate lens transparency while interfering with cataractogenic aggregation of soluble lens proteins at the membrane surface. Taken together, these analyses provide examples of the complex roles that sterol-rich domains may have in mammalian plasma membranes. Combined with the findings from various other laboratories, these data support a model of the membrane in which cholesterol aggregates into structurally distinct regions that regulate the function of the cell membrane and may contribute to mechanisms of disease.

References

Bach D, Borochov N, Wachtel E (1998) Phase separation of cholesterol in dimyristoyl phosphatidylserine cholesterol mixtures. Chem Phys Lipids 92:71–77

Bialecki RA, Tulenko TN (1989) Excess membrane cholesterol alters calcium channels in arterial smooth muscle. Am J Physiol 257:C306–C314

Blaurock AE (1971) Structure of the nerve myelin membrane: proof of the low-resolution profile. J Mol Biol 56:35–52
Blaurock AE (1982) Evidence of bilayer structure and of membrane interactions from X-ray diffraction analysis. Biochim Biophys Acta 650:167–207
Blaurock AE, Wilkins MH (1972) Structure of retinal photoreceptor membranes. Nature 236:313–314
Bloom M, Thewalt JL (1995) Time and distance scales of membrane domain organization. Mol Membr Biol 12:9–13
Bolotina V, Gericke M, Bregestovski P (1991) Kinetic differences between Ca(2+)-dependent K^+ channels in smooth muscle cells isolated from normal and atherosclerotic human aorta. Proc Biol Sci 244:51–55
Borchman D, Cenedella RJ, Lamba OP (1996) Role of cholesterol in the structural order of lens membrane lipids. Exp Eye Res 62:191–197
Bretscher MS, Munro S (1993) Cholesterol and the Golgi apparatus. Science 261:1280–1281
Broderick R, Bialecki R, Tulenko TN (1989) Cholesterol-induced changes in rabbit arterial smooth muscle sensitivity to adrenergic stimulation. Am J Physiol 257:H170–H178
Broekhuyse RM, Soeting WJ (1976) Lipids in tissues of the eye. XV. Essential fatty acids in lens lipids. Exp Eye Res 22:653–657
Brown DA, London E (2000) Structure and function of sphingolipid- and cholesterol-rich membrane rafts. J Biol Chem 275:17221–17224
Byrdwell WC, Borchman D (1997) Liquid chromatography/mass-spectrometric characterization of sphingomyelin and dihydrosphingomyelin of human lens membranes. Ophthalmic Res 29:191–206
Byrdwell WC, Borchman D, Porter RA, Taylor KG, Yappert MC (1994) Separation and characterization of the unknown phospholipid in human lens membranes. Invest Ophthalmol Vis Sci 35:4333–4343
Cenedella RJ, Bierkamper GG (1979) Mechanism of cataract production by 3-beta(2-diethylaminoethoxy) androst-5-en-17-one hydrochloride, U18666A: An inhibitor of cholesterol biosynthesis. Exp Eye Res 28:673–688
Chandrasekher G, Cenedella RJ (1995) Protein associated with human lens 'native' membrane during aging and cataract formation. Exp Eye Res 60:707–717
Chang HM, Reitstetter R, Mason RP, Gruener R (1995) Attenuation of channel kinetics and conductance by cholesterol: An interpretation using structural stress as a unifying concept. J Membr Biol 143:51–63
Chen M, Mason RP, Tulenko TN (1995) Atherosclerosis alters the composition, structure and function of arterial smooth muscle cell plasma membranes. Biochim Biophys Acta 1272:101–112
Corless JM (1972) Lamellar structure of bleached and unbleached rod photoreceptor membranes. Nature 237:229–231
Craven BM (1976) Crystal structure of cholesterol monohydrate. Nature 260:727–729
Dupont Y, Hasselbach W (1973) Structural changes in sarcoplasmic reticulum membrane induced by SH reagents. Nat New Biol 246:41–44
Edidin M (1997) Lipid microdomains in cell surface membranes. Curr Opin Struct Biol 7:528–532
Engelman DM, Rothman JE (1972) The planar organization of lecithin-cholesterol bilayers. J Biol Chem 247:3694–3697
Epand RM, Maekawa S, Yip CM, Epand RF (2001) Protein-induced formation of cholesterol-rich domains. Biochemistry 40:10514–10521
Franks NP, Levine YK (1981) Low-angle X-ray diffraction. Mol Biol Biochem Biophys 31:437–487
Geng YJ, Phillips JE, Mason RP, Casscells SW (2003) Cholesterol crystallization and macrophage apoptosis: Implication for atherosclerotic plaque instability and rupture. Biochem Pharmacol 66:1485–1492
Gleason MM, Medow MS, Tulenko TN (1991) Excess membrane cholesterol alters calcium movements, cytosolic calcium levels, and membrane fluidity in arterial smooth muscle cells. Circ Res 69:216–227

Harris JS, Epps DE, Davio SR, Kezdy FJ (1995) Evidence for transbilayer, tail-to-tail cholesterol dimers in dipalmitoylglycerophosphocholine liposomes. Biochemistry 34:3851–3857

Houslay MD, Stanley KK (1982) Dynamics of biological membranes: influence on synthesis, structure and function. John Wiley, New York

Hui SW (1995) Geometry of domains and domain boundaries in monolayers and bilayers. Mol Membr Biol 12:45–50

Jacob RF, Mason RP (2005) Lipid peroxidation induces cholesterol domain formation in model membranes. J Biol Chem 280:39380–39387

Jacob RF, Cenedella RJ, Mason RP (1999) Direct evidence for immiscible cholesterol domains in human ocular lens fiber cell plasma membranes. J Biol Chem 274:31613–31618

Jacob RF, Aleo MD, Self-Medlin Y, Doshna CM, Mason RP (2013) 1,2-Naphthoquinone stimulates lipid peroxidation and cholesterol domain formation in model membranes. Invest Ophthalmol Vis Sci 54:7189–7197

Janes PW, Ley SC, Magee AI, Kabouridis PS (2000) The role of lipid rafts in T cell antigen receptor (TCR) signalling. Semin Immunol 12:23–34

Katz SS, Small DM, Smith FR, Dell RB, Goodman DS (1982) Cholesterol turnover in lipid phases of human atherosclerotic plaque. J Lipid Res 23:733–737

Kellner-Weibel G, Yancey PG, Jerome WG, Walser T, Mason RP, Phillips MC, Rothblat GH (1999) Crystallization of free cholesterol in model macrophage foam cells. Arterioscler Thromb Vasc Biol 19:1891–1898

Kirschner DA, Caspar DL (1972) Comparative diffraction studies on myelin membranes. Ann N Y Acad Sci 195:309–320

Knutton S, Finean JB, Coleman R, Limbrick AR (1970) Low-angle X-ray diffraction and electron-microscope studies of isolated erythrocyte membranes. J Cell Sci 7:357–371

Langlet C, Bernard AM, Drevot P, He HT (2000) Membrane rafts and signaling by the multichain immune recognition receptors. Curr Opin Immunol 12:250–255

Lau YT (1994) Cholesterol enrichment inhibits Na^+/K^+ pump in endothelial cells. Atherosclerosis 110:251–257

Leonard A, Dufourc EJ (1991) Interactions of cholesterol with the membrane lipid matrix: a solid state NMR approach. Biochimie 73:1295–1302

Li LK, So L, Spector A (1985) Membrane cholesterol and phospholipid in consecutive concentric sections of human lenses. J Lipid Res 26:600–609

Li LK, So L, Spector A (1987) Age-dependent changes in the distribution and concentration of human lens cholesterol and phospholipids. Biochim Biophys Acta 917:112–120

Libby P (2002) Inflammation in atherosclerosis. Nature 420:868–874

Liscum L, Underwood KW (1995) Intracellular cholesterol transport and compartmentation. J Biol Chem 270:15443–15446

Maekawa S, Sato C, Kitajima K, Funatsu N, Kumanogoh H, Sokawa Y (1999) Cholesterol-dependent localization of NAP-22 on a neuronal membrane microdomain (raft). J Biol Chem 274:21369–21374

Mason RP, Walter MF, Day CA, Jacob RF (2006) Active metabolite of atorvastatin inhibits membrane cholesterol domain formation by an antioxidant mechanism. J Biol Chem 281:9337–9345

McIntosh TJ (1978) The effect of cholesterol on the structure of phosphatidylcholine bilayers. Biochim Biophys Acta 513:43–58

Moody MF (1963) X-ray diffraction pattern of nerve myelin: a method for determining the phases. Science 142:1173–1174

Mukherjee S, Chattopadhyay A (1996) Membrane organization at low cholesterol concentrations: a study using 7-nitrobenz-2-oxa-1,3-diazol-4-yl-labeled cholesterol. Biochemistry 35:1311–1322

Ostermeyer AG, Beckrich BT, Ivarson KA, Grove KE, Brown DA (1999) Glycosphingolipids are not essential for formation of detergent- resistant membrane rafts in melanoma cells. Methyl-beta-cyclodextrin does not affect cell surface transport of a GPI-anchored protein. J Biol Chem 274:34459–34466

Phillips MC, Johnson WC, Rothblat GH (1997) Mechanism and consequences of cellular cholesterol exchange and transfer. Biochim Biophys Acta 906:223–276

Phillips JE, Geng YJ, Mason RP (2001) 7-Ketocholesterol forms crystalline domains in model membranes and murine aortic smooth muscle cells. Atherosclerosis 159:125–135
Rafferty NS (1985) Lens morphology. In: Maisel H (ed) The Ocular Lens. Marcel Dekker, New York, pp 1–60
Rice PA, McConnell HM (1989) Critical shape transitions of monolayer lipid domains. Proc Natl Acad Sci U S A 86:6445–6448
Ruocco MJ, Shipley GG (1984) Interaction of cholesterol with galactocerebroside and galactocerebroside-phosphatidylcholine bilayer membranes. Biophys J 46:695–707
Schroeder F, Wood WG (1995) Lateral lipid domains and membrane function. In: Sperelakis N (ed) Cell Physiology Source Book. Academic, New York, pp 36–44
Schroeder F, Jefferson JR, Kier AB, Knittel J, Scallen TJ, Wood WG, Hapala I (1991) Membrane cholesterol dynamics: cholesterol domains and kinetic pools. Proc Soc Exp Biol Med 196:235–252
Schroeder F, Woodford JK, Kavecansky J, Wood WG, Joiner C (1995) Cholesterol domains in biological membranes. Mol Membr Biol 12:113–119
Self-Medlin Y, Byun J, Jacob RF, Mizuno Y, Mason RP (2009) Glucose promotes membrane cholesterol crystalline domain formation by lipid peroxidation. Biochim Biophys Acta 1788: 1398–1403
Shinitzky M, Inbar M (1976) Microviscosity parameters and protein mobility in biological membranes. Biochim Biophys Acta 433:133–149
Shrivastava S, Pucadyil TJ, Paila YD, Ganguly S, Chattopadhyay A (2010) Chronic cholesterol depletion using statin impairs the function and dynamics of human serotonin(1A) receptors. Biochemistry 49:5426–5435
Simons K, Ikonen E (1997) Functional rafts in cell membranes. Nature 387:569–572
Simons K, Ikonen E (2000) How cells handle cholesterol. Science 290:1721–1726
Simons K, Toomre D (2000) Lipid rafts and signal transduction. Nat Rev Mol Cell Biol 1:31–39
Slotte JP (1995a) Lateral domain formation in mixed monolayers containing cholesterol and dipalmitoylphosphatidylcholine or N-palmitoylsphingomyelin. Biochim Biophys Acta 1235:419–427
Slotte PJ (1995b) Effect of sterol structure on molecular interactions and lateral domain formation in monolayers containing dipalmitoyl phosphatidylcholine. Biochim Biophys Acta 1237:127–134
Small DM (1988) Progression and regression of atherosclerotic lesions. Insights from lipid physical biochemistry. Arterioscler Thromb Vasc Biol 8:103–129
Tocanne JF (1992) Detection of lipid domains in biological membranes. Comments Mol Cell Biophys 8:53–72
Tulenko TN, Chen M, Mason PE, Mason RP (1998) Physical effects of cholesterol on arterial smooth muscle membranes: evidence of immiscible cholesterol domains and alterations in bilayer width during atherogenesis. J Lipid Res 39:947–956
Worthington CR, Liu SC (1973) Structure of sarcoplasmic reticulum membranes at low resolution (17Å). Arch Biochem Biophys 157:573–579
Yeagle PL (1985) Cholesterol and the cell membrane. Biochim Biophys Acta 822:267–287
Yeagle PL, Young J, Rice D (1988) Effects of cholesterol on (Na^+, K^+)-ATPase ATP hydrolyzing activity in bovine kidney. Biochemistry 27:6449–6452

Chapter 16
Role of Lipid-Mediated Effects in β_2-Adrenergic Receptor Dimerization

Xavier Prasanna, Amitabha Chattopadhyay, and Durba Sengupta

Introduction

G protein-coupled receptors (GPCRs) constitute the largest and most diverse family of the mammalian membrane receptors (Pierce et al. 2002). Members of the GPCR superfamily are involved in mediating several physiological processes and respond to a wide range of ligands (Rosenbaum et al. 2009). As a result, more than half of the current clinical drugs target GPCRs (Heilker et al. 2009). Members of the GPCR family show a low sequence similarity but they share common structural features (Venkatakrishnan et al. 2013). All GPCRs comprise of seven transmembrane helices that traverse the bilayer (Lagerstroem and Schioeth 2008; Katritch et al. 2012). Recent structural characterization of GPCRs have provided insights into their function (Cherezov et al. 2007; Rosenbaum et al. 2007; Rasmussen et al. 2007; Jaakola et al. 2008; Xu et al. 2011; Chien et al. 2010; Haga et al. 2012; Granier et al. 2012), but the dynamics related to their function are still largely unknown.

The β_2-adrenergic receptor, an important GPCR involved in muscle relaxation, has been the focus of several seminal studies. Until recently, the β_2-adrenergic receptor was hypothesized to exist as monomers and the reconstituted monomeric receptor was shown to be functional (Whorton et al. 2007). Previous studies have proposed that β_2-adrenergic receptors dimerize in the membrane (Angers et al. 2000; Hébert et al. 1996). Advances in spectroscopic and imaging techniques have confirmed

X. Prasanna • D. Sengupta (✉)
CSIR-National Chemical Laboratory, Council of Scientific and Industrial Research, Dr. Homi Bhabha Road, Pune 411 008, India
e-mail: d.sengupta@ncl.res.in

A. Chattopadhyay (✉)
CSIR-Centre for Cellular and Molecular Biology, Council of Scientific and Industrial Research, Uppal Road, Hyderabad 500 007, India
e-mail: amit@ccmb.res.in

A. Chakrabarti, A. Surolia (eds.), *Biochemical Roles of Eukaryotic Cell Surface Macromolecules*, Advances in Experimental Medicine and Biology 842,
DOI 10.1007/978-3-319-11280-0_16

the existence of dimers and higher order oligomers in related receptors (Paila et al. 2011a; Ganguly et al. 2011). Single particle tracking methods have shown that the GPCRs exist in a dynamic equilibrium between the different associated species (Kasai et al. 2011). Recent single molecule studies on the β_2-adrenergic receptor suggested a distinct dynamics and organization of the receptor (Calebiroa et al. 2012). Interestingly, crystal structures of the β_1-adrenergic receptor, revealed two distinct dimer interfaces, corroborating the existence of associated GPCR states (Huang et al. 2013).

The spatio-temporal organization of GPCRs has been shown to be affected by changes in membrane composition, especially cholesterol concentrations (Paila et al. 2011a; Ganguly et al. 2011). Further, lipid composition has been shown to influence organization, stability and function of several GPCRs (Soubias and Gawrisch 2012; Yao and Kobilka 2005; Paila et al. 2011b; Pucadyil and Chattopadhyay 2004, 2007; Gibson and Brown 1993; Brown 1994; Saxena and Chattopadhyay 2012). The effect of the membrane composition on GPCR dynamics has been suggested to be due to either specific direct interactions (such as binding sites) or indirect effects (such as the changes in membrane physical properties) (Paila and Chattopadhyay 2009). The lipid and cholesterol binding sites that have been resolved in several GPCR crystal structures (Cherezov et al. 2007; Liu et al. 2012), point toward a direct effect. In contrast, spectroscopic studies have suggested a modulation of receptor organization by indirect effects such as hydrophobic mismatch and curvature changes (Botelho et al. 2006; Oates and Watts 2011; Alves et al. 2005).

The molecular details of receptor-lipid interactions are being increasingly probed at the atomistic resolution by computational methods, such as molecular dynamics simulations. With increase in computational power, longer timescale atomistic simulations of GPCR monomers have been performed (Grossfield et al. 2006, 2008; Dror et al. 2009; Lyman et al. 2009). The simulations have probed detailed GPCR dynamics and identified several important protein-lipid interactions. Specific cholesterol binding sites have also been proposed based on μs timescale simulations (Sengupta 2012a; Cang et al. 2013; Lee and Lyman 2012). To understand the more complex higher-order organization of GPCRs, coarse-grain methods have been used (Periole et al. 2007; Mondal et al. 2009) that suggest hydrophobic mismatch as an important driving force. Additionally, we have recently shown that membrane composition can directly modulate the dimer interface of the β_2-adrenergic receptor (Prasanna et al. 2014). Dimerization profiles have also been calculated by biased molecular dynamics for a few interfaces, but focused mainly on the protein energetics (Periole et al. 2012; Johnston et al. 2012). It is becoming increasingly clear that GPCR association is modulated by the membrane environment, although a more detailed investigation of the receptor-lipid interactions is still missing. Importantly, the significance of cholesterol in modulating receptor organization has been established, but the role of the phospholipids in modulating organization remains largely unexplored.

In this work, we study the interaction of the β_2-adrenergic receptor with the membrane lipids and explore the direct and indirect membrane effects that could be important in receptor dimerization. We have recently carried out coarse-grain molecular dynamics simulations of the β_2-adrenergic receptor in membranes of varying cholesterol composition (Prasanna et al. 2014). We have shown that cholesterol "hot-spots" present on the receptor surface could modulate the receptor dimerization. Here, we examine the role of the membrane lipids in driving and modulating receptor dimerization. We first calculate the hydrophobic mismatch in the monomeric and dimeric regimes, and compare its role in driving association. We also explore direct lipid binding sites in the receptor monomers and dimers. We have identified a putative lipid-binding site between transmembrane helices I and VII that shows a high occupancy by a lipid molecule. Our results show that both direct and indirect membrane effects contribute toward the dimerization of the receptor.

Methods

System Setup: Molecular dynamics simulations of the membrane embedded β_2-adrenergic receptor were carried out in 1-palmitoyl-2-oleoyl-*sn*-glycero-3-phosphocholine (POPC) bilayers in the absence and presence of 50 % cholesterol. The systems were represented using the MARTINI coarse-grain force-field (version 2.1) (Marrink et al. 2007; Monticelli et al. 2008). We have used the MARTINI force-field in our study since it has been shown to be suitable for applications such as membrane protein association (Periole et al. 2012; Sengupta et al. 2009; Sengupta and Marrink 2010; Prasanna et al. 2013) and partitioning of membrane proteins between membrane domains of varying compositions (Schäfer et al. 2011; Sengupta 2012b). A homology model of β_2-adrenergic receptor (amino acid residues 29–342) was generated from crystal structure (PDB: 2RH1) using the software SWISS-MODEL (Arnold et al. 2006). Bilayers containing POPC and with 50 % cholesterol concentration were generated from an initial conformation of randomly placed POPC, cholesterol and water molecules. For the simulations of the monomeric receptor, a single copy of the receptor in its coarse-grain representation was embedded at the middle of an equilibrated bilayer and simulated for 5 μs. For the dimer simulations, two copies of the receptor, in its coarse-grain representation, were embedded into the equilibrated membrane, such that the inter-receptor distance (centre of mass) was at least 6 nm (minimum distance at least 3 nm). After the receptors associated, the dimer regime was simulated for a further 20 μs. Further details of the simulations are described in our previous study (Prasanna et al. 2014).

Simulation parameters: All simulations were performed using the GROMACS simulation package, version 4.5.4 (Van Der Spoel et al. 2005). The cut-off for non-bonded interactions was 1.2 nm with electrostatic interactions shifted to 0 in the range 0–1.2 nm and Lennard-Jones interactions shifted to 0 in the range 0.9–1.2 nm. A relative electrostatic screening of 15 was used. The temperature for each group was weakly coupled using Berendsen thermostat algorithm with a coupling constant of 0.1 ps to maintain a constant temperature of 300 K during simulation (Berendsen et al. 1984). Semi-isotropic pressure was maintained using Berendsen barostat algorithm with a pressure of 1 bar independently in the plane of the membrane and perpendicular to the membrane with a coupling constant of 0.5 ps and a compressibility of 3×10^{-5} bar^{-1}. The time step used in the simulations was 20 fs. Simulations were rendered using VMD software (Humphrey et al. 1996).

Analysis: Local Membrane Thickness: The local membrane thickness was calculated from the difference in the *z*-position of the PO_4 bead of the POPC molecule. The values were calculated by binning the bilayer into 0.5 nm bins and averaging over the trajectory. To compare between different membrane compositions, a normalized membrane thickness has been defined: $X_{norm} = X/X_{av}$, where X is the local bilayer thickness and X_{av} the average bilayer thickness in the bulk membrane. Correspondingly, the bilayer thickness far from the receptor is 1, and local thickening or thinning will be denoted by a value greater than or less than 1, respectively. A snapshot of the receptor has been superimposed on the 2D local thickness profile.

Spatial density function (*SDF*): The spatial distribution of the phospholipid molecules around the β_2-adrenergic receptor was calculated as the 3D spatial distribution function of the lipid beads. The SDF was calculated from the last 5 μs trajectory by using the g_spatial program in the Gromacs package. The voxel element was set to 0.1 nm in each direction. In general, the SDF reflects the average 3D density distribution of the lipids and therefore points toward the locations where lipid molecules reside with higher probability. The calculated 3D SDFs were averaged over the extracellular and intracellular leaflets by projecting onto the upper and lower membrane planes, respectively. A snapshot of the receptor has been superimposed on the projected SDFs.

Energetics: The protein-protein, protein-lipid, and lipid-lipid interaction energies were calculated by summing the Lennard Jones and Coulomb terms. For POPC-cholesterol bilayers, the contributions from POPC and cholesterol were summed. The values were binned at a bin size of 0.1 nm inter-receptor distance.

Results

Coarse-grain molecular dynamics simulations of β_2-adrenergic receptors were performed in POPC bilayers in the absence and presence (50 %) of cholesterol. We have previously shown that cholesterol modulates the dimer interfaces of the β_2-adrenergic receptor via cholesterol "hot-spots" (Prasanna et al. 2014). Here, we extend our previous work to analyze the effects mediated by the lipids, both the direct and indirect effects, and probe their relation to receptor association.

Indirect Effects: Hydrophobic Mismatch Around the Receptor

We analyzed the local membrane thickness around a β_2-adrenergic receptor, embedded in POPC bilayers containing 0 and 50 % cholesterol (see Fig. 16.1). Since the thickness of the POPC-cholesterol bilayer is larger than the POPC bilayer (Nezil and Bloom 1992), we compare a normalized thickness between the two

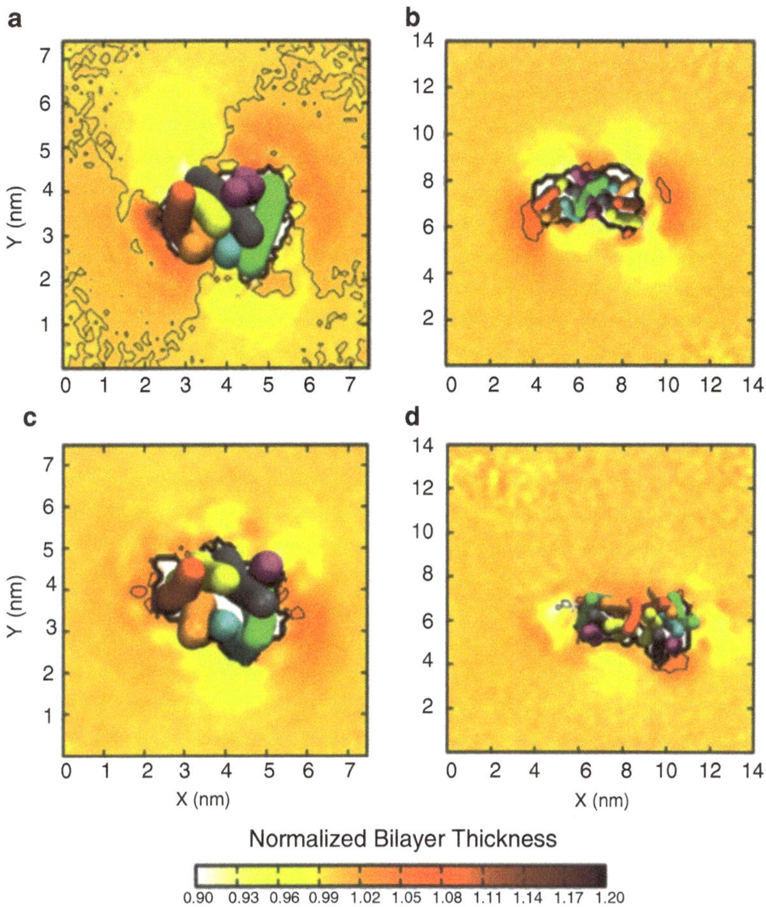

Fig. 16.1 Bilayer thickness profile around the β_2-adrenergic receptor in POPC bilayers with 0 % (**a** and **b**) and 50 % (**c** and **d**) cholesterol concentration. The profiles correspond to the monomer (*left*) and dimer (*right*) regimes of the receptor. To compare the local variations in the bilayer thickness, a normalized bilayer thickness is plotted. The top views of the receptor monomer and dimers are superimposed on the plots and transmembrane helices I and IV are labeled. The transmembrane helices have been color coded as follows: I (red), II (yellow), III (gray), IV (purple), V (green), VI (cyan), VII (orange). Thickness profiles were generated with previously developed tools (Castillo et al. 2013). For further details see Methods

bilayers. The normalized thickness is defined as $X_{norm} = X/X_{av}$, where X is the local bilayer thickness and X_{av} the average bilayer thickness in the bulk membrane. This measure allows us to directly compare the local variations in the thickness compared to the bulk membrane. In POPC bilayers, the thickness profile shows two distinct areas of increased thickness (Fig. 16.1a). The first site is localized around helix I and VII and the second site is located at helices IV and V. At the other faces of the receptor, membrane thinning was observed around the grooves formed by helices I, II and helices VI, VII. The same two sites show a positive mismatch in POPC-cholesterol bilayers (Fig. 16.1c), but the magnitude of the mismatch is less. The decrease in the

Table 16.1 Length of the hydrophobic segment of the transmembrane helices

Transmembrane Helix	Helix length (nm)	Helix length along the bilayer normal (nm)
I	3.7	2.6
II	3.6	2.9
III	3.8	3.4
IV	3.1	3.1
V	3.5	3.3
VI	3.7	3.5
VII	3.8	3.7

The hydrophobic segment of each transmembrane helix was calculated by measuring the average distance between the terminal backbone beads. To account for helix tilting, a second measure of the length component of transmembrane helix parallel to the bilayer normal was also calculated. The hydrophobic region of the bilayer, calculated as the average distance between the beads representing Sn2 glycerol ester carbon of POPC, is 3.25 nm

magnitude of the mismatch is due to the increased membrane thickness in the presence of cholesterol. It is difficult to correlate residue-based hydrophobic length of the helices to the bilayer perturbations due to helix tilting, apolar flanking residues and asymmetric distribution of apolar residues on different helix faces (Table 16.1).

To analyze the membrane profile around the receptor in the dimer regime, two dimer states corresponding to the most sampled states in our simulations were considered (see Rasmussen et al. 2007). The dimer structure most populated in POPC bilayers with 0 % cholesterol is defined by helices IV and V at the interface. At 50 % cholesterol concentration, the most populated dimer structure comprised of helices I and II at the interface. A distinct relationship is seen between the helices that define the dimer interface and those that show a positive hydrophobic mismatch with the bilayer. The differences in the membrane thickness were less pronounced in the dimer regime, for both dimer structures. Interestingly, in the dimers observed in POPC bilayers, the thickness changes around helix I were reduced (Fig. 16.1b), although the helix was not in the dimer interface. In the dimer states in POPC bilayers with 50 % cholesterol, an asymmetric thickening around helix I persisted, though it was involved at the dimer interface (Fig. 16.1d). Although there appears to be a distinct connection between hydrophobic mismatch and dimer interfaces, the relationship between the two is not straightforward. The decreased populations of helix I in the dimer interface in POPC bilayers, despite a high hydrophobic mismatch in the monomer regime, coupled with a complex mismatch pattern in the dimer states, points toward more complex dimerization behavior of the receptor.

Direct Effects: Specific Protein-Lipid Interactions

The spatial distribution function (SDF) of lipid molecules around the monomer was calculated for a representative simulation set and is plotted in Fig. 16.2. The density profiles of the POPC molecules was calculated over several *z*-slices and averaged

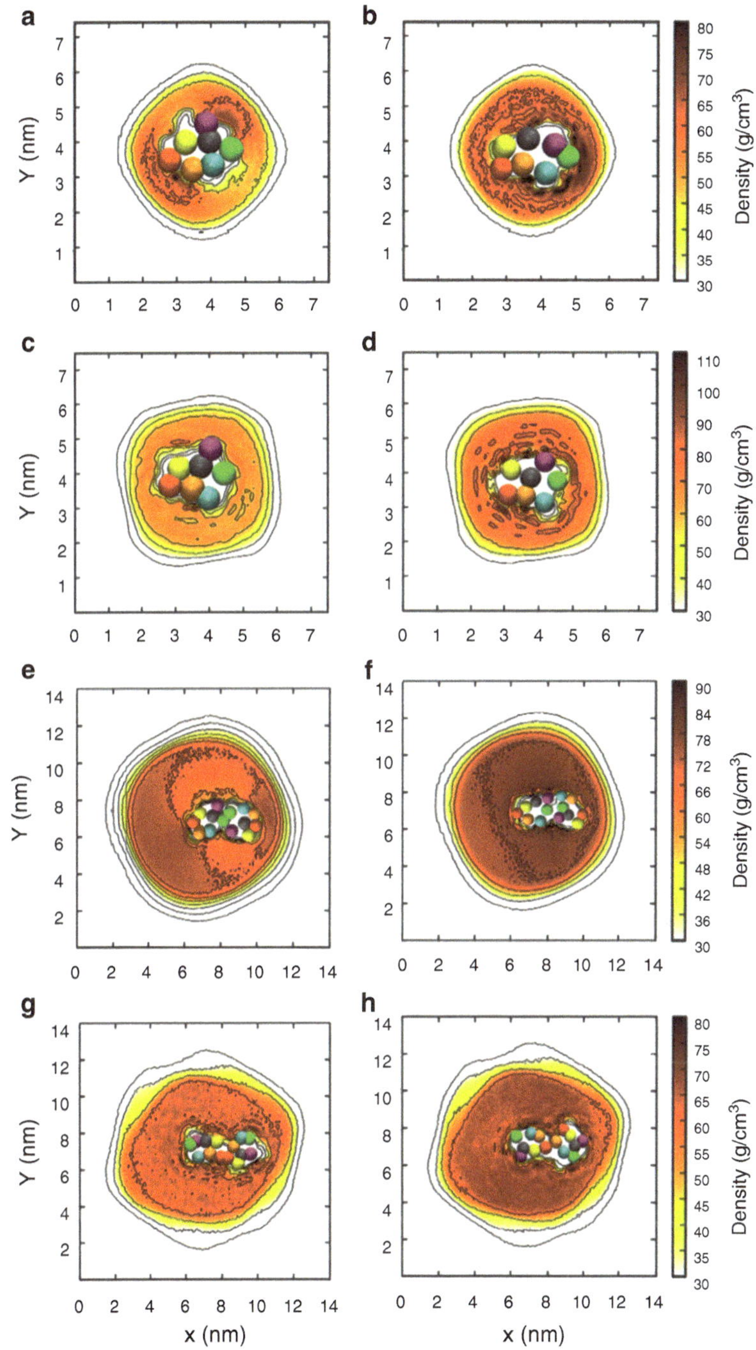

Fig. 16.2 The spatial distribution function (SDF) of POPC around the β_2-adrenergic receptor monomer in POPC bilayers with 0 % (**a**, **b**) and 50 % (**c**, **d**) cholesterol. The SDF is represented for the extracellular (**a**, **c**) and intracellular (**b**, **d**) leaflets separately. The SDF around the receptor dimer is shown for the extracellular (**e**, **g**) and intracellular (**f**, **h**) leaflets in POPC bilayers with 0 % (**e**, **f**) and 50 % (**g**, **h**) cholesterol concentration. The top views of the receptor monomer and dimers are superimposed on the plots. For further details see Methods

over the extracellular and the intracellular leaflets separately. The regions of high density correspond to sites with the highest probability of finding a POPC molecule. The SDF profile shows one distinct site between helices I and VII in the extracellular leaflet (Fig. 16.2a). In the inner leaflet, a more spread-out density was observed (Fig. 16.2b). The first site was centered around helix IV, and was close to the putative cholesterol binding site (CCM). The remaining sites were spread out over helices V, VI and VII. In the presence of cholesterol, a competition was observed between POPC and cholesterol molecules and at most sites POPC binding persisted. On the extracellular site, the SDF is high in the same groove, i.e. between I and VII, although the location appears to be shifted (Fig. 16.2c). The magnitude of the SDF in this groove is similar to that in POPC bilayers. In the intracellular site, the main region of high lipid distribution is located around helix IV, although the magnitude is reduced (Fig. 16.2d). The remaining regions of high density observed in POPC bilayers are further reduced in magnitude. Upon dimerization, the sites of high lipid density persist (Fig. 16.2e–h). Taken together, these results suggest a competition between cholesterol and POPC molecules at the putative cholesterol binding site at helix IV. In contrast, the high SDF site at helix VII appears to be stable and we propose it to be a putative lipid binding site.

To test the robustness of the density data and to exclusively account for the specific binding events, we calculated the maximum occupancy time of POPC around each of the transmembrane helices during the simulation (Fig. 16.3). We defined the maximum occupancy time as the maximum time a given PO_4 bead of the POPC molecule was continuously bound to a given site, normalized to the simulation length. A value of 1 implies that the lipid molecule was present at the given site throughout the entire simulation time and 0 implies it was never present at that site. The values were averaged over ten simulations and the error bars denote the standard deviation between the simulations. A similar trend to the density profiles is observed in maximum occupancies. The maximum occupancy of POPC was the highest at helix VII at the extracellular leaflet and the highest at helix IV on the intracellular leaflet. In presence of cholesterol, the occupancy at both the sites decreases although the trend persists. Combining our data on the SDFs and occupancies, it is clear that POPC binding to some of these sites is specific.

Characterization of the Lipid Site at Helix I

To explore the molecular details of the POPC occupancy site at transmembrane helix VII, we calculated a residue-based distance map between the bound POPC molecule and the amino acid residues on transmembrane helix I and VII. A representative snapshot is shown in Fig. 16.4. The headgroup beads are observed to interact with a charged residue, Glu306. Several aromatic and apolar residues line the lipid occupancy groove. Interestingly, the same site has been identified as a putative lipid binding site in a recent high resolution A_{2A}-adenosine receptor structure (Liu et al. 2012). The lipid binding groove between helices I and VII from the crystal structure of the A_{2A}-adenosine receptor is also depicted in Fig. 16.4.

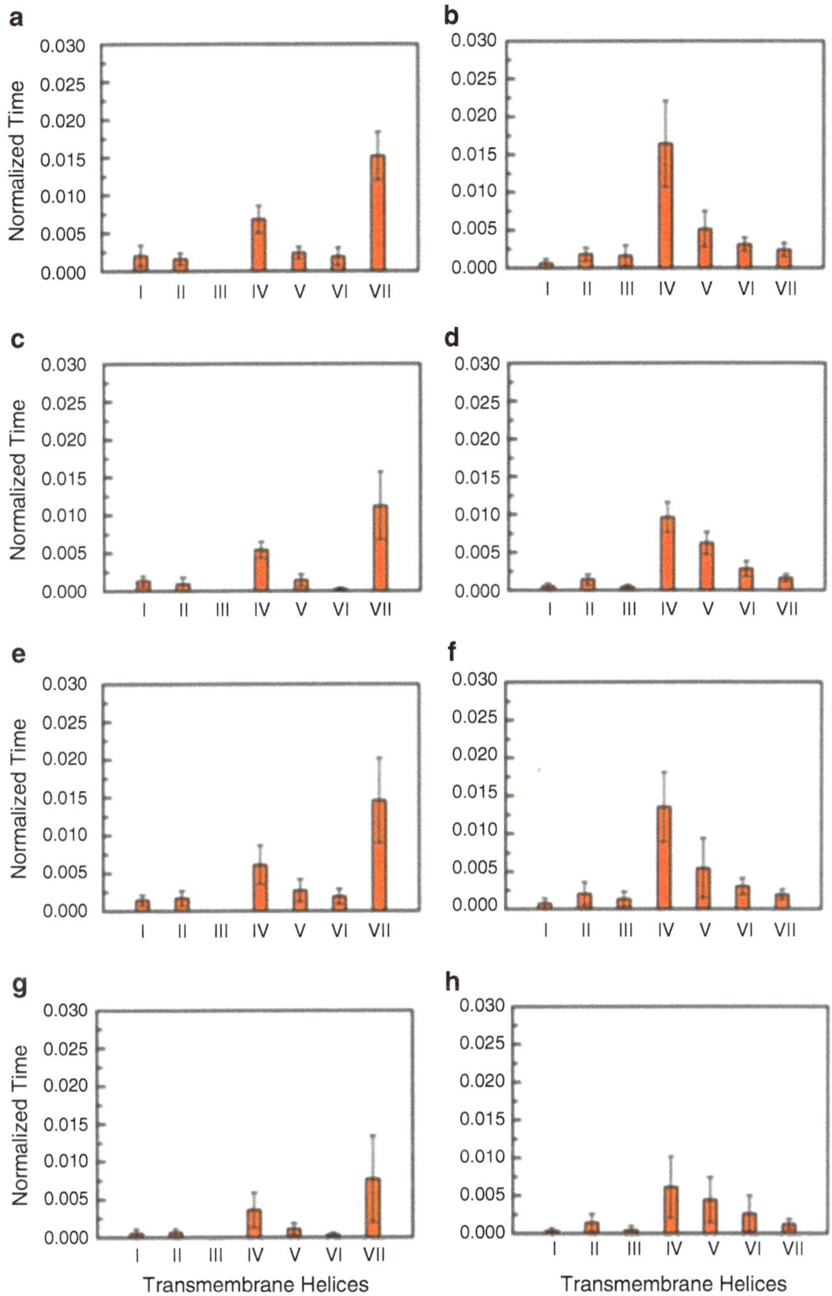

Fig. 16.3 Maximum occupancy of POPC molecules around the β_2-adrenergic receptor monomer in POPC bilayers with 0 % (**a**, **b**) and 50 % (**c**, **d**) cholesterol. The occupancy is shown for the extracellular (**a**, **c**) and intracellular (**b**, **d**) leaflets separately. The occupancy around the receptor dimer is shown for the extracellular (**e**, **g**) and intracellular (**f**, **h**) leaflets in POPC bilayers with 0 % (**e**, **f**) and 50 % (**g**, **h**) cholesterol concentration. For further details see Methods

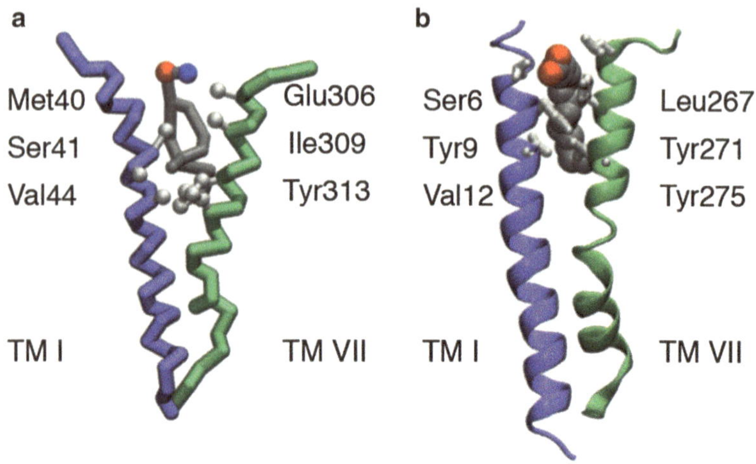

Fig. 16.4 A schematic representation of (**a**) the high lipid occupancy site in seen in coarse-grain simulations and (**b**) the putative lipid binding site in the crystal structure of A_{2A}-adenosine receptor. Only transmembrane helices I and VII are shown for clarity. The side chains that are involved at the site are shown in light grey. The POPC molecule is represented in dark grey and the head-group PO_4 and NC3 beads are depicted as spheres

Exploring the Energetics of Association

The membrane effects described above are a consequence of the interplay between the energetics of the protein and the lipid. We have calculated the interaction energies as a function of inter-receptor separation (shown in Fig. 16.5). The values shown are averaged over ten simulations, totaling about 130 μs of total simulation time and are distinct from the free energy of dimerization. In POPC and POPC-cholesterol bilayers, the protein-protein interaction energy decreases expectedly as the receptors approach each other (Fig. 16.5a, b). A minimum could not be discerned at low inter-receptor distances, possibly due to the lack of sampling of unfavorable close contacts. Multiple smaller minima could be discerned along the pathway that could correspond to meta-stable states along the dimerization pathway. The protein-lipid interaction energy increased as the receptors approach each other due to the de-lipidation of the protein (Fig. 16.5c, d). The lipid-lipid interaction energy decreases correspondingly as the receptors associate and interact with each other (Fig. 16.5e, f). To fully understand the energetics of receptor association, free-energy calculations sampling over all possible receptor interfaces is required, that still remains a challenge within current computational methods.

Discussion

GPCR organization is a critical factor in cellular signaling (Saxena and Chattopadhyay 2011) and understanding these processes within heterogeneous membrane compositions gives rise to new challenges and complexities. It is becoming clear that cellular

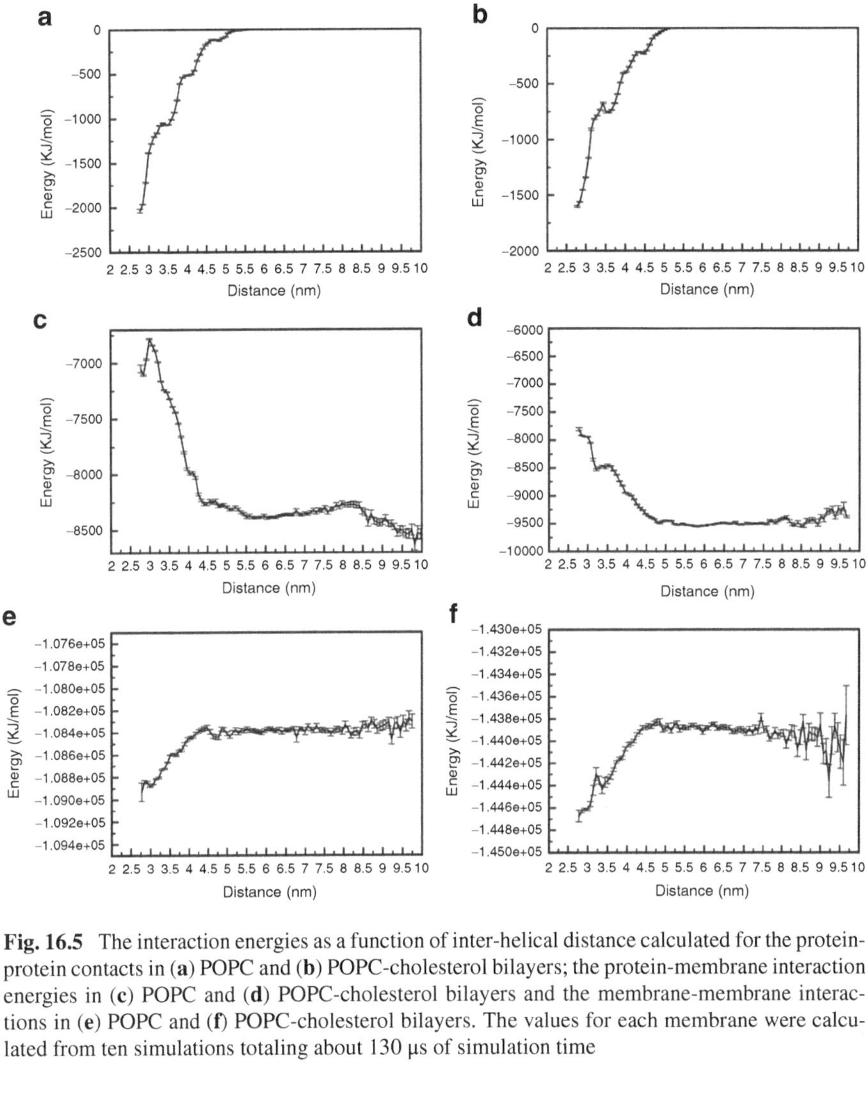

Fig. 16.5 The interaction energies as a function of inter-helical distance calculated for the protein-protein contacts in (**a**) POPC and (**b**) POPC-cholesterol bilayers; the protein-membrane interaction energies in (**c**) POPC and (**d**) POPC-cholesterol bilayers and the membrane-membrane interactions in (**e**) POPC and (**f**) POPC-cholesterol bilayers. The values for each membrane were calculated from ten simulations totaling about 130 μs of simulation time

signaling in general and GPCR function in particular has to be considered in the context of membrane organization and composition. In specific, the interplay between the receptor and the constituent membrane molecules needs to be probed. In this work, we have analyzed the membrane effects around the β_2-adrenergic receptor. Both direct and indirect effects have been probed, together with the estimation of the membrane-protein energetics. We have analyzed multiple μs time scale coarse-grain simulations that allows us to explore the membrane effects in the presence and absence of cholesterol.

Hydrophobic mismatch as a driving force for GPCR association has been proposed previously for β_2-adrenergic receptor (Mondal et al. 2009), rhodopsin (Botelho et al.

2006; Periole et al. 2007) and the opioid receptor (Alves et al. 2005). Although there appears to be a distinct connection between the helices with the maximum hydrophobic mismatch and those present subsequently at the dimer interfaces, the relationship between the two is not straightforward. The decreased populations of helix I in the dimer interface is observed in POPC bilayers, despite a high hydrophobic mismatch in the monomer regime. This, coupled with a complex mismatch pattern in the dimer states, points toward more complex driving forces for receptor association. A more direct effect of the membrane environment on GPCR stability and organization is suggested based on the lipid and cholesterol binding sites that have been resolved in several GPCR crystal structures (Cherezov et al. 2007; Liu et al. 2012). Although cholesterol binding sites have been probed in detail (Sengupta 2012a; Cang et al. 2013; Lee and Lyman 2012), the lipid binding sites remain less explored. Here, using multiple μs time scale coarse-grain simulations we propose a high occupancy site of POPC at helix VII of the β_2-adrenergic receptor. A similar site has been observed in previous atomistic simulations of the β_2-adrenergic receptor, but was not further characterized (Cang et al. 2013). Interestingly, the same site is also proposed to be a lipid binding site based on the recent high-resolution crystal structure of the A_{2A}-adenosine receptor (Liu et al. 2012). We suggest that this site could play a role in the subsequent organization of the receptor and modulate the population of the dimer structures with helices I/VII at the interface.

In conclusion, using multiple coarse-grain simulations of the β_2-adrenergic receptor in POPC bilayers in the absence and presence (50 %) of cholesterol, we have characterized lipid-protein interactions that could play an important role in the stabilization and organization of the receptor. We show the presence of hydrophobic mismatch at helices I/VII and IV/V, that is reduced in the presence of cholesterol. Although there is a relation between the helices with maximum hydrophobic mismatch and its subsequent presence in the dimer interface, the relationship is not straightforward. We have further suggested a putative lipid binding site at helix VII that could play an important role in modulating the dimer interfaces. Based on our results, it appears that lipid-protein interactions play an important role in receptor dimerization. Understanding the underlying membrane-protein interactions will help us appreciate the complex nature of GPCR organization and its link to GPCR function in health and disease.

Acknowledgements This work was supported by the Council of Scientific and Industrial Research, Govt. of India. D.S. gratefully acknowledges the support of the Ramalingaswami Fellowship from the Department of Biotechnology, Govt. of India. X.P. thanks the University Grants Commission (India) for the award of a Junior Research Fellowship. A.C. gratefully acknowledges J.C. Bose Fellowship (Department of Science and Technology, Govt. of India). A.C. is an Adjunct Professor at the Special Centre for Molecular Medicine of Jawaharlal Nehru University (New Delhi) and Indian Institute of Science Education and Research (Mohali), and Honorary Professor of the Jawaharlal Nehru Centre for Advanced Scientific Research (Bangalore). We acknowledge the CSIR Fourth Paradigm Institute (Bangalore) and the Multi-Scale Simulation and Modeling project - MSM (CSC0129) for computational time.

References

Alves ID, Salamon Z, Hruby VJ, Tollin G (2005) Ligand modulation of lateral segregation of a G-Protein-Coupled receptor into lipid microdomains in sphingomyelin/phosphatidylcholine solid-supported bilayers. Biochemistry 44:9168–9178

Angers S, Salahpour A, Joly E, Hilairet S, Chelsky D, Dennis M, Bouvier M (2000) Detection of β2-adrenergic receptor dimerization in living cells using bioluminescence resonance energy transfer (BRET). Proc Natl Acad Sci U S A 97:3684–3689

Arnold K, Bordoli L, Kopp J, Schwede T (2006) The SWISS-MODEL workspace: a web-based environment for protein structure homology modelling. Bioinformatics 22:195–201

Berendsen HJC, Postma JPM, van Gunsteren WF, Dinola A, Haak JR (1984) Molecular dynamics with coupling to an external bath. J Chem Phys 81:3684–3690

Botelho VA, Huber T, Sakmar TP, Brown MF (2006) Curvature and hydrophobic forces drive oligomerization and modulate activity of rhodopsin in membranes. Biophys J 91:4464–4477

Brown MF (1994) Modulation of rhodopsin function by properties of the membrane bilayer. Chem Phys Lipids 73:159–180

Calebiroa D, Riekena F, Wagner J, Sungkaworna T, Zabela U, Borzid A, Cocuccie E, Zürna A, Lohse MJ (2012) Single-molecule analysis of fluorescently labeled G-protein-coupled receptors reveals complexes with distinct dynamics and organization. Proc Natl Acad Sci U S A 110:743–748

Cang X, Du Y, Mao Y, Wang Y, Yang H, Jiang H (2013) Mapping the functional binding sites of cholesterol in β_2-adrenergic receptor by long-time molecular dynamics simulations. J Phys Chem B 117:1085–1094

Castillo N, Monticelli L, Barnoud J, Tieleman DP (2013) Free energy of WALP23 dimer association in DMPC, DPPC, and DOPC bilayers. Chem Phys Lip 169:95–105

Cherezov V, Rosenbaum DM, Hanson MA, Rasmussen SGF, Thian FS, Kobilka TS, Choi H-J, Kuhn P, Weis WI, Kobilka BK, Stevens RC (2007) High-Resolution crystal structure of an engineered human β_2-adrenergic G protein-coupled receptor. Science 318:1258–1265

Chien EYT, Liu W, Zhao Q, Katritch V, Won Han G, Hanson MA, Shi L, Newman AH, Javitch JA, Cherezov V, Stevens RC (2010) Structure of the human dopamine D_3 receptor in complex with a D2/D3 selective antagonist. Science 330:1091–1095

Dror RO, Arlow DH, Borhani DW, Jensen MA, Piana S, Shaw DE (2009) Identification of two distinct inactive conformations of the β_2-adrenergic receptor reconciles structural and biochemical observations. Proc Natl Acad Sci U S A 106:4689–4694

Ganguly S, Clayton AH, Chattopadhyay A (2011) Organization of higher-order oligomers of the serotonin$_{1A}$ receptor explored utilizing homo-FRET in live cells. Biophys J 100:361–368

Gibson NJ, Brown MF (1993) Lipid headgroup and acyl chain composition modulate the MI-MII equilibrium of rhodopsin in recombinant membranes. Biochemistry 32:2438–2454

Granier S, Manglik A, Kruse AC, Kobilka TS, Thian FS, Weis WI, Kobilka BK (2012) Structure of the d-opioid receptor bound to naltrindole. Nature 485:400–404

Grossfield A, Feller SE, Pitman MC (2006) A role for direct interactions in the modulation of rhodopsin by w-3 polyunsaturated lipids. Proc Natl Acad Sci U S A 103:4888–4893

Grossfield A, Pitman MC, Feller SE, Soubias O, Gawrisch K (2008) Internal hydration increases during activation of the G-protein-coupled receptor Rhodopsin. J Mol Biol 381:478–486

Haga K, Kruse AC, Asada H, Yurugi-Kobayashi T, Shiroishi M, Zhang C, Weis WI, Okada T, Kobilka BK, Haga T, Kobayashi T (2012) Structure of the human M_2 muscarinic acetylcholine receptor bound to an antagonist. Nature 482:547–551

Hébert TE, Moffett S, Morello J-P, Loisel TP, Bichet DG, Barret C, Bouvier M (1996) A peptide derived from a β_2-adrenergic receptor transmembrane domain inhibits both receptor dimerization and activation. J Biol Chem 271:16384–16392

Heilker R, Wolff M, Tautermann CS, Bieler M (2009) G-protein-coupled receptor-focused drug discovery using a target class platform approach. Drug Discov Today 14:231–240

Huang J, Chen S, Zhang JJ, Huang X-Y (2013) Crystal structure of oligomeric β_1-adrenergic G protein-coupled receptors in ligand-free basal state. Nat Struct Mol Biol 20:419–425
Humphrey W, Dalke A, Schulten K (1996) VMD: visual molecular dynamics. J Mol Graph 14:33–38
Jaakola V-P, Griffith MT, Hanson MA, Cherezov V, Chien EYT, Lane JR, IJzerman AP, Stevens RC (2008) The 2.6 Å crystal structure of a human A_{2A}-adenosine receptor bound to an antagonist. Science 322:1211–1217
Johnston JM, Wang H, Provasi D, Filizola M (2012) Assessing the relative stability of dimer interfaces in G protein-coupled receptors. PLoS Comput Biol 8:e1002649
Kasai R, Suzuki K, Prossnitz E, Koyama-Honda I, Nakada C, Fujiwara T, Kusumi A (2011) Full characterization of GPCR monomer-dimer dynamic equilibrium by single molecule imaging. J Cell Biol 192:463–480
Katritch V, Cherezov V, Stevens RC (2012) Diversity and modularity of G protein-coupled receptor structures. Trends Pharmacol Sci 33:17–27
Lagerstroem MC, Schioeth HB (2008) Structural diversity of G protein-coupled receptors and significance for drug discovery. Nat Rev Drug Discov 7:339–357
Lee JY, Lyman E (2012) Predictions for cholesterol interaction sites on the A_{2A} Adenosine receptor. J Am Chem Soc 134:16512–16515
Liu W, Chun E, Thompson AA, Chubukov P, Xu F, Katritch V, Han GW, Roth CB, Heitman LH, IJzerman AP, Cherezov V, Stevens RC (2012) Structural basis for allosteric regulation of GPCRs by sodium ions. Science 337:232–236
Lyman E, Higgs C, Kim B, Lupyan D, Shelley JC, Farid R, Voth GA (2009) A Role for a specific cholesterol interaction in stabilizing the Apo configuration of the Human A_{2A}- Adenosine Receptor. Structure 17:1660–1668
Marrink SJ, Risselada HJ, Yefimov S, Tieleman DP, de Vries AH (2007) The MARTINI forcefield: coarse grained model for biomolecular simulations. J Phys Chem B 111:7812–7824
Mondal S, Johnston JM, Wang H, Khelashvili G, Filizol M, Weinstein H (2009) Membrane driven spatial organization of GPCRs. Sci Rep 3:2909
Monticelli L, Kandasamy SK, Periole X, Larson RG, Tieleman DP, Marrink S-J (2008) The MARTINI coarse grained forcefield: extension to proteins. J Chem Theory Comput 4:819–834
Nezil F, Bloom M (1992) Combined influence of cholesterol and synthetic amphiphillic peptides upon bilayer thickness in model membranes. Biophys J 61:1176–1183
Oates J, Watts A (2011) Uncovering the intimate relationship between lipids cholesterol and GPCR activation. Curr Opin Struct Biol 21:802–807
Paila Y, Chattopadhyay A (2009) The function of G-protein coupled receptors and membrane cholesterol: specific or general interaction? Glycoconj J 26:711–720
Paila YD, Kombrabail M, Krishnamoorthy G, Chattopadhyay A (2011a) Oligomerization of the serotonin$_{1A}$ receptor in live cells: A time-resolved fluorescence anisotropy approach. J Phys Chem B 115:11439–11447
Paila YD, Jindal E, Goswami SK, Chattopadhyay A (2011b) Cholesterol depletion enhances adrenergic signalling in cardiac myocytes. Biochim Biophys Acta 1808:461–465
Periole X, Huber T, Marrink S-J, Sakmar TP (2007) G Protein-coupled receptors self-assemble in dynamics simulations of model bilayers. J Am Chem Soc 129:10126–10132
Periole X, Knepp AM, Sakmar TP, Marrink SJ, Huber T (2012) Structural determinants of the supramolecular organization of G protein-coupled receptors in bilayers. J Am Chem Soc 134:10959–10965
Pierce KL, Premont RT, Lefkowitz RJ (2002) Seven-transmembrane receptors. Nat Rev Mol Cell Biol 3:639–650
Prasanna X, Praveen PJ, Sengupta D (2013) Sequence dependent lipid-mediated effects modulate the dimerization of ErbB2 and its associative mutants. J Phys Chem B 15:19031–19041
Prasanna X, Chattopadhyay A, Sengupta D (2014) Cholesterol modulates the dimer interface of the β_2-Adrenergic receptor via cholesterol occupancy sites. Biophys J 106:1290–1300
Pucadyil TJ, Chattopadhyay A (2004) Cholesterol modulates ligand binding and G-protein coupling to serotonin$_{1A}$ receptors from bovine hippocampus. Biochim Biophys Acta 1663:188–200

Pucadyil TJ, Chattopadhyay A (2007) Cholesterol depletion induces dynamic confinement of the G protein-coupled serotonin$_{1A}$ receptor in the plasma membrane of living cells. Biochim Biophys Acta 1768:655–668

Rasmussen SGF, Choi H-J, Rosenbaum DM, Kobilka TS, Thian FS, Edwards PC, Burghammer M, Ratnala VRP, Sanishvili R, Fischetti RF, Schertler GFX, Weis WI, Kobilka BK (2007) Crystal structure of the human β_2-adrenergic G protein-coupled receptor. Nature 450:383–387

Rosenbaum DM, Cherezov V, Hanson MA, Rasmussen SGF, Thian FS, Kobilka TS, Choi H-J, Yao X-J, Weis WI, Stevens RC, Kobilka BK (2007) GPCR engineering yields high-resolution structural insights into β_2-adrenergic receptor function. Science 318:1266–1273

Rosenbaum DM, Rasmussen SGF, Kobilka BK (2009) The structure and function of G protein-coupled receptors. Nature 459:356–363

Saxena R, Chattopadhyay A (2011) A Membrane organization and dynamics of the serotonin$_{1A}$ receptor in live cells. J Neurochem 116:726–733

Saxena R, Chattopadhyay A (2012) Membrane cholesterol stabilizes the human serotonin$_{1A}$ Receptor. Biochim Biophys Acta 1818:2936–2942

Schäfer LV, de Jong DH, Holt A, Rzepiela AJ, de Vries AH, Poolman B, Killian JA, Marrink SJ (2011) Lipid packing drives the segregation of transmembrane helices into disordered lipid domains in model membranes. Proc Natl Acad Sci U S A 108:1343–1348

Sengupta D (2012a) Chattopadhyay A (2012) Identification of cholesterol binding sites in the serotonin$_{1A}$ receptor. J Phys Chem B 116:12991–12996

Sengupta D (2012b) Cholesterol modulates the structure binding modes and energetics of caveolin-1 membrane interactions. J Phys Chem B 116:14556–14564

Sengupta D, Marrink SJ (2010) Lipid-mediated interactions tune the association of glycophorin A helix and its disruptive mutants in membranes. Phys Chem Chem Phys 12:12987–12996

Sengupta D, Rampioni A, Marrink SJ (2009) Simulations of the c-subunit of ATPsynthase reveal helix rearrangements. Mol Membr Biol 26:422–434

Soubias O, Gawrisch K (2012) The role of the lipid matrix for structure and function of the GPCR rhodopsin. Biochim Biophys Acta 1818:234–240

Van Der Spoel D, Lindahl E, Hess B, Groenhof G, Mark AE, Berendsen HJC (2005) GROMACS: Fast, flexible, and free. J Comput Chem 26:1701–1718

Venkatakrishnan AJ, Deupi X, Lebon G, Tate CG, Schertler GF, Babu MM (2013) Molecular signatures of G-protein-coupled receptors. Nature 494:185–194

Whorton MR, Bokoch MP, Rasmussen SGF, Huang B, Zare RN, Kobilka B, Sunahara RKA (2007) Monomeric G protein-coupled receptor isolated in a high-density lipoprotein particle efficiently activates its G protein. Proc Natl Acad Sci U S A 104:7682–7687

Xu F, Wu H, Katritch V, Han GW, Jacobson KA, Gao Z-G, Cherezov V, Stevens RC (2011) Structure of an agonist-bound human A_{2A}-adenosine receptor. Science 332:322–327

Yao Z, Kobilka B (2005) Using synthetic lipids to stabilize purified β_2-adrenergic receptor in detergent micelles. Anal Biochem 343:344–346

Chapter 17
Effect of Temperature on the Phase Behaviour of Fully Saturated DAPC Lipid Bilayer: A Comparative Molecular Dynamics Simulation Study

Ipsita Basu and Chaitali Mukhopadhyay

The plasma membrane is made of phospholipid bilayer in which membrane proteins are embedded. The physiochemical properties of the cell membrane are determinant in many important cellular processes. The lipid bilayer has a fluid like consistency and the fundamental structure and dynamic properties are dependent upon temperature. Depending upon temperature lipid bilayer has many phases namely gel phase (L_β), liquid-crystalline phase (L_α), subgel phase (L_c), and ripple phase (P_β). We have investigated the structural properties of DAPC lipid bilayer at two different temperatures, one above the phase transition temperature (at 350 K) and another below the phase transition temperature (300 K). We have been able to discriminate the two phases at two different temperatures and determine the phase behavior of DAPC lipid. We have found that change in temperature may have serious consequences for the structural properties of lipid bilayer systems. The reduced area per lipid and the corresponding ordering of the lipid acyl chain lead to phase change of the bilayer. Energy calculation also supports that the system at temperature 300 K is at gel phase. Preliminary studies of dynamic quantities like diffusion coefficient showed that this parameter is also sensitive to temperature and shows lower value at lower temperature indicating ordering at lower temperature.

Introduction

The features of membranes have been broadly explored for a long time, and extensive information about the physicochemical characteristics of membrane systems has been provided (Garcia-Manyes et al. 2005a; Büldt et al. 1978; Pak et al. 1987;

I. Basu • C. Mukhopadhyay (✉)
Department of Chemistry, University of Calcutta, 92, A. P. C. Road, Kolkata 700009, India
e-mail: chaitalicu@yahoo.com; cmchem@caluniv.ac.in

A. Chakrabarti, A. Surolia (eds.), *Biochemical Roles of Eukaryotic Cell Surface Macromolecules*, Advances in Experimental Medicine and Biology 842,
DOI 10.1007/978-3-319-11280-0_16

Pastor et al. 1991; Schwyzer 1991). However, it is often difficult to obtain an atomic level description of the phenomena taking place in lipid bilayers by experiments only. Therefore, atomistic molecular dynamics simulation has become a useful tool for studies of biomembrane systems at the molecular level (Tieleman et al. 1997; Feller 2000; Saiz and Klein 2002). Many fundamental structure and dynamical properties of lipid bilayer such as area per lipid, lipid tail order parameter, and lipid thickness are dependent upon temperature (Garcia-Manyes et al. 2005b). One of the most important properties of lipid bilayer is the relative mobility of lipid molecules. The response of this mobility to temperature is known as phase behavior of the bilayer. Depending upon temperature lipid bilayer has many phases namely gel phase (L_β), liquid-crystalline phase (L_α), subgel phase (L_c), and ripple phase (P_β) (Goto et al. 2008). The first one is found at low temperature in which the lipid tails are tilted with respect to the membrane normal. The second one i.e. liquid-crystalline phase exists at higher temperature at which the lipid tails are parallel to membrane plane. Liquid-crystalline phase (L_α) is the most common phase of lipid bilayers under which most of the biological processes occur and therefore it has been of focus in many works (Zubrzycki et al. 2000). Subgel phase exist between gel and liquid crystalline phase. For the common phospholipids the low temperature phase is the subgel phase (L_c) where the lipid tails are highly ordered and tilted with the bilayer normal (Koynova and Caffrey 1998). Upon heating it is converted to liquid-crystalline phase (L_α) through some intermediate phases depending upon the type of lipid head group (Goto et al. 2008). At moderate temperatures ripple phase (P_β) of lipid bilayer is found. At this phase the tails remain tilted but behave as rotationally symmetric (Lewis and McElhaney 1992). Systematic study of lipid bilayers at different temperatures will elaborate how the structure and mechanical properties change with the temperature dependent phase transition.

Gel to liquid-crystalline phase transition may be induced by different factors such as temperature, hydration, pressure, pH etc. (Jacobson and Papahadjopoulos 1975). In this work we will concentrate thermally induced phase transition as it has been extensively studied and are of direct biological significance (Oldfield and Chapman 1972). Most of the phospholipids are studied well in different phases at different temperatures which then add to the important conclusions drawn on the physiochemical properties of lipid membranes. Fully saturated PC (phosphatidylcholine) lipid containing 20 CH2 groups in the acyl tail named 1, 2-Diarachidoyl-sn-glycero-3-phosphocholine (DAPC) is difficult to study experimentally as it is readily oxidizable lipid. The structure and dynamic properties of DAPC lipid bilayer is not well understood though it has significant biological significance. It can alter the structure and function of the ion channel Gramicidin A in its gel phase due to hydrophobic mismatch (Basu et al. 2014; Kelkar and Chattopadhyay 2007). Due to its high chain length and high phase transition temperature it forms asymmetric lipid bilayer domain when mixed with lipids with lower chain length and low melting temperature (Rittera et al. 2013). As the phase behavior of lipid bilayers is reflected from the structure and dynamics of phospholipids, we focus on the structural and dynamic properties of this lipid bilayer which can correlate with its temperature dependent phase behavior.

In this work, we aim to study the dependence of lipid structure and dynamics on temperature which may lead to change of phase of the bilayer. The known phase transition temperature of DAPC lipid is 66 °C (Bagatolli and Gratton 2000). To this aim, we have performed Molecular Dynamics Simulation at two different temperatures, one at above the phase transition temperature (at 350 K) and another at below the phase transition temperature (300 K) and try to characterize lipid structural and dynamic properties. We found different structural and dynamic properties at different temperatures which are further explained as the phase change of the lipid bilayer due to change in temperature.

Methodology

System Setup

The atomistic simulations were performed using NAMD_2.7 (Phillips et al. 2005) package with the standard CHARMM27 force field (Mackerell et al. 1998) including dihedral cross term corrections (CMAP) (Mackerell et al. 2004). At first a single DAPC lipid was built using the software 'Insight II' (Accelrys Inc., USA) running on a silicon Graphics O2. The bilayer consisting of 128 lipids was prepared by replicating, translating, rotating the lipids. Then water molecules were added. KCl was added to maintain physiological salt concentration of 150 mM. These processes were carried out using CHARMM. The resulting bilayer was then energy minimized to remove unwanted overlaps and equilibrated for 30 ns. This equilibrated structure is considered as the starting structure for production run. TIP3P water model was used (Jorgensen et al. 1983). We carried out the simulations for 150 ns at two different temperatures one at 300 K and another at 350 K.

Simulation Protocol

All MD simulations were carried out under the isobaric-isothermal (NPT) ensemble with imposed 3D periodic boundary conditions. A time step of 2 fs was used to integrate the equation of motion. The temperatures were maintained for the simulations using Langevin dynamics, while the pressure was kept constant at 1 bar using a Nośe -Hoover-Langevin piston (Feller et al. 1995). The smooth particle mesh Ewald method was used to calculate long range electrostatic calculations (Darden et al. 1993). Short range interactions were cutoff at 10 Å. All bond lengths involving hydrogen atoms were held fixed using the RATTLE (Anderson 1983) and SETTLE (Miyamoto and Kollman 1992) algorithm. The trajectory analysis was performed with CHARMM (Chemistry at Harvard Macromolecular Mechanics) (Brooks et al. 1983) and the snapshots were generated by VMD (Humphrey et al. 1996).

Results and Discussion

The phase behavior of lipids is dependent on their structural properties and hence the properties were calculated. The distinct phase behavior of the lipid bilayer at two different temperatures is clearly observed from the final snapshot at Fig. 17.1 where the left side represents the final structure at 300 K and the right side stands for the final structure at 350 K. We observed that at 300 K temperature, the lipids in the bilayer are tilted with respect to the bilayer normal, and the individual lipid tail is extended and ordered. The state is reminiscent to Lβ· which is the common gel phase of PC lipids. In contrast to this, at temperature 350 K, the lipids are perpendicular to the membrane plane and random which is a characteristic of liquid crystalline phase. So, it may be the fact that a phase transition occurs in DAPC lipid bilayer by increasing the temperature from 300 to 350 K. It is known that for hydrocarbon chains containing equal or more than 20 carbon atoms, pre-transition state is not observed (Kranenburg and Smit 2005). So instead of analyzing the phase transition process, we focus to distinguish the structural and dynamical behavior of DAPC lipid bilayer at two different temperatures.

Area Per Lipid

One of the basic quantities used in describing lipid bilayer is the average area per lipid (Klauda et al. 2006). Figure 17.2 shows that area per lipid strongly depends upon temperature. At 300 K, the area initially fluctuates and then become stable

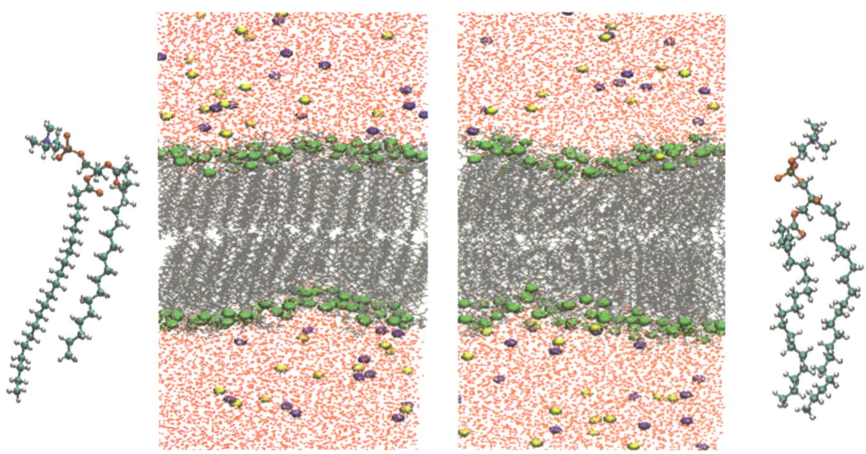

Fig. 17.1 The snapshots showing the DAPC lipid bilayer after 150 ns of simulation at 300 K (*left*) and at 350 K (*right*). The Phosphorus atoms of phosphate group of lipids are shown as green spheres. Individual lipid at both temperatures is highlighted separately beside the corresponding bilayer snapshot in bonds. The image rendering is done with VMD

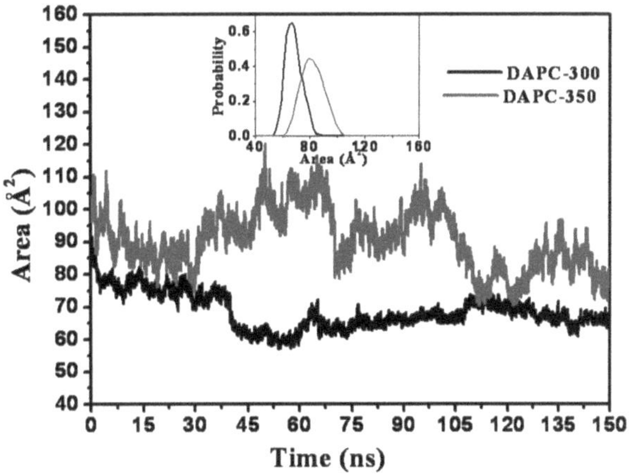

Fig. 17.2 Headgroup area per lipid as a function of time and probability of area (*Inset*) at 300 K (*black line*) and 350 K (*grey line*). The calculations were done from the 150 ns simulation trajectory after the equilibration

after 60 ns to a value of 65 Å² and at 350 K the area is always fluctuating and gives a value of 80 Å². At 300 K temperature, the system is more compact and the lower value of area per lipid can be attributed to tight packing of the lipids whereas at 350 K temperature the system is in disordered state showing higher value of area per lipid. The starting configuration of both the simulations being the same, so the difference in nature of area per lipid is arising due to the effect of temperature and not as a consequence of initial condition.

The probability distribution for the area per lipid molecule (*Inset* of Fig. 17.2) reveals that the minimum area per lipid is above 50 Å². Further we find that though the shapes of the distributions are similar, but the distribution is sharper at temperature 300 K showing a peak at 65 Å² while the distribution broad at temperature 350 K indicating the system flexibility and the peak shifts towards 81 Å².

Atom Density Distribution and Lipid Hydrophobic Thickness

The atom density distributions of each component of the systems were estimated and the lipid hydrophobic thickness was calculated. The atom density distribution for water and different component of lipids along the membrane normal for final 20 ns are shown in Fig. 17.3a, b at temperature 300 K and 350 K respectively. The curves are sharper at lower temperature indicating more ordered state. The lipid acyl tail density at the membrane centre is higher at 300 K temperature which indicate that the lipid tails are more stretched and rigid giving higher density at the membrane centre while at temperature 350 K, the tails are disordered.

The lipid hydrophobic thickness is defined as the distance between the acyl chain C2 carbon atoms of the two opposing bilayer leaflets (Harroun et al. 1999). From Fig. 17.4, we observe that at temperature 300 K, the value rises initially then merges to 45 Å for the rest of the simulation. At temperature 350 K, the same rise is observed.

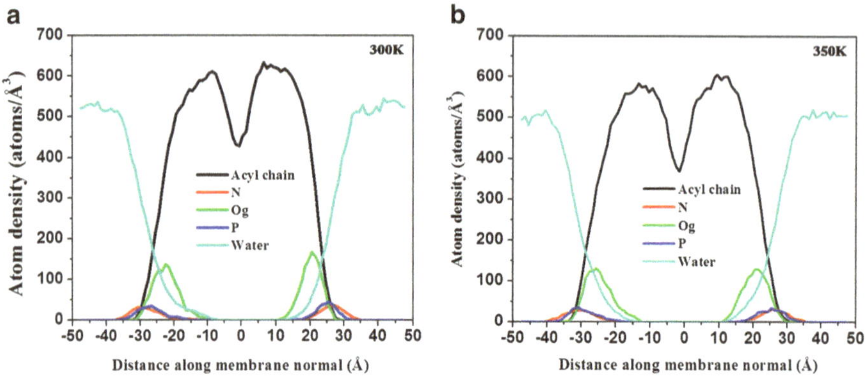

Fig. 17.3 Time average of atom density distribution of individual parts of lipid along membrane normal (*Z* axis) for the system at 300 K (**a**) and at 350 K (**b**)

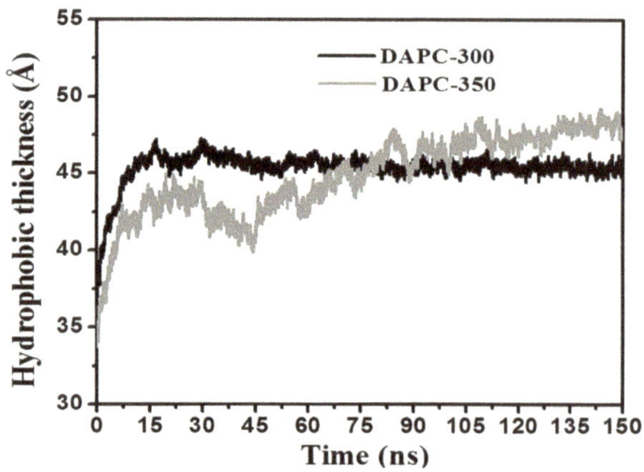

Fig. 17.4 Hydrophobic widths at temperature 300 K (*black line*) and at 350 K (*light grey line*) of lipid bilayer as a function of time. The calculation was done from the last 150 ns simulation trajectory after the equilibration

As lower temperature leads to a decreased area per lipid, the expected hydrophobic thickness should be high. But after 100 ns, the value becomes higher in case of higher temperature than that of lower temperature. So the ordering and rigidity of the lipids are not reflected from the lipid hydrophobic thickness.

To understand the above discrepancy we have calculated the tilt angle between the hydrocarbon tail and bilayer normal for the two systems and the probability function of the tilt angle is plotted in Fig. 17.5. It is very much clear that the tails are tilted to the normal as the peak shows a value of 27° at 300 K while the tails are almost parallel at 350 K which is exactly observed from Fig. 17.1. This calculation along with atom density distribution together suggests that at 300 K the system remains at gel phase and at 350 K attains fluid phase. And due to this tilting the lipid hydrophobic thickness is not as high as expected.

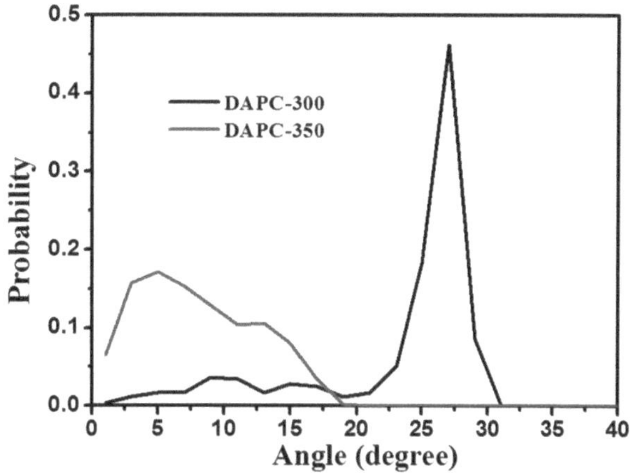

Fig. 17.5 The probability distribution of tilt angle between lipid acyl tail and bilayer normal at 300 K (*black line*) and at 350 K (*light grey line*). The calculation was done from the last 150 ns simulation trajectory after the equilibration

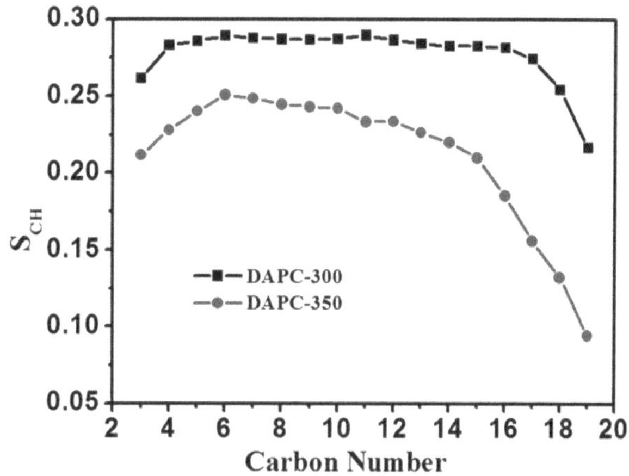

Fig. 17.6 The order parameter profile of lipid acyl tail for the last 40 ns of simulation trajectory at 300 K (*black line* with points) and at 350 K (*grey line* with points)

Order Parameter

Lipid tail order parameter is one of the most important parameters to understand the lipid structure. It gives the information about the degree of rigidity in the lipid membrane interior (Moore et al. 2001). The computed S_{CH} for all carbon atoms in the lipid acyl tail are shown in Fig. 17.6. The change in the value of S_{CH} in the two systems at two different temperature will give the altered rigidity of the lipid tails due to change in temperature (Seelig and Seelig 1974). The higher temperature yields an order parameter profile having higher value than that at temperature 350 K. The value is higher near the glycerol group for both the system, but it decreases very much at the end of the tail at temperature 350 K whereas at temp 300 K, the value is still high at the tail end. This implies that the tails are reasonably ordered at the headgroup region but conformational disorder comes more and more apparent toward the bilayer centre at higher temperature, whereas at lower temperature the tails are still ordered. The S_{CH} of pure POPC lipid molecules at temperature 300 K indicate that POPC lipid bilayer is at their fluid phase at that temperature, as phase transition temperature is different for lipids with different chain length (Leekumjorn and Sum 2007).

Headgroup Orientation of Lipids

We consider the radial distribution function (RDF) (Chiu et al. 2002) between the phosphate groups in a DAPC molecule. The RDFs for the P-P pair in the headgroup are shown in Figure S1 of Supporting Information. High temperature yields a radial distribution function which has a narrow peak at 0.7 nm and essentially no structure beyond r = 1.0 nm. The RDF at lower temperature is little bit different. Though at lower temperature the radial distribution function has a narrow peak at almost same distance as of at higher temperature, it has widely oscillating long range component and this oscillation persists beyond 1.0 nm. This result indicates ordering of the system at lower temperature.

To observe the headgroup orientation of the lipids in a system, we have calculated the angle between the P-N vector of the headgroup with the bilayer normal and the probability of the angle is shown in Fig. 17.7. This plot shows that the orientation of P-N vector is parallel with the bilayer normal, however the angle shows a shift towards a higher value in case of lower temperature indicating more parallel orientation than at higher temperature. This is expected as we find more ordering at lower temperature. Lower head group area of lipid at 300 K temperature is also reflected from the head group orientation of lipids. As the orientation is more parallel, lower area is occupied by the headgroup of lipids at lower temperature than that of higher temperature.

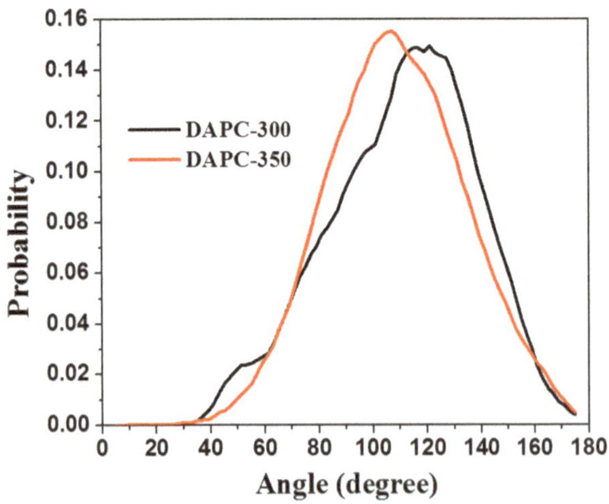

Fig. 17.7 The probability of the angle between the P-N vector of the headgroup with the bilayer normal at 300 K and at 350 K temperatures

Rotational Autocorrelation Function

The amount of rigidity can be understood from the rotational autocorrelation function of P-N vector of headgroup of lipids, plotted at Fig. 17.8, is calculated from the given equation (Rocchi et al. 1998)

$$C(t) = < P_2\left[\mu(0)\mu(t)\right] > \tag{17.1}$$

where μ(0) and μ(t) are the P-N vectors at beginning and time t respectively. P2 is the second-order Legendre polynomial. The calculation is averaged over last 20 ns of simulation. The figure shows that the decay becomes slower at temperature 300 K. To assess the restriction in rotation we have calculated the time of decay by fitting a bi exponential decay with a fixed part as follows:

$$C(t) = A_0 + A_1 e^{-\left(t/\tau_1\right)} + A_2 e^{-\left(t/\tau_2\right)} \tag{17.2}$$

where A_1 and A_2 are fractions of the fast and slow components of motions to the decay and A_0 is a constant indicating the stable component of decay and indicative of restriction of rotation (Mondal and Mukhopadhyay 2008). The reorientational correlation time τ_1 and τ_2 along with their relative contributions A_1, A_2, and A_0 are shown in Table 17.1.

It is observed from the table that A_0 is three times higher at T=300 K which shows the restriction of rotation is higher at lower temperature. The shorter component is much faster (τ_1=0.129 ns with A_1=0.109) at 300 K temperature than at 350 K (τ_1=0.189 ns with A_1=0.297) where as the longer component is slower

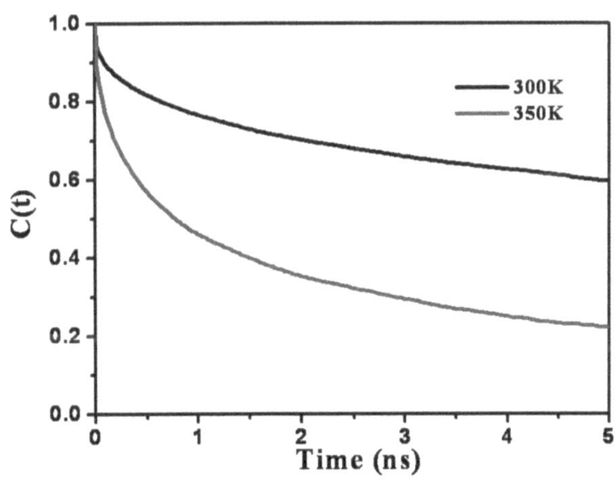

Fig. 17.8 Decay of rotational relaxation of P-N vector of DAPC lipid molecule at 300 K (*black line*) and at 350 K (*grey line*)

Table 17.1 Fitting parameters for the rotational autocorrelation function of lipid P-N vector

System (K)	A_0	A_1	τ_1 (ns)	A_2	τ_2 (ns)
300	0.530	0.109	0.129	0.332	3.04
350	0.183	0.297	0.189	0.454	2.06

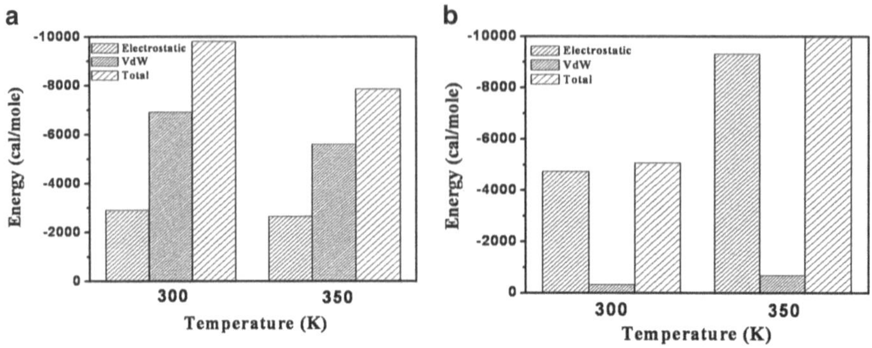

Fig. 17.9 Interaction energy diagram for lipid-lipid interaction (**a**) and for lipid-water interaction (**b**)

(τ_2=3.04 ns with A_2=0.332) at temperature 300 K than at higher temperature (τ_2=2.06 ns with A_2=0.454). The higher constant component and longer decay time at 300 K temperature indicate that the head groups of lipids are more restricted at temperature 300 k.

Energy

The phase behavior of lipid bilayer is largely determined by the attractive van der waals interaction between adjacent lipid molecules. The extent of this interaction is governed by the lipid tail length and the way of packing. Longer tailed lipids have more space to interact, consequently increasing the strength of this attraction and decreasing the lipid mobility. So at a fixed temperature a short tailed lipid will be more fluid (Rawicz et al. 2000). On the other hand, if a system is at gel phase then one lipid is more closely packed with another, thus should give more attractive van der waal interaction than that of the system which is at fluid phase. To observe this behavior we have calculated lipid-lipid and lipid-water interaction energies and their distributions (Cordomi and Perez 2007), which are shown in Fig. 17.9a, b respectively.

From Fig. 17.9a, we observed that the electrostatic and van der waal interaction is more negative i.e. the lipid-lipid interactions are more attractive at temperature 300 K than at temperature 350 K. Specially the van der waal interaction is more attractive indicating the nature of close packing at lower temperature and which is in agreement with the previous findings at gel phase (Seddon et al. 1997). Thus this analysis again confirms that at temperature 300 K the system is more closely packed indicating gel phase. Again the lateral mobility of the lipid molecules can be obtained from their translational diffusion coefficient (Czub and Baginski 2006). The diffusion coefficient which is estimated from the slope of the time series of the mean square displacement of the DAPC lipids in the membrane plane, is given in Table 17.2. We find the lateral

Table 17.2 Diffusion coefficient of lipid

Systems	Temperature (K)	Lateral diffusion coefficients (10^{-8} cm²/s)
DAPC-300	300	4.963
DAPC-350	350	8.057

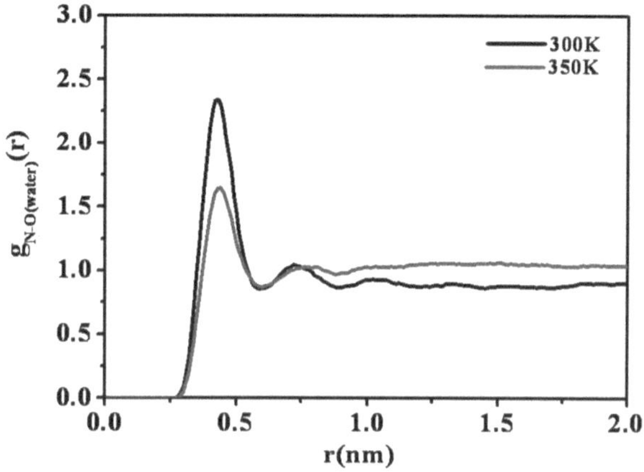

Fig. 17.10 The RDF plot of oxygen atom of water around nitrogen atom of choline group of lipids for the last 40 ns of simulation trajectory at 300 K (*black line*) and at 350 K (*grey line*)

diffusion coefficient is almost double at temperature 350 K than at temperature 300 K that means restricted mobility of the lipids at temperature 300 K.

Interaction with Water

On the other hand lipid-water interaction is more attractive at temperature 350 K that at 300 K temperature. As the lipids are less tightly packed at 350 K and they are freer to move than at 300 K temperature, so the lipids have better chance to interact with the water molecules more and thereby giving more negative interaction energy at temperature 350 K.

Another critical issue is to understand how the properties of water are affected by the ordering of lipids. The difference in the reorientation behavior of the water near and far from the interface can be explained by the orientational behavior of water molecules around headgroup of lipids. The radial distribution of the oxygen atoms of the water molecules around the nitrogen atom of choline group is determined and shown in Fig. 17.10. The larger peak at 0.5 nm revealed that the water molecules are strongly oriented around the head group of lipids at 300 K than at 350 K. The loss of water structure beyond the second layer hydration beyond ~ 0.75 nm at 350 K indicates that hydration waters are more mobile at higher temperature. The narrow peak around 0.35 nm at temperature 300 K, observed in the radial distribution profile of the oxygen of water around the phosphorus atom of the phosphoryl group (shown in Figure S2 of the supporting Information) also supports such orientation.

Table 17.3 Hydrogen bond lifetime between lipid head group and water

Systems	H-bond lifetime (ps)
DAPC-300	98.28
DAPC-350	85.24

Table 17.4 Diffusion coefficient of water

Systems	Temperature (K)	Lateral diffusion coefficients (10^{-5} cm^2/s)
DAPC-300	300	8.15
DAPC-350	350	19.9

But at temperature 350 K, no such orientation is observed around and beyond the phosphorus atom. This is due to the hydrogen bonding of the oxygen atoms of the phosphoryl group with the hydrogen atoms of water molecules. To analyze this, the survival time of H-bond between the phosphoryl group and water molecules for the two systems are calculated and shown in Table 17.3. The increased affinity of water towards the lipid bilayer at lower temperature is clearly evidenced from the higher value of H-bond lifetime between the lipid head group and water at that temperature. Another study is that the average number of hydrogen bonds between the acyl group and the water molecules is higher in case of the system at temperature 350 K (shown in table S1 of Supporting Information) indicating more water penetration after the head group region giving more system flexibility i.e. less ordering at that temperature (also evidenced from Fig. 17.1). These results conclude that the packing of lipids affect the properties of interfacial water.

We have also considered the reorientation dynamics of water near the bilayer interface and of the bulk. To do this water is separated into two categories; if any oxygen atom of water is located within 6 Å from the P atom of lipid for a time period, then it is considered as local water and the rest are bulk. These are shown in Figure S3a and S3b in the Supporting Information.

Figure S3a and S3b stand for the local and bulk waters respectively. The difference in the reorientational behavior can be noticed from the fitting parameters which are shown in Table S2. The constant parameter is zero for all the systems. In all the cases the contributions for the faster and slower decay time are equal. The slowest longer component is observed for the local water molecules at temperature 300 K (τ_2=8.32 ns with A_2=0.499). This is higher than that of the bulk at the same temperature (τ_2=6.40 ns with A_2=0.499). In contrast, there is no such difference in this component at 350 K (τ_2=5.30 and 5.05 for the local and bulk respectively). These results clearly demonstrate that the ordering of lipid molecules at 300 K imposed some restriction on the local water molecules and that's why they possessed slow rotational relaxation than the bulk water. While at 350 K temperature the lipids don't have significant restriction effect on the local water molecules.

The lateral diffusion coefficient of water is also calculated from the slope of mean square displacement of water and shown in Table 17.4. The calculated diffusion coefficient is in well correlation with previously obtained data (Marrink and Berendsen 1994). The diffusion coefficient of water is 2.5 times higher at temperature 350 K which again proofs the above interpretations regarding water dynamics.

Conclusion

We have investigated the effect of temperature on the properties of DAPC lipid bilayer which reflect their phase behavior at two different temperatures and how this affect the properties of the water molecules at the vicinity of the bilayer-water interface. Here we are able to distinguish different structural and dynamic characteristics at two different temperatures. The reduced area per lipid and the corresponding ordering of the acyl chain lead to phase change of the lipid bilayer. Energy calculation also supports that the system at temperature 300 K is at gel phase. Preliminary studies of dynamic quantities like diffusion coefficient have been shown to be sensitive to temperature and shows lower value at lower temperature indicating ordering at lower temperature. All the results found here indicated that change in temperature by 50 K is responsible for phase change due to change in lipid lateral mobility and lipid chain packing. With increasing temperature the van der Waal interaction that favours a crystal like packing of lipids is overcome by thermally induced rotational excitation of lipids and hence leads to phase transition.

Acknowledgements This work is partially supported by the Department of Science and Technology, Government of India, [Project number: No. SR/ S1/PC-60/2009] and a fellowship to IB through UGC-NET. We are also thankful to the NANO Project (CONV/002/NANORAC/2008) of the Department of Chemistry, University of Calcutta, Kolkata, India.

Supporting Information Data regarding the calculation of number of hydrogen bonds between lipid head group and acyl region separately with water molecules at two temperatures and the rotational reorientation decay for local and bulk water molecules with their fitting parameters are shown.

References

Anderson HC (1983) Rattle: a "velocity" version of the shake algorithm for molecular dynamics calculations. J Chem Phys 52:24–34

Bagatolli LA, Gratton E (2000) A correlation between lipid domain shape and binary phospholipid mixture composition in free standing bilayers: a two-photon fluorescence microscopy study. Biophys J 79:434–447

Basu I, Chattopadhyay A, Mukhopadhyay C (2014) Ion channel stability of Gramicidin a in lipid bilayers: effect of hydrophobic mismatch. Biochim Biophys Acta 1838:328–338

Brooks BR, Bruccoleri RE, Olafson BD, States DJ, Swaminathan S, Karplus M (1983) CHARMM: a program for macromolecular energy, minimization, and dynamics calculations. J Comput Chem 4:187–217

Büldt G, Gally HU, Seelig A, Seelig J, Zaccai G (1978) Neutron diffraction studies on selectively deuterated phospholipid bilayers. Nature 271:182–184
Chiu SW, Jakobsson E, Mashl RJ, Scott HL (2002) Cholesterol-induced modifications in lipid bilayers: a simulation study. Biophys J 83:1842–1853
Cordomi A, Perez JJ (2007) Molecular dynamics simulations of rhodopsin in different one-component lipid bilayers. J Phys Chem B 111:7052–7063
Czub J, Baginski M (2006) Comparative molecular dynamics study of lipid membranes containing cholesterol and ergosterol. Biophys J 90:2368–2382
Darden T, York D, Pedersen L (1993) Particle mesh Ewald: an $N \cdot \log(N)$ method for Ewald sums in large systems. J Chem Phys 98:10089–10092
Feller SE (2000) Molecular dynamics simulations of lipid bilayers. Curr Opin Colloid In 5:217–223
Feller SE, Zhang Y, Pastor RW, Brooks BR (1995) Constant pressure molecular dynamics simulation: the Langevin piston method. J Chem Phys 103:4613–4621
Garcia-Manyes S, Oncins G, Sanz F (2005a) Effect of ion-binding and chemical phospholipid structure on the nanomechanics of lipid bilayers studied by force spectroscopy. Biophys J 89:1812–1826
Garcia-Manyes S, Oncins G, Sanz F (2005b) Effect of temperature on the nanomechanics of lipid bilayers studied by force spectroscopy. Biophys J 89:4261–4274
Goto M, Kusube M, Tamai N, Matsuki H, Kaneshina S (2008) Effect of hydrostatic pressure on the bilayer phase behavior of symmetric and asymmetric phospholipids with the same total chain length. Biochim Biophys Acta 1778:1067–1078
Harroun TA, Heller WT, Weiss TM, Yang L, Huang HW (1999) Theoretical analysis of hydrophobic matching and membrane-mediated interactions in lipid bilayers containing gramicidin. Biophys J 76:937–945
Humphrey W, Dalke A, Schulten K (1996) VMD: visual molecular dynamics. J Mol Graph 14:33–38
Jacobson K, Papahadjopoulos D (1975) Phase transitions and phase separations in phospholipid membranes induced by changes in temperature, pH, and concentration of bivalent cations. Biochemistry 1(4):152–161
Jorgensen WL, Chandrasekhar J, Madura JD, Impey RW, Klein ML (1983) Comparison of simple potential functions for simulating liquid water. J Chem Phys 79:926–935
Kelkar DA, Chattopadhyay A (2007) Modulation of gramicidin channel conformation and organization by hydrophobic mismatch in saturated phosphatidylcholine bilayers. Biochim Biophys Acta 1768:1103–1113
Klauda JB, Kučerka N, Brooks BR, Pastor RW, Nagle JF (2006) Simulation-based methods for interpreting X-ray data from lipid bilayers. Biophys J 90:2796–2807
Koynova R, Caffrey M (1998) Phases and phase transitions of the phosphatidylcholines. Biochim Biophys Acta 1376:91–145
Kranenburg M, Smit B (2005) Phase behavior of model lipid bilayers. J Phys Chem B 109:6553–6563
Leekumjorn S, Sum AK (2007) Molecular characterization of gel and liquid-crystalline structures of fully hydrated POPC and POPE bilayers. J Phys Chem B 111:6026–6033
Lewis RNAH, McElhaney RN (1992) The mesomorphic phase behaviour lipid bilayers. In: Yeagle P (ed) The structure of biological membranes, 2nd edn. CRC, Boca Raton, pp 53–71
Mackerell AD Jr, Bashford D, Bellott M, Dunbrack RL Jr, Evanseck JD, Field MJ, Fischer S, Gao J, Guo H, Ha S, Joseph-McCarthy D, Kuchnir L, Kuczera K, Lau FTK, Mattos C, Michnick S, Ngo T, Nguyen DT, Prodhom B, Reiher WE III, Roux B, Schienkrich M, Smith JC, Stote R, Straub J, Watanabe M, Wiorkiewicz-Kuczera J, Yin D, Karplus MJ (1998) All-atom empirical potential for molecular modeling and dynamics studies of proteins. J Phys Chem B 102:3586–3616
Mackerell AD Jr, Feig M, Brooks CL III (2004) Extending the treatment of backbone energetics in protein force fields: limitations of gas-phase quantum mechanics in reproducing protein conformational distributions in molecular dynamics simulations. J Comput Chem 25:1400–1415
Marrink SJ, Berendsen HJC (1994) Simulation of water transport through a lipid membrane. J Phys Chem B 98:4155–4168

Miyamoto S, Kollman PA (1992) Settle: an analytical version of the SHAKE and RATTLE algorithm for rigid water models. J Comput Chem 13:952–962
Mondal S, Mukhopadhyay C (2008) Molecular level investigation of organization in ternary lipid bilayer: a computational approach. Langmuir 24:10298–10305
Moore PB, Lopez CF, Klein ML (2001) Dynamical properties of a hydrated lipid bilayer from a multinanosecond molecular dynamics simulation. Biophys J 81:2484–2494
Oldfield E, Chapman D (1972) Dynamics of lipids in membranes: heterogeneity and the role of cholesterol. FEBS Lett 23:285–297
Pak JH, Bork VP, Norberg RE, Creer MH, Wolf RA, Gross RW (1987) Disparate molecular dynamics of plasmenylcholine and phosphatidylcholine bilayers. Biochemistry 26:4824–4830
Pastor RW, Venable RM, Karplus M (1991) Model for the structure of the lipid bilayer. Proc Natl Acad Sci U S A 88:892–896
Phillips C, Braun R, Wang W, Gumbart J, Tajkhorshid E, Villa E, Chipot C, Skeel RD, Kale L, Schulten K (2005) Scalable molecular dynamics with NAMD. J Comput Chem 26:1781–1802
Rawicz W, Olbrich KC, McIntosh T, Needham D, Evans E (2000) Effect of chain length and unsaturation on elasticity of lipid bilayers. Biophys J 79:328–339
Rittera M, Schmidta S, Jakaba M, Paulmichlb M, Henderson R (2013) Evidence for the formation of symmetric and asymmetric DLPC-DAPC lipid bilayer domains. Cell Physiol Biochem 32:46–52
Rocchi C, Bizzarri RA, Cannistraro S (1998) Water dynamical anomalies evidenced by molecular-dynamics simulations at the solvent-protein interface. Phys Rev E 57:3315–3325
Saiz L, Klein ML (2002) Computer simulation studies of model biological membranes. Acc Chem Res 35:482–489
Schwyzer R (1991) Peptide–membrane interactions and a new principle in quantitative structure—activity relationships. Biopolymers 31:785–792
Seddon JM, Templer RH, Warrender NA, Huang Z, Cevc G, Marsh D (1997) Phosphatidylcholine–fatty acid membranes: effects of headgroup hydration on the phase behaviour and structural parameters of the gel and inverse hexagonal (H_{II}) phases. Biochim Biophys Acta 1327:131–147
Seelig A, Seelig J (1974) The dynamic structure of fatty acyl chains in a phospholipid bilayer measured by deuterium magnetic resonance. Biochemistry 13:4839–4845
Tieleman DP, Marrink SJ, Berendsen HJC (1997) A computer perspective of membranes: molecular dynamics studies of lipid bilayer systems. Biochim Biophys Acta 1331:235–270
Zubrzycki IZ, Xu Y, Madrid M, Tang P (2000) Molecular dynamics simulations of a fully hydrated dimyristoylphosphatidylcholine membrane in liquid-crystalline phase. J Chem Phys 112:3437–3441

Chapter 18
Biophysical Characterization of the Interaction of *O*-acylcholines with the Major Bovine Seminal Plasma Protein, PDC-109

Rajani S. Damai, Pradip K. Tarafdar, Bhanu Pratap Singh, S. Thirupathi Reddy, and Musti J. Swamy

Introduction

Sperm-egg fusion is the first major event in the reproductive cycle of mammals. The seminal plasma, which carries sperm cells to the female uterus, contains several proteinaceous factors that modulate the fertilizing capacity of the spermatozoa (Shivaji et al. 1990; Yanagimachi 1994). In this regard, a group of acidic proteins of the bovine seminal plasma, namely PDC-109 (BSP-A1/-A2), BSP-A3 and BSP-30 kDa have been investigated in detail. These proteins, collectively referred to as *bovine seminal plasma proteins* or *BSP proteins* constitute the major protein fraction of bull seminal plasma (about 65 % of total proteins) (Manjunath and Thérien 2002). The primary structure of PDC-109 is made up of 109 amino acids, with two tandemly repeating fibronectin type-II (FnII) domains, connected to an *N*-terminal segment (Esch et al. 1983; Baker 1985; Seidah et al. 1987). The crystal structure of PDC-109/phosphorylcholine (PrC) complex shows that each FnII domain can bind one choline phospholipid molecule, through its specific interaction with the PrC moiety of the head group (Wah et al. 2002). At ejaculation, binding of PDC-109 to sperm surface by such interaction results in an efflux of choline phospholipids and cholesterol—a process referred to as *cholesterol efflux*—which appears to be an important step in sperm capacitation (Thérien et al. 1998; Moreau et al. 1998; Manjunath and Thérien 2002; Swamy 2004). In light of this, the interaction of PDC-109 with membranes made up of different phospholipids has been investigated by a variety of biophysical approaches (Müller et al. 1998; Gasset et al. 1997; Ramakrishnan et al. 2001; Greube et al. 2001; Swamy et al. 2002; Thomas et al. 2003; Anbazhagan and Swamy 2005; Anbazhagan et al. 2008; Damai et al. 2009;

R.S. Damai • P.K. Tarafdar • B.P. Singh • S.T. Reddy • M.J. Swamy (✉)
School of Chemistry, University of Hyderabad, Hyderabad 500 046, India
e-mail: mjssc@uohyd.ernet.in; mjswamy1@gmail.com

A. Chakrabarti, A. Surolia (eds.), *Biochemical Roles of Eukaryotic Cell Surface Macromolecules*, Advances in Experimental Medicine and Biology 842,
DOI 10.1007/978-3-319-11280-0_18

Damai et al. 2010). More recently, it has been shown that PDC-109 exhibits chaperone like activity (CLA) against different target proteins *in vitro*, suggesting that it may protect other proteins of the seminal plasma against misfolding and assist in their proper folding (Sankhala and Swamy 2010). Importantly, it has been demonstrated that binding of phosphatidylcholine (PC) to PDC-109 increases its CLA significantly (Sankhala et al. 2011).

Although PDC-109 exhibits high affinity towards PC and other choline phospholipids (Desnoyers and Manjunath 1992, 1993), it also recognizes several other phospholipids such as phosphatidylglycerol and phosphatidylserine, albeit with a considerably lower affinity (Ramakrishnan et al. 2001). The high specificity for PC is due to a faster association rate constant and a slower dissociation rate constant for choline phospholipids as compared to phospholipids with other head groups (Thomas et al. 2003). Earlier studies have shown that PrC, the head group moiety of PC and lysophosphatidlycholine (Lyso-PC) bind to PDC-109 and that the binding affinity for the interaction of PDC-109 with Lyso-PC is 250-fold higher than the affinity of PrC (Desnoyers and Manjunath 1992, 1993; Müller et al. 2002; Anbazhagan and Swamy 2005). This suggests that, in spite of the obligatory requirement of the choline head group in phospholipids, addition of the glycerol backbone and acyl chains enhance the strength of binding of PDC-109 for choline containing phospholipids.

Isothermal titration calorimetric studies have shown that PDC-109 exhibits two types of binding sites for Lyso-PC, a high affinity site and a low affinity site (Anbazhagan et al. 2011). This is most likely due to the coexistence of Lyso-PC micelles and the monomer. It was not possible to carry out experiments at low concentration where Lyso-PC would be in the monomeric form because of the very low CMC of this lipid. Therefore, we have synthesized *O*-acylcholines (OACs), viz., *O*-lauroylcholine (OLC) and *O*-myristoylcholine (OMC), in which the acyl chain is directly connected to the choline group without the phosphate and glycerol units. The interaction of OLC and OMC with PDC-109 was investigated at various temperatures by ITC under conditions where the OACs exist in the monomer form. Effect of OAC binding on the structure and thermal stability of PDC-109 was studied by circular dichroism (CD) spectroscopy and differential scanning calorimetry, respectively.

Materials and Methods

Materials

Choline chloride (Ca^{2+} salt), tris (hydroxymethyl)-aminomethane (Tris base), acrylamide, phosphoryl choline, bis acrylamide and TEMED were obtained from Sigma (St. Louis, MO, USA). Sephadex G-50 (Superfine) and DEAE Sephadex A-25 were from Pharmacia Biotech (Uppsala, Sweden). OLC and OMC were synthesized by condensing *N,N*-dimethylethanolamine with lauric acid or myristic acid, followed by quaternization of the product with methyl iodide and characterized as described in Tarafdar et al. (2013). OMC iodide was converted to the corresponding chloride salt by ion exchange chromatography (Tarafdar et al. 2013). Sodium chloride, EDTA and sodium azide were obtained from local suppliers and were of the highest purity available.

Isothermal Titration Calorimetry (ITC)

The interaction of PDC 109 with OLC and OMC was investigated by ITC using a MicroCal VP-ITC instrument (Northampton, MA, USA) at different temperatures between 20 and 40 °C. Both lipid and protein solutions were degassed under vacuum prior to use. A 360 μM solution of PDC-109 was taken in the syringe and injected in 12 μL aliquots into the ITC cell filled with 1.445 mL of 50 μM OLC (or OMC). Injections were made at 5 min intervals and a constant stirring speed of 300 rpm was maintained during the experiment in order to ensure proper mixing after each injection. To minimise contribution to binding heat from dilution, the final dialysate obtained after the protein dialysis was used to prepare lipid solution (TBS-I, pH 7.4). The Control experiments were performed by titrating PDC-109 into TBS-I and the resulting heats were subtracted from measured heats from *O*-acylcholine-PDC-109 binding. The binding data were analysed using the Origin software provided by MicroCal. The equilibrium association constants (K_a), the enthalpy of binding (ΔH°), the entropy change (ΔS°) and the stoichiometry (n), were obtained from the curve fitting of the experimental data to '*one set of sites*' model in MicroCal origin. The values of K_a and ΔH° obtained from curve fitting were used to calculate the standard free energy change (ΔG°) and the standard entropy change (ΔS°) for the binding using the equation:

$$\Delta G^\circ = -RT \ln K_a = \Delta H^\circ - T\Delta S^\circ$$

where R is the molar gas constant. Heat capacity change (ΔCp°) for the binding was also determined from the enthalpy changes determined at different temperatures.

Fluorescence Spectroscopy

Steady state fluorescence measurements were performed using a Spex model Fluoromax-3 spectrofluorimeter at 25 °C. The excitation and emission band pass filters were set at 3 nm. PDC-109 in TBS-I (pH 7.4) buffer with $OD_{280nm} < 0.1$ was excited at 280 nm and emission spectra were recorded between 300 and 400 nm. Titrations were performed by adding small aliquots of 1 mM solution of OMC to PDC-109 in TBS-I buffer taken in a $1 \times 1 \times 4.5$ cm quartz fluorescence cuvette.

Differential Scanning Calorimetry

DSC measurements were carried out on a MicroCal VP-DSC apparatus (MicroCal LLC, Northampton, MA, USA). Protein samples were dialyzed extensively against TBS-I and the dialysate was used as the reference. Protein samples of 0.6 mg/mL were heated from 10 to 90 °C at a scan rate of 30°/h. To investigate the effect of OAC binding on the protein thermal unfolding, PDC-109 was preincubated with 1 mM OMC before the DSC scans. Sample and reference solutions were properly degassed prior to the DSC experiment in order to eliminate bubbling effects. Reproducibility of baselines was verified by multiple scans, and reversibility of protein unfolding was investigated by scanning the sample twice. Buffer scans were subtracted from thermograms obtained with PDC-109 in the absence and also in the presence of OMC before further analysis. The DSC data were analyzed by non-two-state fits of the thermograms using the Origin 7.0 software provided with the instrument.

Circular Dichroism Spectroscopy

Circular dichroism (CD) spectra were recorded at 25 °C on a Jasco-J-810 spectropolarimeter equipped with a thermostatted water bath. Samples were taken in a 0.2 cm path length quartz cell and spectra were recorded at a scan speed of 20 nm/min. Protein concentrations of 0.1 mg/mL and 1.0 mg/mL were used in the far-UV (200–250 nm) and near-UV (250–300 nm) regions, respectively. Data were collected with a response time of 2 s and a slit width of 1 nm. Each spectrum reported was the average of 20 scans from which the buffer scans, recorded under the same conditions, were subtracted. The observed ellipticities were converted to mean residual ellipticities (Θ) using a mean residue mass of 117 (Gasset et al. 2000).

Results

Isothermal Titration Calorimetry

In the present study the interaction of PDC-109 with *O*-lauroylcholine and *O*-myristoylcholine, two synthetic derivatives of choline, has been investigated by isothermal titration calorimetry and the thermodynamic parameters characterising the binding have been determined. Figure 18.1 shows a representative ITC profile for the binding of OLC to PDC-109 at 25 °C. Qualitatively similar profiles were obtained for this interaction at other temperatures as well as for the interaction of OMC and PDC-109 at different temperatures. Injection of 12 μL aliquots of 360 μM PDC-109 (from syringe) into 50 μM solution of OLC (in sample cell) resulted in exothermic peaks, indicating that the binding process is associated with a negative

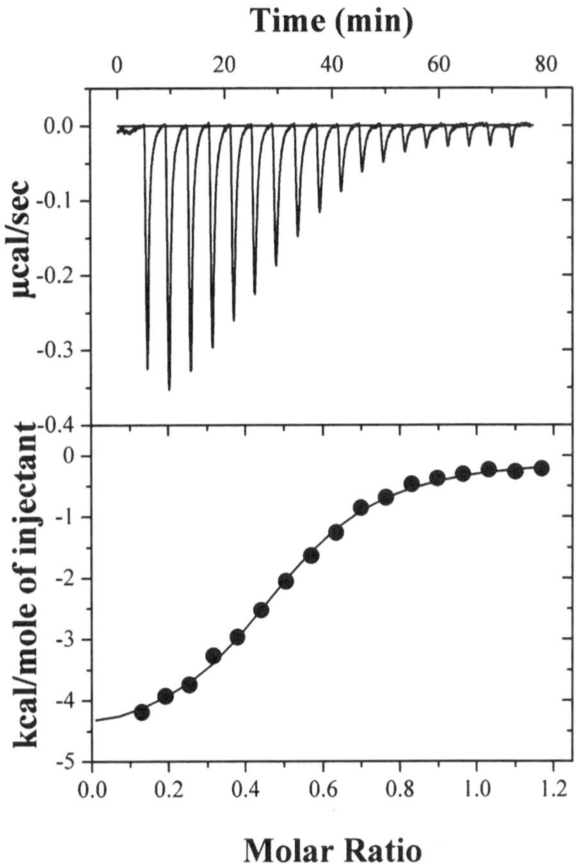

Fig. 18.1 ITC profile of PDC-109 titration with *O*-lauroylcholine at 25 °C. The upper panel shows the raw data for the titration of 50 µM *O*-lauroylcholine with 360 µM of PDC-109. The lower panel shows the integrated data obtained from the raw data shown in the upper panel, after subtracting the dilution experiments. The *solid line* in the bottom panels represents the best curve fit to the experimental data, using the '*one set of sites*' model from MicroCal Origin

enthalpy change ($\Delta H^\circ < 0$). The magnitude of the exothermic peaks decreased with subsequent injections, showing saturation of the binding sites at high protein concentration. The data could be fit satisfactory by nonlinear least squares method to '*one set of sites*' binding model from MicroCal Origin. Data obtained at other temperatures for the binding of OLC to PDC-109 as well as for the interaction of OMC with PDC-109 at various temperatures could also be fitted satisfactorily to the '*one set of sites*' model in a similar fashion.

Values of stoichiometry (n), the equilibrium association constant (K_a) and thermodynamic parameters, ΔH° and ΔS° for the binding of OLC and OMC to PDC-109 at different temperatures were obtained from the above analysis and listed in Table 18.1. The association constant (K_a), stoichiometry (n) and the thermodynamic parameters, ΔH° and ΔS° obtained for the binding of OLC at 25 °C are 5.1×10^5 M^{-1}, 0.47, −4.46 kcal mol^{-1}, and 11.13 cal mol^{-1} K^{-1}, respectively. These values indicate that the binding of PDC-109 to *O*-lauroylcholine is governed by a negative enthalpy contribution. The entropic contribution is positive, which augments the favourable enthalpy and makes the binding stronger.

Table 18.1 Temperature dependence of the thermodynamic parameters associated with the binding of PDC-109 to *O*-lauroylcholine and *O*-myristoylcholine

Temperature (°C)	n	$K_a \times 10^{-5}$ (M^{-1})	$\Delta H°$ (kcal mol^{-1})	$\Delta S°$ (cal mol^{-1} K^{-1})
O-lauroylcholine				
25	0.47(±0.02)	5.1(±0.7)	−4.46(±0.27)	11.13(±1.17)
30	0.42(±0.04)	3.8(±1.0)	−7.74(±0.44)	−0.09(±0.09)
35	0.49(±0.04)	2.6(±0.1)	−10.19(±0.01)	−8.31(±0.075)
40	0.45(±0.08)	1.3(±0.1)	−16.27(±1.1)	−28.65(±3.75)
O-myristoylcholine				
30	0.29(±0.01)	19.9(±10)	−7.01(±0.48)	5.69
35	0.30(±0.04)	8.9(±2.1)	−11.4(±0.30)	−9.97(±1.4)
40	0.36(±0.001)	4.3(±0.4)	−16.6(±0.35)	−27.4

The values listed are averages of duplicate experiments

The data presented in Table 18.1 show that the enthalpy of binding decreases with increase in temperature, i.e., the magnitude of $\Delta H°$ increases while retaining the negative sign. This would be expected to result in an increase of the association constants with temperature. However, the K_a values decrease with increase in temperature, due to a larger negative change in the entropy of binding. The stoichiometry of binding (n) ranges between 0.42 and 0.49, which is consistent with each molecule of PDC-109 binding two molecules of OLC, through the two FnII domains.

At 30 °C the K_a value obtained for the binding of OMC to PDC-109 is 19.9×10^5 M^{-1}, which is considerably larger than the value of 3.8×10^5 M^{-1}, obtained for the OLC/PDC-109 interaction. This is clearly due to the larger positive entropy of binding observed with OMC (5.69 cal mol^{-1} K^{-1}) as compared to that observed with OLC (−0.09 cal mol^{-1} K^{-1}). This suggests that the interaction of the longer acyl chain in OMC with the protein exhibits a larger hydrophobic component which leads to larger entropy of binding.

Fluorescence Studies on the Interaction of PDC-109 with O-lauroylcholine

The binding of *O*-lauroylcholine to PDC-109 was also investigated by fluorescence spectroscopy at 25 °C. Fluorescence emission spectra obtained for the titration of OLC with PDC-109 are shown in Fig. 18.2a. Spectrum 1 corresponds to PDC-109 alone and spectra 2–20 correspond to PDC-109 in presence of increasing concentrations of OLC. The binding of OLC to PDC-109 resulted in a gradual decrease in the fluorescence intensity of the protein. The emission maximum for PDC-109 was observed at 340.5 nm, which upon addition of OLC is blue shifted to 333 nm. A plot of ΔF *vs* concentration of added OLC gives the binding curve for the PDC-109/OLC interaction. A typical binding curve thus obtained is given in Fig. 18.2b.

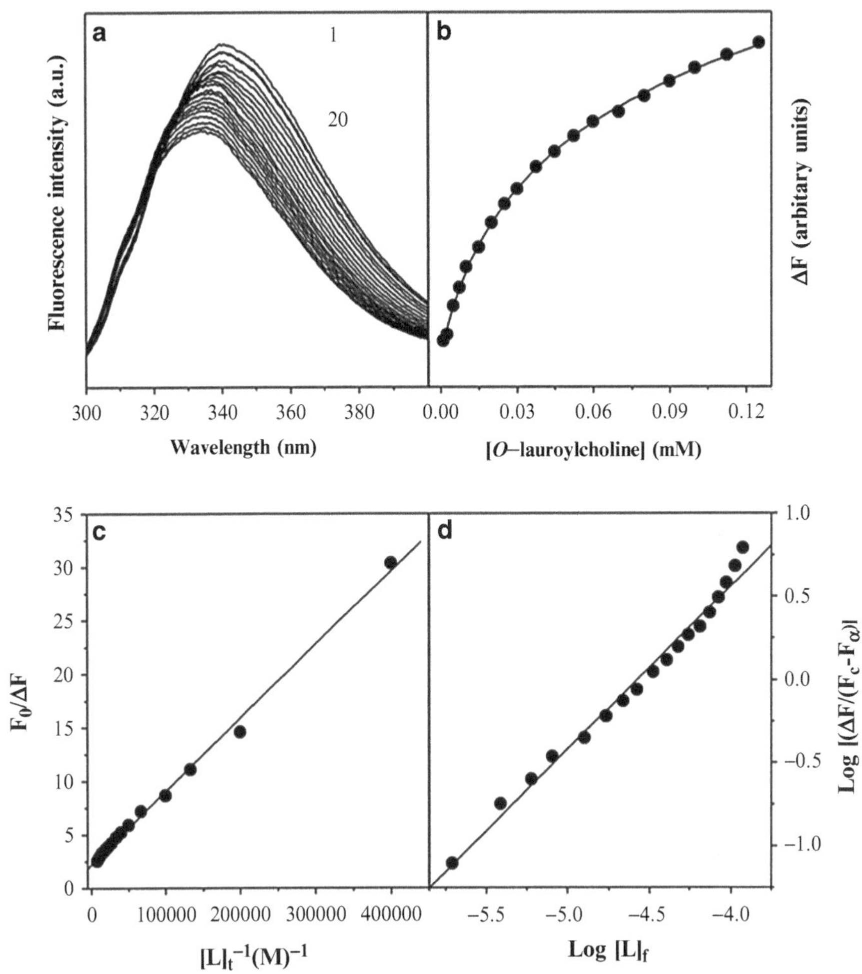

Fig. 18.2 Determination of the association constant for the interaction of OLC with PDC-109. (**a**) Fluorescence emission spectra of PDC-109 in the absence and presence of OLC at 25 °C. Spectrum 1 corresponds to native PDC-109 and spectra 2–20 correspond to PDC-109 in presence of increasing concentrations of OLC. (**b**) Binding curve for the titration of PDC-109 with OLC. The change in fluorescence intensity (ΔF) was plotted as a function of the concentration of added OLC. (**c**) Plot of ($F_o/\Delta F$) *vs* ($1/[L]_t$). From the Y-intercept of this plot fluorescence intensity of the protein at saturation binding was determined. (**d**) Chipman plot for the binding of OLC to PDC-109. The abscissa of the plot yields pK_a, from which the association constant, K_a was determined

This binding curve shows that the change in fluorescence intensity of the protein decreases with increased ligand concentration, displaying saturation behaviour.

A plot of ($F_o/\Delta F$) *vs* ($1/[L]_t$), where F_o refers to the fluorescence intensity of the protein in the absence of any ligand and $[L]_t$ is the total ligand (OLC) concentration, yielded a straight line (Fig. 18.2c). From the ordinate intercept of the plot, the fluorescence intensity of the protein at infinite concentration of OLC (F_∞) was obtained. The maximal change in fluorescence intensity (ΔF_∞) corresponding to the complete saturation was calculated using the expression, $\Delta F\infty = (F_0 - F_\infty)$. Now, the p$K_a$ for the PDC-109/OLC interaction was obtained from the abscissa intercept of a plot of $[L]_f$ *vs* log $[\Delta F/(F_c - F_\infty)]$ according to the relationship (Chipman et al. 1967):

$$\mathrm{Log}\left[\Delta F/\left(F_c - F_\infty\right)\right] = \log K_a + \log\left[L\right]_f \quad (18.1)$$

where, F_c is the fluorescence intensity of the sample at any point during the titration and $[L]_f$ is the free ligand concentration and is given by:

$$\left[L\right]_f = \{\left[L\right]_t \quad \left(\left[P\right]_t \times \left(\Delta F/\Delta F_\infty\right)\right)\} \quad (18.2)$$

where $[P]_t$ is the total protein concentration. A plot of log$[\Delta F/(F_c - F_\infty]$ *vs* $[L]_f$ is shown in Fig. 18.2d. From the abscissa of the plot which gives pK_a, the association constant, K_a, at 25 °C was estimated to be 5.46×10^4 M^{-1}.

CD Spectroscopy

The far-UV and near-UV circular dichroic (CD) spectra of PDC-109 alone and in the presence of *O*-lauroylcholine are shown in Fig. 18.3a and b, respectively. The far-UV CD spectrum of PDC-109 (thick line, Fig. 18.3a) is characterised by a broad positive band with a maximum at 222.2 nm and a shoulder at 211.8 nm which is consistent with previous reports (Gasset et al. 1997; Anbazhagan and Swamy 2005).

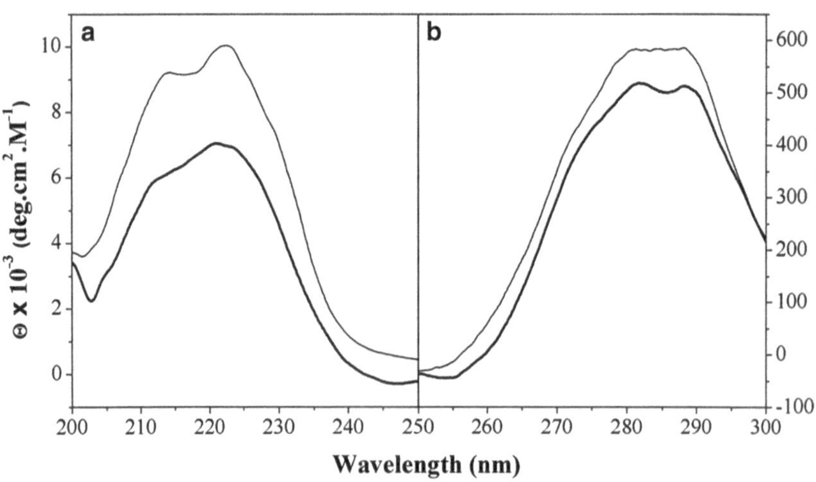

Fig. 18.3 Far and near-ultraviolet CD spectra of PDC-109 in presence and absence of *O*-lauroylcholine. (**a**) Far-UV spectra. (**b**) Near-UV CD spectra. PDC-109 alone (*thick line*), PDC-109 + 1 mM OLC (*thin line*)

Binding of OLC resulted in a considerable increase in the intensity of the CD spectrum of PDC-109. However, there was no significant change in the position of broad band maximum which is observed at 222.4 nm, although the shoulder has moved closed to major band with maximum at 213.8 nm (Fig. 18.3a). The near-UV CD spectrum of PDC-109 was characterized by a positive band with discrete maxima at 281.8 and 288.4 nm. Binding of OLC leads to an increase in the spectral intensity, with the two maxima merging somewhat, resulting to a relatively flat region at the top of the band.

Differential Scanning Calorimetry

The effect of *O*-acylcholine binding on the thermal stability of PDC-109 was investigated by differential scanning calorimetry. Thermograms corresponding to PDC-109 alone and in the presence of various concentrations of *O*-myristoylcholine, corrected for buffer baseline and normalized to the peak height are shown in Fig. 18.4a. From this figure it is clear that binding of OMC increases the thermal stability of PDC-109 and shifts the unfolding transition to higher temperatures in a concentration dependent manner. A plot of the shift in the unfolding temperature (ΔT) against the concentration of OMC exhibits a steep increase at low concentrations, increases more gradually at intermediate concentrations and levels off at high

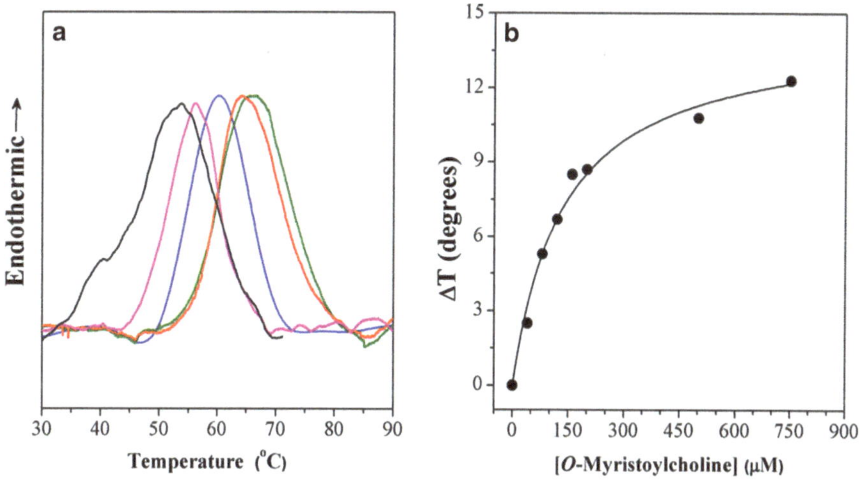

Fig. 18.4 Effect of *O*-myristoylcholine binding on the thermal unfolding of PDC-109. (**a**) DSC scans of PDC-109 in the absence and in the presence of different concentrations of OMC are shown. The concentration of OMC in different scans shown are: *black*, 0 μM; *purple*, 40 μM; *blue*, 120 μM; *red*, 500 μM; *green*, 750 μM. (**b**) A plot of the thermal unfolding temperature versus the concentration of OMC. The data clearly show that OMC binding increases the thermal unfolding temperature of PDC-109 in a concentration dependent manner and exhibits saturation behavior. Thermogram of PDC-109 alone was taken from our earlier study (Sankhala et al. 2011)

concentrations, exhibiting saturation behavior (Fig. 18.4b). These observations are qualitatively in agreement with the previous DSC studies of Gasset et al. (1997) who reported that binding of PrC stabilizes PDC-109 against thermal unfolding.

Discussion

Previous work has clearly shown that the binding of PDC-109 to membranes containing choline phospholipids such as phosphatidylcholine is mediated by its specific interaction with the choline moiety in the head group of choline phospholipids. Therefore, it is of considerable interest to investigate its interaction with synthetic, choline containing molecules. Such studies can lead to the development of new synthetic molecules with high affinity to PDC-109. In preliminary work done in this direction, the interaction of two *O*-acyl derivatives of choline, namely *O*-lauroylcholine and *O*-myristoylcholine with PDC-109 was investigated by ITC, DSC, fluorescence spectroscopy and CD spectroscopy and the results obtained are discussed here.

Although the binding of OLC to PDC-109 is favoured by both enthalpy and entropy at 25 °C, at higher temperatures the entropic factors do not favour binding. Thus the $\Delta S°$ value which is positive at 25 °C becomes negative at 30 °C, although very small in magnitude (−0.09 cal mol^{-1} K^{-1}). As the temperature is increased further, the magnitude of the (negative) entropy of binding increases further. These values outweigh the larger (negative) values of enthalpy of binding, which results in a decrease in the association constants.

From the temperature dependence of the enthalpy of binding for the interaction of OLC and OMC with PDC-109, we have obtained the change in heat capacity (ΔC_p) of the binding process (Fig. 18.5a). The ΔC_p values obtained are −258 cal mol^{-1} K^{-1} and −233 cal mol^{-1} K^{-1}, for OLC and OMC, respectively. Negative values of ΔC_p have been reported earlier for a number of protein-ligand systems including the binding of chitobiose (−347 J mol^{-1} K^{-1}) and chitotriose (−498 J mol^{-1} K^{-1}) to lysozyme (García-Hernández et al. 2003) and the association of mono- and disaccharides with the *Momordica charantia* lectin (Sultan and Swamy 2005). In these two studies, the decrease in ΔC_p has been attributed to the dehydration of polar groups that accompanies the binding process. If no protonation/deprotonation takes place during the binding process, the ΔC_p values can be attributed to hydration and dehydration of apolar and polar groups (Gómez et al. 1995). Conformational changes induced by the binding reaction may also be responsible for a change in the heat capacity upon binding (Baxa et al. 2001). Because the CD spectra shown in Fig. 18.3 indicate changes in the secondary structure of PDC-109 upon binding of OLC, at least part of the change in heat capacity associated with the binding could be due to conformational changes in the protein. Indeed, comparison of the structure of domain B of PDC-109—determined by solution NMR studies (Constantine et al. 1992)—with the structure of the corresponding region in the three-dimensional structure of full length PDC-109 complexed with PrC, determined by X-ray diffraction studies, revealed a significant reorientation of a loop near the PrC binding site,

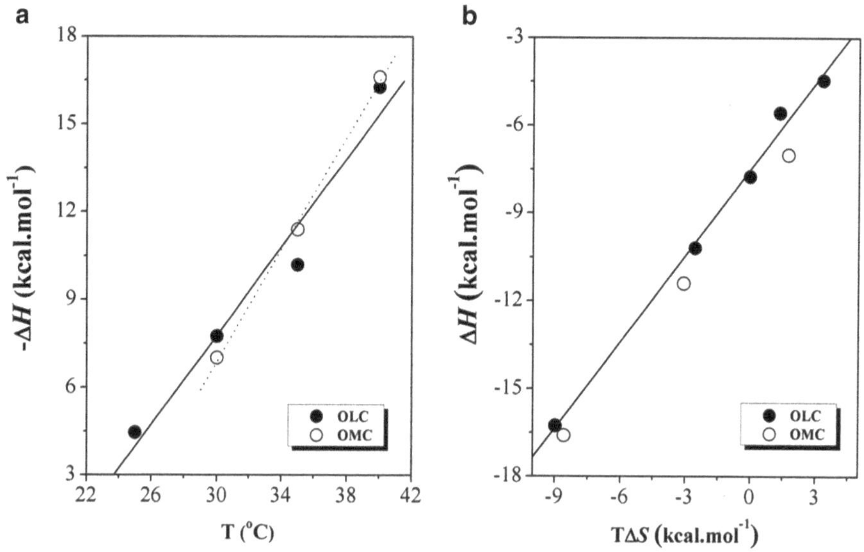

Fig. 18.5 Determination of the change in heat capacity (ΔC_p) and enthalpy-entropy compensation for the association of *O*-acylcholines with PDC-109. (**a**) A plot of ΔH^o versus *T*. From the slope of the linear least squares fits, the ΔC_p values were obtained. (**b**) Enthalpy-entropy compensation plot for the interaction of PDC-109 with *O*-acylcholines. The *straight line* corresponds to a linear least squares fit

whereas the rest of the polypeptide chain in the domain B essentially overlaps with the corresponding polypeptide segment in the full length protein (Wah et al. 2002). Additionally, dehydration of the binding site resulting from the binding of OLC/OMC could also be responsible for the negative heat capacity changes observed as fluorescence studies have suggested that water molecules are most likely removed from the choline binding site upon binding of PrC and choline phospholipids (Anbazhagan et al. 2008).

The thermodynamic data presented in Table 18.1 show that the increase in (negative) enthalpy of binding is compensated by a corresponding (negative) change in entropy. This is clearly seen in the enthalpy-entropy compensation plot of ΔH *vs* $T\Delta S$ for the interaction of PDC-109 with OLC and OMC at different temperatures (Fig. 18.5b). This plot shows that the enthalpy and entropy of binding for the binding of OLC and OMC are very closely compensated with a slope of 0.985. An exact compensation of enthalpy by entropy would be expected to yield a slope of 1.0 (Eads et al. 1998).

It has been shown previously that binding of ligands containing the choline moiety, e.g., choline chloride, phosphorylcholine, Lyso-PC and diacyl PC to PDC-109 results in an increase in the emission intensity of the protein fluorescence (Müller et al. 1998; Gasset et al. 1997; Anbazhagan and Swamy 2005; Anbazhagan et al. 2008). In contrast, it was observed in the present study that binding of OLC to PDC-109 results in a decrease in the protein intrinsic fluorescence. While this appeared surprising at first, a careful analysis suggested that since OLC used in these titrations has iodide as the counterion, the decrease in the fluorescence intensity is likely to be due to quenching of the protein intrinsic fluorescence by the iodide ion. Consistent with this, titration of PDC-109 with OMC chloride led to a significant increase in the emission intensity of the protein fluorescence which was associated with a concomitant blue shift in the emission λ_{max}. At 0.15 mM concentration of OMC.Cl the emission intensity increased by about 56 % with a concomitant 8 nm blue shift in the emission λ_{max}. These observations are in good agreement with the results obtained earlier for the binding of Lyso-PC to PDC-109 (Anbazhagan et al. 2008).

The CD spectra presented in Fig. 18.4 show that binding of *O*-lauroylcholine results in an increase in the intensity of the spectral bands of PDC-109 in the near UV and far UV region, which is similar to the changes observed upon binding of phosphorylcholine or Lyso-PC with previous results (Gasset et al. 1997; Anbazhagan and Swamy 2005). These changes could be due to the changes in the secondary and tertiary structure of the protein, as PDC-109 which exists as a polydisperse aggregate in solution dissociates into dimeric form upon ligand binding (Gasset et al. 1997).

The K_a value of 5.1×10^5 M^{-1} for the binding of OLC to PDC-109 at 25 °C is significantly higher than the K_a value estimated at the same temperature for the low affinity site (1.19×10^5 M^{-1}) but somewhat lower than that for the high affinity site (7.01×10^5 M^{-1}) for the Lyso-PC/PDC-109 interaction (Anbazhagan et al. 2011). It should be possible to increase the binding affinity further by modifying the structure of the ligands suitably. The three-dimensional structure(s) of complex(es) of single chain, choline-containing amphiphiles such as OLC or Lyso-PC would greatly help in designing suitable structures for such studies.

References

Anbazhagan V, Swamy MJ (2005) Thermodynamics of phosphorylcholine and lysophosphatidylcholine binding to the major protein of bovine seminal plasma, PDC-109. FEBS Lett 579: 2933–2938

Anbazhagan V, Damai RS, Paul A, Swamy MJ (2008) Interaction of the major protein from bovine seminal plasma, PDC-109 with phospholipid membranes and soluble ligands investigated by fluorescence approaches. Biochim Biophys Acta 1784:891–899

Anbazhagan V, Sankhala RS, Singh BP, Swamy MJ (2011) Isothermal titration calorimetric studies on the interaction of the major bovine seminal plasma protein, PDC-109 with phospholipid membranes. PLoS One 6:e25993

Baker ME (1985) The PDC-109 protein from bovine seminal plasma is similar to the gelatin-binding domain of bovine fibronectin and a kringle domain of human tissue-type plasminogen activator. Biochem Biophys Res Commun 130:1010–1014

Baxa U, Cooper A, Weintraub A, Pfeil W, Seckler R (2001) Enthalpic barriers to the hydrophobic binding of oligosaccharides to phage P22 tailspike protein. Biochemistry 40:5144–5150

Chipman DM, Grisaro V, Sharon N (1967) The binding of oligosaccharides containing *N*-acetylglucosamine and *N*-acetylmuramic acid to lysozyme. The specificity of binding subsites. J Biol Chem 242:4388–4394

Constantine KL, Madrid M, Bányai L, Trexler M, Patthy L, Llinás M (1992) Refined solution structure and ligand-binding properties of PDC-109 domain b. A collagen-binding type II domain. J Mol Biol 223:281–298

Damai RS, Anbazhagan V, Rao KB, Swamy MJ (2009) Fluorescence studies on the interaction of choline-binding domain B of the major bovine seminal plasma protein, PDC-109 with phospholipid membranes. Biochim Biophys Acta 1794:1725–1733

Damai RS, Sankhala RS, Anbazhagan V, Swamy MJ (2010) ^{31}P-NMR and AFM studies on the destabilization of cell and model membranes by the major bovine seminal plasma protein, PDC-109. IUBMB Life 62:841–851

Desnoyers L, Manjunath P (1992) Major proteins of bovine seminal plasma exhibit novel interactions with phospholipids. J Biol Chem 267:10149–10155

Desnoyers L, Manjunath P (1993) Interaction of a novel class of phospholipid-binding proteins of bovine seminal fluid with different affinity matrices. Arch Biochem Biophys 305:341–349

Eads JC, Mahoney NM, Vorobiev S, Bresnick AR, Wen KK, Rubienstein PA, Haarer BK, Almo SC (1998) Structure determination and characterization of *Saccharomyces cerevisiae* profilin. Biochemistry 37:11171–11181

Esch FS, Ling NC, Bohlen P, Ying SY, Guillemin R (1983) Primary structure of PDC-109, a major protein constituent of bovine seminal plasma. Biochem Biophys Res Commun 113:861–867

García-Hernández E, Zubillaga RA, Chavelas-Adame EA, Vázquez-Contreras E, Rojo-Domínguez A, Costas M (2003) Structural energetics of protein–carbohydrate interactions: insights derived from the study of lysozyme binding to its natural saccharide inhibitors. Protein Sci 12: 135–142

Gasset M, Saiz JL, Sanz L, Gentzel M, Töpfer-Petersen E, Calvete JJ (1997) Conformational features and thermal stability of bovine seminal plasma protein PDC-109 oligomers and phosphorylcholine-bound complexes. Eur J Biochem 250:735–744

Gasset M, Magdaleno L, Calvete JJ (2000) Biophysical study of the perturbation of model membrane structure cased by seminal plasma protein PDC-109. Arch Biochem Biophys 374:241–247

Gómez J, Hilser VJ, Xie D, Freire E (1995) The heat capacity of proteins. Proteins 22:404–412

Greube A, Müller K, Töpfer-Petersen E, Herrmann A, Müller P (2001) Influence of the bovine seminal plasma protein PDC-109 on the physical state of membrane. Biochemistry 40:8326–8334

Manjunath P, Thérien I (2002) Role of seminal plasma phospholipid-binding proteins in sperm membrane lipid modification that occurs during capacitation. J Reprod Immunol 53:109–119

Moreau R, Thérien I, Lazure C, Manjunath P (1998) Type II domains of BSPA1/-A2 proteins: binding properties, lipid efflux and sperm capacitation potential. Biochem Biophys Res Commun 246:148–154

Müller P, Erlemann K-R, Müller K, Calvete JJ, Töpfer-Petersen E, Marienfeld K, Herrmann A (1998) Biophysical characterization of the interaction of bovine seminal plasma protein PDC-109 with phospholipid vesicles. Eur Biophys J 27:33–41

Müller P, Greube A, Töpfer-Petersen E, Herrmann A (2002) Influence of the bovine seminal plasma protein PDC-109 on cholesterol in the presence of phospholipids. Eur Biophys J 31:438–447

Ramakrishnan M, Anbazhagan V, Pratap TV, Marsh D, Swamy MJ (2001) Membrane insertion and lipid–protein interactions of bovine seminal plasma protein, PDC-109 investigated by spin label electron spin resonance spectroscopy. Biophys J 81:2215–2225

Sankhala RS, Swamy MJ (2010) The major protein of bovine seminal plasma, PDC-209 is a molecular chaperone. Biochemistry 49:3908–3918

Sankhala RS, Damai RS, Swamy MJ (2011) Correlation of membrane binding and hydrophobicity to the chaperone-like activity of PDC-109, the major protein of bovine seminal plasma. PLoS One 6:e17330

Seidah NG, Manjunath P, Rochemont J, Sairam MR, Cheretian M (1987) Complete amino acid sequence of BSP-A3 from bovine seminal plasma. Homology to PDC-109 and to the collagen-binding domain of fibronectin. Biochem J 243:195–203

Shivaji S, Scheit K-H, Bhargava PM (1990) Proteins of seminal plasma. Wiley, New York

Sultan NAM, Swamy MJ (2005) Energetics of carbohydrate binding to *Momordica charantia* (bitter gourd) lectin. An isothermal titration calorimetric study. Arch Biochem Biophys 437:115–125

Swamy MJ (2004) Interaction of bovine seminal plasma proteins with model membranes and sperm plasma membranes. Curr Sci 87:203–211

Swamy MJ, Marsh D, Anbazhagan V, Ramakrishnan M (2002) Effect of cholesterol on the interaction of seminal plasma protein, PDC-109 with phosphatidylcholine membranes. FEBS Lett 528:230–234

Tarafdar PK, Reddy ST, Swamy MJ (2013) Effect of Hofmeister series anions on the thermotropic phase behavior of bioactive *O*-acylcholines. J Phys Chem B 117:9900–9909

Thérien I, Moreau R, Manjunath P (1998) Major proteins of bovine seminal plasma and high-density lipoprotein induce cholesterol efflux from epididymal sperm. Biol Reprod 59: 768–776

Thomas CJ, Anbazhagan V, Ramakrishnan M, Sultan N, Surolia I, Swamy MJ (2003) Mechanism of membrane binding by the bovine seminal plasma protein, PDC-109. A surface plasmon resonance study. Biophys J 84:3037–3044

Wah DA, Fernández-Tornero C, Sanz L, Romero A, Calvete JJ (2002) Sperm coating mechanism from the 1.8 Å crystal structure of PDC-109-phosphorylcholine complex. Structure 10:505–514

Yanagimachi R (1994) Mammalian fertilization. In: Knobil E, Neill JD (eds) The physiology of reproduction, 2nd edn. Raven, New York, pp 189–317

Chapter 19
Crystal Structure of Apo and Ligand Bound *Vibrio cholerae* Ribokinase (Vc-RK): Role of Monovalent Cation Induced Activation and Structural Flexibility in Sugar Phosphorylation

Rakhi Paul, Madhumita Dandopath Patra, and Udayaditya Sen

Introduction

D-Ribose is a ubiquitously found essential pentose sugar, which not only serving as an energy source but also is a constituent of RNA, DNA and many other cofactors. However, D-ribose must be phosphorylated before entering into the metabolic pathways for further use and also in the synthesis of nucleotides and amino acids like tryptophan and histidine (Anderson and Cooper 1969; Lopilato et al. 1984). Ribokinase (RK; EC 2.7.1.15) catalyzes the phosphorylation reaction of ribose to ribose-5-phosphate, a key ingredient of the pentose phosphate pathway and nucleotide synthesis (Anderson and Cooper 1970; Karasevich and Ivoĭlov 1975). During phosphorylation, ribokinase specifically transfers the γ-phosphate of adenosine tri-phosphate (ATP) to O5′ of D-ribose to form D-ribose-5-phosphate.

RK belongs to a well known protein family called Ribokinase superfamily whose folding patterns and substrate specificity have been studied extensively during the last decade (Bork et al. 1993). Apart from ribokinase (RK), this family includes fructokinase (FK), 1-phosphofructokinase (PFK), 6-phosphofructo-2-kinase (PfkB or PFK2), tagatose-6-phosphate kinase (LacC), 4-Amino-5-Hydroxymethyl-2-Methylpyrimidine Phosphate Kinase (HMPP kinase), inosine kinase (INGK), adenosine kinase (AK) and 2-dehydro-3-deoxygluconokinase (KDG kinase) and pyridoxal kinase (PK) (Bork et al. 1993; Chua et al. 2010; Cheng et al. 2002; Park and Gupta 2008; Li et al. 2002). The key structural features of this family consist of highly conserved sequence motifs near the N and C terminals and in the vicinity of the active site (Park and Gupta 2008). Structural studies (Sigrell et al. 1998, 1999)

R. Paul • M.D. Patra • U. Sen, Ph.D. (✉)
Crystallography and Molecular Biology Division, Saha Institute of Nuclear Physics, 1/AF Bidhan Nagar, WB, Kolkata 700 064, India
e-mail: udayaditya.sen@saha.ac.in

A. Chakrabarti, A. Surolia (eds.), *Biochemical Roles of Eukaryotic Cell Surface Macromolecules*, Advances in Experimental Medicine and Biology 842,
DOI 10.1007/978-3-319-11280-0_19

have revealed that most of the members of RK superfamily act as dimer where each monomer consists of two distinct domains (Chua et al. 2010; Cheng et al. 2002; Li et al. 2002; Sigrell et al. 1997, 1998, 1999). A large catalytic domain that provides the binding site for sugar and ATP comprises of a twisted nine stranded β-sheet flanked by α-helices, five on each side. A smaller domain made of β-strands is responsible for crucial contacts at the dimerization interface and also acts as a lid over the catalytic domain to bury the sugar during complex formation. A crucial Asp residue at the α/β domain is believed to act as a base responsible for activating O5′ of ribose to make it a better nucleophile in attacking the γ-phosphate of ATP. Beside the substrate and ATP, this class of proteins requires divalent cations like Mg^{2+}, Ca^{2+}, Mn^{2+} during catalysis to stabilize the multiple negative charges of the nucleotide (Park and Gupta 2008; Chuvikovsky et al. 2006; Schimmel et al. 1974). In addition to the divalent cations, catalytic activity of RK is strongly dependent on the presence of monovalent cations which activate the enzyme before phosphoryl group transfer (Suelter 1970). For *E coli* RK, monovalent cations with ionic radii greater than sodium works well (apparent K_d of 5 and 17 mM for K^+ and Cs^+, but not Na^+ or Li^+ up to 140 mM) (Andersson and Mowbray 2002). Although the structure of *E. coli* RK in complex with Cs^+ was determined, no structure of ribokinase in apo-form has ever been reported to asses the structural changes associated with the activation process (Andersson and Mowbray 2002). Re-evaluation of the crystallization conditions and electron density maps of previously reported structures of *E. coli* RK strongly suggests that a water molecule that was inadvertently placed closed to the K^+/Cs^+ binding site should have actually been an NH_4^+. This NH_4^+ bound RK structure is similar to the Cs^+ bound RK structure as NH_4^+ is also capable of activating RK (Chuvikovsky et al. 2006; Andersson and Mowbray 2002). Recently the crystal structure of Staphylococcus aureus RK has been reported in the apo-form (Li et al. 2012) and upon comparison with the *E. coli* RK it has been possible to shed light on the activation mechanism by monovalent cations. This comparison indicates that significant conformational changes of the large ATP loop occurred via binding of the monovalent cation which facilitates the formation of nucleotide binding pocket. Apart from the role of metal ion, crystal structure of an aminoimidazole riboside (AIR) kinase from *Salmonella enterica* (Zhang et al. 2004) reports the implications of the β-strand domain for the evolution of the ribokinase superfamily.

Vibrio cholerae shows a biphasic life style with a periodic transition between environmental and host-associated niches (Sack et al. 2004). During this process, metabolic reactions and gene expressions of *V. cholerae* are modified according to ambient environmental factors. However, the exact condition under which ribose metabolism is initiated remains unknown. RK from *V. cholerae* O395 (Vc-RK) (UniProtKb accession code A5F1B7, gene name *rbsk*) (Lopilato et al. 1984; Anderson and Cooper 1970; Iida et al. 1984), possesses 306 amino acids with a molecular weight of 32 kDa. In order to gain insight into the catalytic functions of Vc-RK at the atomic level we have solved the crystal structural of Vc-RK in the apo form (3.4 Å), sugar + ADP bound form (1.75 Å) and sugar + ADP + Cs^+ bound form (2.37 Å) and compared them with *E. coli* RK and Sa239 RK. Vc-RK exists as dimer and each monomer has two domains, a large catalytic α/β domain consisting of a central nine stranded twisted β-sheet flanked on both faces by five α-helices and a small four stranded β-sheet region protruding from the α/β domain. Comparison with Cs^+ bound *E. coli* RK structure shows that Vc-RK structure in sugar + ADP bound form is activated and this activation is caused by Na^+. The location of the Na^+ is confirmed by the sugar + ADP + Cs^+ bound Vc-RK structure where Cs^+ occupies the same position as Na^+. Comparisons of the apo and ligand bound Vc-RK structures have allowed us to identify not only the conformational changes associated with its activation (in the presence of Na^+ ion) but also sugar induced structural changes and the mechanism of the catalytic phosphorylation reaction in Vc-RK.

Materials and Methods

Cloning, Expression and Purification

Vc-RK protein was purified according to the previously described protocols elaborated earlier (Paul and Sen 2014). Briefly, the *rbsK* gene encoding Vc-RK was amplified from *V. cholerae* O395 genomic DNA and cloned into pET28a$^+$ vector. After transformation in *E. coli* BL21 (DE3), cells were grown at 37 °C until the optical density at 600 nm (OD_{600}) reached 0.4–0.6. Protein expression was induced by the addition of IPTG (isopropyl-D-thiogalactopyranoside) to a final concentration of 0.1 mM. The cells were harvested by centrifugation and the re-suspended pellet was lysed by sonication in presence of PMSF (phenylmethanesulfonyl fluoride) followed by centrifugation (12,000*g* for 50 min) at 4 °C. The 6×His tagged protein was isolated from the supernatant using Ni^{2+}–NTA affinity chromatography (Qiagen) and eluted with lysis buffer containing 150 mM imidazole. The eluted fractions were checked by 10 % SDS–PAGE, pooled and dialyzed overnight against the thrombin cleavage buffer (0.05 M Tris–HCl pH 8.0, 100 mM NaCl) and the 6×His tag was cleaved with 1 U thrombin by overnight incubation at 4 °C. The proteins were further purified by gel filtration chromatography using a Sephacryl S-100 (GE-Healthcare) column (78×1.4 cm) pre-equilibrated with thrombin cleavage buffer containing 0.02 % sodium azide at 4 °C.

Crystallization and Data Collection

The previously reported crystallization condition (5 % PEG 6 K, 0.1 M MES pH 6.0 as precipitant and 35 % (w/v) PEG 6 K as reservoir) (Paul and Sen) were used to obtain single crystals suitable for X-ray diffraction, for both apo Vc-RK and ribose + ADP bound Vc-RK. For apo Vc-RK, a low-salt condition (<100 mM NaCl) was maintained during crystallization to make sure that Vc-RK is not activated. For the co-crystals, Vc-RK was incubated with ribose (5 mM) and ADP (10 mM) before crystallization in the presence of a buffer of high ionic strength (>200 mM NaCl).

For data collection, crystals were harvested from the crystallization drops using a 20 mm nylon loop and flash-cooled in a stream of nitrogen (Oxford Cryosystems) at 100 K. Diffraction data were collected on a MAR 345 CCD on BM14 at the

European Synchrotron Radiation Facility (ESRF), Grenoble. The incident X-ray beam had a wavelength of 0.97 Å and 360 frames were recorded with non-overlapping 1° oscillations. For apo Vc-RK crystals 12 s exposure per image was used whereas for its co-crystals exposure time was 1 s only. Cs^+ bound data were collected at home source by soaking ribose + ADP bound Vc-RK crystals with a cryoprotectant containing 20 mM CsCl. Data were processed and scaled using *iMosflm* (Battye et al. 2011), *POINTLESS* and *SCALA* (Evans 2006) respectively. Apo Vc-RK crystals diffracted poorly and we could process the data to a resolution of 3.4 Å. Co-crystals of Vc-RK, however, produced excellent quality diffraction data to a resolution of 1.75 Å. Cs^+ soaked crystals diffracted up to 2.37 Å. Data-collection and processing statistics are detailed in Table 19.1.

Structure Determination and Refinement

Since the apo Vc-RK crystals diffracted to a lower resolution (3.4 Å) we first solved the sugar + ADP bound structure of Vc-RK using the high resolution (1.75 Å) diffraction data. For phasing, the coordinates of the 1.8 Å data of *E. coli* RK (PDB code 1RK2) (Sigrell et al. 1999), which has 54 % amino acid sequence identity with Vc-RK as determined by BLAST (Altschul et al. 1990) against PDB database (Fig. 19.1) was used for molecular replacement with Phaser (McCoy et al. 2007) in CCP4 (Potterton et al. 2003). Phaser located four molecules in the asymmetric unit with LLG value 12,686 and R-factor of 43.2 %. The MR model was built with COOT (Emsley and Cowtan 2004) and refinement was carried out with Phenix refine (Adams et al. 2010). TLS refinement was performed during the final stages of refinement (Painter and Merritt 2006) to an R-work of 18.4 % (R_{free} = 22.2 %). Coordinates of the D-chain of this refined Vc-RK molecule (that contains only ADP but no sugar) was used to solve the apo Vc-RK structure. Sixteen molecules of this modified model in the asymmetric unit produced an LLG value of 7,531 with an R-factor of 45.5 % for the data between 30.0 and 3.4 Å. Analysis of the packing ruled out the presence of additional molecules in the asymmetric unit. Structure of Cs^+ soaked crystals were refined using the 1.75 Å sugar + ADP bound Vc-RK structure. The structures were refined with several cycles of model building in COOT (Emsley and Cowtan 2004) and refinement with PHENIX (Adams et al. 2010). The quality of the final model was analyzed with the program PROCHECK (Laskowski et al. 1993). The data processing and structure determination statistics are listed in Table 19.1.

Structural Analysis

Average B-factors for each residue were calculated using B average in CCP4 (Potterton et al. 2003). PISA webserver (Krissinel and Henrick 2007) was used for the analysis of the oligomeric state. Sequence alignment of Vc-RK with other RKs was done using ClustalW (Goujon et al. 2010; McWilliam et al. 2013). Figures were prepared using Pymol (http://www.pymol.org).

Table 19.1 Data collection and refinement statistics for Vc-RK ctystals

	Apo Vc-RK	Vc-RK+ribose+ADP+Na	Vc-RK+ribose+ADP+Cs
Space group	*P1*	*P1*	*P1*
Unit-cell parameters (Å, °)	a=129.22, b=130.85, c=145.69	a=59.42, b=70.70, c=79.91	a=59.30, b=69.55, c=78.31
	α=110.52, β=90.00, γ=119.59	α=106.31, β=97.66, γ=98.69	α=106.99, β=98.76, γ=98.46
Resolution (Å)	3.40 (43.57–3.40)[a]	1.75 (27.47–1.75)	2.37 (26.64–2.37)
Oscillation range (°)	1	1	1
Molecules/a.u.	16	4	4
Matthews coefficient V_M (Å^3 Da^{-1})	3.9	2.4	2.3
Solvent content	68	50	48
No. of unique reflections	97,045	117,357	42,241
Multiplicity	2.1 (2.5)	4.0 (3.9)	2.07 (2.11)
Mosaicity (°)	1.5	0.8	0.7
Completeness (%)	92.9 (85.6)	95.1 (95.1)	93.8 (93.4)
R_{merge} (%)	15.1 (45.9)	9.3 (56.7)	6.4 (28.37)
Average I/σ(I)	4.7 (2.0)	8.5 (2.4)	5.1 (1.5)
Refinement			
Resolution range (Å)	43.57–3.40	27.47–1.75	26.64–2.37
R-factor/R free (%)	23.86/30.60	18.46/22.20	21.13/25.48
Average B-factor/r.m.s.d. (Å²)			
Main chain (306 residues)	81.2	26.1	45.5
Side chain	82.6	33.1	48.4
Solvent (molecules)		38.5	44.2
r.m.s.d. from standard geometries			
Bond length (Å)	0.012	0.011	0.007
Bond angles (°)	1.840	1.278	1.173
Ramachandran plot (%)			
Most favored	81.31	97.62	96.00
Allowed	12.73	1.80	3.59
Outliers	5.96	0.57	0.41

[a]Values in parenthesis are for the highest resolution shell

Results

Overall Structure

Each monomer of Vc-RK consists of two domains: one large globular α/β domain and a β-sheet region that distinctly protrudes from the former (Fig. 19.2a). The catalytic α/β domain consists of a central nine stranded twisted β-sheet flanked on both faces

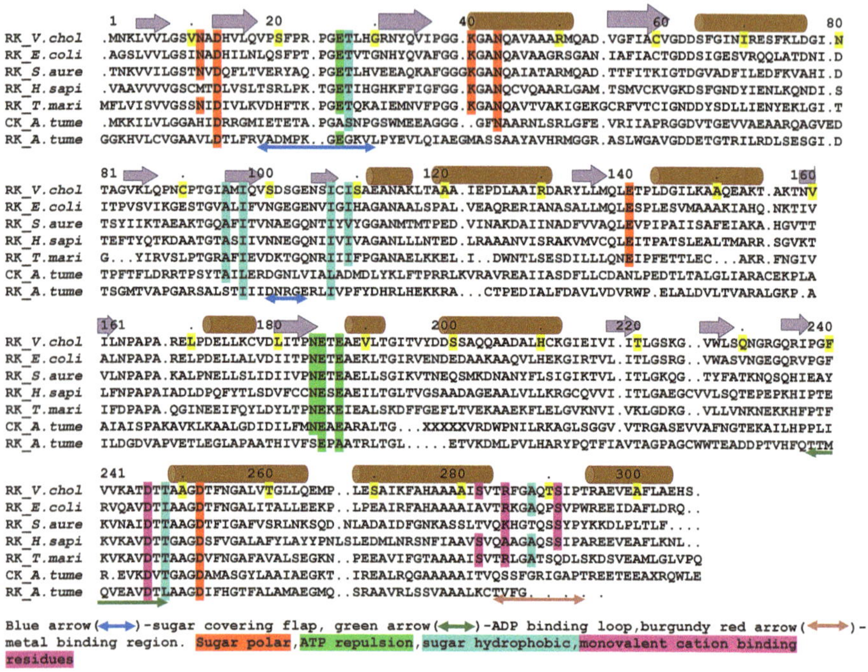

Fig. 19.1 Sequence alignment of VcRK with other sugar kinases RK from *Escherechia coli*, *Sachharomyces serevaci*, *Homo sapiens*, *Thermotoga maritima*, *Agrobacterium tumefaciens*, and a carbohydrate kinase (CK) from *Agrobacterium tumefaciens* are used in MultAlin. The secondary structure of the first sequence (i.e.,Vc-RK) is shown overhead by *brown helices* (for **α**-helices) and *violet arrows* (for β-strands). The conserved residues of the ribose-binding site are highlighted in *red* and the flap, which encloses the sugar, is indicated in *blue arrow* at the bottom of sequence. The ADP binding loop is shown in *green arrow* and the loop near metal binding site is shown as *burgundy red arrow*. The *green* highlighted conserved hydrophobic residues repel ADP and *cyan* highlighted residues make hydrophobic interaction with ribose. Residues interacting with monovalent cation are highlighted in *pink*

by five α-helices which dominate the surface of the molecule. The first six β-strands of this nine stranded sheet together with the associated α helices have the topology of a typical Rossmann fold. The ribokinase fold is extended by three additional β strands (β11–β13) that are connected by short reverse turns. Both faces of this extension are also shielded by helices, α7 and α8 on one side and α9 and α10 on the other (Fig 19.2a). The protruding β-sheet domain is formed by two insertions into the central α/β fold, each of which forms two pairs of antiparallel β-strands: β2–β3 (residue 13–36) and β6–β7 (residue 95–107). The first pair connects β1 and α1, and the other is located between β5 and α3. A long loop (residues 18–31) joins β2 and β3 which bends at Pro19 and Pro24.

Packing of the Vc-RK molecules inside the crystal shows that they form a spiral type of arrangement enclosing a prominent cavity at the centre, accounting for the high solvent content (68 %) of the crystals (Table 19.1). Sixteen Vc-RK molecules,

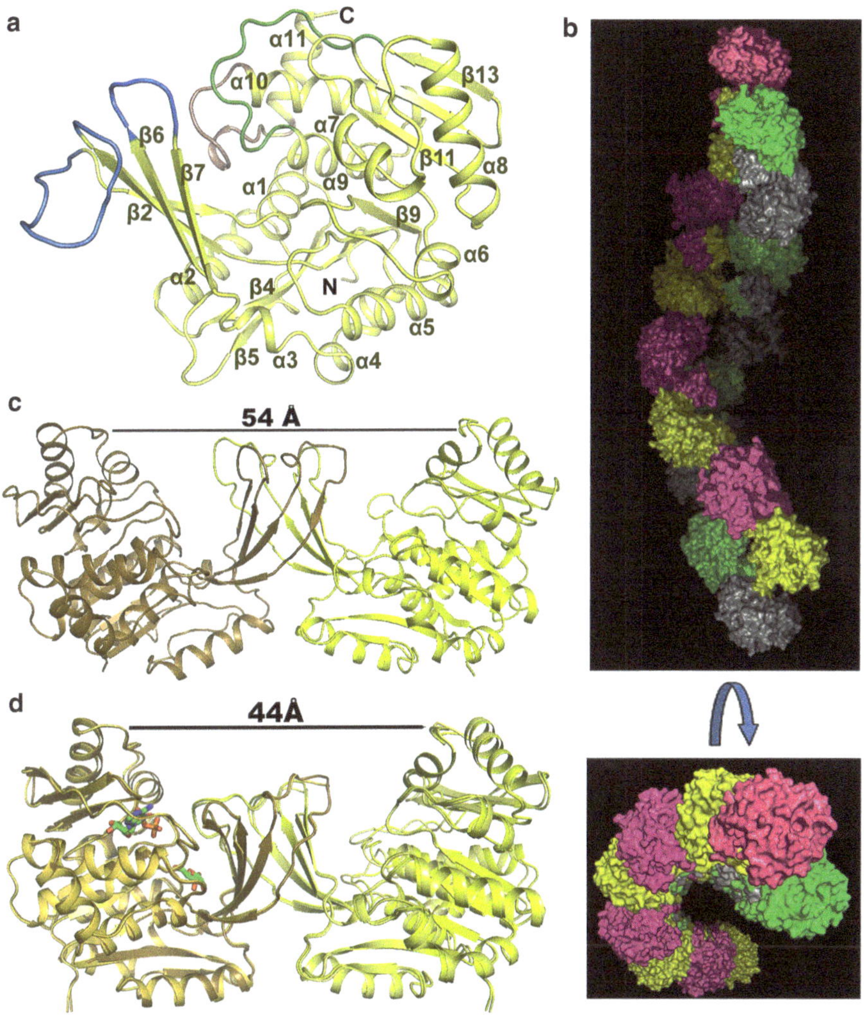

Fig. 19.2 Secondary structure and oligomeric state of Vc-RK (**a**) secondary structure of ribokinase monomer shown in cartoon. The dimerization loop (between β2–β3 and β6–β7) is shown in blue, ADP binding loop (between β12 and α9) is shown in green and the metal binding loop (between α10 and α11) is shown in burgundy color. (**b**) Arrangement of 16 monomers in the form of eight dimers forming a double-stranded helical pattern; upper strand composed of four '*yellow* and *magenta*' dimers and the lower strand consists of four '*grey* and *green*' dimers (*top*), perpendicular view of the helix axis where spiral like look is evident (*bottom*). (**c**, **d**) Ribokinase dimer in apo and substrate bound form; apo dimer is symmetric about β-clasp axis and more open than the ternary complex structure. The distances indicated were measured between Cα atoms of residue Thr195 from the two subunits. Ternary complex is more closed towards ADP-ribose bound chain than towards ADP bound chain

in the form of eight dimers, are arranged in the asymmetric unit in a spiral like geometry (Fig. 19.2b). In a close view, it looks like a twisted double stranded helix (Fig. 19.2b) where four dimers make the upper strand of the helix and the remaining four dimers make the lower strand.

Dimerization

Size exclusion chromatography and dynamic light scattering (DLS) experiments indicate that Vc-RK exists as dimer in solution. The dimeric form is also evident in the crystal structure of apo and sugar+ADP bound Vc-RK where the protruding β-sheet region, consisting of four β-strands from each neighboring monomers, interacts with each other to form the dimer (Fig. 19.2c, d). Eight β-strands along with their connecting loops pack around a twofold axis in the form of a flattened barrel like structure. Several hydrophobic residues interact with each other at the dimeric interface. This barrel-like structure has been termed as β-clasp (handshake motif) and described in detail in case of *E. coli* RK structure (Sigrell et al. 1998, 1999). Eight β-strands from two chains (A and B) are arranged in the order β3A–β2A–β6A–β7A–β3B–β2B–β6B–β7B to form the symmetric dimer. The large loop between β3 and β2 bends about 90° near Pro19 to complete the 'clasp'.

'Open' Apo Dimer

The apo form of Vc-RK crystallized in space group P1 and contains eight dimers in each asymmetric unit (Table 19.1). According to PDBePiSA program (Krissinel and Henrick 2007), the average buried surface area per dimer is about 2,350 $Å^2$ which is about 9 % of the total surface area with ΔG_{int} value of ~22 kcal/mol and ΔG_{disso} value of ~14 kcal/mol implying that the dimer is energetically stable. In the course of dimer formation, residues Leu28, Gly30, Tyr33 from the larger β-turn of one monomer comes close to the residues Ile109, Ser110 and Glu112 from the shorter β-turn of another monomer. The inner part of the clasp region contains mainly hydrophobic residues like Met96, Val35, Pro22, Tyr33 and Phe21. The apo dimer adopts an open structure where α/β domain stays away from the β-clasp region (Fig. 19.2c). The distance between the α/β domains of two monomers is about 54 Å (Fig. 19.2c). This open form of the apo Vc-RK structure allows entry of the ribose into the sugar binding pocket.

'Half-Way Closed' Substrate Dimer

Crystals of Vc-RK complexed with ribose and ADP contain four molecules in the asymmetric unit (Table 19.1). The four monomers (chains A, B, C, D) are arranged in the form of two dimers A:B and C:D. Out of the four monomers chain A and chain C bind both ribose and ADP, chain B binds two ADP molecules and chain D binds only one ADP molecule without any sugar. It is well known that upon ribose binding, the protruding lid (composed of β2–β3 and β6–β7 and the associated loops) acquires a closed conformation and folds back to the α/β domain with the sugar trapped underneath (Sigrell et al. 1999). Therefore, chain A and chain C assume a closed conformation due to ribose binding but, chain D remains in the open form as no sugar is bound at the active site. For chain B, the ribose site is occupied by ADP where the sugar part of ADP matches with the site normally occupied by ribose keeping the phosphates towards the α/β domain and the adenine base towards the lid. Larger size of ADP and/or its non-cognate nature may keep the lid in the open conformation. Therefore, for each dimer one chain (A or C) is in the closed conformation while the other chain (B or D) is in the open conformation leading to a 'half-way closed' asymmetric dimers about their symmetry axis (Fig. 19.2d).

The Active Site

The active site is located near a shallow trench between the β-flap and the central α/β domain. The region consists of a ribose binding site, an ATP or analog binding site and an adjacent anion hole. The β-flap participates in both dimerization and trapping of ribose in the active site before the enzymatic reaction proceeds. Among the loop regions, two sugar binding loops (residues 18–34 and 100–105, shown in blue), monovalent cation binding loop (residues 284–295, shown in flesh colour) and large ATP loop (residues 237–250, shown in green) are crucial for catalysis (Fig. 19.2a). The green loop creates a compartment, which separates the monovalent cation from ribose and ADP and therefore makes it an allosteric effector. These regions are quite flexible in apo form but become structurally rigidified upon sugar binding.

Ribose Binding

The hydrogen bonding pattern with the nearby residues (Fig. 19.3a) and examination of the electron density map (Fig. 19.3b) clearly indicate that when bound to Vc-RK, D-ribose adopts a five membered α-furanose form. Although less prevalent in solution, this form allows the O5′ atom of ribose to be available for phosphorylation. This form of ribose projects the ring oxygen O4′ and O5′ on one side of the plane made of four carbon atoms of the ring while all the remaining non-ring oxygen atoms remain on the other side of the plane (Fig. 19.3a, b). D-ribose molecule binds on top of the conserved GGK motif (Fig. 19.1) projecting all non-ring 'O' atoms towards the α/β domain forming altogether 12 hydrogen bonds with the side chains of Asn11, Asp13, Lys40, Asn43, Glu140 and Asp252 and the main-chain amide atoms of Gly39 and Asn11 (Fig. 19.3b). These sugar binding residues are highly conserved (Fig. 19.1). Asp252 is believed to act as the base responsible for ionizing the O5′ of ribose to make it better a nucleophile in attacking the γ-phosphate of ATP. The hydrophobic part of ribose, constituted by four ring carbons and C5′,

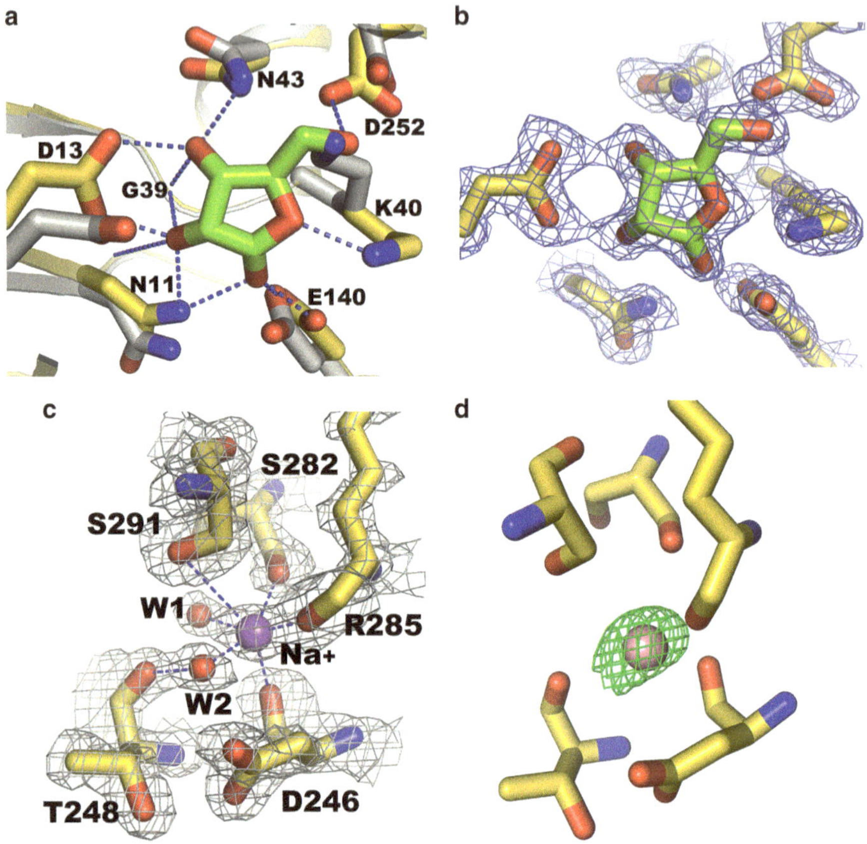

Fig. 19.3 Sugar and metal ion binding in RK. (**a**) Residues of Vc-RK (*yellow sticks*) involved in binding sugar (*green sticks*). Hydrogen bonds between sugar and Vc-RK are shown as dashed line (*blue*). Amide nitrogen atoms of G39 and N11 are not shown. The structure of apo Vc-RK (grey) is overlaid on it to show the local structural change upon sugar binding. (**b**) 2Fo-Fc map contoured at 1.5σ around ribose and the surrounding residues. (**c**) 2Fo-Fc map contoured at 1.6σ (*light blue mesh*) around the monovalent cation (Na^+ ion, *violet sphere*) binding site. Residues interacting with the metal ion (*yellow sticks*) and two water molecules (W1 and W2, small *red* spheres) are also shown (**d**) Fo-Fc map contoured at 6.5σ (*green mesh*) with Cs^+ (CsCl) soaked crystal showing Cs^+ (*flesh sphere*) occupying roughly the same position as Na^+

faces Ile97, Ile107, Ile109, Ala95 and Ala288. These conserved hydrophobic residues (Fig. 19.1) are located at the bottom of the lid and pack with the hydrophobic face of the ribose.

Monovalent Cation Binding

Phosphorylation of sugar by RK is catalyzed by divalent cations such as Mg^{2+}, Mn^{2+}, Co^{2+}, Ca^{2+}, Ni^{2+} and Cu^{2+} (Chuvikovsky et al. 2006; Park and Gupta 2008) whereas monovalent cations like Na^+, K^+, Cs^+ and NH_4^+ ions serve as an activator of

RK (Chuvikovsky et al. 2006; Andersson and Mowbray 2002). Although we used Ca^{2+} during crystallization no electron density was observed for this ion. As the usual binding site for divalent cations is located between β and γ phosphate (Andersson and Mowbray 2002), density for the divalent metal ion was not observed probably due to the absence of γ phosphate in ADP. However, all the four chains of RK are seen to be activated through monovalent cation binding and from a scrutiny of salts used during crystallization this cation was assigned as Na^+. 2Fo-Fc map around the Na^+ ion is shown along with the metal ion binding residues of Vc-RK and two water molecules (Fig. 19.3c). Na^+ is seen to be hexa-coordinated and is within polar contact with Asp246, Ser282, Arg285, Ser291 and two water molecules. Na^+ interacting residues are disposed roughly in an octahedral geometry and the interacting residues are roughly same as *E. coli* RK. To further confirm the position of Na^+, we soaked the crystals of Vc-RK+ribose+ADP with a cryo buffer containing 40 mM CsCl and solved the structure to a resolution of 2.37 Å. A Fo–Fc map calculated with the refined structure (with no monovalent cation at this site) shows a clear density for Cs^+ located roughly in the same position of Na^+ (Fig. 19.3d).

Discussion

RK catalyzes the phosphorylation of ribose in the presence of ATP and divalent cations. For *E. coli* RK, it has been reported that monovalent cations with ionic radii greater than sodium acts as an activator (Andersson and Mowbray 2002). Structural basis of this activation has been proposed for *E. coli* RK but since no apo RK structure was known for *E. coli* exact structural changes associated with metal binding was not known. Crystal structure of Sa239 RK solved in apo-form when compared with the *E. coli* RK sheds light on the activation mechanism by monovalent cations. This comparison indicates that significant conformational changes of the large ATP loop occurred via binding of the monovalent cation which facilitates the formation of nucleotide binding pocket.

The structure of apo Vc-RK solved at 3.4 Å (with low Na^+ concentration), when compared with its ribose+ADP bound form, provides an opportunity not only to decipher the underlying activation mechanism but also to identify the necessary structural changes required for catalysis. Although atomistic structural details are not expected from the low resolution structure of apo Vc-RK, the 16-fold NCS map is of high quality for most of the regions. Superposition of the 16 chains in the asymmetric unit provides some structural details. The helices and strands belonging to the α/β domain along with their connecting loops superpose well excepting for a few loops near the active site which are highly flexible (Fig. 19.4a). A superposition of ribose+ADP complex structure with the apo Vc-RK provides the location of these flexible loops (Fig. 19.4a). These are the large ATP loop (shown in green), monovalent metal binding loop (loop connecting α10–α11; flesh colored loop) and the two connecting loops which join the four β-strands of the lid (shown in blue). Different conformation of the large ATP loop is clearly seen to encroach upon the ATP binding site (Fig. 19.4a). However if we compare the four chains of ribose+ADP

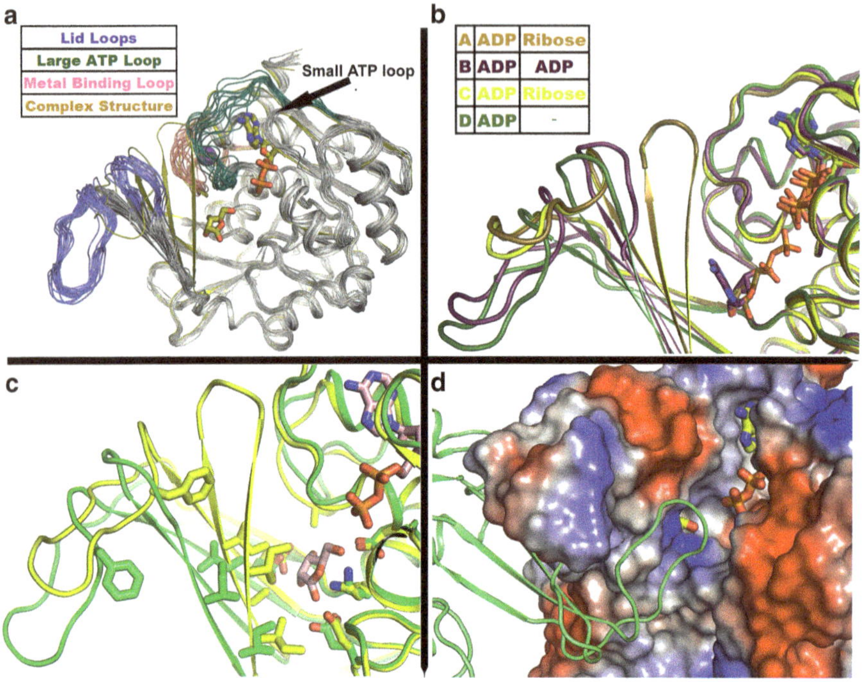

Fig. 19.4 (**a**) Superposition of 16-chains of apo structure (*grey*) overlaid with the ternary complex (*sand color*). ADP and sugar molecule are shown in sticks and the metal ion as *violet sphere*. Structural flexibility of ADP loop, metal binding loop and β loops are evident. (**b**) Relative movement of β-clasp region for different chains. β-clasp is closed for chain A and chain C (*yellow shades*) while it is open for chain B (*magenta*) and chain D (*green*). ADP and ribose is shown in sticks. (**c**) Superposition of chain A and chain D show lid movement and residues involved in sugar binding. A Phe residue (F21) is shown as marker for both apo (*green*) and sugar + ADP bound (*yellow*) structure to show the movement of the loop. (**d**) Covering of the residues (ADP and ribose) involved in phosphoryl group transfer by lid residues of the same monomer (surface) and lid residues from the other monomer (*green cartoon*). This also implies the role of dimerisation in RKs to proceed the phosphorylation reaction with high fidelity

bound Vc-RK we see that all of them possess the same conformation for the large ATP loop and monovalent metal binding loop and all of them bind Na^+ (Fig. 19.4b). Upon binding, Na^+ pulls these two loops and sequesters them to shape-up for ATP binding. This pulling of large ATP loop removes the physical blockage for the incoming ATP molecule and explains the structural basis of its activation.

Pairwise superposition of the four chains of ribose + ADP bound Vc-RK shows that chain A and chain C superposes well (with an r.m.s.d of 0.19 Å for 262 Cα atoms) whereas the r.m.s.d. value of superposition between chain B and Chain D is 0.3 Å for 244 Cα atoms. Chain B when superposed with Chain A (or Chain C) produces an r.m.s.d value between 0.33 (or 0.38) Å for 244 Cα atoms while Chain D when superposed with Chain A (or Chain C) produces an r.m.s.d value between

0.38 (or 0.46) Å for 238 Cα atoms. Chain B and Chain D when superposed with Chain A of Chain C their α/β domain matches with each other but the flap regions varies (Fig. 19.4b). Out of the four chains chain A and chain C binds ribose. One side of ribose is anchored with the α/β domain through several strong H-bonds (Figs. 19.3a and 19.4c) whereas the other side is tethered through hydrophobic residues which are highly conserved in the family (Figs. 19.1 and 19.4c). Therefore the ribose molecule mediated interactions bring the β-sheet domain and α/β domain close to each other to trap ribose (Fig. 19.4b, c). Therefore, the β-sheet domain along with the associated connecting loops, experience a huge structural change as they fold back towards the α/β domain (Fig. 19.4c). These regions also show flexibility in their apo form (Fig. 19.4a). Therefore, each region showing structural flexibility in the apo structure has different signature in the catalytic process either binding monovalent cation followed by activation (flesh and green region) or trapping the sugar (blue region).

Chain D does not bind sugar and the pocket is filled with water molecules and no closure of the lid is observed in this case. Chain B has a unique feature as it binds two ADP molecules, one ADP molecule binds at its usual binding site while the other ADP binds such that its ribose counterpart matches the position of ribose while the adenosine ring face the lid and the diphosphate group face the α/β domain. Among the sugar kinases only RK has this unique feature to bind ADP in place of its usual substrate as its cognate sugar is same as the sugar part of ADP. Surprisingly, these extra phosphate groups do not disturb other nearby residues of the α/β domain but the lid remains in the open position. The phosphate group of ADP is now stabilized by Lys40 that was interacting with the O4′ of ribose (Fig. 19.3a). Therefore, upon phosphorylation of ribose Lys40 may play a crucial role to in stabilization of phosphate.

Attempt to crystallize Vc-RK in presence of ANP-PNP and sugar produced a crystal where the terminal group is not seen in the structure. We presume that at the pH of the crystallizing condition the terminal phosphate is probably hydrolyzed. In the ADP+sugar bound structure the distance between the O5′ of ribose and the β-phosphate group of ADP is 7 Å and it is expected that ATP instead of ADP would dispose the γ-phosphate in close proximity to O5′ of ribose for catalysis. The intervening position is, however, occupied by some water molecules that might be the position of γ-phosphate. The ribose molecule is covered by the residues from the lid where the O5′ atom to be phosphorylated is projected through a small opening to attack the γ-phosphate of ATP. Moreover, one side of the active site is covered by the lid residues from other monomer that cover the region between the sugar and ADP/ATP (Fig. 19.4d). This covering is required to shield the groups involved in phosphoryl transfer from the bulk solvent in order to prevent abortive hydrolysis. This also implies the necessity of dimerisation in catalysis of RKs.

Acknowledgements The laboratory of US is supported by the MSACR project, DAE, Government of India and SINP. This work is supported by the Department of Biotechnology (DBT), Government of India (The BM14 beamline project for synchrotron data collection at the ESRF, Grenoble). M.D.P. thanks the DBT for a DBT-RA fellowship.

References

Adams PD, Afonine PV, Bunkóczi G, Chen VB, Davis IW, Echols N, Headd JJ, Hung L, Kapral GJ, Grosse-Kunstleve RW, McCoy AJ, Moriarty NW, Oeffner R, Read RJ, Richardson DC, Richardson JS, Terwilliger TC, Zwart PH (2010) PHENIX: a comprehensive python-based system for macromolecular structure solution. Acta Crystallogr D Biol Crystallogr 66: 213–221

Altschul SF, Gish W, Miller W, Myers EW, Lipman DJ (1990) Basic local alignment search tool. J Mol Biol 215(3):403–410

Anderson A, Cooper RA (1969) The significance of ribokinase for ribose utilization by *Escherichia coli*. Biochim Biophys Acta 177:163–165

Anderson A, Cooper RA (1970) Biochemical and genetical studies on ribose catabolism in *Escherichia coli* K12. J Gen Microbiol 62:335–339

Andersson CE, Mowbray SL (2002) Activation of ribokinase by monovalent cations. J Mol Biol 315:409–419

Battye TG, Kontogiannis L, Johnson O, Powell HR, Leslie AG (2011) iMOSFLM: a new graphical interface for diffraction-image processing with MOSFLM. Acta Crystallogr D Biol Crystallogr 67(4):271–281

Bork P, Sander C, Valencia A (1993) Convergent evolution of similar enzymatic function on different protein folds: the hexokinase, ribokinase, and galactokinase families of sugar kinases. Protein Sci 2:31–40

Cheng G, Bennett EM, Begley TP, Ealick SE (2002) Crystal structure of 4-amino-5-hydroxymethyl-2- methylpyrimidine phosphate kinase from *Salmonella typhimurium* at 2.3 Å resolution. Structure 10:225–235

Chua TK, Seetharaman J, Kasprzak JM, Ng C, Patel BK, Love C, Bujnicki JM, Sivaraman J (2010) Crystal structure of a fructokinase homolog from *Halothermothrix orenii*. J Struct Biol 171(3):397–401

Chuvikovsky DV, Esipov RS, Skoblov YS, Chupova LA, Muravyova TI, Miroshnikov AI, Lapinjoki S, Mikhailopulo IA (2006) Ribokinase from *E. coli*: expression, purification, and substrate specificity. Bioorg Med Chem 14:6327–6332

Emsley P, Cowtan K (2004) Coot: model-building tools for molecular graphics. Acta Crystallogr D Biol Crystallogr 60:2126–2132

Evans P (2006) Scaling and assessment of data quality. Acta Crystallogr D Biol Crystallogr 62(1):72–82

Goujon M, McWilliam H, Li W, Valentin F, Squizzato S, Paern J, Lopez R (2010) A new bioinformatics analysis tools framework at EMBL-EBI. Nucleic Acids Res 38:695–699

Iida A, Harayama S, Iino T, Hazelbauer GL (1984) Molecular cloning and characterization of genes required for ribose transport and utilization in *Escherichia coli* K-12. J Bacteriol 158(2):674–682

Karasevich I, Ivoĭlov VS (1975) Preliminary metabolism of D-ribose by *Candida bombi*. Mikrobiologiia 44(2):202–205 (Russian)

Krissinel E, Henrick K (2007) Inference of macromolecular assemblies from crystalline state. J Mol Biol 372:774–797

Laskowski RA, MacArthur MW, Moss DS, Thornton JM (1993) PROCHECK—a program to check the stereochemical quality of protein structures. J Appl Crystallogr 26:283–291

Li MH, Kwok F, Chang WR, Lau CK, Zhang JP, Lo SC, Jiang T, Liang DC (2002) Crystal structure of brain pyridoxal kinase, a novel member of the ribokinase superfamily. J Biol Chem 277(48):46385–46390

Li J, Wang C, Wu Y, Wu M, Wang L, Wang Y, Zang J (2012) Crystal structure of Sa239 reveals the structural basis for the activation of ribokinase by monovalent cations. J Struct Biol 177:578–582

Lopilato JE, Garwin JL, Emr SD, Silhavy TJ, Beckwith JR (1984) D-Ribose metabolism in *Escherichia coli* K-12: genetics, regulation, and transport. J Bacteriol 158:665–673

McCoy AJ, Grosse-Kunstleve RW, Adams PD, Winn MD, Storoni LC, Read RJ (2007) Phaser crystallographic software. J Appl Crystallogr 40:658–674

McWilliam H, Li W, Uludag M, Squizzato S, Park YM, Buso N, Cowley AP, Lopez R (2013) Analysis tool web services from the EMBL-EBI. Nucleic Acids Res 41:597–600

Painter J, Merritt EA (2006) Optimal description of a protein structure in terms of multiple groups undergoing TLS motion. Acta Crystallogr D Biol Crystallogr 62:439–450

Park J, Gupta RS (2008) Adenosine kinase and ribokinase—the RK family of proteins. Cell Mol Life Sci 65:2875–2896

Paul R, Dandopath Patra M, Banerjee R, Sen U (2014) Crystallization and preliminary X-ray analysis of a ribokinase from Vibrio cholerae O395. Acta Crystallogr F Struct Biol Commun 70(8):1098–1102

Potterton E, Briggs P, Turkenburg M, Dodson E (2003) A graphical user interface to the CCP4 program suite. Acta Crystallogr D Biol Crystallogr 59:1131–1137

Sack DA, Sack RB, Nair GB, Siddique AK (2004) Cholera. Lancet 363(9404):223–233

Schimmel SD, Hoffee P, Horecker BL (1974) Deoxyribokinase from Salmonella typhimurium: purification and properties. Arch Biochem Biophys 164:560–570

Sigrell JA, Cameron AD, Jones TA, Mowbray SL (1997) Purification, characterization, and crystallization of *Escherichia coli* ribokinase. Protein Sci 6:2474–2476

Sigrell JA, Cameron AD, Jones TA, Mowbray SL (1998) Structure of *Escherichia coli* ribokinase in complex with ribose and dinucleotide determined to 1.8 Å resolution: insights into a new family of kinase structures. Structure 6:183–193

Sigrell JA, Cameron AD, Jones TA, Mowbray SL (1999) Induced fit on sugar binding activates ribokinase. J Mol Biol 290:1009–1018

Suelter CH (1970) Enzymes activated by monovalent cations. Science 168:789–795

Zhang Y, Dougherty M, Downs DM, Ealick SE (2004) Crystal structure of an aminoimidazole riboside kinase from *Salmonella enterica*: implications for the evolution of the ribokinase superfamily. Structure 12(10):1809–1821

Chapter 20
Synthetic Glycolipids and (p)ppGpp Analogs: Development of Inhibitors for Mycobacterial Growth, Biofilm and Stringent Response

Kirtimaan Syal, Krishnagopal Maiti, Kottari Naresh, Dipankar Chatterji, and N. Jayaraman

Abbreviations

(p)ppGpp	Guanosine 5′-(tri)diphosphate, 3′-diphosphate
c-di-GMP	Bis-(3′–5′)-cyclic dimeric guanosine monophosphate
CTD	C-terminal domain
DAP	Meso-2,6-diaminopimelic acid
EPS	Extracellular polymeric substance
GDP	Guanosine-5′-diphosphate
Gly	Glycine
GMP	Guanosine-5′-monophosphate
GPLs	Glycolipids
GTP	Guanosine-5′-triphosphate
KatG	Catalase-peroxidase
LAM	Lipoarabinomannan
LM	Lipomannan
mAGP	Mycolylarabinogalactan–peptidoglycan
PG	Peptidoglycan
PM	Plasma membrane

K. Syal • D. Chatterji (✉)
Molecular Biophysics Unit, Indian Institute of Science, Bangalore 560 012, India
e-mail: dipankar@mbu.iisc.ernet.in

K. Maiti • K. Naresh • N. Jayaraman (✉)
Department of Organic Chemistry, Indian Institute of Science, Bangalore 560 012, India
e-mail: jayaraman@orgchem.iisc.ernet.in

A. Chakrabarti, A. Surolia (eds.), *Biochemical Roles of Eukaryotic Cell Surface Macromolecules*, Advances in Experimental Medicine and Biology 842,
DOI 10.1007/978-3-319-11280-0_20

Introduction

Bacteria are the most abundant and one of the oldest life forms present on earth. Their unique metabolic features and time tested survival strategies allow them to flourish in hostile environments, for example, hot springs, radioactive waste, deep in earth's crust, in soil and on/in living organisms. There are 10^{14} microbial residents on human body outnumbering human cells by 10:1. Some bacterial species are beneficial to humans and play crucial role in immunity and nutrition. On the other hand, some bacteria serve as the major source of human disease and mortality. Bacterial diseases, such as, tuberculosis, diphtheria, typhoid, cholera, dysentery, pneumonia and listeriosis are detrimental diseases affecting human welfare (Roberfroid et al. 2010; Aujoulat et al. 2012; Ursell et al. 2012; Wilson 2012; Kellow et al. 2014). Due to their resistance against antibiotics, bacterial diseases have also become more difficult to treat increasingly. Bacterial pathogens responsible for respiratory diseases, gut and immumodeficiency-associated infections are among deadly and rapidly evolving family of bacteria. In current scenario, treatment options for bacterial pathogens are steadily depleting due to the evolution and horizontal spread of drug resistance. Persistence, stringent response, sporulation, cyst formation, biofilm formation, horizontal transfer of genes, rapid evolution and KatG pulse survival (Chatterji and Ojha 2001; van Schaik et al. 2007; Bassetti and Righi 2013; Wakamoto et al. 2013) are the major survival strategies implicated by bacterial systems to safeguard under difficult environments. Secondary messengers, such as, c-di-GMP and (p)ppGpp play key roles in the cell–cell communication and quorum sensing among bacteria. The most interesting aspect of quorum sensing is the coordinated response of bacteria, which confers multicellularity to the bacterial population and generates responses essential for survival of bacteria (Bharati et al. 2012). The discovery and synthesis of new antibacterials for treatment of bacterial infections have proven to be extremely slow, in contrast to the rapid evolution rate of bacterial systems (Roszak and Colwell 1987; Beceiro et al. 2013). Most antibiotics target the cellular components, such as, ribosomes and cell wall synthesis, thereby interfering bacteria in its growth phase. However, bacteria are present in several other phases. Thus there is a need to develop new antibiotics, which target 'alternate adaption strategies' of bacteria in their dormant phases. This report provides details of our study on the synthesis of a new competent antibacterial agent, targeting the stringent response and biofilm formation of bacteria.

Mycobacterial Cell Wall

The mycobacterial cell wall envelope consists of thick and waxy structure containing diverse lipids (Recht and Kolter 2001). Mycobacterial cell wall is distinct from gram-negative, as well as, gram-positive bacteria. The cell wall once stained with differential stains such as Ziehl Neelsen stain, resist the de-staining by weak acid or alcohol, thus deriving the name acid fast (Mindolli et al. 2013). It lacks outer

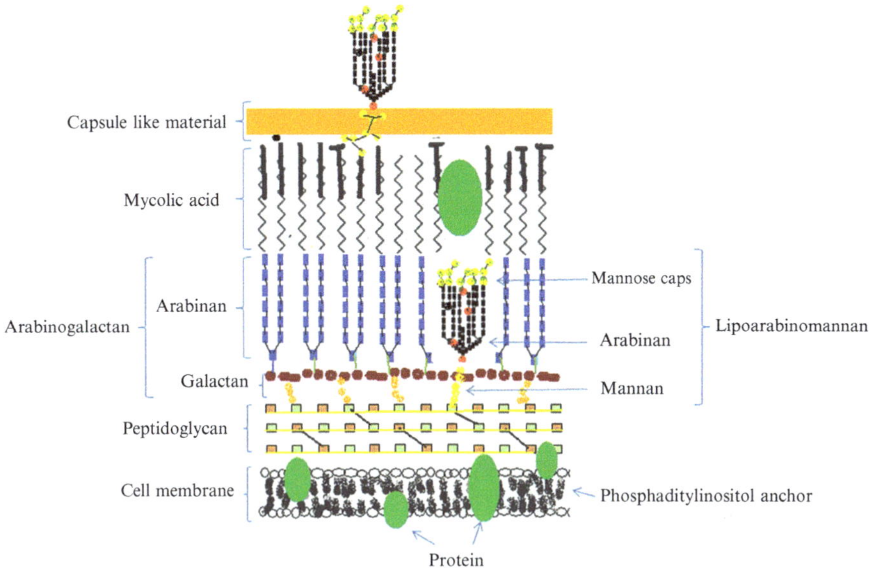

Fig. 20.1 Schematic view of cell wall components of a *Mycobacterium*

membrane similar to one present in gram negative bacteria. Mycobacterial cell wall is composed of significant amount of peptidoglycan–arabinogalactan–mycolic acid which makes the cell wall impermeable to most of the undesirable substances (Fig. 20.1). The cell envelope is compartmentalized in three major compartments: a plasma membrane (PM), similar to PM of other bacteria; the cell wall, a complex structure responsible for the rigidity and the outermost layer comprised of polysaccharides, proteins and lipids.

The cell wall consists of peptidoglycan (PG) covalently attached to arabinogalactan (AG), which is further linked to the mycolic acids. The latter complex is known as the mycolylarabinogalactan–peptidoglycan (mAGP) complex. The cell-wall proteins, the phosphatidylinositol mannosides (PIMs), the *p*-thiocerol-containing lipids and lipoarabinomannan (LAM) constitute the dynamic cell wall of *Mycobacteria* (Brennan 2003; Vats et al. 2012). Cell wall has two major components, namely, peptidoglycan and glycolipid. These are discussed in detail below.

1. The peptidoglycan (murein): It contains glycan backbone made up of disaccharide unit and a peptide moiety. Disaccharide unit consist of *N*-acetyl-D-glucosamine, linked to *N*-acetyl muramic acid by $\beta 1 \rightarrow 4$ linkage. Peptide moiety contains D- and L-alanine, D-glutamic acid and meso-2,6-diaminopimelic acid (DAP). In model organism *Mycobacterium smegmatis*, the N-terminal amino acid of peptide moiety is L-alanine, C-terminal amino acid is D-alanine and that DAP is meso-configured (Brennan 2003).
2. Glycolipid (GPLs): The peptidoglycan of the cell wall is linked to glycolipid containing mycolic acids esterified to an arabinogalactan. The mycolic acid size

varies from C_{60} to C_{90}. The glycan backbone is linked to the arabinogalactan *via* phosphodiester linkage (Brennan 2003). In *M. smegmatis*, GPLs consist of a lipopeptide core unit made of a modified C_{26}–C_{34} fatty acyl chain, which is further linked to a tetrapeptide (Phe-Thr-Ala-alaninol). The hydroxyl groups of threonine and terminal alaninol are glycosylated. A recent study showed C-5 position of the fatty acyl component as the correct position of the hydroxyl group (Vats et al. 2012). Mycolates, such as, trehalosedimycolate and phosphatidylinositolates are the major and essential components of the mycobacterial cell envelope. Mutations of these components were shown to have detrimental effects on cell structure and viability. There are other non-essential glycolipids, that do not have direct effects on cell viability, however, if mutated may change permeability to undesirable substances. Wax D, sulpholipids and cord factors are the other important glycolipids necessary for virulence of *M. tuberculosis*. The sliding motility, biofilm formation and virulence have been reported to be associated with GPLs. Glycolipids have been shown to be essential for mycobacterial survival and persistence. It is also involved in host–pathogen interactions. Their role in mycobacterial pathogenicity and their mode of action is not completely understood (Bharati et al. 2013).

Biofilm

Biofilms are matrix-enclosed microbial cumulation adhered to biotic/abiotic surfaces. The bacterial cell reorganizes itself from planktonic growth to biofilm in response to environmental changes that involves concerted signalling processes bringing regulatory networks to favour the transition. Fossil records confirm the historic presence of biofilms in bacterial lineages. Biofilm is a dynamic phase comprising of active exchanges of products/metabolites for maintenance of the living territory for the inhabitant bacteria and is a fundamental integrant of the prokaryotic life cycle, playing a central role for survival in hostile condition. The composition of the biofilm is a crucial factor which may influence the susceptibility of the resident microorganisms. Bacterial biofilm consists of various steps involving reversible and irreversible stages. Primarily, bacteria come in contact with surface under the influence of Brownian motion, gravitational forces and by surrounding hydrodynamic forces. Once the bacteria is affixed to surface, bacteria starts secreting extracellular polymeric substance (EPS) which forms the major component of biofilm to seclude itself from environment. EPS keeps the biofilm together and protect it from hostile environment (O'Toole et al. 2000; Hall-Stoodley et al. 2004; Beloin et al. 2008; Kostakioti et al. 2013). Oxygen gradients combined with nutrient controls metabolism and growth. As biofilm matures, there is a decrease in entry of new cells. Decrease in antimicrobial activity ensures due to the less permeability of the antibiotics, which makes biofilm increasingly resistant (Hall-Stoodley et al. 2004).

With maturation, the cells start dispersing in surroundings. Biofilms can disperse actively as well as passively. Passive dispersion generally happens due to shear tension and active one is initiated with stimulus from environment. Bacteria can sense environmental changes and assess the scope of residing inside the biofilm. Sustainability of bacteria to reside inside biofilm depends on various factors including nutrition and environment (O'Toole et al. 2000; Hall-Stoodley et al. 2004; Kostakioti et al. 2013).

Biofilm and Human Pathogens

Many bacterial pathogens can persist inside the host by formation of bacterial communities with biofilm-like properties. In addition to the action as barrier to obstruct antibiotic entry, biofilm extracellular matrix also facilitates the horizontal transmission of resistance genes within the biofilm community. Chances of survival are further increased by inactivation of antibiotics by high metal ion concentration and the presence of persistent bacteria. Thus, the need for new biofilm inhibitors has become indispensable (Kostakioti et al. 2013).

Pathogenic bacteria, such as, *Mycobacterium tuberculosis*, *Staphylococcus epidermidis*, *S. aureus* and the *enterococci* have been reported to benefit from biofilm formation. Capacity of a strain to cause disease is linked with prowess of the strain to form a biofilm in microtiter (Ackart et al. 2014).

In *S. epidermidis*, biofilm formation is achieved on abiotic surfaces which have been isolated with a microtiter dish-based assay. Crucial factors, mainly capsular polysaccharide and many other surface proteins help cell in binding to both biological and non-biological surfaces, detailed characterization of which remains to be elucidated (Xu and Siedlecki 2014).

Biofilms of dental microbes are well-known example of interspecies cooperation. Despite of harsh environmental conditions like nutrient availability and exposure to cleansing agents, bacteria are able to stay developed. During plaque formation, primary colonizers, mostly *Streptococci*, attached to the surfaces. These organisms are capable of interacting with both their own and the other genera. This is followed by settlement of actinomycetes in the vicinity of *Streptococci* (O'Toole et al. 2000; Jakubovics et al. 2014). Growth rate increases many fold once the threshold density of biofilm is reached. It is known that microbes secrete extracellular matrix and build up a safe niche. Molecular details of detachment are not known in any of the organisms. Though we know the role of protease is crucial, regulation, signalling and other factors are largely unknown (O'Toole et al. 2000).

Synthetic Glycolipids—Potential Inhibitors of Biofilm Formation

The arabinofurnanoside forms the crucial component of bacterial cell wall. Arabinan biosynthetic pathway is targeted by many life-line anti-tuberculosis drugs, such as, ethambutol. Enzyme arabinofuranosyltransferase adds arabinan to a growing arabinofuranoside oligomer by concomitant utilization of substrate decaprenolphosphoarabinose (DPA), the synthetic pathway of which is one of the important drug targets. For the purpose of drug discovery, analogue of the substrate, as well as, arabinofuranoside oligomers have been suggested. Many inhibitors are based on arabinofuranoside glycolipids, constituted with decaprenol alkyl chains (Bharati et al. 2013).

Novel synthetic oligoarabinomannans were developed as potential inhibitors of mycobacterial cellular assembly pathways. As stringent response inhibitors are known to control the maintenance of a pathogenic organism, such inhibitors are presumed to affect the growth, impair sliding motilities and biofilm formation. Synthetic lipomannan (LM), lipoarabinomannan (LAM) and mannose-capped lipoarabinomannan (Man-LAM) can modulate the cytokine response by binding to macrophages and dendritic cells. The structural patterns of saccharide and lipid moieties, for example, stereochemistry, acyl chain length and acyl branch patterns are proved to be important to elicit the desired biological responses (Recht and Kolter 2001; Vats et al. 2012). The following discussion pertains to the development of synthetic glycolipid inhibitors.

Arabinofurnaoside oligomer forms important integral component of both mycolylarabinogalactan-peptidoglycan complex (mAGP) and lipoarabinomannan (LAM) of the cell wall. A tight packing of the cell wall results from the presence of conformationally flexible furanoside units, in both mAGP and LAM. Tight packing, in turn, acts as a barrier to the entry of most hydrophilic drugs inside the bacterium. Thus, the arabinan structures in mycobacterial cell walls were targeted not only to delineate the functions of arabinofuranosyltransferase enzymes (Centrone and Lowary 2002; Zhang et al. 2007), but also to construct arabinofuranosyltransferase inhibitors (Rose et al. 2002; Davis et al. 2007) and highly immunogenic lipoarabinomannan oligosaccharides protein conjugates as potential TB vaccines (Hamasur et al. 1999). The particular component of LAM was studied in a greater detail, in the light of the importance that the virulence factor of the pathogen strongly depends on the LAM component, for example, association of pathogen to host receptor proteins, such as, pulmonary surfactant proteins, present in alveolar macrophages (Gaynor et al. 1995; Sidobre et al. 2000). The structure of LAM has been elucidated by Brennan and co-workers in a series of studies, as shown in Fig. 20.2. The lipidic portion in LAM retain the structural integrity of the cell wall through hydrophobic interactions and aggregate formation upon receptor binding (Sidobre et al. 2002). The repeating units of LAM are constituted by arabinose, mannose and long aliphatic chain. This repeating nature of the constituents of LAM was taken advantage in studies focused on synthetic variants of LAM. Synthetic variants provide an opportunity by which a diversity of structures, within major constituents of arabinose, mannose and lipids, could be constructed synthetically. The biological efficacies of such synthetic variants to interfere functions of a mycobacterium and conversely, to bind to a receptor protein provide important clues as to the functions of LAMs. Figure 20.3 shows example of oligoarabinofuranoside containing glycolipid synthetic variants. These variants were synthesized by chemical methods. Upon synthesis, these structural variants consisting of arabinan, mannan and lipid portions were subjected to biological evaluations, in an assay identifying the efficacies

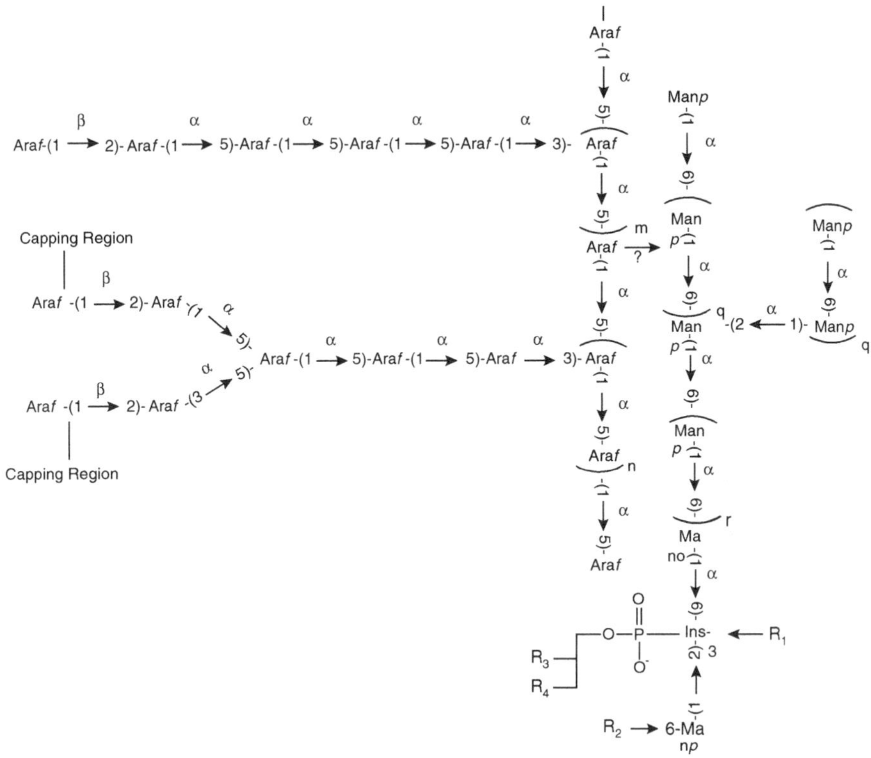

Fig. 20.2 A schematic representation of Lipoarabinomannan (Brennan 2003)

of these variants to affect the mycobacterial growth. In the event, the studies showed a significant reduction in the mycobacterial growth, for example, that of *M. smegmatis*, resulted in the presence of these variants (Bharati et al. 2013). That the reduction in the growth of the mycobacterium was specific to arabinofuranoside-containing glycolipids **1** and **2** was ascertained by replacing the sugar moieties to, for example, a maltose disaccharide. Further, the lipidic constituent in **1** and **2** was essential, as arabinofuransides without lipids alone did not elicit a reduction in the mycobacteria growth. Figure 20.3 also shows LAM synthetic variants incorporated with mannose residues along with **1** arabinofuranosidic residues. Among **3** and **4**, the tetrasaccharide glycolipid was more effective in the reduction of mycobacterial growth, illustrating that glycolipid containing more arabinofuranosidic residues are more effective in inhibiting mycobacterial growth (Naresh et al. 2008, 2010).

In order to assess the interaction, synthetic glycolipids were subjected to a surface plasmon resonance study, involving mycobacterial cell lysate as the analyte. A covalent functionalization of the arabinomannans on to a sensor surface was preferred as the method of choice, in order to assess the interaction of synthetic glycolipids with bacterial cell lystates. Covalent functionalization provides a better control of the extent of functionalization of the sensor surface with the sugar ligands.

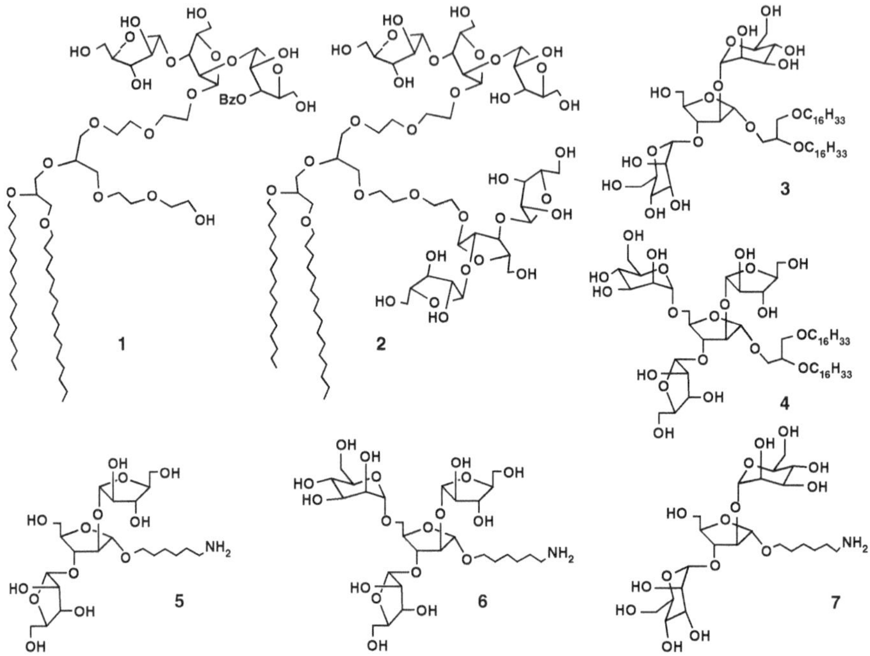

Fig. 20.3 Molecular structures of few synthetic variants of arabino- and arabinomannan glycolipids (Naresh et al. 2008)

Installation of reactive amine functionality onto the synthetic sugar ligands was accomplished by incorporating aminohexyl moiety at the reducing end of synthetic oligosaccharides, in place of the lipidic portion. Synthetic arabinomannans **5–7** (Fig. 20.3) with amine tether were then immobilized on a carboxylic acid functionalized sensor chip. Cell lysates from motile and non-motile bacteria, grown with 2 and 0.02 % glucose nutrient, respectively, were chosen as analytes. Cell lysates were extracted at two stages of the growth: one was at exponential phase of the growth, *i.e.* after 36 h of cell growth, and other at late stationary phase of the growth, *i.e.* after 88 h of growth. The cell lysates were passed for 60 s during the association phase. The dissociation phase was monitored for 60 s by passing the buffer alone over the surface. Sensorgrams of the binding of motile and non-motile strains with arabinan compound **5** are shown in Fig. 20.4(i). Cell lysates from motile and non-motile strain at exponential and stationary phases, grown with 2 % of glucose nutrient were taken for the studies. In order to identify the relative affinities between different cell lysates, uniform concentrations were used for the binding analysis. In general, it was found that the motile strain was binding stronger than the non-motile strain. In the case of non-motile strain, cell lysates from the stationary phase exhibited a higher affinity than the exponential phase, whereas for the motile strain, cell lysates from both the phases showed similar binding profile. Binding affinities of cell lysates from motile and non-motile strains, grown with a glucose concentration

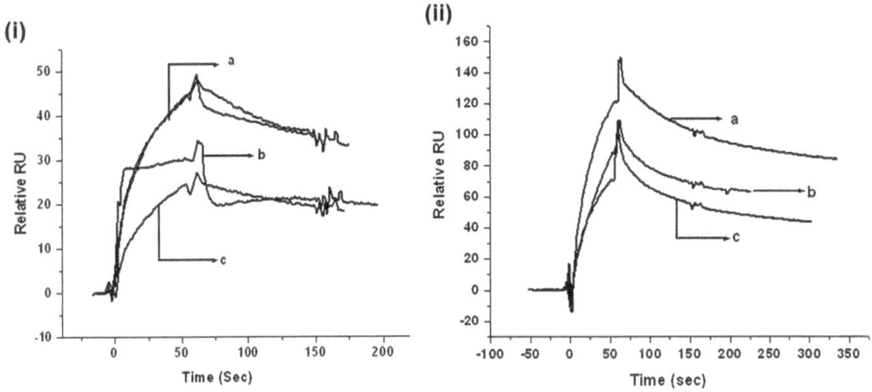

Fig. 20.4 Sensorgrams of the binding of **5** with cell lysates from motile and non-motile bacteria, with glucose (2 %), after 36 and 88 h of growth. (**i**) (*a*) Motile strain with glucose (2 %), after 36 and 88 h of growth; (*b*) non-motile strain with glucose (2 %), after 88 h of growth; (*c*) non-motile strain with glucose (2 %), after 36 h of growth. (**ii**) Sensorgrams of the binding of the cell extracts of motile strain with the glucose (2 %) after 36 h of growth: (*a*) **7**; (*b*) **6** and (*c*) **5**

of 0.02 %, with **5** was found to be low. It was shown previously that *M. smegmatis* shows significant differences in colony morphology, growth rate and cellular appearance when grown under nutrient-deprived conditions. Under nutrient-deprived conditions, *M. smegmatis* exhibits an increased resistance to acid, osmotic stress, and oxidative stress due to the GASP (growth advantage in stationary phase) response (Ojha et al. 2002; Chowdhury et al. 2010). We presume that this altered colony morphology and other factors contribute to the overall reduced effect of the synthetic glycolipids interacting with cell lysates grown under nutrient-starved conditions.

The relative affinities between arabinan and arabinomannans were also evaluated. For this purpose, cell lysate of motile strain, with 2 % glucose after 36 h of the growth was passed over the sensor surface of **5**–**7**. Sensorgrams of relative binding between cell lysates and **5**–**7** are shown in Fig. 20.4(ii). From the sensograms, the binding affinity was relatively higher with **7** containing two mannose units, than the **5** and **6** having no or one mannose unit, respectively.

In addition to α-Ara*f* linkages, Ara*f*-β-(1 → 2)-Ara*f* linkages also exist at the non-reducing end of the arabinan core of LAM and mAG, the biosynthesis of which involves enzyme, namely, AftB (Seidel et al. 2007). Synthetic variants of arabinofuranoside glycolipids, constituted with β-glycosidic linkage (Fig. 20.5), were thus prepared, through chemical glycosylation techniques. Further studies of these glycolipids in mycobacterial growth assays showed inhibitions of growth to the extent of 15–25 % in both the log and stationary phases, that were lower than the glycolipids constituted with α-glycosidic linkages (Naresh et al. 2012).

In order to increase the number sugar residues in synthetic glycolipids and their effects on mycobacterial growth, several glycolipids with oligomeric arabinofuranoside and mannopyranoside were synthesized. Few such oligomeric glycolipids are shown in Fig. 20.5. Structural variations in the oligomer arise as a result of diversity

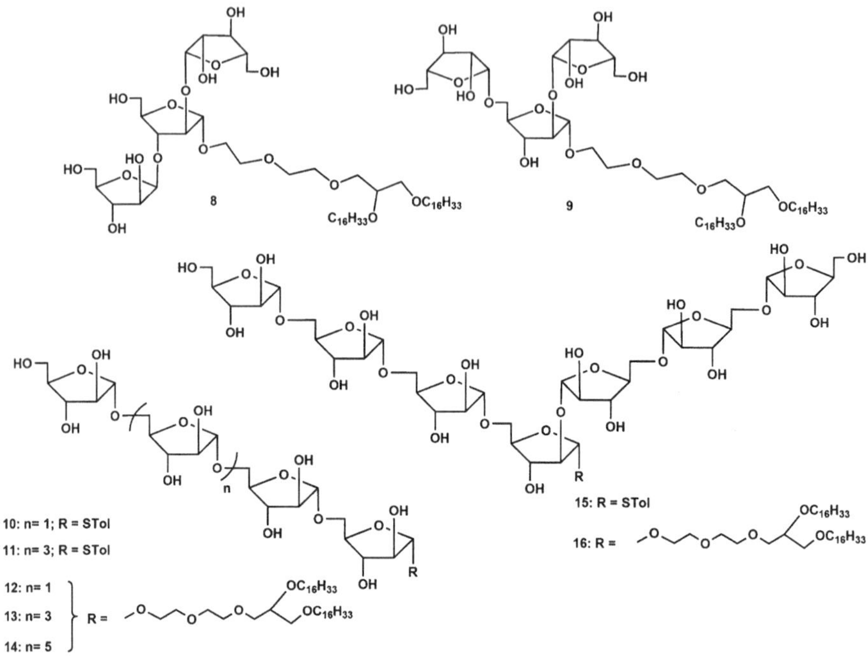

Fig. 20.5 Molecular structures of glycolipids with β-arabinofuranoside linkages and few linear and branched oligoarabinan compounds

which the chemical methods facilitate. When these glycolipids were subjected to evaluate their effects on *M. smegmatis* growth, ~45 % inhibition of mycobacterial growth was observed, at a concentration of 100 μg/mL. Whereas significant difference in the percentage inhibition efficacies was not observed between linear and branched arabinofuranosides, the linear oligosaccharide glycolipid was found to exhibit inhibition activity at a glycolipid concentration of 50 μg/mL. In these instances too, the lipidic portion in the glycolipids was essential, without which the inhibition efficacies were only marginal (Naresh et al. 2011).

Whereas above discussion pertained to the effect of synthetic glycolipids on mycobacterial growth, pertinent queries relating to the effect of such glycolipids on sessile, self-produced extracellular polymeric matrix of a mycobacteria species, namely, the biofilm, were also investigated. Biofilm acts as a protective coat to the survival of the organism against harsh environmental conditions and antibiotics. Biofilm formation is a heavily coordinated group behaviour regulated by the quorum sensing, which detects the density of other bacteria in the environment (Nadell et al. 2008). In the light of role of biofilms on the survival mechanisms of the bacteria, studies on the effect of synthetic glycolipids on biofilm formation arise further interest. Synthetic glycolipids **10–16**, when tested at concentrations of 100–200 μg/mL, were found to affect biofilm formation and maturation over a period of 7–10 days. For biofilm formation, *Mycobacterium smegmatis* mc^2155 was grown in Sauton's media containing 2 % glycerol and 0.05 % tween-80. 1 % of this primary culture was used for secondary culture on petri-plates incubated at 37 °C under humid conditions for growth of biofilms (Naresh et al. 2010). Quantitative assays showed that inhibition of biofilm by synthetic glycolipids varied, the highest of ~65 % reduction in the biofilm formation was observed with the branched heptasaccharide glycolipid, at a concentration of 100 μg/mL. In addition to inhibiting biofilm formation, the glycolipids prevented bacterial attachment on abiotic surfaces. Bacterial attachment to the abiotic surface propagates biofilm growth, as could be observed with wild-type bacterial growth. Further observation was the imminent requirement of the lipidic portion in the glycolipids, as oligosaccharide portion alone did not show a considerable effect in the phenotype of the biofilm, as in the case with control bacteria grown without the presence of synthetic glycolipids (Naresh et al. 2011).

Sliding Motility

Mycobacteria are acid fast bacteria categorized as gram-positive. Due to lack of flagella and cilia, it was thought to be non-motile and can spread across the surface of growth medium by the sliding activity. It has been proposed that *M. smegmatis* spread as a monolayer arranged in pseudo filaments. The inherent details of the cell-to-cell contactin monolayer are largely unknown (Martinez et al. 1999).

A subsequent query is how a bacterium is able to overcome its interaction with substratum and neighbouring cells, when they do not have flagella to facilitate the movement. For translocation to occur, interaction with base and neighbouring cells should be weakened. It has been suggested that GPLs play crucial role in sliding motility. *M. smegmatis* cells, that have rough colony morphology and deficient in capsular GPL, are inefficient in spreading over the surface. GPLs make the bacterial surface more hydrophobic and therefore decrease association with the agarose surface, slowing cells to spread. GPLs role in sliding motility may be explained on the basis of its ability to reduce hydrophilic interactions. Mycobacterial GPLs are reported to be the major components of the superficial layer of smooth variants of *M. smegmatis* (Martinez et al. 1999; Recht et al. 2000; Bharati et al. 2013). Synthetic glycolipids, shown in Figs. 20.3 and 20.5 were shown to inhibit the sliding motility of mycobacterial cells. When tested on a moist Petri plates imbedded with mycobacteria, synthetic glycolipids severely affected motility of the bacteria, the zone spreading did not reach the periphery of the plates for over period of several days, as opposed to control bacteria without the glycolipids, which was found to spread the plates within very few days of bacterial growth. In these instances, it was observed that the glycolipid constitution is essential, since the saccharide portion alone without the lipidic chain did not affect the sliding motility (Naresh et al. 2010, 2011). Mechanism of sliding motility is not known in detail (Recht et al. 2000; Etienne et al. 2002). For these studies, MB7H9 base medium (Difco) supplemented with 2 % glucose (solidified on petriplates) was inoculated at the centre with 10 μL cultures (OD_{600} to 0.5). Spreading was observed after incubation for 5–7 days at

37 °C in humid conditions (Naresh et al. 2010). It is presumed currently that synthetic glycolipids affect the normal profile of glycopeptidolipids or proteins involved in the biosynthesis of membrane bound glycolipids.

Growth Phases of Bacteria and Inhibitors

Most antibiotics target the cellular components such as ribosomes and cell wall. In this manner, they are able to counter act the bacteria in its growth phase primarily. The bacterial growth involves four different phases, namely, lag (preparatory), log (exponential), stationary and death phase. In lag phase, bacteria prepare for the rapid replication and acclimatize to the environment, exponential phase characterizes self-replication. Whereas in the stationary phase, where due to nutrient limitation and other factors, rate of growth and rate of cell death are almost equal. In death phase, lysis of cells is more than the replication. Most of the antibiotics target the growth phase of bacteria, however bacteria lodge in nature in several other phases too (dormant/spores/biofilm) (Bharati et al. 2013). Antimicrobial agents act on almost all the steps of central dogma that is replication, transcription and translation. However, stress induced stringent response pathways have not been thought of as a good antimicrobial target, although such an approach offers a unique possibility.

Stringent Response and Persistence

In stress conditions, all organisms initiate rescue pathway which facilitates change in pattern of gene expression and activation of signaling cascades. Mostly proteins, small polynucleotides and secondary messenger, such as, (p)ppGpp and c-di-GMP are responsible for the regulation of gene expression. (p)ppGpp, a modified nucleotide, has been discovered as a magic spot on autoradiograms derived from the extracts of *E. coli* stressed with amino acid starvation (Cashel and Gallant 1969; Kostakioti et al. 2013). It is synthesized and hydrolyzed by bi-functional Rel protein in gram positive bacteria, including *Mycobacteria* and Rel/SpoT in gram negative bacteria. In *E. coli*, RelA and SpoT have been found to play key role in production of (p)ppGpp. Further, RelA has been dissected both functionally and physically into two domains: N-terminal domain (NTD) containing the catalytic domain of RelA and the C-terminal domain (CTD) is involved in regulating RelA activity. Overexpression of CTD has been shown to negatively affect the production of (p)ppGpp (Gropp et al. 2001). In addition to amino acid starvation, the (p)ppGpp mediated stringent response has also been found in other types of nutrient, such as, glucose/fatty acid starvation and factors causing growth arrest in variety of organisms (Pal et al. 2011). The rates of synthesis and conversion of pppGpp and ppGpp in various spoT+and spoT- strains indicate that the ppGpp concentration indirectly controls the rate of (p)ppGpp synthesis (Fiil et al. 1977). In case of bifunctional RelA/SpoT homolog, conformational antagonism between the two active sites has been observed (Roszak and Colwell 1987; Hogg et al. 2004). Survival of pathogenic bacteria inside the host has been shown to be (p)ppGpp dependent. It is now known that (p)ppGpp has a substantial role in regulation of many physiological processes including transcription, translation, replication, GTP homeostasis and virulence (Jain et al. 2006a, b; Sharma and Chatterji 2010; Dalebroux and Swanson 2012). In *B subtilis*, (p)ppGpp null mutant failed to survive in normal conditions due to dysregulated GTP homeostasis. Role of stringent response in antibiotic resistance, long term persistence, survival in nutrient deprived conditions, sporulation and biofilm formation is well proved. Recently, Relacin, an analog of (p)ppGpp, has been synthesized in which di-peptide (Gly-Gly) is located at both 3′ and 5′ positions of the deoxyguanosine with anisobutyryl group at position 2 (Wexselblatt et al. 2012), which showed an inhibition of the synthesis of (p)ppGpp by RelA.

Interestingly, (p)ppGpp is ubiquitous in bacteria and absent in mammals (Wexselblatt et al. 2013). Thus, blocking (p)ppGpp synthesis would impede the survival of bacteria without having an effect on humans. However, there are certain limitations with analogs, such as, high dosage requirement, permeability across cell membranes *etc.* that required to be overcome.

GMP: Intrinsic Regulator of pppGpp Synthesis

In late stationary phase, cell faces acute shortage of energy sources, a cell has to conserve energy for survival. GTP is a key energy source required for diverse cellular processes. The deficiency of GTP has been shown to negatively regulate the transcription of rRNA and elicit signalling cascades of sporulation in *B. subtilis* (Krasny and Gourse 2004).

The starvation-inducible ((p)ppGpp) in bacteria and plants is crucial for bacterial survival, persistence, virulence and development. (p)ppGpp synthesis utilizes GTP/GDP and ATP as substrates, which results in the decrease of cellular levels of GTP (Kriel et al. 2012).

The direct regulation of cellular GTP levels by (p)ppGpp involves the control of enzymes, Gmk and HprT, that catalyze the synthesis of GMP and GDP. Therefore, it was of interest to identify whether GMP could competitively inhibit the pppGpp synthesis. We found GMP at very high concentration can inhibit pppGpp synthesis. At low concentration, it does not have any effect. Michelis Menten kinetics for pppGpp synthesis by Rel_{MSM} (Full length Rel enzyme in *M. smegmatis*) in the presence of 100 μM GMP has been followed to understand the level of inhibition (Fig. 20.6). Substrate concentration was varied from 0 to 2,000 μM for the kinetics study. It is possible that in acute conditions, GMP may inhibit the pppGpp synthesis so as to conserve the energy currency.

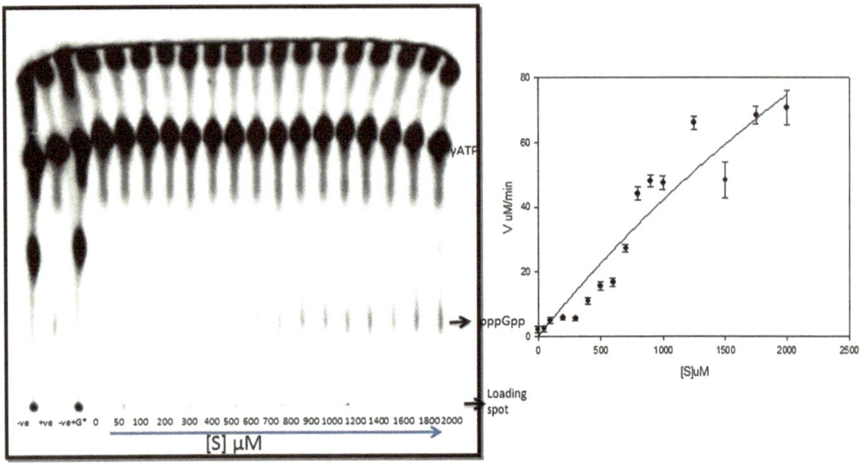

*100µM GMP

-ve: Negative control (Reaction mix without enzyme)
+ve: Positive control (Reaction mix with enzyme)
-ve+G: Negative control + 100µM GMP
0-2000: Varying substrate concentration in presence of 100µM GMP

	Km (µM)	Vmax (µM/min)
With 100µM GMP	890 ± 69.5	75 ± 5.9
Rel_{MSM} control	584 ± 231.9	94.3 ± 14.7

Fig. 20.6 Michelis Menten kinetics of pppGpp synthesis by Rel_{MSM} in presence of 100 µM GMP

Cyclic-di-GMP

C-di-GMP is a ubiquitous secondary messenger that regulates essential biological processes, such as, dormancy, biofilm formation, quorum sensing and virulence in bacteria. It is synthesized by diguanylatecyclases (DGC) and degraded by phosphodiesterases (PDE). C-di-GMP synthesis and hydrolyzing enzymes are wide spread in bacteria. Recently, it has been shown that c-di-GMP helps *M. tuberculosis* in coming out of dormancy and initiates the signalling cascades necessary for active metabolism and virulence (Hong et al. 2013). Both secondary messenger, (p)ppGpp and c-diGMP are involved in stress response and thus it was of interest to follow whether there is any cross talk between the two pathways. We checked the binding of cyclic-di-GMP with Rel_{MSM} enzyme and also followed the activity in presence of c-di-GMP. Enzyme purification was done as per the referred protocol (Fig. 20.7b) (Jain et al. 2006b). Further, we confirmed the identity of the protein by ingel trypsin digestion and mass spectrometry analysis. We found that the binding of c-di-GMP is weak, although specific as determined by isothermal titration calorimetry (Fig. 20.7a and c). Michelis Menten kinetics for pppGpp synthesis in presence of 100 µM c-di-GMP has been followed to understand the level of inhibition. The kinetics was followed for range of substrate concentration (0–2,000 µM) with purified enzyme (Fig. 20.8). pppGpp activity was inhibited in presence of c-di-GMP at high concentrations. Such a high concentration is unrealistic in *in vivo* conditions,

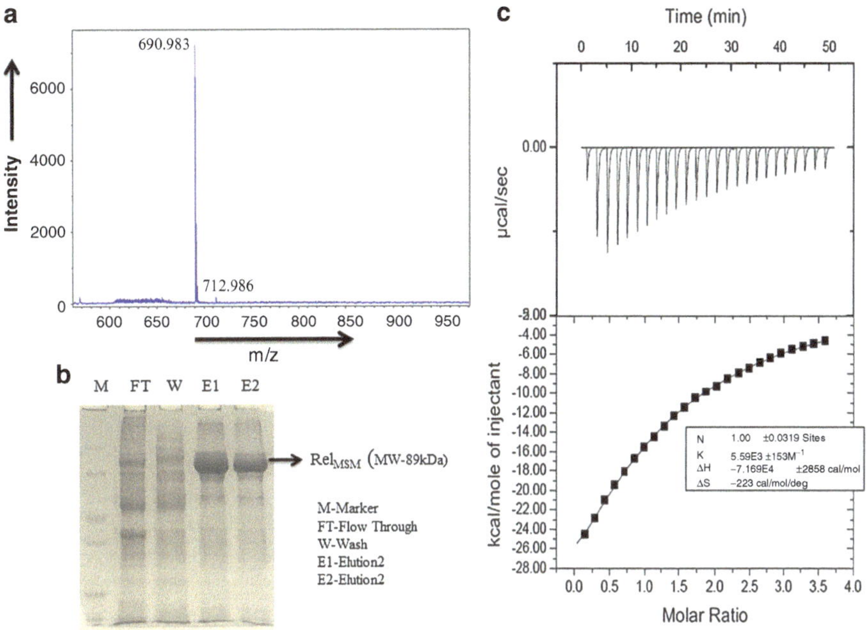

Fig. 20.7 (**a**) Mass spectrum of purified c-di-GMP. (**b**) SDS-Polyacrylamide gel electrophoresis of fractions collected in Ni-NTA chromatography for purification of His-tagged Rel_{MSM}. (**c**) Isothermal titration calorimetry analysis of c-di-GMP binding with Rel_{MSM} (K_D = 178 mM)

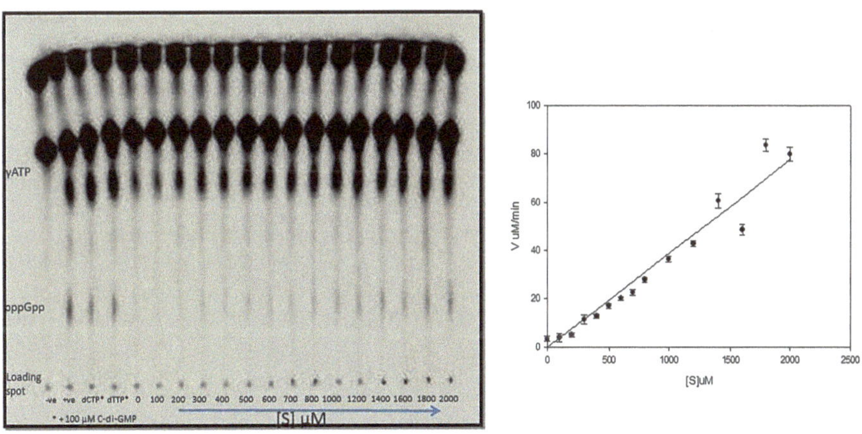

-ve: Negative control (Reaction mix without enzyme)
+ve: Positive control (Reaction mix with enzyme)
dCTP: +ve with dCTP and 100µM c-di-GMP
dTTP: +ve with dCTP and 100µM c-di-GMP
0-2000: Varying substrate concentration in presence of 100µM c-di-GMP

	Km (µM)	Vmax (µM/min)
Rel_{MSM}	584±231.9	94.2±14.7
Rel_{MSM}+c-di-GMP (100µM)	1030±63.8	78.5±4.9

Fig. 20.8 Michelis Menten kinetics of pppGpp synthesis by Rel_{MSM} in presence of 100 µM c-di-GMP. dCTP and dTTP are the controls confirming specific inhibition by c-di-GMP

although a local high concentration could be achieved under a microenvironment. We predict that c-di-GMP may play a significant role in the regulation of pppGpp synthesis and stringent response. pppGpp and c-di-GMP participate in many common metabolic processes such as long term survival and biofilm formation. More investigation is awaited to uncover common features of both the secondary messengers.

Conclusion

There is an ever-increasing demand for identifying new anti-bacterial in order to overcome multiple resistant bacteria. Glycolipids and pppGpp analogs hold promise to target putative mycobacterial cell wall structures for antimicrobial therapies. Specifically, pppGpp analogs can interfere with the activation of the stringent response. For example, (p)ppGpp analogs block *B. subtilis* sporulation, proving the importance of the stringent response. (p)ppGpp analogs were also shown to impede biofilm formation and long term persistence, thus negatively affecting *in vivo* survivability. Synthetic glycolipids act as inhibitors of mycobacterial growth, sliding motility and biofilm formation. However, their nanomolar inhibition concentrations remain to be achieved. Rupturing biofilms and reducing the sliding motilities of growing bacteria by synthetic glycolipids provide additional approaches to the on-going search for newer antimicrobial agents. Issues surrounding the thick, waxy mycobacterial cell wall structures will continue to be the focus in manifold approaches to mitigate detrimental effects of mycobacterial pathogens.

References

Ackart DF, Hascall-Dove L, Caceres SM, Kirk NM, Podell BK, Melander C, Orme IM, Leid JG, Nick JA, Basaraba RJ (2014) Expression of antimicrobial drug tolerance by attached communities of *Mycobacterium tuberculosis*. Pathog Dis 70:359–369

Aujoulat F, Roger F, Bourdier A, Lotthe A, Lamy B, Marchandin H, Jumas-Bilak E (2012) From environment to man: genome evolution and adaptation of human opportunistic bacterial pathogens. Genes (Basel) 3:191–232

Bassetti M, Righi E (2013) Multidrug-resistant bacteria: what is the threat? Hematology Am Soc Hematol Educ Program 2013:428–432

Beceiro A, Tomas M, Bou G (2013) Antimicrobial resistance and virulence: a successful or deleterious association in the bacterial world? Clin Microbiol Rev 26:185–230

Beloin C, Roux A, Ghigo JM (2008) *Escherichia coli* biofilms. Curr Top Microbiol Immunol 322:249–289

Bharati BK, Sharma IM, Kasetty S, Kumar M, Mukherjee R, Chatterji D (2012) A full-length bifunctional protein involved in c-di-GMP turnover is required for long-term survival under nutrient starvation in *Mycobacterium smegmatis*. Microbiology 158:1415–1427

Bharati BK, Naresh K, Chatterji D, Jayaraman N (2013) Synthetic arabinan, arabinomannan glycolipids and their effects on mycobacterial growth, sliding motility and biofilm formation. In: Rauter AP, Lindhorst TK (eds) Carbohydrate chemistry, vol 39. The Royal Society of Chemistry, Cambridge, UK, p 58–77

Brennan PJ (2003) Structure, function, and biogenesis of the cell wall of *Mycobacterium tuberculosis*. Tuberculosis (Edinb) 83:91–97

Cashel M, Gallant J (1969) Two compounds implicated in the function of the RC gene of *Escherichia coli*. Nature 221:838–841

Centrone CA, Lowary TL (2002) Synthesis and antituberculosis activity of C-phosphonate analogues of decaprenolphosphoarabinose, a key intermediate in the biosynthesis of mycobacterial arabinogalactan and lipoarabinomannan. J Org Chem 67:8862–8870

Chatterji D, Ojha AK (2001) Revisiting the stringent response, ppGpp and starvation signaling. Curr Opin Microbiol 4:160–165

Chowdhury RP, Saraswathi R, Chatterji D (2010) Mycobacterial stress regulation: the Dps "twin sister" defense mechanism and structure–function relationship. IUBMB Life 62:67–77

Dalebroux ZD, Swanson MS (2012) ppGpp: magic beyond RNA polymerase. Nat Rev Microbiol 10:203–212

Davis CB, Hartnell RD, Madge PD, Owen DJ, Thomson RJ, Chong AK, Coppel RL, von Itzstein M (2007) Synthesis and biological evaluation of galactofuranosyl alkyl thioglycosides as inhibitors of mycobacteria. Carbohydr Res 342:1773–1780

Etienne G, Villeneuve C, Billman-Jacobe H, Astarie-Dequeker C, Dupont MA, Daffé M (2002) The impact of the absence of glycopeptidolipids on the ultrastructure, cell surface and cell wall properties, and phagocytosis of Mycobacterium smegmatis. Microbiology 148:3089–3100

Fiil NP, Willumsen BM, Friesen JD, von Meyenburg K (1977) Interaction of alleles of the relA, relC and spoT genes in Escherichia coli: analysis of the interconversion of GTP, ppGpp and pppGpp. Mol Gen Genet 150:87–101

Gaynor CD, McCormack FX, Voelker DR, McGowan SE, Schlesinger LS (1995) Pulmonary surfactant protein A mediates enhanced phagocytosis of *Mycobacterium tuberculosis* by a direct interaction with human macrophages. J Immunol 155:5343–5351

Gropp M, Strausz Y, Gross M, Glaser G (2001) Regulation of Escherichia coli RelA requires oligomerization of the C-terminal domain. J Bacteriol 183:570–579

Hall-Stoodley L, Costerton JW, Stoodley P (2004) Bacterial biofilms: from the natural environment to infectious diseases. Nat Rev Microbiol 2:95–108

Hamasur B, Kallenius G, Svenson SB (1999) Synthesis and immunologic characterisation of *Mycobacterium tuberculosis* lipoarabinomannan specific oligosaccharide–protein conjugates. Vaccine 17:2853–2861

Hogg T, Mechold U, Malke H, Cashel M, Hilgenfeld R (2004) Conformational antagonism between opposing active sites in a bifunctional RelA/SpoT homolog modulates (p)ppGpp metabolism during the stringent response [corrected]. Cell 117:57–68

Hong Y, Zhou X, Fang H, Yu D, Li C, Sun B (2013) Cyclic di-GMP mediates *Mycobacterium tuberculosis* dormancy and pathogenecity. Tuberculosis (Edinb) 93:625–634

Jain V, Kumar M, Chatterji D (2006a) ppGpp: stringent response and survival. J Microbiol 44:1–10

Jain V, Saleem-Batcha R, China A, Chatterji D (2006b) Molecular dissection of the mycobacterial stringent response protein Rel. Protein Sci 15:1449–1464

Jakubovics NS, Yassin SA, Rickard AH (2014) Community interactions of oral streptococci. Adv Appl Microbiol 87:43–110

Kellow NJ, Coughlan MT, Reid CM (2014) Metabolic benefits of dietary prebiotics in human subjects: a systematic review of randomised controlled trials. Br J Nutr 111:1147–1161

Kostakioti M, Hadjifrangiskou M, Hultgren SJ (2013) Bacterial biofilms: development, dispersal, and therapeutic strategies in the dawn of the postantibiotic era. Cold Spring Harb Perspect Med 3:a010306

Krasny L, Gourse RL (2004) An alternative strategy for bacterial ribosome synthesis: Bacillus subtilis rRNA transcription regulation. EMBO J 23:4473–4483

Kriel A, Bittner AN, Kim SH, Liu K, Tehranchi AK, Zou WY, Rendon S, Chen R, Tu BP, Wang JD (2012) Direct regulation of GTP homeostasis by (p)ppGpp: a critical component of viability and stress resistance. Mol Cell 48:231–241

Martinez A, Torello S, Kolter R (1999) Sliding motility in mycobacteria. J Bacteriol 181: 7331–7338

Mindolli PB, Salmani MP, Parandekar PK (2013) Improved diagnosis of pulmonary tuberculosis using bleach microscopy method. J Clin Diagn Res 7:1336–1338
Nadell CD, Xavier JB, Levin SA, Foster KR (2008) The evolution of quorum sensing in bacterial biofilms. PLoS Biol 6:e14
Naresh K, Bharati BK, Jayaraman N, Chatterji D (2008) Synthesis and mycobacterial growth inhibition activities of bivalent and monovalent arabinofuranoside containing alkyl glycosides. Org Biomol Chem 6:2388–2393
Naresh K, Bharati BK, Avaji PG, Jayaraman N, Chatterji D (2010) Synthetic arabinomannan glycolipids and their effects on growth and motility of the *Mycobacterium smegmatis*. Org Biomol Chem 8:592–599
Naresh K, Bharati BK, Avaji PG, Chatterji D, Jayaraman N (2011) Synthesis, biological studies of linear and branched arabinofuranoside-containing glycolipids and their interaction with surfactant protein A. Glycobiology 21:1237–1254
Naresh K, Avaji PG, Maiti K, Bharati BK, Syal K, Chatterji D, Jayaraman N (2012) Synthesis of beta-arabinofuranoside glycolipids, studies of their binding to surfactant protein-A and effect on sliding motilities of *M. smegmatis*. Glycoconj J 29:107–118
O'Toole G, Kaplan HB, Kolter R (2000) Biofilm formation as microbial development. Annu Rev Microbiol 54:49–79
Ojha AK, Varma S, Chatterji D (2002) Synthesis of an unusual polar glycopeptidolipid in glucose-limited culture of Mycobacterium smegmatis. Microbiology 148:3039–3048
Pal RR, Das B, Dasgupta S, Bhadra RK (2011) Genetic components of stringent response in Vibrio cholerae. Indian J Med Res 133:212–217
Recht J, Kolter R (2001) Glycopeptidolipid acetylation affects sliding motility and biofilm formation in *Mycobacterium smegmatis*. J Bacteriol 183:5718–5724
Recht J, Martinez A, Torello S, Kolter R (2000) Genetic analysis of sliding motility in *Mycobacterium smegmatis*. J Bacteriol 182:4348–4351
Roberfroid M, Gibson GR, Hoyles L, McCartney AL, Rastall R, Rowland I, Wolvers D, Watzl B, Szajewska H, Stahl B, Guarner F, Respondek F, Whelan K, Coxam V, Davicco MJ, Leotoing L, Wittrant Y, Delzenne NM, Cani PD, Neyrinck AM, Meheust A (2010) Prebiotic effects: metabolic and health benefits. Br J Nutr 104(Suppl 2):S1–S63
Rose JD, Maddry JA, Comber RN, Suling WJ, Wilson LN, Reynolds RC (2002) Synthesis and biological evaluation of trehalose analogs as potential inhibitors of mycobacterial cell wall biosynthesis. Carbohydr Res 337:105–120
Roszak DB, Colwell RR (1987) Survival strategies of bacteria in the natural environment. Microbiol Rev 51:365–379
Seidel M, Alderwick LJ, Birch HL, Sahm H, Eggeling L, Besra GS (2007) J Biol Chem 282:14729
Sharma UK, Chatterji D (2010) Transcriptional switching in *Escherichia coli* during stress and starvation by modulation of sigma activity. FEMS Microbiol Rev 34:646–657
Sidobre S, Nigou J, Puzo G, Riviere M (2000) Lipoglycans are putative ligands for the human pulmonary surfactant protein A attachment to mycobacteria. Critical role of the lipids for lectin-carbohydrate recognition. J Biol Chem 275:2415–2422
Sidobre S, Puzo G, Riviere M (2002) Lipid-restricted recognition of mycobacterial lipoglycans by human pulmonary surfactant protein A: a surface-plasmon-resonance study. Biochem J 365:89–97
Ursell LK, Clemente JC, Rideout JR, Gevers D, Caporaso JG, Knight R (2012) The interpersonal and intrapersonal diversity of human-associated microbiota in key body sites. J Allergy Clin Immunol 129:1204–1208
van Schaik W, Prigent J, Fouet A (2007) The stringent response of Bacillus anthracis contributes to sporulation but not to virulence. Microbiology 153:4234–4239
Vats A, Singh AK, Mukherjee R, Chopra T, Ravindran MS, Mohanty D, Chatterji D, Reyrat JM, Gokhale RS (2012) Retrobiosynthetic approach delineates the biosynthetic pathway and the structure of the acyl chain of mycobacterial glycopeptidolipids. J Biol Chem 287:30677–30687
Wakamoto Y, Dhar N, Chait R, Schneider K, Signorino-Gelo F, Leibler S, McKinney JD (2013) Dynamic persistence of antibiotic-stressed mycobacteria. Science 339:91–95

Wexselblatt E, Oppenheimer-Shaanan Y, Kaspy I, London N, Schueler-Furman O, Yavin E, Glaser G, Katzhendler J, Ben-Yehuda S (2012) Relacin, a novel antibacterial agent targeting the Stringent Response. PLoS Pathog 8:e1002925

Wexselblatt E, Kaspy I, Glaser G, Katzhendler J, Yavin E (2013) Design, synthesis and structure–activity relationship of novel Relacin analogs as inhibitors of Rel proteins. Eur J Med Chem 70:497–504

Wilson DJ (2012) Insights from genomics into bacterial pathogen populations. PLoS Pathog 8:e1002874

Xu LC, Siedlecki CA (2014) Staphylococcus epidermidis adhesion on hydrophobic and hydrophilic textured biomaterial surfaces. Biomed Mater 9:035003

Zhang J, Khoo KH, Wu SW, Chatterjee D (2007) Characterization of a distinct arabinofuranosyltransferase in *Mycobacterium smegmatis*. J Am Chem Soc 129:9650–9662

Chapter 21
Regulations of Glycolipid: XI. Glycosyltransferase (GSL: GLTs) Genes Involved in SA-LeX and Related GSLs Biosynthesis in Carcinoma Cells by Biosimilar Apoptotic Agents: Potential Anticancer Drugs

Subhash Basu, Rui Ma, Joseph R. Moskal, and Manju Basu

Introduction

The term "biosimilar" refers to products marketed after expiration of patents and claimed to have similar properties to the existing biologic products (Bull and Taylor 2014; Abraham 2013; Lopes et al. 2013; Zajdel and Zajdel 2013). Our laboratory is looking for the potential anticancer drugs or agents to kill breast and colon cancers by initiating apoptosis in these cells (Basu et al. 2004a, b). The degree of apoptosis was evaluated by the fluorescence microscopic studies using PSS-380 dye binding to the phosphatidyl serine excluded on the outer leaflet from the inside of the cells (Ma et al. 2004). Various glycolipid: glycosyltransferases (GLTs) were modulated differently during induced apoptosis (Basu et al. 2004a, c, 2012a, b; Ma et al. 2004, 2009; Ma 2008; Boyle 2005; Boyle et al. 2006). However, different agents could not use these parameters for biosimilar comparison because of the unknown mechanisms of gene regulations. Simple DNA laddering experiments after treatment with biosimilar drugs could be used for quick comparisons of these drugs. In addition to the above agents, the apoptotic killing-effect of at least three common disialosylgangliosides (GD3, GD1a, and GD1b) was also compared (Ma et al. 2004;

S. Basu (✉) • M. Basu
Department of Chemistry and Biochemistry and Cancer Drug Delivery Research Foundation, University of Notre Dame, Notre Dame, IN 46556, USA
e-mail: sbasu@nd.edu

R. Ma
Diagnostic Division, Siemens Corporation, Shanghai, China

J.R. Moskal
The Falk Center for Molecular Therapeutics, Northwestern University, Evanston, IL 60201, USA

A. Chakrabarti, A. Surolia (eds.), *Biochemical Roles of Eukaryotic Cell Surface Macromolecules*, Advances in Experimental Medicine and Biology 842,
DOI 10.1007/978-3-319-11280-0_21

Basu et al. 2004c; Ma 2008) for biosimilar effect. On the other hand one of these biosimilar agents (cis-platin) inactivated both DNA polymerase-alpha and Helicase-III in breast carcinoma cells (Boyle 2005; Boyle et al. 2006).

Biosimilar Drugs in Cancer Chemotherapy

Biosimilar agents offer prospect of providing efficient and safe treatment option for many diseases, including cancers (1). The treatment of many diseases, particularly cancer, has been highly impacted by the introduction of biologics (biological therapies) (1). These biologics are large molecular weight structurally complex proteins that are produced by complex manufacturing processes (2). The regulatory evaluations of biosimilar biologics in the clinical trial are sometimes confusing (3). While the goal of developments to demonstrate that the biosimilar product is highly similar to the reference biologic product, Biosimilars should not be considered as "generic" biologics (4) generic drugs are small chemical compounds that are identical to the patent-expired "reference" small-molecule drugs. On the other hand use of relatively small molecular weight apoptotic agents gives us opportunity to treat cancer patients by relatively simple compounds and almost similar biological regulation and gene control (Ma et al. 2011; Basu et al. 2012a, b).

Lewis Antigen Epitopes on Cancer Cell Surfaces

Several *in vivo* studies suggested that tumor metastasis depends on the expression carbohydrate Lewis structures (Table 21.1). Lewis antigens and their derivatives such as LewisX (LeX), Sialyl LewisX (SA-LeX), Sialyl Lewis a (SA-Lea), Sialyl Leb (SA-Leb); Lewis Y (LeY), were identified as tumor-associated antigens by Koprowski and associates (Fukushi et al. 1984a) almost three decades ago. All these epitopes are expressed as glycosphingolipids (Fukushi et al. 1984a, b) as well as glycoproteins (Ashizawa et al. 2003; Silva et al. 2011; Zhang et al. 2000, 2002; Matsumoto et al. 2002; Mitoma et al. 2003; Schuldes et al. 2003; Fuster et al. 2003; Kashiwagi et al. 2004; Kannagi 2004). Both LeX and SA-LeX are ligands for

Table 21.1 Structures of type-2 chain containing fuco-glycolipids

Name	Structures of Lewis antigen epitopes
nLcOse4Cer(nLc4) (Type-2 chain)	Galb1-4GlcNAcb1-3Galb1-4Glc-Cer
LeX (Type-2 chain)	Galb1-4(Fuca1-3)GlcNAcb1-3Galb1-4Glc-R
SA-LeX (Tyoe-2)	NeuAca2-3Galb1-4(Fuca1-3)GlcNAcb1-3Galb1-4Glc-R
SA-Lea(Type-1 chain)	NeuAca2-3Galb1-3(Fuca1-3)GlcNAcb1-3Galb1-4Glc-R
LeY (Type-2 chain)	Fuca1-2Galb1-4(Fuca1-3)GlcNAcb1-3Galb1-4Glc-R

Mono-, di-, and tri-fucosyl alpha1-3 type-2 structures are also present in human embryonic cells and cancer cells; R-ceramide or glycoproteins

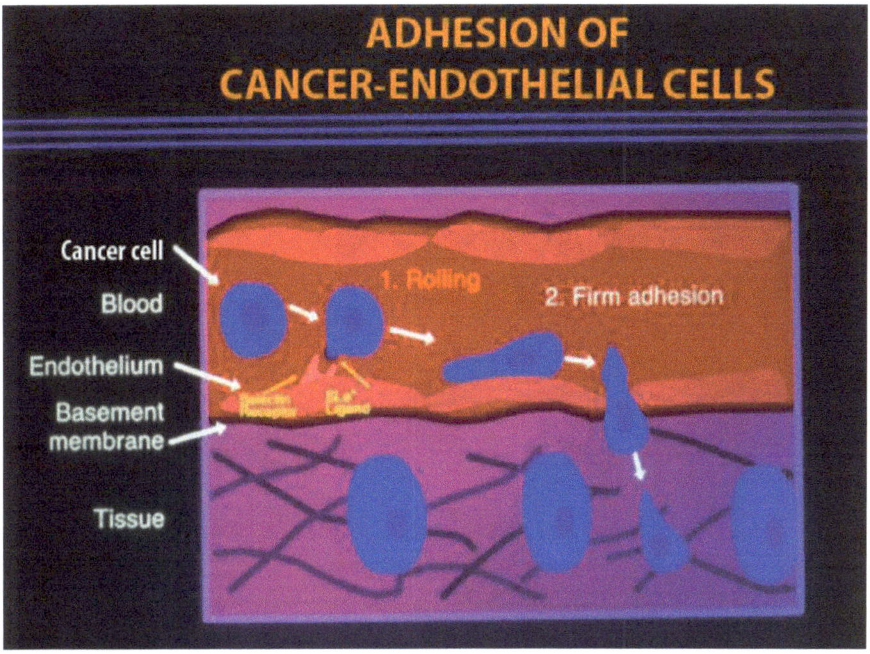

Fig. 21.1 Hypothesis of adhesion of cancer-endothelial cells (Lewis antigen vs selectin binding)

E- and P-selectin binding, respectively. Both E- and P-selectins are vascular receptors and are expressed on endothelial cells and activated on endothelial cells. L-lectins are expressed on leucocytes. Researchers have proposed that metastatic migration of cancer cells is almost similar to leucocytes migration using interaction between selectin receptor and ligand interaction (Fig. 21.1). Cancer cell surface Lewis antigens are believed to be involved in the *in vivo* metastatic propagation in different organs. Polyfucosyl-Type-2 LeX and LeY families of glycolipids have been established by Hakomori and his associates (Fukushi et al. 1984a). Research on biosynthesis (Basu and Basu 1972, 1973; Higashi et al. 1985; Basu et al. 1991, 1999, 2000; Holmes et al. 1985, 1986 and regulation (Ma et al. 2009, 2011; Basu et al. 2012a, b; Radhakrishnan et al. 2007; Julien et al. 2007; Barthel et al. 2008; Li et al. 2010a; Sugiarto et al. 2011; Chachadi et al. 2011; Shirue et al. 2011) of LeX, SA-LeX biosynthesis in cancer cells is the exploding field at the present time.

Induction of Apoptosis in Cancer Cells

Apoptosis is a naturally occurring process by which normal cells are directed to programmed death activating several signaling pathways by some inducer molecules from outside. It is a distinctive mode of cell death (through self-destruction) with characteristic changes in morphologic features, which is regulating the size of animal tissues. In contrast, the process of necrosis in cancer cells is a progressive

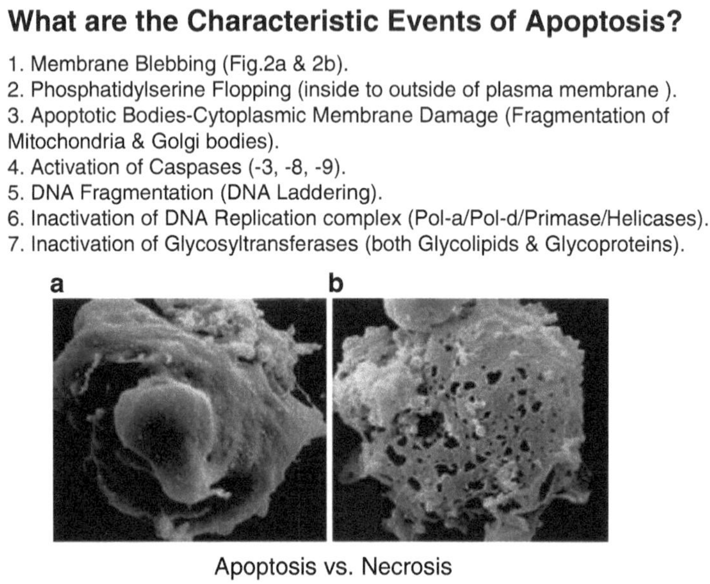

Fig. 21.2 The characteristic events during apoptosis

disintegration of cells, which affects nearby viable cells to disintegrate. Viable cells discriminate apoptosis and necrosis targets via distinct cell surface receptors (Fig. 21.2) (Weedon et al. 1979; Kostrzewa 2000; Blagosklonny 2000; Patel et al. 2009). Unlike necrosis, apoptotic cell death is less damaging (Fig. 21.2) in the patients carrying highly metastatic breast carcinomas. Studies of induction (or initiation) and regulation of apoptotic cascades (Fig. 21.3) in breast and colon cancer cells are of prime interest in anticancer drug discovery. If we search this topic in the PubMed, we will obtain at least 11,000 reports regarding apoptosis in breast cancer cells. Induction of apoptosis can be classified into four categories: (1) internal mitochondrial-caspase activation pathway (IMCAP), (2) external Bad-receptor activating pathway (EBRAP), (3) NFkappaB activation pathway (NFkBP), and (4) cascades of protein kinase activation Pathways (CPKAP). Using highly metastatic breast carcinoma cells (SKBR-3, MDA-468, and MCF-7) (Fig. 21.4a), we have reported in recent years (Basu et al. 2004a, b; Ma et al. 2004; Ma 2008; Boyle 2005; Boyle et al. 2006) apoptotic induction by D-PPMP, D-PDMP (inhibitors of GSL biosynthesis), *cis*-platin (inhibitor of DNA biosynthesis), Betulinic acid (a triterpinoid used as an anticancer agent in China as an alternative medicine), Tamoxifen (a common anticancer agent used for the treatment of breast cancers today), GD3, GD1b, and Melphalan (a scrambler for Golgi bodies) in the range of 2–16 μM concentrations (Fig. 21.4b) (Basu et al. 2004b, c; Ma et al. 2004; Ma 2008; Boyle 2005; Boyle et al. 2006). Inhibition of cell growth (IC-50) was different with three different metastatic breast cancer cell lines when tested with a single apoptotic agent (L-PPMP) for 24 h (Boyle 2005; Boyle et al. 2006). All these chemicals (Fig. 21.4b) induce apoptosis by activating IMCAP (Fig. 21.3). On the other hand, *cis*-platin (used in the treatment of testicular cancers) induced apoptosis (Fig. 21.5a) occurs in

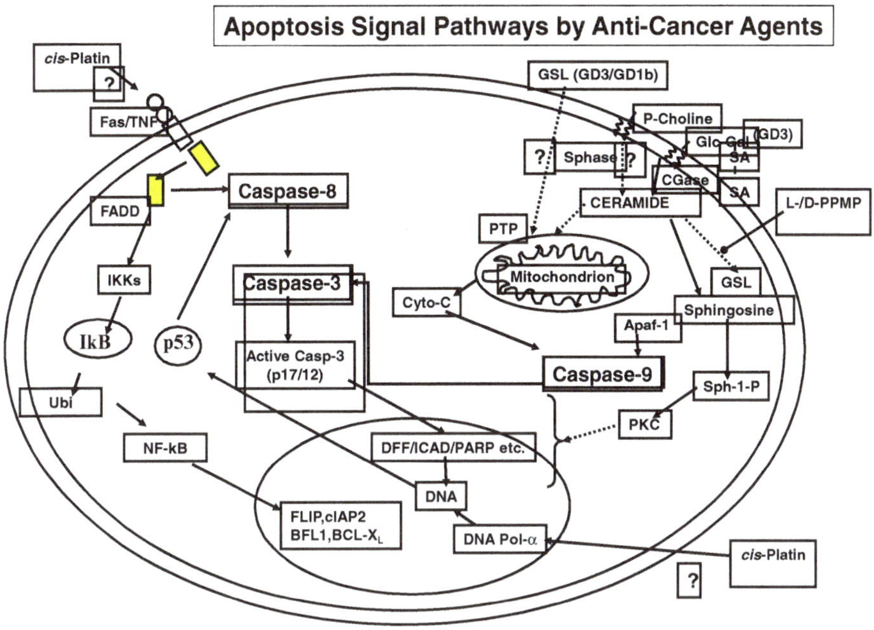

Fig. 21.3 Apoptosis signal pathways induced by anti-cancer agents

those cell lines at much higher concentrations (50–150 mM) (Boyle 2005; Boyle et al. 2006) by EBRAP followed by activation of caspase-8 (Boyle et al. 2006). Several new chemicals have been tested in recent years for induction of the apoptotic process by activating IMCAP (internal mitochondrial caspase activating pathway), NF-kappaB (NFKBAP), EBRAP (external Bad-receptor activating pathway), or CPKAP pathway (Yuan et al. 2011; Leung et al. 2011; Ullah et al. 2011; Marchetti et al. 2011; Kim et al. 2010; Laezza et al. 2010; Wesierska et al. 2011a; Chou et al. 2010; Zhang et al. 2010; Patil et al. 2010; Banerjee et al. 2010; Shirure et al. 2011; Cazet et al. 2010). In addition to five different compounds (Fig. 21.5) we tried to induce apoptosis in three different breast carcinoma cells (SKBR-3, MDA-468, and MCF-7) with at least six different brain gangliosides (GM3, GM2, GM1a, GDia, GD1b, and GD3). Only disialogangliosides containing NeuAc (alpha2,8)NeuAc (a2, 3)-linked di- or tetraglycosyl gangliosides (GD3 and GD1b) activated Casoase-3 (Ma et al. 2004; Ma 2008).

Induction of Apoptosis in Breast and Colon Carcinoma Cells by Simple Apoptotic Agents

Unlike necrosis, apoptotic cell death is less damaging (Fig. 21.2; Scanning electron microscopic pictures were obtained from internet display) in the patients carrying highly metastatic breast carcinomas. Characteristic events during apoptotic cell

a

MCF-7, MDA-468, and SKBR-3

Derived from pleural effusion of breast carcinoma

	Cells	ER	PgR	p53	Caspase-3
MCF-7	Normal	+	+	Normal	Mutant
MDA-468	Normal	-	-	Mutant	Normal
SKBR-3	High	-	-	Mutant	Normal

- HER2 (c-erbB-2/neu) protein

  HER2 comes from a proto-oncogene encoding a transmembrane glycoprotein of 185 kDa (p185(HER2)) with intrinsic tyrosine kinase activity. HER2 gives the cells different responsiveness to anti-cancer drugs versus HER2 negative breast cancers cells

- Estrogen receptor (ER) and progesterone receptor (PgR)

b

Structures of Anti-Cancer Apoptotic Agents

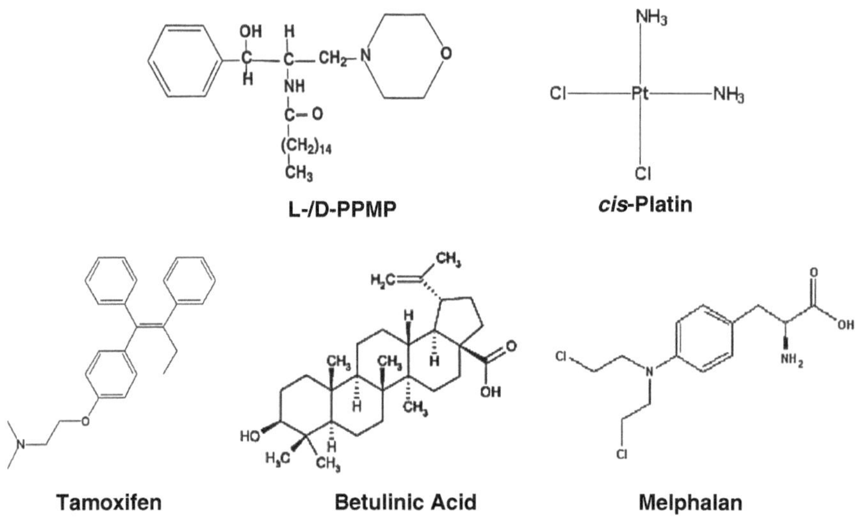

Fig. 21.4 (**a**) Expression of different receptors on three different metastatic carcinoma cells (MCF-7, MDA-468, and SKBD-3) derived from pleural effusion of breast carcinomas. (**b**) Structures of anti-cancer apoptotic agents (Tested in Basu Laboratory)

death could be monitored (Fig. 21.2) by: (1) membrane blebbing (by scanning electron microscopy); (2) flopping out of phosphatidyl serine on the outer layer of the plasma membrane (by fluorescent dye binding microscopy); (3) fragmentation of mitochondrial and other inner membranes; (4) activation of Caspases (-3, -6, -9 etc.), by Westernblot gel analysis of activated products of Caspases (Caspase Cascades) (Fig. 21.3; Fig. 21.5a, b); (5) DNA fragmentation (DNA laddering gel

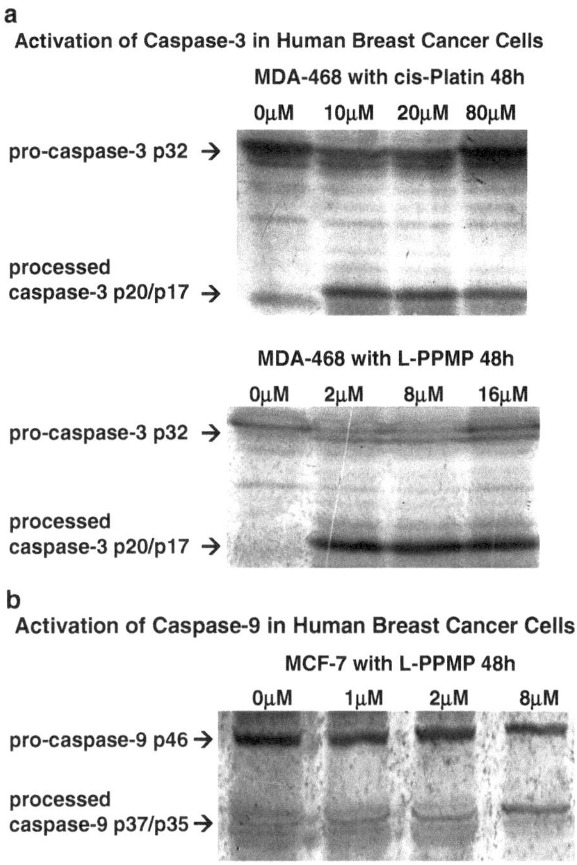

Fig. 21.5 (**a**) Activation of Caspase-3 in apoptotic MDA-468 cancer cells treated with *cis*-platin 10–80 μM and L-PPMP (2–16 μM) (Western Blot Examination). (**b**) Activation of Caspase-9 in MCF-7 cancer cells treated with L-PPMP (1–8 μM) (Westernblot Examination)

analysis; (6) inactivation of DNA replication complex (containing DNA polymerase-alpha, -delta, primase, helicases and 30 other proteins); (7) Inactivation of Golgi bodies (containing Glycosyltransferases for glycolipids and glycoproteins biosyntheses (Fig. 21.8).

Studies of induction (or initiation) and regulation of apoptotic cascades in breast and colon cancer cells are of prime interest in the field of anti-cancer drug research. If we search this topic in the Internet we would obtain at least 11,000 reports regarding apoptosis in cancer cells. Induction of apoptosis could be classified in four categories (Fig. 21.3): (1) Internal mitochondrial-caspase activation pathway (IMCAP); (2) external bad receptor activating pathway (EBRAP); (3) NFkappaB activation pathway (NFKBP; and (4) Cascades of protein kinase activation Pathways (CPKAP). Using highly metastatic breast carcinoma cells (Fig. 21.4a) (SKBR-3, MDA-468, and MCF-7) (we have reported in recent years (Ma et al. 2004; Basu et al. 2004c, 2012a; 2012b; Ma 2008; Boyle 2005; Boyle et al. 2006; Ma et al. 2009, 2011). These cultured cells grown in T-flasks or plastic petri dishes showed varied expression of ER (estrogen receptor), PgR (Progesterone receptor), p53 protein expression,

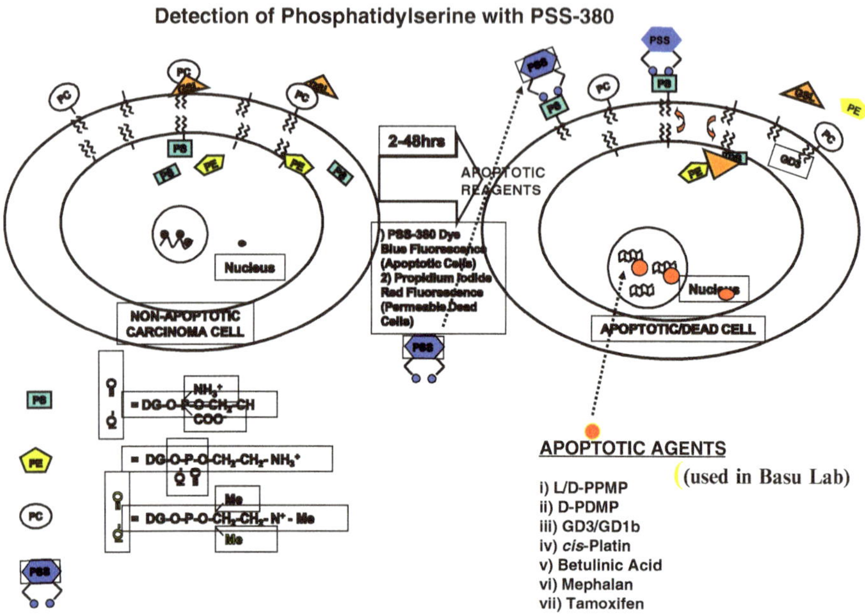

Fig. 21.6 Detection of phosphatidyl serine on the outer leaflet of apoptotic cells with the New Dye PSS-380. (Schematic drawing)

or Caspase-3 or, Caspase-8 cascade pathways. Using five simple apoptotic inducers (Fig. 21.4b: D-/L-PPMP, D-PDMP (inhibitors of GSL biosynthesis), Betulinic acid (a triterpinoid used as anti-cancer agent in China as an alternative medicine), Tamoxifen (a common anti-cancer agent used for the treatment of breast cancers today), and Melphalan (a scrambler for Golgi bodies) in the range of 2–16μM concentrations for L-PPMP showed apoptosis cell death (Figs. 21.6 and 21.7a). On the other hand, *cis*-platin showed apoptotic killing at higher 20–160 μM concentrations (Fig. 21.6b). These two chemicals induce apoptosis by activating IMCAP (Fig. 21.3). These chemicals activate both Caspases-3 (Fig. 21.5a) and Caspase-9 (Fig. 21.5b). On the other hand *cis*-platin (used in the treatment of testicular cancers) induced-apoptosis occurs in those cell lines at much higher concentrations (50–150 μM) by EBRAP followed by activation of Caspase-8 as we published before (Boyle et al. 2006). Several new chemicals have been tested in recent years for induction of the apoptotic process by activating IMCAP, NF-kappaB (NFKBAP) pathway, EBRAP or CPKAP.

Human MCF-7 breast cancer cells are resistant to pro-apoptotic stimuli due to Caspase-3 inactivation. These cell lines activate Caspase-9 only (Fig. 21.7c) in the presence of L-PPMP. Reconstitution of human MCF-7 breast cancer cells with Caspase-3 gene does not sensitize these cells to the inhibitory action of ROSC (roscovitine) and OLO (Olomoucine) (Wesierska et al. 2011b). However, apoptotic mechanisms of MCF-7cells by *cis*-platin or Betulinic acid are still unknown.

Table 21.2 Enzymatic levels of GSL: glycosyltransferases in breast cancer cells

Enzymatic Level of GSL Glycosyltransferase in Breast Cancer Cells

Enzyme(Substrate)	SK-BR-3	MDA-468	MCF-7
GlcT(Ceramide)	N.T.	N.T.	-
GalT-2(Glc-Cer)	-	N.T.	N.T.
GalT-3(GM2)	-	N.T.	N.T.
GalT-4(Lc3)	+++	+++	+
GalT-5(nLc4)	++	+	+
SAT-1 (Lc2)	-	-	-
SAT-2 (GM3)	-	-	+
SAT-3 (nLc4)	+	-	-
SAT-4 (GM1)	+	+	+
SAT-4' (Gg4)	++	++	++

"-" < 200 CPM / "+" 200 CPM - 1000 CPM / "++" 1000-5000 CPM / "+++" > 5000 CPM / "N.T." Not tested.
1 CPM Count = 10^{-9} μmol nucleotide sugar incorporated per μg protein of cell lysate in 4 hours

Quercetin (a natural polyphenolic compound) induced apoptosis in many human cancer cells, including MCF-7 human breast cancer cells (Chou et al. 2010). The involvement of possible signaling pathways and the roles of quercetin in the apoptosis of other cancer cells remain undefined. The recent results (Chou et al. 2010) suggest that quercetin may induce apoptosis by direct activation of Caspase cascade through a noncannonical IMCAP bypassing the Caspase-3 activation process or without involving Caspase-canonical pathway.

Betulin and Betulinic Acid as Apoptotic Agents in Cancer Cells

Several reports are available where Betulin [(lup-20)-ene-3beta28-diol], the natural occurring triterpene, triggers apoptosis (Basu et al. 2004b; Ma 2008; Fulda et al. 1997; Li et al. 2010b; Chaouki et al. 2010; Mullauer et al. 2011; Kessler et al. 2007) in human cancer cells through IMCAP (intrinsic mitochondrial Caspase activation pathways. The results showed Betulin significantly inhibited cell viability in cervix

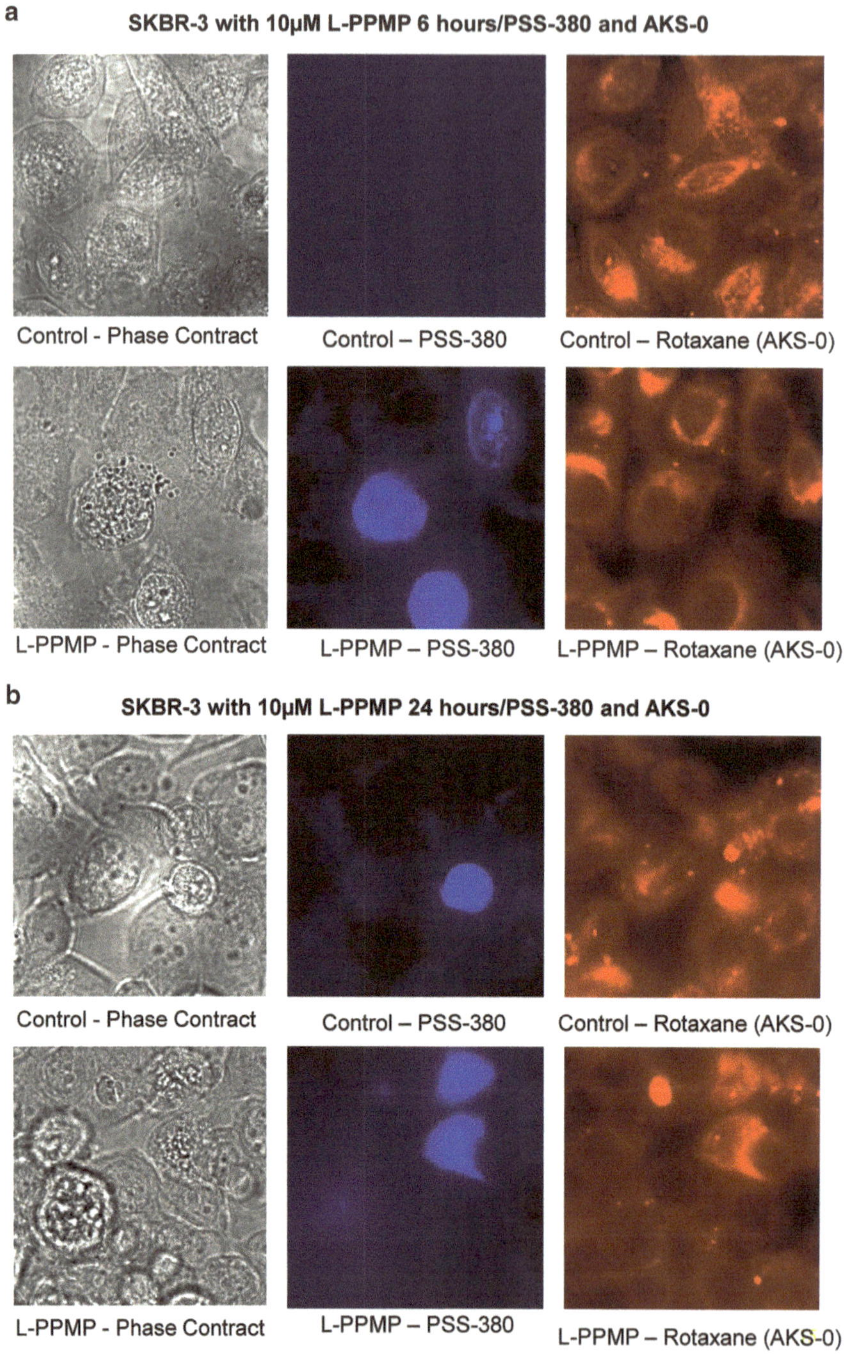

Fig. 21.7 (a) Fluorescent Microscopic Examination (with PSS-380/ Rotaxane (AKS-0)) of Apoptotic SKBR-3 Cells after 6 h treatment with L-PPMP (10 μM). (**b**) Fluorescent Microscopic Examination (with PSS-380/Rotaxane (AKS-0)) of Apoptotic SKBR-3 cells after 24 h treatment with L-PPMP (10 μM)

carcinoma HeLa cells, hepatoma HepG2 cells, lung adenocarcinoma A549 cells, prostate carcinoma PC3, and lung carcinoma NCI-H460, with IC50 values ranging 20–60 μg/mL It also showed a minor growth inhibitor in human erythroleukimic K562 cells (IC50 > 100 μg/mL). Activation of mitochondria and release of mitochondrial apoptogenic factors by Betulinic acid (BA), a melanoma-specific cytotoxic agent in neuroectodermal tumors such as neuroblastoma, medulloblastoma, and Ewing's sarcoma was first recognized by Fulda and his associates (Fulda et al. 1997). In recent years we have demonstrated that BA activates the apoptotic pathway via IMCAP in human breast carcinoma cells: SKBR-3, MDA-468, and MCF-7 also (Fukushi et al. 1984b; Shirue et al. 2011; Weedon et al. 1979; Kostrzewa 2000). Whether these cells also regulate cell surface GSL biosynthesis is not known but is under investigation. Apoptotic cell death by membrane phosphatidylserine translocation was observed. However, exact mechanism of apoptosis induced by Betulinic acid is not understood as yet (Basu et al. 1979, 1982; Presper et al. 1978; Radin 1999).

Proteins and Peptides

Decorin (a protein core that directly modulates collagen fibrillogenesis and matrix assembly) is a member of the small leucine-rich proteoglycan gene family (Koulov et al. 2003), It down-regulates members of the ErbB-receptor, tyrosine kinase family and regulates their signaling pathway, leading to growth inhibition. The effect of Decorin on the overexpression of ErbB2 in mammary carcinoma cells suggests it is an effective therapeutic agent against tumor growth of breast cancer and its metastatic spreading to other organs. Decorin inhibits MTLn3 cell proliferation, in a dose-dependent manner, as well as anchorage-independent cell growth and colony formation (Basu et al. 1971). Decorin also slows cell motility and stops cell invasion through a three-dimensional extracellular matrix formation. Anti-cancer activity of targeted proapoptotic peptides has also been reported (Arunkumar et al. 2006). An inhibitor of glycoprotein biosynthesis, Tunicamycin (an apoptotic agent) produce unfolded protein also inhibited in Nu/Nu mice microvasculature is suggested for human breast cancer treatment (Oskouian and Saba 2010). However, Survivin, is a small protein (142 amino acids; 16.5 kDa), belongs to the inhibitor of apoptosis.

Disialogangliosides

A short chain ganglioside, GD3 (Sialic-alpha2-8Sialic-alpha2-3Galactose-beta1-4Glc-beta1-1ceramide) occurs in the central nervous system (CNS) as an intermediate in the long chain ganglioside biosynthesis (Fig. 21.8; (Basu et al. 2000; Radhakrishnan et al. 2007; Sugiarto et al. 2011; Chachadi et al. 2011; Basu et al. 1987)). It also occurs in the optic nerve (Basu and Basu 1982; Kroes et al. 2006) and is a minor ganglioside in the normal tissue also (Kroes et al. 2011). It has been detected as a major GSL in meningiomas (Oskouian and Saba 2010),

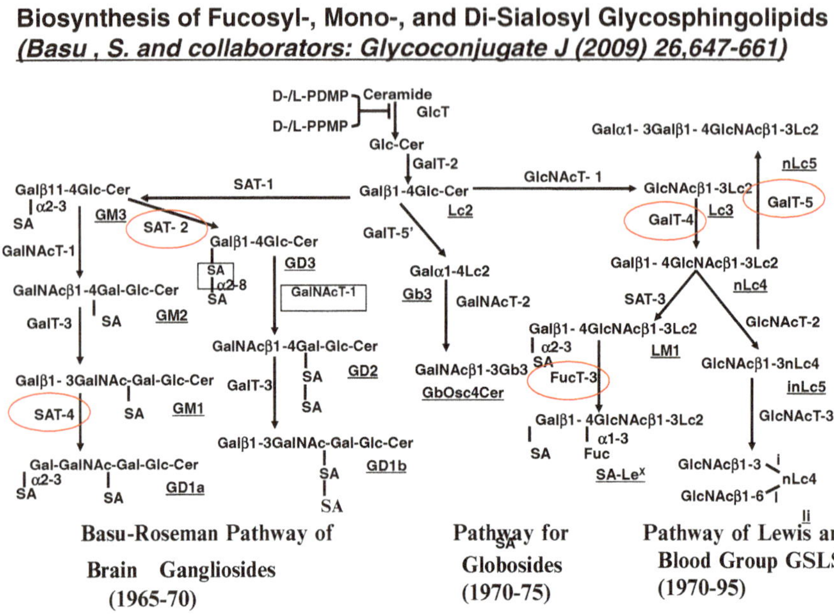

Fig. 21.8 Proposed biosynthetic pathways for disialosyl gangliosides in brain tissues (Basu Roseman Pathway); Globoside GSL, Lewis X, SA-LewisX, and Other Blood Group Active GSLs (B, Ii) biosynthesis in Cancer Cells were completed in Basu-Lab (1965–1995)

gliomas (Flemming and Saltzman 2001), melanomas (Masserini et al. 2002), colorectal carcinomas (Pecheur E-I Hoekstra 2002), and breast cancer cells (Ghosh and Bell 2002). Increased GD3 concentration during neuronal differentiation and on growth rate of CHO-K1 cells has been reported recently (Basu and Basu 2002). GD3 also sensitizes human hepatoma cells to cancer therapy (Aziz and Qiu 2014). Chimeric anti-GD3 monoclonal antibody by KM871 is proved to enhance *in vitro* antibody-dependent cellular cytotoxicity (Basu and Basu 2002) and to inhibit the proliferation of human malignant glioma cells *in vitro* (Ma 2008), GD3 has been recognized in recent years as an apoptotic agent in oligodendrocytes [97] and neuronal cell apoptosis in culture [98]. The pathways for biosynthesis of GD3 (Fig. 21.4; (Radhakrishnan et al. 2007; Basu et al. 1987)) and LD1a (Radhakrishnan et al. 2007) were established before in embryonic chicken brain cells. GD3 ganglioside as a proapoptotic agent has been established in recent years (Fukushi et al. 1984a, b; Mao et al. 1999; Shirue et al. 2011; Weedon et al. 1979; Kostrzewa 2000) [99, 100]. We have employed the disialosyl gangliosides (GD3 and GD1b) to induce apoptosis (Fukushi et al. 1984a, b; Mao et al. 1999; Shirue et al. 2011; Weedon et al. 1979; Kostrzewa 2000) in human breast cancer cells, SKBR-3 grown in culture. Apoptosis induction was monitored by the concomitant appearance of activated Caspase-3 and by binding of PS-380 to the outer leaflet of phosphatidyl serine (Fig. 21.6; Shirue et al. 2011; Weedon et al. 1979; Kostrzewa 2000). These results indicated that, in addition to many unknown GSLs on the cancer cell surfaces disialosylgangliosides

(GD3 or GD1b) could be employed as a breast cancer killing therapeutic agent (Weedon et al. 1979; Kostrzewa 2000). Post-translational and transcriptional regulation of GSL biosynthesizing genes during the induction of apoptosis by L-PPMP in breast cancer cells has been published in recent years (Shirue et al. 2011; Weedon et al. 1979; Kostrzewa 2000). However, exact regulations of GLT-genes in disialosylganglioside are not known [Ma et al. 2004, 2009; Ma 2008].

Apoptotic Membrane Damages Monitored by Fluorescence Microscopic Studies

Fluorescence Staining of Apoptotic Carcinoma Cells

Previous publications from other laboratories used Annexin-V (phosphatidyl specific blood clotting protein) for staining flopping of phosphatidyl serine (as a mark of apoptosis) on the outer leaflet of the apoptotic cells. However, this method had several drawbacks. This binding process needed a high concentration of added calcium in the medium to alter the apoptotic processes; (ii) to have also the commercial source of pure Annexin-v is very costly. Instead of this method with the availability of cheaper PSS-380 dye-binding assay (synthesized by our colleague Prof. Bradely Smith) was used for our carcinoma cells apoptotic studies as given below.

PSS-380/Propidium Iodide (PropI) Staining

The breast carcinoma cells were cultured and synchronized on the Nunc Lab-Tek 16-well Chamber Slide system at the starting concentration of 1×10^4 cells/200 μL per well. After synchronization and drug treatment, each well was washed once with 200 μL TES buffer mixture (5 mM N-[Tris (hydroxymethyl) methyl]-2-aminoethanesulfonic acid, 140 mM NaCl, 2 mM $MgCl_2$, 1 mM KCl). The cells were then incubated in 100 μL TES buffer containing 25 μM PSS-380 (Fig. 21.7a and b) (the dye has been synthesized and patented by Professor Bradely smith of the University of Notre Dame) plus 0.25 μg/mL Propidium iodide at 37 °C for 5 min. The staining buffer was removed; the wells were washed with 200-μL TES buffer mixture once and soaked in 100 μL of fresh TES buffer mixture for fluorescence observation (Figs. 21.7a and b).

PSS-380/or AKS-0 Staining

The breast carcinoma cells were cultured and synchronized on the Nunc Lab-Tek 8-well Chambered #1.0 Borosilicate Coverglass system at the starting concentration of 1×10^4 cells/200 μL per well. After synchronization and drug treatment, 10 μL of

1 mM AKS-0 stock (the complex dye which binds to both outer and inner organelle membrane; synthesized by Prof. Bradley Smith and his coworkers in our department at the University of Notre Dame; Fig. 12a and b) in DMSO was added to the medium (~200 μL) and incubated at 37 °C for 30 min. Then each well was washed with 200 μL TES buffer mixture and stained with 25 μM PSS-380 as described above. The cells were washed with 200 μL TES buffer mixture once and soaked in 100 μL of fresh TES buffer mixture for fluorescence observation.

Image Capture and Processing

The fluorescence was visualized under the Zeiss Axiovert S100TV confocal microscope with Chroma Cy3 (for Propidium iodide and/orAKS-0) and DAPI (for PSS-380) filters. The 100x objective lens with mineral oil was used for cell observation. The gray-scale image, which represents the single wavelength through different filters, was acquired in 16-bit TIF format with MetaMorph software developed by Molecular Devices Co. (Sunnyvale, CA). The 16-bit TIF file was transferred to 8-bit with MetaMorph. For color processing, the tif file was transferred to RGB mode with Photoshop, and artificial colors of blue or red were added in Photoshop by removing signals in other channels (e.g. to add red color, blue and green channels were removed in the RGB mode file). The processed image was saved in GIF or JPG format for data presentation (Fig. 21.7a, b).

Using these fluorescent dyes induction of apoptosis in all these three-breast carcinoma (SKBR-3, MDA-468, MCF-7) and colon carcinoma (Colo-205) was tested and reported from our laboratory during last decade (Zajdel and Zajdel 2013; Basu et al. 2004a, b, c, 2012a, b; Ma et al. 2004, 2009, 2011; Ma 2008; Boyle 2005; Boyle et al. 2006). Studies with AKS-0 showed the time dependent damage of inner membranes of the apoptotic cells within 24 h of treatment (Ma 2008; Ma et al. 2011).

Biosynthesis of Lewis and Blood Group Active Glycosphingolipids in Carcinoma Cells

Breast cancer cells adhesion to vascular endothelium is a critical process (Fig. 21.1) in metastatic MDA-468 and BT-20 breast cancer cells (BCC) adhered to cytokine-activated human umbilical cord vein endothelial cells (HUVECVs). The same is not true for anti-E selectin monoclonal antibody-treated HUVECs: BT. It is suggested that BT-20 cells express sialosyl-LewisX (SA-LeX) and sialosyl Lewis A (SA-Lea), but MDA-MB-468 BCC has novel unidentified E-selectin-binding epitopes (Nohara et al. 1998). Biosynthetic pathways (Fig. 21.8) for both SA-Lex and SA-Lea have been established in colon carcinoma cells by Basu et al. (Basu and Basu 1973; Higashi et al. 1985; Basu et al. 1991, 1999, 2000) and Hakomori and his associates (Holmes et al. 1985) by 1985–1991.

The disialoganglioside GD3 (NeuAcalpha2, 8NeuAcalpha2, 3Galbeta1, 4Glc-Ceramide) is overexpressed in 50 % of invasive ductal breast carcinoma; and the SAT-2 (or ST8SIAT) (Fig. 21.8) displays higher expression among estrogen receptor-negative breast cancer tumors, associated with a decreased survival of BC patients. It was shown previously that overexpression of SAT-2 in MDA-MB-231 acquires a proliferative phenotype in the absence of serum when grown in culture. Using two animal models (leghorn chicken and C57BL/6 mice) in human breast cancer cells increased NeuGcGM3 expression has also been reported. SAT-2 or GD3 synthase Overexpression enhances proliferation and migration of MDA-MB-231 breast cancer cells (Nohara et al. 1998).

Analysis of glycosphingolipid composition of MDA-MB-231 and MCF-7 human BCCs showed abundant presence of GM3, GM2, GM1, and GD1a in both the cell lines. The 18-fold increased amount of GM3 ganglioside suggests some role for this simple ganglioside in the growth regulation in MDA-MB-231 BCCs. However, insertion of GM3 ganglioside into the plasma membrane of MCF-7 cells blocked the growth stimulatory effect of EGF. Biosynthesis of glycosphingolipids in all Ganglio (Gg)-, Globo-, and Lacto (Lc)-series pathways (Fig. 21.8) is catalyzed by at least 18 different glycolipid glycosyltransferases (GLTs) expressed in normal embryonic tissues (Basu et al. 1965, 1968, 1971, 1973; Kaufman et al. 1968), rat liver Golgi bodies (Keenan et al. 1974), tumor tissues (Kijimoto and Hakomori 1971; Basu et al. 1980; Jenis et al. 1982) and cancer cells (Yeung et al. 1974; Moskal et al. 1974; Basu et al. 1979; Presper et al. 1978) have been characterized in last three decades in our and Dr. Hakomori's laboratories (Fig. 21.8) and have also been cloned in recent years by many laboratories around the world. GLTs involved in the syntheses of sialo-LeX in the non-apoptotic breast cancer cells have also been investigated in our laboratory (Fig. 21.8; (Basu et al. 2004a; Basu et al. 2004b; Ma et al. 2004; Basu et al. 2004c; Ma 2008; Boyle 2005; Boyle et al. 2006; Ma et al. 2009; Ma et al. 2011; Basu et al. 2012a; Basu et al. 2012b)) as well as in other laboratories.

Functions of glycosphingolipids on the eukaryotic cell plasma membrane during the onset of oncogenic processes and cell death are not well understood. Several inhibitors of glycosphingolipid biosynthesis were recently found to trigger apoptosis in many carcinoma cells including breast cancer SKBR-3, MCF-7, and MDA-468 cells through either intrinsic or extrinsic apoptotic pathways (Fig. 21.3) as we previously reported (Basu et al. 2004a, b, c, 2012a, b; Ma et al. 2004, 2009, 2011; Ma 2008; Boyle 2005; Boyle et al. 2006). These inhibitors (L-/day-PPMP) of glucosylceramide biosynthesis (Basu et al. 1982) could increase ceramide concentration () by blocking the functions of glycolipid glycosyltransferases (GLTs; Fig. 21.8). Using three novel fluorescent dyes PSS-380 (Figs. 21.9) (Radin 1999) and ASK-0 (Koulov et al. 2003) our recent studies revealed (Fig. 21.7a and b) (Ma et al. 2011) the damage of cell organelle membranes during apoptosis by the inhibitor of glucosylceramide biosynthesis (L-PPMP). Inhibition of GalT-2 (Fig. 21.8) by L- and D-PDMP has also been reported. The drug- and cell-dependent regulation of MAPKs was also found by *cis*-platin and L-PPMP when inducing apoptosis in SKBR-3, MCF-7, and MDA-468 cells. A summary of our protein kinase studies with the apoptotic BCCs is published (Basu et al. 2012b). In the presence of L-PPMP, both MDA-468, and MCF-7 cell lines all three pathways (ERK, JNK/SAPK and p38)

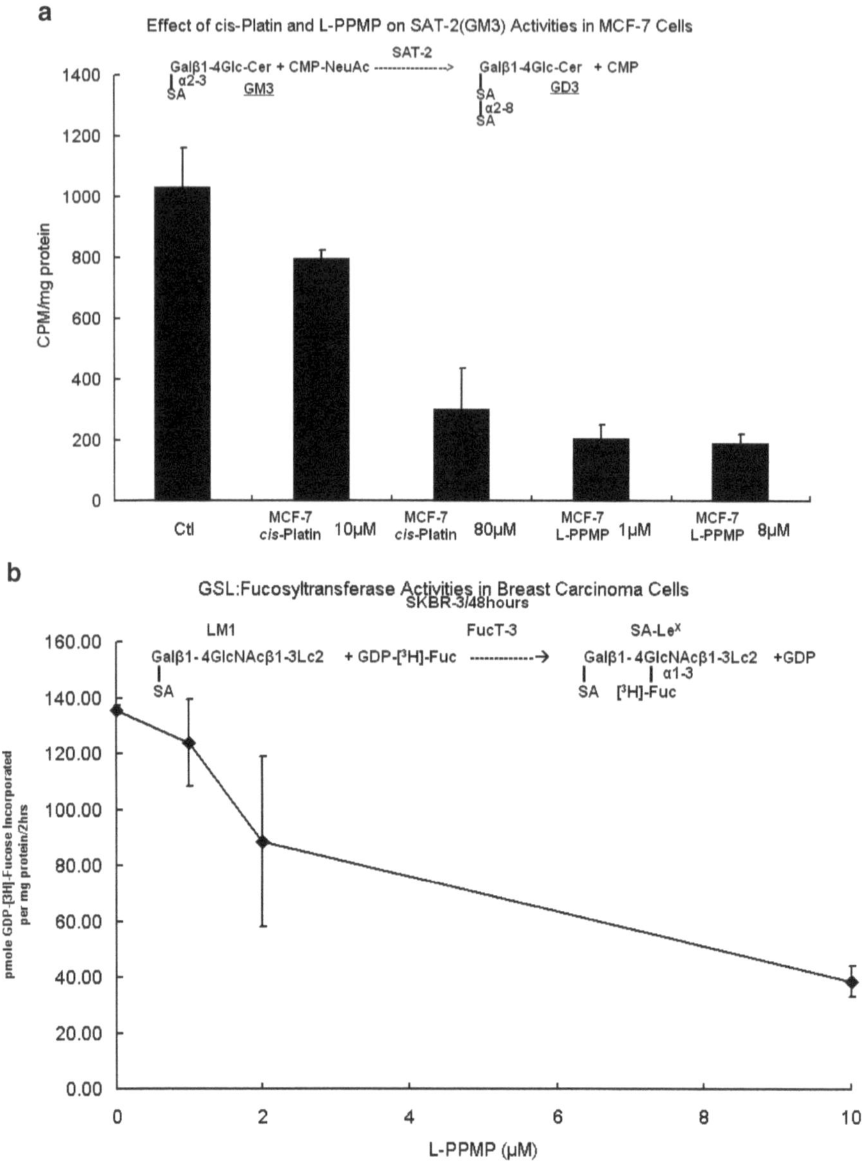

Fig. 21.9 (**a**) GSL: Alfa2, 8Sialyltransferase (SAT-2) activities in MCF-7 cells after treatment with cis-Platin (10–80 μM) and L-PPMP (1–8 μM) for 48 h. (**b**) GSL: Alfa1, 4 Fucosyltransferase (FucT-3) activities in SKBE-3 cells after treatment with L-PPMP (2–10 μM) for 48 h [12]

were activated whereas in SKBR-3 cell lines these pathways were inhibited. Further study is needed to implicate these pathways in the apoptotic breast cancer cells (MDA-468, MCF-7, and SKBR-3) induced by L-PPMP and cis-platin.

Regulation of Lewis Glycosphingolipid (LeX, SA-LeX) Biosynthetic Genes in Normal Tissues and Apoptotic Breast Cancer Cells

Starting from ceramide, the four *in vitro* steps (Fi, 8) for biosynthesis of neolactotetraosylceramide (nLcOse4-Cer) and globoside (GbOse4Cer) were established in embryonic chicken brains (Basu et al. 1991, 1999, 2000; Chien et al. 1973), rabbit bone marrow tissues (Basu and Basu 1972, 1973), bovine spleen Golgi preparations (Basu et al. 1999, 2000), Neuroblastoma cells (Presper et al. 1978), Colo-205 cells, and breast carcinoma cells (Basu et al. 2004a, b, c, 2012a, b; Ma et al. 2004, 2009, 2011; Ma 2008; Boyle 2005; Boyle et al. 2006) were established during the last three decades. Conversion of nLcOse4-Cer to Sialyl-nLcOse4-Cer (LM1) (Fig. 21.8) was established in embryonic chicken brains (Basu et al. 1991) as well as in bovine spleen (Basu et al. 1999, 2000), and its conversion *in vitro* to SA-LeX almost two decades ago. During process of normal growth and differentiation, the cell surface glycosphingolipids (GSLs) are proved to be regulated by the interaction of small molecules to the cell signaling systems. Using breast carcinoma lines (SKBR-3, MDA-468, and MCF-7), we at first used external chemicals, which induce apoptosis, may regulate expression of macromolecules on the cell surfaces and may control directly at the gene level in the production of catalytic proteins such as glycosyltransferases which in turn regulate expression of cell surface GSLs.

The glycosyltransferases (GLTs) (Fig. 21.8; (Ma et al. 2011; Basu et al. 2012a; Basu et al. 2012b)) catalyzing their synthesis have been characterized in Golgi-bodies (Keenan et al. 1974). Very little is known about gene-regulation of these GLTs either during embryonic development (Basu et al. 2012a) or during metastatic processes (Table 21.3). We know the complete biosynthetic pathways of GSLs during embryonic development or onset of oncogenic processes, but its regulation during apoptosis is unknown. Inhibitors of GLTs (L-PPMP and D-PDMP) and DNA (*cis*-platin) trigger apoptosis in Colo-205, SKBR-3, MCF-7, and MDA-468 through either intrinsic or extrinsic apoptotic pathways. These inhibitors regulate GLT gene expression post-translationally (Table 21.3) as well as post-transcriptionally (Ma et al. 2011; Basu et al. 2012a, b). Apoptotic effects initiate activation of Caspases (-3, -8, and -9). Using novel DNA-microarrays specifically designed for screening over 359 Glyco-related genes, transcriptional-regulation of several glycosyltransferases involved in the biosyntheses of Sialo-Le^X and Sialo-Le^a (cancer cell surface antigens) was observed with L-PPMP. Down-regulation of GLT activities (Table 21.3) and up-regulation of some GLT mRNA (Tables 21.4 and 5) suggest a tight regulation of these enzymes by signal transduction pathways. A total of 359 genes with four copies of unique sense 45 mer oligonucleotides were individually synthesized in the laboratory of Dr. Moskal at Northwestern University (Kroes et al. 2006, 2011). These DNA microarrays were prepared in his laboratory. Total RNA from MCF-7, MDA-468, and SKBR-3 cells treated with L-PPMP 2 μM for 2 h (Table 21.4) and 24 h were reversely transcribed (Table 21.5) and used as the substrate for RNA amplification and labeling using the Ambion Amino Allyl Message

Table 21.3 Overall conclusion of post-translational activities of glycosphingolipid: GLTs

Overall conclusion of post-translational activities of glycosphingolipid : GLTs (cell/agents-time)		
GSL-GLT	Catalyzed reaction	Enzymatic activity
GalT-4	Lc3 (GlcNAc-Gal-Glc-Cer) → Galβ-Lc3	Decrease
		(MCF-7/L-PPMP-2 h,6 h; SKBR-3/*cis*P, L-PPMP and MDA-468/L-PPMP, 48 h)
GalT-5	Lc4) (Gal-GlcNAc-Gal-Glc-Cer) → Galα-Lc4	Decrease
		(SKBR-3/L-PPMP-2 h,6 h; MCF-7,MDA-468/L-PPMP-6 h; SKBR-3/*cis*P and MDA-468/L-PPMP-48 h)
SAT-2	GM3 → GD3	Decrease
		(MCF-7/*cis*P,L-PPMP-48 h)
SAT-4	GM1 → GD1a	Decrease
		(MCF-7/L-PPMP, SKBR-3/*cis*P and MDA-468/L-PPMP-48 h)
SAT-4'	Gg4 (Gal-GalNAc-Gal-Glc-Cer) → GM1b	Decrease
		(SKBR-3/*cis*P and MDA-468/L-PPMP-48 h)
FucT-3	LM1 → SA-Le^X	Decrease
		(SKBR-3/L-PPMP-48 h)

Table 21.4 Changes in GLT expression in apoptotic breast cancer cells (DNA microarray/2 μM L-PPMP/2 h)

Cell line	GLT gene name	Linkage formed	Fold change
MCF-7	B3GALT5	Galβ1-3GlcNAc-R1	1.19–1.33
	B3GNT3	GlcNAcβ1-3 Gal-R2	1.20
	B3GNT1	GlcNAcβ1-3 Gal-R2	−1.19
MDA-468	UGCGL2	Glcβ1-1 Cer	1.28
	B3GNT4	GlcNAcβ1-3 Gal-R2	1.34
	B3GNT1	GlcNAcβ1-3 Gal-R2	−1.34
SKBR-3	B3GNT4	GlcNAcβ1-3 Gal-R2	1.35
	ST6GAL1	NeuAcα2-6Galβ1-3GlcNAc-R3	−1.55

Core Blood Type: Galβ1-3/4GlcNAcβ1-3Galβ1-4-R1/R2
R1 = Glcβ1-1 Cer / R2 = OligoN-Protein (values more than one means-transcriptional stimulation; negative values means inhibition of transcription)

AMP/TM II aRNA amplification kit. Each labeled sample with pooled reference human RNA was hybridized with the three arrays. The scanned arrays in acceptable quality were transferred to Blue Fuse format and quantified with Gene Traffic (DUO) v3.2-11 Comprehensive statistical analysis with data from Gene Traffic was performed with SAM (Significance Analysis of Microarrays) at 10 % false discovery (Kroes et al. 2011). The significantly changed genes were listed in Table 21.4 (Glycosyltransferase involved in SA-LeX biosynthesis) and Table 21.5 (Glyco-related nonrelated enzymes involved in carbohydrate metabolism). The values in

Table 21.5 Transcriptional regulation of glyco-related genes in breast carcinoma cells (2 h/24 h with L-PPMP:2 μM)

Gene name	MCF-7 2 h	MCF-7 24 h	Symbol	Gene name
	2 h	24 h		
NM_000188	1.2	0.67	HK1	HEXOKINASE 1
NM_000194	1.19	0.6	HPRT1	HYPOXANTHINE PHOSPHORIBOSYLTRANSFERASE 1 (LESCH-NYHAN SYNDROME)
NM_001069	1.45	0.38	TUBB2A	TUBULIN, BETA 2A
NM_002629	1.11	1.2	PGAM1	PHOSPHOGLYCERATE MUTASE 1 (BRAIN)
NM_005573	1.4	0.46	LMNB1	LAMIN B1
NM_033170	1.19	1.16	B3GALT5	UDP-Gal:betaGlcNAc beta 1,3-galactosyltransferase, polypeptide 5 (B3GALT5), transcript variant 2
NM_170707	1.35	0.68	LMNA	LAMIN A/C
	SKBR-3	SKBR-3		
	2 h	24 h		
NM_152932	1.35	0.87	GLT8D1	GLYCOSYLTRANSFERASE 8 DOMAIN CONTAINING 1
	MDA-468	MDA-468		
	2 h	24 h		
NM_006082	0.78	0.77	TUBA6	TUBULIN, ALPHA, UBIQUITOUS

the columns of "Fold Change" are the mRNA level of the gene in L-PPMP-treated sample divided by the control cell sample without L-PPMP treatment. For example, when the fold change is 1.55, L-PPMP increased the gene expression by 55 % compared to the control. When the value is 0.73, it means L-PPMP decreases the expression by 27 %. With the given Gene short name the full name and function can be found at the Genbank database.

In this study, the normalized data from the Gene Traffic (DUO) were processed with the two classes, unpaired analysis on a minimum of 500 permutations. The comparison was between the data derived from control cells and 2 μM L-PPMP-TREATED cells.

A dose- and time-dependent down-regulation of GLTs was investigated by GLT enzymatic assays (Table 21.2) (Basu et al. 1987, 1991; Basu and Basu 1982) and DNA microarray analyses (Kroes et al. 2006). The GLTs are involved in biosynthesis of Le^X (neolactosyl-ceramide series) such as GalT-4 (UDP-Gal: LcOse3cer beta-galactosyltransferase (Fig. 21.8) and LeA(UDP-Gal: LcOse3-Cer beta1,3 Galactosyltransferase (Basu et al. 1987); SAT-3(CMP-NeuAc: nLcose4Cer alpha 2,3 Sialyltransferase, SAT-3 (Fig. 21.8 (Basu et al. 1982); Fig. 21.8, and FucT-3 (GDP-Fucose: LM1 alpha1, 4fucosyltransferase) (Fig. 21.8) (Basu et al. 1987, 1991, 1999, 2000; Presper et al. 1978). The reaction steps are clearer in the chart, Fig. 21.8 (Basu et al. 2012a, b). A similar effect was observed with the GLTs involved in the biosyntheses of Gg-series gangliosides, such as SAT-4 (CMP-NeuAc: GgOse4Cer alpha2, 3sialyltransferase), and SAT-3 (CMP-NeuAc: nLcOse4-cer alpha2, 3sialyltransferase) published previously (Basu et al. 2012a, b). The glyco-related gene DNA-microarrays (containing more than 359 well selected different genes, also suggested (Tables 21.4 and 5) modulation of the transcriptional regulation (many were stimulated-when vaues given in the tables are more than 1.0) of several GLTs involved in the biosynthesis of neolactosylceramide containing cell-surface antigens in these apoptotic breast carcinoma cells. In the early apoptotic stages (2–6 h after L-PPMP treatment) in addition to the GlcT-1 gene, several genes (betaGalTs and betaGlcNAcTs) in the SA-Le^a pathway were stimulated (Tables 21.4 and 5). Transcriptional regulation of different glyco-related and nonrelated genes during apoptosis of breast cancer cells have been reported (Oskouian and Saba 2010). Overexpression of ST6GalNacV, a ganglioside-specific alpha-2,6 sialyltransferase in glioma growth (Kroes et al. 2011) is also reported in recent years.

Novel Drug Delivery Systems for Cancer Treatments

These apoptotic agents could be employed as a new generation of anti-cancer drugs. Proper drug delivery system (Liposome Magic Bullet containing cis-platin) is discussed previously (Basu et al. 2012b).

The tissues in human bodies contain 70–90 % water. Drug molecules (soluble or suspended fine nono-structure) can be introduced into the body of patients in a variety of ways: topical, intravenous injection, intravenous infusion, subcutaneous injection, submuscular injection, or by controlled release from any transplant. The effectiveness of a drug therapy depends on the rate and extent to which drug molecules can move through structures to their targeted site of action in breast cancer tumors. Diffusion is the basic process by which migration of drug molecule occurs in the cells (normal or carcinoma). The rate of diffusion (i.e., a diffusion constant) depends on the structure of the diffusing molecules. An average diffusion coefficient of 10^{-7} cm^2/s is desirable for an effective therapeutic drug (Flemming and Saltzman 2001).

However, this diffusion process can be enhanced when a therapeutic drug is targeted by the aid of a special molecule present on the cancer cell surfaces. The search for better therapeutics (e.g., apoptotic agents) includes search for its proper strategies to cross the cancer cell surfaces without damaging normal cells by simple diffusion process. Modern anti-cancer therapeutic science is a developing field. Properties of the lipid membranes are critically important in regulating the movement of the molecules between these aqueous spaces, from blood to the intracellular space of cancer cells. The relationship between liposome structure, stability, and penetration through plasma membranes is an area of active, ongoing study in our present research also. Much of the effort in drug design and drug delivery devoted to overcoming the membrane diffusional barriers of the cancer cells (Masserini et al. 2002; Pecheur E-I Hoekstra 2002; Ghosh and Bell 2002; Basu and Basu 2002) could be adopted as an efficient drug delivery system. An ideal drug delivery system should have two things: (1) The physicochemical properties of a drug must be well controlled during the delivery inside a cancer cell and, (2) the drug should be targeted to the specific cancer cells avoiding any normal cell. A tentative model of a targeted drug delivery system (Basu et al. 2012b) is under study in our laboratory using advantage of the antigens of cancer cell surfaces.

Summary

Our present studies suggest induced apoptosis occur *in vitro* in highly metastatic breast carcinoma cells (SKBR-3, MDA-468, and MCF-3) and colon carcinoma cells (Colo-205) in the presence of biosimilar simple compounds (L-/day-PPMP, *cis*-platin, Betulinic acid, Tamoxifen, Melphalan, GD3 ganglioside, and GD1b ganglioside). Except for *cis*-platin all other apoptotic agents (as mentioned above) activate through Caspase-3 and Caspase-9 pathways, perhaps follow the canonical "internal mitochondrial pathway (IMAP)" where as *cis*-platin follows the "external activation pathway (EXAP)" through Caspase-8 activation or through some noncannonical pathway. All these activations at early stages (in first 2 h) induce transcriptional regulation of a large number of Glyco genes whereas after 6 h they regulate genes differently, ultimately down-regulating all those glyco-genes tested for SA-LeX biosynthesis. An accurate correlation between availability of SA-LeX or LeY (Aziz and Qiu 2014) on the surfaces of carcinoma cells and their metastatic properties have not been established as yet and would be the goal for future research. Using availability of Lewis antigen on the tumor cell surfaces, proper-targeted drug delivery is under study in different laboratories and would be beneficial for cancer patients.

Acknowledgments We thank Mrs. Dorisanne Nielsen and Mr. Eric Kuehner for their help during preparation of this manuscript. Our sincere thanks to Dr. Sipra Banerjee of the Cleveland Clinic Foundation who initially (decade ago) provided us the clones of the breast carcinoma cell: SKBR-3, MDA-468, and MCF-7, which we have maintained continuously for this work. This research work was supported by the Jacob Javits Research Award from NIH-NINDS, NCI, and the Coleman Cancer Foundation (continuous long-term support) to S. Basu; and a grant-in-aid from Siemens Corporation to M. Basu.

References

Abraham J (2013) Developing oncology biosimilars: an essential approach for the future. Semin Oncol 40(Suppl 1):5–24

Arunkumar E, Fu N, Smith BD (2006) Squarine-derived rotaxanes: highly stable, fluorescent near-IR dyes. Chemistry 12(17):4684–4690

Ashizawa T, Aoki T, Yanazaki T, Katayanagi S, Shinizu H, Koyanagy Y (2003) The clinical significance of sialyl-Lewis antigen expression in the spread of gastric cancer. J Exp Clin Cancer Res 22(1):91–98

Aziz F, Qiu Y (2014) The role of anti-LeY antibody in the downregulation of MAPKs/COX-2 pathway in gastric cancer. Curr Drug Targets 15:465–472

Banerjee M, Singh P, Panda D (2010) Curcumin suppresses the dsynamic instability of microtubules activates the mitotic checkpoint and induces apoptosis in MCF-7 cells. FEBS J 277(16): 3437–3448

Barthel SR, Gavino JD, Wiese GK, Jaynes JM, Siddiqui J, Dimittrof CJ (2008) Analysis of glycosyltransferase expression in metastatic prostate cancer cells: capable of rolling activity on microvascular endothelial (E) selectin. Glycobiology 18(19):806–817

Basu M, Basu S (1972) Enzymatic synthesis of a tetraglycosylceramide by a galactosyltransferase from rabbit bone marrow. J Biol Chem 247(5):1489–1495

Basu M, Basu S (1973) Enzymatic synthesis of a blood group B relatesd pentaglycosyl-ceramide by an alpha-galactosyltransferase from rabbit bone marrow. J Biol Chem 248(5):1700–1706

Basu S, Basu M (1982) Expression of glycolipid glycosyltransferases in development and transformation. In: Horowitz M (ed) Glycoconjugates, vol 3. Academic, New York, pp 265–285

Basu M, Basu S (2002) Micelles and liposomes in metabolic enzymes and glycolipid glycosyltransferase assays. In: Basu SC, Basu M (eds) Liposome methods and protocols. Humana, Totowa, pp 107–130

Basu S, Kaufman B, Roseman S (1965) Conversion of Tay-Sachs ganglioside to monosialoganglioside by brain uridine diphosphate D-galactose: glycolipid galactosyltransferase. J Biol Chem 240:4114–4117

Basu S, Kaufman B, Roseman S (1968) Enzymatic synthesis of ceramide-glucose and ceramide lactose by glycosyltransferase from embryonic chicken brain. J Biol Chem 243:5802–5804

Basu S, Schultz A, Basu M, Roseman S (1971) Enzymatic synthesis of galactocerebroside by a galactosyltransferase from embryonic chicken brain. J Biol Chem 243:4272–4279

Basu S, Kaufman B, Roseman S (1973) Enzymatic synthesis of glucocerebroside by a glucosyltransferase from embryonic chicken brain. J Biol Chem 248:1388–1394

Basu M, Presper KA, Basu S, Hoffman LM, Brooks SE (1979) Differential activities of glycolipid glycosyltransferase in Tay-Sachs disease: studies in cultured cells from cerebrum. Proc Natl Acad Sci U S A 76:4270–4274

Basu M, Basu S, Potter M (1980) Biosynthesis of blood group related glycosphingolipids in T- and B-lymphomas and neuroblastoma cells. In: Sweeley CC (ed) Cell surface glycolipids. American chemical society symposium, vol. 128. p 187–212

Basu M, Basu S, Stoffyn A, Stoffyn P (1982) Biosynthesis in vitro of Sialyl-alpha2,3-neolactotetraosylceramide by a sialyltransferase from embryonic chicken brain. J Biol Chem 257:12765–12769

Basu M, De T, Das K, Kyle JW, Chon HC, Schaeper RJ, Basu S (1987) Glycosyltransferases involved in glycolipid biosynthesis. In: Ginsburg V (ed) Methods in enzymology, vol 38. Academic, New York, pp 575–607

Basu M, Hawes JW, Li Z, Ghosh S, Khan FA, Zhang BJ, Basu S (1991) Biosynthesis *in vitro* of SA-LeX and SA-diLeX by alpha 1-3 fucosyltransferases from colon carcinoma cells and embryonic brain tissues. Glycobiology 1(5):527–535

Basu S, Basu M, Dastgheib S, Hawes JW (1999) Biosynthesis and regulation of glycosphingolipids. In: Meth-Cohn O, Pinto BM, Barton DHR, Nakanishi K (eds) Comprehensive natural products chemistry. Pergamon, New York, pp 107–128

Basu S, Das K, Basu M (2000) Glycosyltransferase in glycosphingolipid biosynthesis. In: Ernst B, Sinay P, Hart G (eds) Oligosaccharides in chemistry and biology—a comprehensive handbook. Wiley-VCH Verlag Gmbh, Weinheim, pp 329–347

Basu S, Ma R, Mikulla B, Bradley M, Moulton C, Basu M, Banerjee S, Inokuchi J (2004a) Apoptosis of human carcinoma cells in the presence of inhibitors of Glycosphingolipid biosynthesis: I. Treatment of Colo-205 and SKBR3 cells with isomers of PDMP and PPMP. Glycoconj J 20(3):157–168

Basu S, Ma R, Boyle PJ, Mikulla B, Bradley M, Smith B, Basu M, Banerjee S (2004b) Apoptosis of human carcinoma cells in the presence of potential anti-cancer drugs: III. Treatment of Colo-205 and SKBR3 cells with: *cis*-Platin, Tamoxifen, Melphalan, Betulinic acid, L-PDMP, L-PPMP, and GD3 ganglioside. Glycoconj J 20(9):563–577

Basu S, Ma R, Basu M, Goodson H, Smith B, Banerjee S (2004a) Glycosphingolipid metabolism and signaling in apoptotic cancer cells, lipids: sphingolipid metabolizing enzymes. In: Haldar DK, Das SK, (eds) Research Signpost, Kerala. p 81–100

Basu S, Ma R, Moskal JR, Basu M (2012a) Ganglioside biosynthesis in developing brains and apoptotic cancer cells. X. Regulation of glyco-genes involved in GD3 and Sialyl-LeX/a syntheses. Neurochem Res 37:1245–1255

Basu S, Ma R, Moskal JR, Basu M, Banerjee S (2012b) Apoptosis of breast cancer cells: XI. Modulation of genes of glycoconjugate biosynthesis and targeted drug delivery. In: Sudhakaran PR, Surolia A (eds) Proceedings of 9th international symposium, biochemical roles of eukaryotic cell surface macromolecules. Adv Exp Med Biol 749:233–255

Blagosklonny MV (2000) Cell death beyond apoptosis. Leukemia 14(8):1502–1508

Boyle PJ (2005) Characterization of DNA Helicase-III in replication complexes isolated from embryonic chicken brains and breast carcinoma cells. PhD Thesis, University of Notre Dame, Notre Dame, IN: 1–186

Boyle PJ, Ma R, Tuteja N, Banerjee S, Basu S (2006) Apoptosis of human breast carcinoma cells in the presence of *cis*-platin and L-/D-PPMP: IV. Modulation of replication complexes and glycolipid: glycosyltransferases. Glycoconj J 23(3–4):175–187

Bull LA, Taylor C (2014) Developing clinical trials for biosimilar. Semin Oncol 41(Suppl 1):S15–S25

Cazet A, Lefebvre J, Adriaenssens E, Julien S, Bobowski H, Grigoriads A, Tutt A, Tulasne D, LeBourhis X, Delannoy P (2010) GD3 synthase expression enhances proliferation and tumor growth of MDA-MB-231 breast cancer cells through c-met activation. Mol Cancer Res 8(11):1526–1535

Chachadi VB, Cheng H, Klinkebiel D, Christman JK, Cheng PW (2011) 5-Aza-2′-deoxycytidine increases sialyl Lewis X on MUC1 stimulating beta-galactoside:alpha2,3-sialyltransferase 6 gene. Int J Biochem Cell Biol 43(4):L586–L593

Chaouki W, Leger DY, Eliastimi JL, Hmamouchi M (2010) Antiproliferative effect of extracts from Aristolochia baetica and Origanum compactumm on human breast cancer cell line MCF-7. Pharm Biol 48(3):269–274

Chien JL, Williams T, Basu S (1973) Biosynthesis of a globoside-type glycosphingolipid by an beta-N-acetygalactosaminyltransferase from embryonic chicken brain. J Biol Chem 248:1778–1785

Chou CC, Yang JS, Lu HF, Ip SW, Lo C, Wu CC, Lin JP, Tang NNY, Chung JG, Chou MJ, Teng YH, Chen DR (2010) Quercetin-mediated cell cycle arrest and apoptosis involving activation of a Caspase cascade through mitochondrial pathway in human breast cancer MCF-7 cells. Arch Pharm Res 33(8):1181–1191

Flemming AS, Saltzman WM (2001) Simultaneous delivery of an active protein and neutralizing antibody: creation of separated regions of biological activity. J. Control Release, 70(1-2):29–36

Fukushi Y, Hakomori S, Nudelman E, Cochran N (1984a) Novel fucolipids accumulating in human adenocarcinoma. II. Selective isolation of Hybridoma antibodies that differentially recognize mono-, di-, and trifucosylated type 2 chain. J Biol Chem 259(7):4681–4685

Fukushi Y, Hakomori S, Shepard T (1984b) Localization and alteration of mon-, di-, and trifucosyl alpha1,3 type3 chain structure in human embyogenesis and human cancer. J Exp Med 160(2):506–520

Fulda S, Friesen C, Los M, Mier W, Benedict M, Nunez G, Krammer PH, Peter ME, Debatin KM (1997) Betulinic acid triggers CD95 (APO-1/Fas)- and p53-independent apoptosis via activation of caspases in neurroectodermal tumors. Cancer Res 37:4956–4964

Fuster MM, Brown JR, Wang L, Esko JD (2003) A disaccharide precursor of sialyl X inhibits metastatic potential of tumor cells. Cancer Res 63(11):2775–2781

Ghosh S, Bell R (2002) Liposomes: applications and protein–lipid interaction studies. In: Basu SC, Basu M (eds) Liposome methods and protocols. Humana, Totowa, pp 49–60

Higashi H, Basu M, Basu S (1985) Biosynthesis *in vitro* of disialosylneolacto-tetraosyl-ceramide by a solubilized sialyltransferase from embryonic chicken brain. J Biol Chem 260(2):824–828

Holmes EH, Ostander GK, Hakomori S (1985) Enzymatic basis for the accumulation of glycolipids with X and dimeric X determinants in human lung cancer cells (NCI-H69). J Biol Chem 260(12):7619–7762

Holmes EH, Ostander GK, Hakomori S (1986) Biosynthesis of the sialyl-Lex determinant carried by the type 2 chain glycosphingolipid in humanhuman lung carcinoma PC9 Cells. J Biol Chem 261(8):3737–3743

Jenis DM, Basu S, Pollard M (1982) Increased activity of a beta-galactosyltransferase in tissues of rats bearing prostate and mammary adenocarcinomas. Cancer Biochem Biophys 6:37–45

Julien S, Grimshaw MJ, Simon-Smith M, Coleman J, Morris HR, Dell A, Taylor-Papadimitiou J, Burchell JM (2007) Sialyl Lewis X on P-selectin glycoprotein ligand-1 is regulated during differentiation and maturation of dendritic cells" a mechanism involving the glycosyltransferases C2GnT1 and STGalI. J Immunol 179(9):5701–5710

Kannagi R (2004) Molecular mechanism for cancer-associated induction of sialyl LewisX and sialyl Lewis A expression. The Warburg effect revisited. (Review). Glycoconj J 20(5): 353–364

Kashiwagi H, Kijima H, Dowaki S, Obtani Y, Tobita K, Yamazaki H, Nakamura M, Ueyama Y, Tanaka M, Inokuchi S, Imaizumi T, Makuuchi H (2004) Clinicopathology significance of sialyl LeX expression in human gallbladder carcinoma. Oncol Rep 11(6):1139–1143

Kaufman B, Basu S, Roseman S (1968) Enzymatic synthesis of disialogangliosides from monosailogangliosides by sialyltransferases from embryonic chicken brain. J Biol Chem 243:5804–5806

Keenan TW, Morre JD, Basu S (1974) Ganglioside biosynthesis: concentration of glycosphingolipid glycosyltransferase in golgi apparatus from rat liver. J Biol Chem 249:310

Kessler JH, Mullauer FB, de Roo GM, Medema JP (2007) Broad in vitro efficacy of plant-derived Betulinic acid against cell lines derived from the most prevent human cancer types. Cancer Lett 251(1):132–145

Kijimoto S, Hakomori S (1971) Enganced glycolipid:alpha-galactosyltransferase activity in contact-inhibited hamster cells and loss of this response in polyoma transformation. Biochem Biophys Res Commun 44L:557–563

Kim DY, Kang SH, Ghil SH (2010) Circium japonicum extract induces apoptosis and anti-proliferation in the human breast cancer cell line MCF. Mol Med Rep 3(3):427–432

Kostrzewa RM (2000) Review of apoptosis vs. necrosis of substantia nigra pars compacta in Parkinson's disease. Neurotox Res 2(2–3):239–250

Koulov AV, Stucker KA, Lakshmi C, Robinson JP, Smith BD (2003) Detection of apoptotic cells using a synthetic fluorescent sensor for membrane surfaces that contain phosphatidylserine. Cell Death Differ 10(12):1357–13595

Kroes RA, Panksepp J, Burgdorf J, Otto NJ, Moskal JR (2006) Modeling depression: social dominance-submission gene expression patterns in rat neocortex. Neuroscience 137:37–49

Kroes RA, He H, Emmett MR, Nilsson CL, Leach FE III, Amster IJ, Marshall AG, Moskal JR (2011) Overexpression of ST6GalNAcV, a gangliposide-specific alpha-2,6-sialyltransferase, inhibits glioma growth in vivo. Proc Natl Acad Sci USA 107(28):12646–12651

Laezza C, Malfitano AM, DiMatola T, Ricchi P, Bifulco M (2010) Involvement of Akt/NF-kappaB pathway in N6-isopentenyladenosine-induced apoptosis in human breast cancer cells. Mol Carcinog 49(10):892–901

Leung E, Kim JE, Rewcastle GW, Finlay GJ, Baguley BC (2011) Comparison of the effects of the P13K/mTOR inhibitors NVP-BEZ235 and GSK2126458 on tamoxifen-resistant breast cancer cells. Cancer Biol Ther 11(11):938–946

Li W, Zhang W, Luo J, Cao A, Zhang D, Sheng W, Cai S, Li J (2010a) Alpha1,3 fucosyltransferase VII plays a rile in colorectal carcinoma by promoting the carbohydration of glycoprotein CD24. Oncol Rep 23(6):1609–1617

Li Y, He K, Huang Y, Zheng D, Gap C, Jin YH (2010b) Betulin induces mitochondrial cytochrome c release associated apoptosis in human cancer cells. Mol Carcinog 49(7):630–640

Lopes GL, de Souza JA, Barrios C (2013) Access to cancer medication in low and middle-income countries. Nat Rev Clin Oncol 10(6):314–322

Ma R (2008) Apoptosis of breast and colon cancer cells by inhibitors of glycolipid and DNA biosynthesis. PhD Thesis, University of Notre Dame, Notre Dame, IN: 1–271

Ma R, Koulov A, Moulton C, Basu M, Banerjee S, Goodson H, Basu S (2004) Apoptosis of human breast carcinoma cells in the presence of Disialosyl gangliosides: II. Treatment of SKBR3 cells with GD3 and GD1b gangliosides. Glycoconj J 20(5):319–330

Ma R, Decker N, Matthew AV, Moskal JR, Bergdorf J, Johnson J, Basu M, Banerjee S, Basu S (2009) Post-translational and transcriptional regulation of glycolipid glycosyltransferase genes in apoptotic breast carcinoma cells: VII. After treatment with L-PPMP. Glycoconj J 26:647–661

Ma R, Hopp EA, Decker M, Loucks A, Johnson JP, Moskal JR, Basu M, Banerjee S, Basu S (2011) VIII. Regulation of glycosyltranferase genes in apoptotic breast cancer cells by inhibitors of glycolipid and DNA biosynthesis. In: Wu A (ed) Immunology of complex carbohydrates: advances in experimental biology, vol 705. p 621–642

Mao S, Gao C, Lo CH, Wirsching P, Wong CH, Janda KD (1999) Phage-display library selection of high-affinity human single-chain antibodies to tumor associated carbohydrate antigebv sialyl LewisX and Lewis Z. Proc Natl Acad Sci U S A 96(12):6953–6958

Marchetti M, Russo L, Balducci D, Falanga A (2011) All trans-retinoic acid modulates the procoagulant activity of human breast cancer cells. Thromb Res 128(4):368–7

Masserini M, Palestini P, Pitto M, Chigorno V, Sonnino S (2002) Preparation and use for the study of sphingolipid segregation in membrane model systems. In: Basu SC, Basu M (eds) Liposome methods and protocols. Humana, Totowa, pp 17–27

Matsumoto S, Imaeda Y, Umemoto S, Kobayashi K, Suzuki H, Okamoto T (2002) Cimetidine increases survival of colorectal cancer patients with high levels of sialyl Lewis X and sialyl Lewis A epitope expression on tumor cells. Br J Cancer 86(2):161–167

Mitoma J, Petryniak B, Hiraoka N, Yeh JC, Lowe JB, Fukuda M (2003) Extended core 1 and 2 branched O-glycans differently modulate sialyl Lewis X-type L-selectin ligand activity. J Biol Chem 276(11):9953–9961

Moskal JR, Gardner DA, Basu S (1974) Changes in glycolipid glycosyltransferases and glutamate decarboxylase and their relationship to differentiation in neuroblastoma cells. Biochem Biophys Res Commun 61:751–758

Mullauer EB, van Bloois L, Ten Daalhusin JB, Brink MS, Storm G, Medema JP, Schiffelers RM, Kesseler JH (2011) Betulinic acid delivered in liposomes reduces growth of human lung and colon cancers in mice without causing systemic toxicity. Anticancer Drugs 22(3):223–233

Nohara K, Wang F, Spiegel S (1998) Glycosphingolipid composition of MDA-MB-231 and MCF-7 human breast cancer cells. Breast Cancer Res Treat 48(2):149–157

Oskouian B, Saba JD (2010) Cancer treatment strategies targeting sphingolipid metabolism. Adv Exp Med Biol 688:185–205

Patel VA, Lee DJ, Longacre-Antoni A, Feng L, Lieberthal W, Rauch J, Ucker DS, Levine JS (2009) Apoptotic and necrotic cells as sentinels of local tissue stress and inflammation: response pathways initiated in nearby viable cells. Autoimmunity 42(4):317–3921

Patil JB, Kim J, Jayprakasha GK (2010) Berbarine induces apoptosis in breast cancer cells (MCF-7) through mitochondrial-dependent pathway. Eur J Pharmacol 645(1–3):70–78

Pecheur E-I Hoekstra D (2002) Peptide-induced fusion of liposomes. In: Basu SC, Basu M (eds) Liposome methods and protocols. Humana, Totowa, pp 31–48

Presper KA, Basu M, Basu S (1978) Biosynthesis *in vitro* of fucose-containing glycosphingolipids in human neuroblastoma IMR-32 cells. Proc Natl Acad Sci U S A 75:289–293

Radhakrishnan P, Beum PV, Tan S, Cheng PW (2007) Butyrate induces sLeX synthesis by stimulation of selective glycosyltransferase genes. Biochem Biophys Res Commun 359(3): 457–462

Radin NS (1999) Chemotherapy by slowing glucosphingolipid synthesis. Biochem Pharmacol 57(6):0589–0595

Schuldes H, Schleicher D, Mayer G, Markus BH, Cinati J, Blaheta RA (2003) Bonding of gastrointestinal tumor cells in endothelial E- and P-selectin adhesion receptors leads to transient down regulation of sLeX ligands in vitro. Int J Colorectal Dis 18(4):292–299

Shirue VS, Henson KA, Schnaar RL, Nimrichter L, Burdick MM (2011) Gangliosides expressed on breast cancer cells are E-selectin ligands. Biochem Biophys Res Commun 406(3):423–429

Silva Z, Tong Z, Cabral MG, Martins C, Castro R, Reis C, Trindade H, Konstantopoulos K, Videira PA (2011) Sialyl LewisX-dependent binding of human monocyte-derived dendritic cells to selectins. Biochem Biophys Res Commun 409(3):459–464

Sugiarto G, Lau K, Ya H, Vuong S, Thon V, Li Y, Huang S, Chen X (2011) Cloning and charax-teruzation of a viral alpha2,3 Sialyltransferase (vSTGal-1) for the synthesis of sialyl LewisX. Glycobiology 21(3):387–396

Ullah MF, Ahmad A, Zubair H, Khan HY, Wang Z, Sarkar FH, Hadi SM (2011) Soy isoflavone genisten induces cell death in breast cancer cells through mobilization of endogenous copper ions and generation of reactive oxygen species. Mol Nutr Food Res 55(4):553–559

Weedon D, Searle J, Kerr JF (1979) Apoptosis. Its nature and implications for dermatopathology. Am J Dermatopathol 1(2):133–144

Wesierska J, Hacki S, Zulehner N, Maurer M, Komina O (2011a) Reconstitution of human MCF-7 breast cancer cells with Caspase-3 does not sensitize them to action of CDK inhibitor. J Cell Biochem 112(1):273–288

Wesierska J, Hacki S, Zulehner N, Maurer M, Komina O (2011b) Reconstitution of human MCF-7 breast cancer cells with Caspase-3 does not sensitize them to action of CDK inhibitor. J Cell Biochem 112(1):273–288

Yeung KK, Moskal J, Chien JL, Gardner D, Basu S (1974) Biosynthesis of globoside and Forssman related glycosphingolipid in mouse adrenal Y-1 tumor cells. Biochem Biophys Res Commun 59:252–260

Yuan J, He Z, Wu J, Lin Y, Zhu X (2011) A novel adriamycin analogue derived from marinemi-crobes induces apoptosis by blocking Akt activation in human breast cancer cells. Mol Med Rep 4(2):261–265

Zajdel J, Zajdel R (2013) Brand-name drug, generic drug, orphan drug. Pharmacological therapy with biosimilar drugs: provision of due diligence in the treatment process. Contemp Oncol (Pozn) 17(6):477–483

Zhang B, Chen H, Yao X, Cong W, Wu M (2000) E-selextin and its ligand-sLeX in metastasis of hepatocekkukar carcinoma. Zhonghua Wai Ke Za Zhi 38(7):534–536 (Chinese)

Zhang BH, Chen H, Yao XP, Cong WM, Wu MC (2002) E-selectun and its ligand-sLeX in the metastasis of heptacellular carcinoma. Hepatobiliary Pancreat Dis Int 1(1):L80–L82

Zhang N, Kong X, Yan YC, Yang Q (2010) Huair aqueous extract inhibits proliferation of breast cancer cells by inducing apoptosis. Cancer Sci 101(11):2375–2383

Chapter 22
N-Acetylglucosaminyl 1-Phosphate Transferase: An Excellent Target for Developing New Generation Breast Cancer Therapeutic

Aditi Banerjee, Juan A. Martinez, Maria O. Longas, Zhenbo Zhang, Jesus Santiago, Krishna Baksi, and Dipak K. Banerjee

Introduction

Each year, breast cancer is diagnosed in over one million women worldwide, with a survival rate of approximately 400,000 women (Jemal et al. 2006). The American Cancer Society estimates the following for breast cancer in the United States for 2013:

The original version of this chapter was revised. The correction to this chapter is available at https://doi.org/10.1007/978-3-319-11280-0_25

A. Banerjee
Department of Pediatrics, School of Medicine, University of Maryland, Baltimore, MD 21202, USA

Department of Biochemistry, School of Medicine, University of Puerto Rico, Medical Sciences Campus, San Juan, PR 00936-5067, USA

J.A. Martinez • Z. Zhang • J. Santiago
Department of Biochemistry, School of Medicine, University of Puerto Rico, Nanotherapeutics, Inc., Alachua, FL 32615, USA

M.O. Longas
Department of Chemistry and Physics, Purdue University Calumet Hammond, Hammond, IN 46323-2094, USA

K. Baksi
Department of Anatomy and Cell Biology, School of Medicine, Universidad Central del Caribe, Bayamon, PR 00960-3001, USA

D.K. Banerjee (✉)
Institute of Functional Nanomaterials, University of Puerto Rico-Rio Piedras, San Juan, PR 00931-1907, USA

Department of Biochemistry, School of Medicine, University of Puerto Rico, Medical Sciences Campus, San Juan, PR 00936-5067, USA
e-mail: dipak.banerjee@upr.edu

A. Chakrabarti, A. Surolia (eds.), *Biochemical Roles of Eukaryotic Cell Surface Macromolecules*, Advances in Experimental Medicine and Biology 842,
DOI 10.1007/978-3-319-11280-0_22

232,340 new cases of invasive breast cancer, 64,640 new cases of carcinoma *in situ* (CIS; non-invasive), and nearly 39,620 deaths. The disease is more common among women in upper socioeconomic classes, among women who never have been married and among women living in rural areas. Lower-than-average rates of breast cancer have been recorded for Mexican-Americans, Japanese and Filipino women in Hawaii, American Indians, Seventh-Day Adventists, and Mormons, while Jewish women have a higher-than-average risk. Nuns have a higher risk for breast cancer, presumably because of their usual nulliparous status (Berkowitz and Kelsey 2006a).

Triple Negative Breast Cancer: The triple negative breast cancer (TNBC) accounts for 15 % of all breast cancers and has a disproportionate share of mortality. In TNBC, tumors do not express estrogen receptors, progesterone receptors or Her2 (Cleator et al. 2007). The patients are younger (Rhee et al. 2008), and the disease is more common in pre-menopausal African-American women (Carey et al. 2006). Histology includes a high grade, a high proliferation rate, and necrosis (Livasy et al. 2006; Aksoy et al. 2007). TNBC tends to metastasize hematogenously rather than *via* the lymphatics, and thus shows less axillary lymph node metastasis than non-TNBC (Van Calster et al. 2009). Patients with TNBC unfortunately have a higher risk of recurrence and death than those with non-triple negative tumors. Recurrence risk increases rapidly in the first 2 years, with a peak at 2–3 years, then declines over the next 5 years (Dent et al. 2007). Majority of deaths occurs in the first 5 years (Stockmans et al. 2008).

Both endogenous and exogenous factors contribute to the development and progression of breast cancer. These include reproductive factors, endogenous hormones, exogenous hormones [oral contraceptives, depot-metroxyprogesterone acetate/DMPA, estrogen-replacement therapy, diethyl-stilbestrol/DES], body build, diet, alcohol consumption, lifestyle/medications/electric power use, benign breast conditions, multiple primary neoplasms, familial aggregation and heredity, estrogen receptors, and radiation (Berkowitz and Kelsey 2006b). This is complicated further with the metastatic load due to epithelial-mesenchymal transition (EMT; Kalluri and Weinberg 2009).

The disease was recognized by the Egyptians as early as 1600 BC. But, it has become a major public health problem over the last 50 years, affecting as many as one in eight women during their life time (Ries et al. 1999; Sondik 1994). Furthermore, in many regions of the world breast cancer is the most frequently occurring malignant disease in women (Forbes 1997). When the deaths are aggregated by age, cancer has surpassed heart disease as the leading cause of death for those younger than age 85 since 1999. Delay-adjusted cancer incidence rates stabilized in men from 1995 through 2002, but continued to increase by 0.3 % per year from 1987 through 2002 in women (Jemal et al. 2006). African-American men and women have 40 and 18 % higher death rates from all cancers combined than White men and women, respectively. Furthermore, minority populations are more likely to be diagnosed with advanced stage disease than Whites. At present, breast cancer incidence rates are higher in White women than in African-Americans over age 45; the rates are similar in the 40–44 years age group, and higher in African-American younger than age 40 (Boring et al. 1991; Kelsey and Gammon 1991). In the United

States, the incidence rate for breast cancer has increased steadily by about 1–2 % per year since 1960 (Berkowitz and Kelsey 2006a; Forbes 1997; Lynn and Ries 1995; Hortobagyl and Buzdar 1995). The prognosis of breast cancer depends upon the stage at diagnosis: 5-year survival rate is 100 % for Stage 0, 98 % for Stage I, 88 % for Stage II, 56 % for Stage IIIA, 49 % for Stage IIIB and 16 % for Stage IV.

Breast cancer therapy: The disease is preventable if detected early and treated with appropriate therapy. Although several therapeutic options exist, the treatment of breast cancer is typically expensive and accompanied by a host of adverse side effects that are detrimental to patient's quality of life. In many cases, treatments are effective in only a small percentage of the total patient population. As a consequence, there is a poor patient outcomes, an economic burden on the healthcare system, added costs of the physician's time, wasted drugs, and increased hospitalization. In malignant breast tissue, the intratumoral endothelial cell proliferation rate is 45 times higher than that of the surrounding benign breast. Therefore, therapies have been designed to interrupt the targets of this process (Folkman 1990). These targets are categorized as: endothelial toxins, growth factor antagonists, protease inhibitors, and endogenous anti-angiogenics (Broomhall et al. 2002; Erlichman et al. 2001; Rudek et al. 2001; Fotsis et al. 1994; Klauber et al. 1997). Since, the TNBCs are poorly differentiated, most of them fall into the basal subgroup of breast cancers. Because of the absence of specific treatment guidelines, TNBC patients are managed with standard treatment which leaves them with a high rate of local and systemic relapse.

Breast tumor growth is angiogenesis dependent: The hallmark of cancer follows the core principles of sustainability to proliferative signaling, ability to evade growth suppressors, ability to resistance cell death, ability to enable replicative immortality, angiogenesis induction, and activating invasion and metastasis. Neovascularization, i.e., angiogenesis is a "key" (Uhr et al. 1997; Gastl et al. 1997) to breast cancer progression and occurs from pre-existing vasculature in stages that orchestrate a network of cooperative interactions. These include endothelial cell migration, capillary budding, establishment of capillary loops, and neovascular remodeling. A "angiogenesis switch" thus initiates the formation of new capillary tubes from host vessels (Folkman 1992). Factors helping the "angiogenic switch" may include, but are not limited to, the tumor microenvironment; mutation in oncogenes or tumor-suppressor genes; pro-angiogenic molecules like VEGF, FGF-2, EGF, PDGF, PIGF and MMPs; and anti-angiogenic factors (thrombospondin, angiostatin, tumstatin, and endostatin (Uhr et al. 1997; Gastl et al. 1997).

Angiogenesis depends on asparagine-linked glycoproteins: Asparagine-linked (N-linked) glycoproteins have been found to play an important role in capillary endothelial cell proliferation and differentiation (Banerjee 1988a; Oliveira and Banerjee 1990; Tiganis et al. 1992; Nguyen et al. 1992, 1993; Banerjee and Vendrell-Ramos 1993; Pili et al. 1995; Ingber and Folkman 1989). It has also been suggested that inhibition of hybrid and complex-type N-glycans synthesis inhibits the formation of capillary tubes. In contrast, inhibition of only complex-type N-glycans, but not hybrid-type N-glycans, does not inhibit the tube formation. This suggestion supports that (a) the synthesis of at least hybrid-type glycans is required for capillary tube formation *in vitro*, and (b) that an increase in monosialylated, fucosylated N-linked glycans occurs during capillary tube formation (Nguyen et al. 1992, 1993). In subsequent studies, it has been proposed that 8Br-cAMP treatment upregulates the proliferation of capillary endothelial cells and the lumen formation (Colleoni et al. 2002; Banerjee 1988b). The suggested molecular mechanism is increased N-glycosylation of cellular glycoproteins (e.g., eFactor VIIIc) by accelerating the synthesis and turnover of $Glc_3Man_9GlcNAc_2$-PP-Dol (lipid-linked oligosaccharide, LLO). Enhanced synthesis and turnover ($t_{½}$) of LLO correlates with increased mannosylphospho dolichol synthase (DPMS) activity in the endoplasmic reticulum (ER). cDNA cloning of Dpm1 gene identifies a cAMP-dependent protein kinase (PKA) mediated phosphorylation motif in the capillary endothelial cell DPMS (Baksi et al. 2008).

Study Design: High expression of some glycosyl epitopes promotes invasion and metastasis, leading to shorter 5–10 year survival rates of patients. Whereas, expression of some other glycosyl epitopes suppress tumor progression, leading to higher post-operative survival rates (Hakomori 1996; Muramatsu 1993). The former category of epitopes include β6GlcNAc branching in N-linked structure; and the latter category includes β4GlcNAc competitive with β6GlcNAc. The expression mechanism of these glycosyl epitopes in terms of status of respective glycosyltransferase genes has been extensively studied (Wandall et al. 2010; Taniguchi et al. 2002). Unfortunately, due to the lack of a systematic study these results cannot be used to develop a new therapy.

Aberrant glycosylation occurs in essentially all types of experimental and human cancers, and many glycosyl epitopes constitute tumor-associated carbohydrate antigens (TACA, 46), but N-glycosylation is evolutionary conserved. A long-standing debate has been whether it is a result or a cause of cancer. Many recent studies indicate with no reasoning that aberrant glycosylation is a result of initial oncogenic transformation, as well as a "key" event in induction of invasion and metastasis. In current cancer research, glycosylation promoting or inhibiting tumor cell invasion and metastasis is of crucial importance (Cazet et al. 2010; Fuster and Esko 2005). This area has received much less attention from most cell biologists involved in cancer research, mainly because structural and functional concepts of glycosylation in cancer were more difficult to understand than the functional role of certain proteins and their genes in defining cancer cell phenotypes. The focus is now shifting because of emerging technologies on functional glycomics (http://www.functionalglycomics.org) developed by the Consortium of Functional Glycomics and is being supported further by the initiatives from the US National Research Council, US National Academy of Sciences, and the National Cancer Institute. Therefore, our objective has been to test a transformative idea that inhibiting *N*-acetylglucosaminyl 1-phosphate transferase (GPT) with a 840 Da glucosamine-containing pyrimidine nucleoside (i.e., tunicamycin) we could inhibit angiogenesis and consequently the breast tumor growth. The significance of this innovative study is to target the protein N-glycosylation machinery of the tumor microvasculature and develop a glycotherapy to be used in the clinic.

Results and Discussion

Tunicamycin inhibits angiogenesis in vitro and decreases invasion and chemotaxis: A non-transformed capillary endothelial cell line was used as a model. The cells had a population doubling time of 68 h at 2 % serum but reduced to 56 h in the presence of high serum (i.e., 10 %) or other growth stimulants such as vascular endothelial growth factor (VEGF), basic fibroblast growth factor (FGF-2), 8Br-cAMP, etc. (Martínez et al. 1999; Banerjee et al. 2007, 2011). Flow cytometric analysis of the cell cycle indicated that the change in population doubling was due to reduction of G1 phase. A synchronized culture of cells when treated with tunicamycin (1 μg/mL), it exhibited inhibition of cellular proliferation in a time and dose-dependent manner (Martínez et al. 2000). The IC_{50} for tunicamycin was 438 ng when treated for 48 h, it however changed to 235, 100 and 85 ng when treated for 72, 96 and 120 h, respectively (Fig. 22.1a). There was a considerable loss of cellular morphology. In addition, the cells exhibited reduction in volume (i.e., cell shrinkage, chromatin condensation, picnotic appearance and the cytosol was full of apoptotic bodies (Fig. 22.1b). Interestingly, the cells failed to recover ones exposed to tunicamycin treatment. The recovery may be quantitative if treated for 24 h at concentrations of 10 ng, 1 μg, or 10 μg but it was approximately 33 % if the treatment was for 48 h, and after 72 h the recovery was extremely poor, i.e., between 8 and 10 % (Fig. 22.2a). This was also analyzed by studying their colony forming ability. The cells lost their colony forming ability whether treated for 3 or 32 h with tunicmycin (1 μg/mL) (Fig. 22.2b).

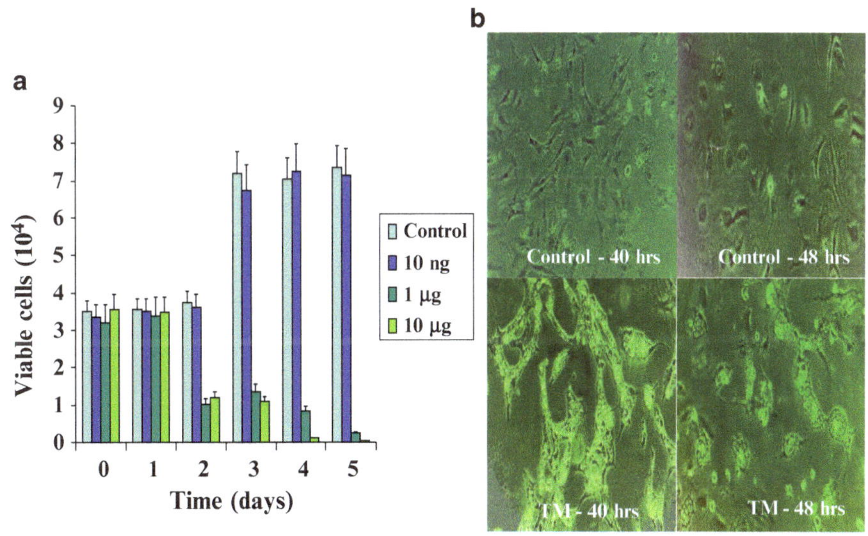

Fig. 22.1 Proliferation of capillary endothelial cells as a function of tunicamycin concentration and the time of treatment (**a**). Morphological changes associated with apoptotic death of microvascular endothelial cells following tunicamycin treatment (**b**)

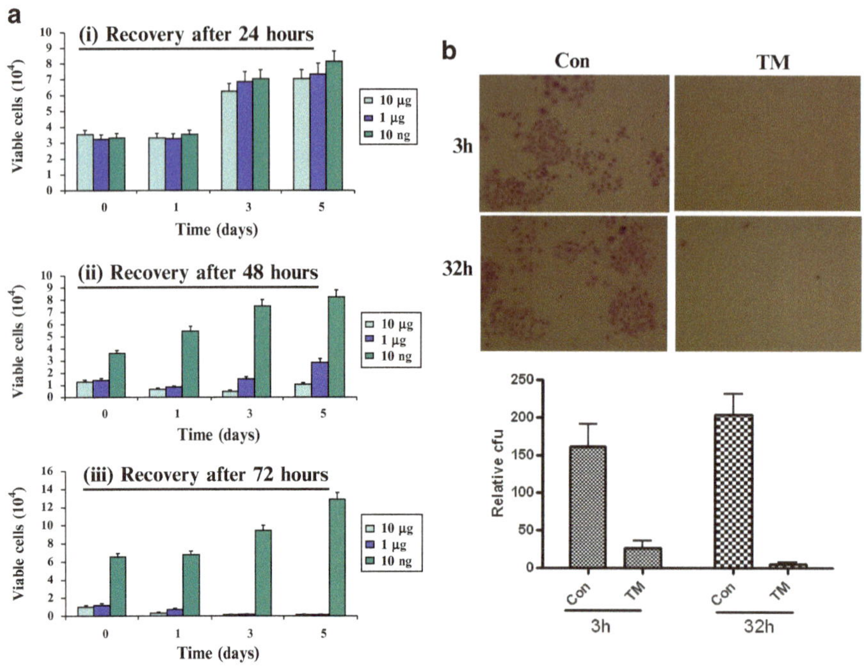

Fig. 22.2 The capillary endothelial cells do not regain their virulence upon withdrawal of tunicamycin. (**a**) Monitoring cell number after 24, 48 and 72 h. (**b**) Clonogenic assay (Longas et al. 2012)

Tunicamycin-treated cells undergo nuclear fragmentation: Reduction of cellular proliferation is associated with either a cell cycle arrest or a cell death or a combination of both. In earlier study (Martínez et al. 1999), we have observed not only a cell cycle arrest in G1 but there was also a loss in cell numbers with time due to cell death in following tunicamycin treatment. To evaluate the nature of cell death, we stained the nuclei with Hoechst 33528 (10 µg/mL in PBS, pH 7.2) and examined under a fluorescence microscope. The photomicrographs of the nuclei from tunicamycin-treated cells exhibited fragmentation (arrows) with no changes in untreated controls (Fig. 22.3). Thus, confirming apoptosis (i.e., programmed cell death).

Tunicamycin inhibition of capillary endothelial cell proliferation is not reversible by FGF-2 or VEGF165: To evaluate the stability of the anti-angiogenic effect of tunicamycin in tumor microenvironment, the proliferation of capillary endothelial cells treated with tunicamycin (1 µg/mL) was examined in the presence of FGF-2 (10 ng/mL) and $VEGF_{165}$ (10 ng/mL). The cells were monitored every 12 h for 108 h (FGF-2) or 24 h for 144 h ($VEGF_{165}$). Exogenous addition of FGF-2 or $VEGF_{165}$ stimulated cell proliferation, but they failed to protect the cells from the growth inhibitory effect of tunicamycin (Fig. 22.4a, b). The inhibition of cell proliferation was less during the first 24 h of FGF-2 or $VEGF_{165}$ treatment but after 48 h neither FGF-2 nor $VEGF_{165}$ showed any protection ($p<0.001$).

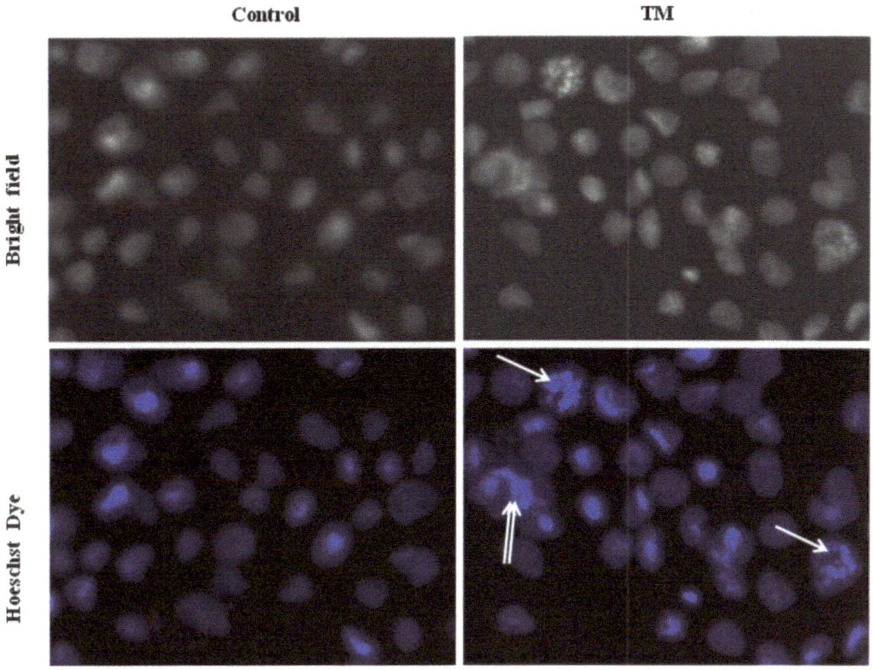

Fig. 22.3 Nuclear fragmentation of capillary endothelial cells following tunicamycin treatment (Longas et al. 2012)

Tunicamycin down-regulates phospho-VEGFR1 and phospho-VEGFR2 receptors: VEGF-induced changes are expected to be seen in cells expressing functional VEGF receptors (Banerjee et al. 2011). Endothelial cells express both VEGFR1 and VEGFR2 receptors although the VEGFR2 receptor has been claimed to be more effective in inducing angiogenesis. The status of total, phosphorylated VEGFR1 and VEGFR2 receptors and their ratios in capillary endothelial cells were evaluated in tunicamycin (1 μg/mL) treated cells. Total VEGFR1 expression was ~97.2, ~83.3, and ~89.4 % at 3, 12, and 32 h, respectively, over the synchronized cells, i.e. 0 h. These values for VEGFR2 were ~110.3, ~91.8, and ~68.3 %, respectively, for the same time period. The level of phosphorylated VEGFR1 was down-regulated by ~30.5, ~33.5, and ~75.7 % in cells treated with tunicamycin for 3, 12, and 32 h, respectively compared with untreated controls. The phosphorylated VEGFR2 was down-regulated by ~44.4 %, and ~72.6 % in cells treated with tunicamycin for 3 and 32 h, respectively but increased by ~3.3 % in cells treated with tunicamycin for 12 h (Fig. 22.5a, b) compared with controls. Interestingly, the reduction of phosphorylated VEGFR1 (i.e., ~30.5 %) and VEGFR2 (i.e., ~44.4 %) was serendipitously very similar to 46.6 % reduction of protein tyrosine kinase activity following $VEGF_{165}$ stimulation in cells treated with tunicamycin for 3 h (Fig. 22.5c). There was also a 1.6–2.4-fold reduction of phosphorylated VEGFR1 and VEGFR2 in cells treated with tunicamycin for 32 h over that of 3 h. On the other hand, there was a

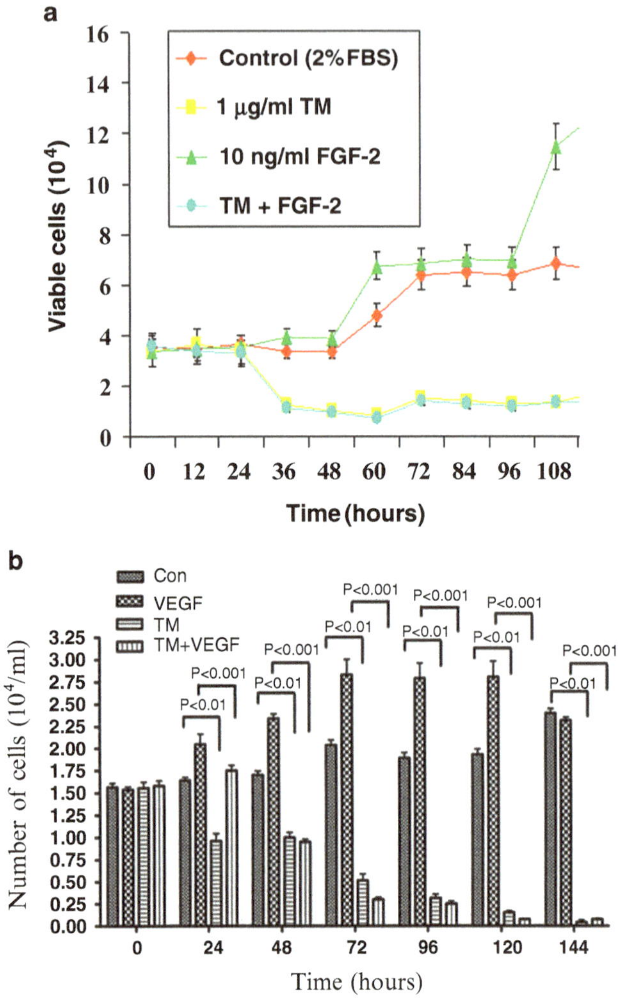

Fig. 22.4 Tunicamycin survives under tumor microenvironment. (**a**) In the presence of FGF-2; and (**b**) in the presence of $VEGF_{165}$ (Banerjee et al. 2011)

2.7–2.9-fold difference between the ratio of down-regulated phospho-VEGF receptors (i.e. VEGFR1 and VEGFR2) and the total VEGFR 1 and VEGFR2 at 3 and 32 h of tunicamycin treatment (Table 22.1).

Tunicamycin down-regulates phosphotyrosine kinase activity in VEGF165-treated capillary endothelial cells: VEGF signals through activation of tyrosine kinase (Zwick et al. 2001; Robinson et al. 2000). We have examined the status of VEGF-stimulated tyrosine kinase activity in cells treated with tunicamycin. Cells stimulated with $VEGF_{165}$ exhibited an increase in tyrosine kinase activity over the basal level. A marked reduction in the enzyme activity ($p < 0.001$) however,

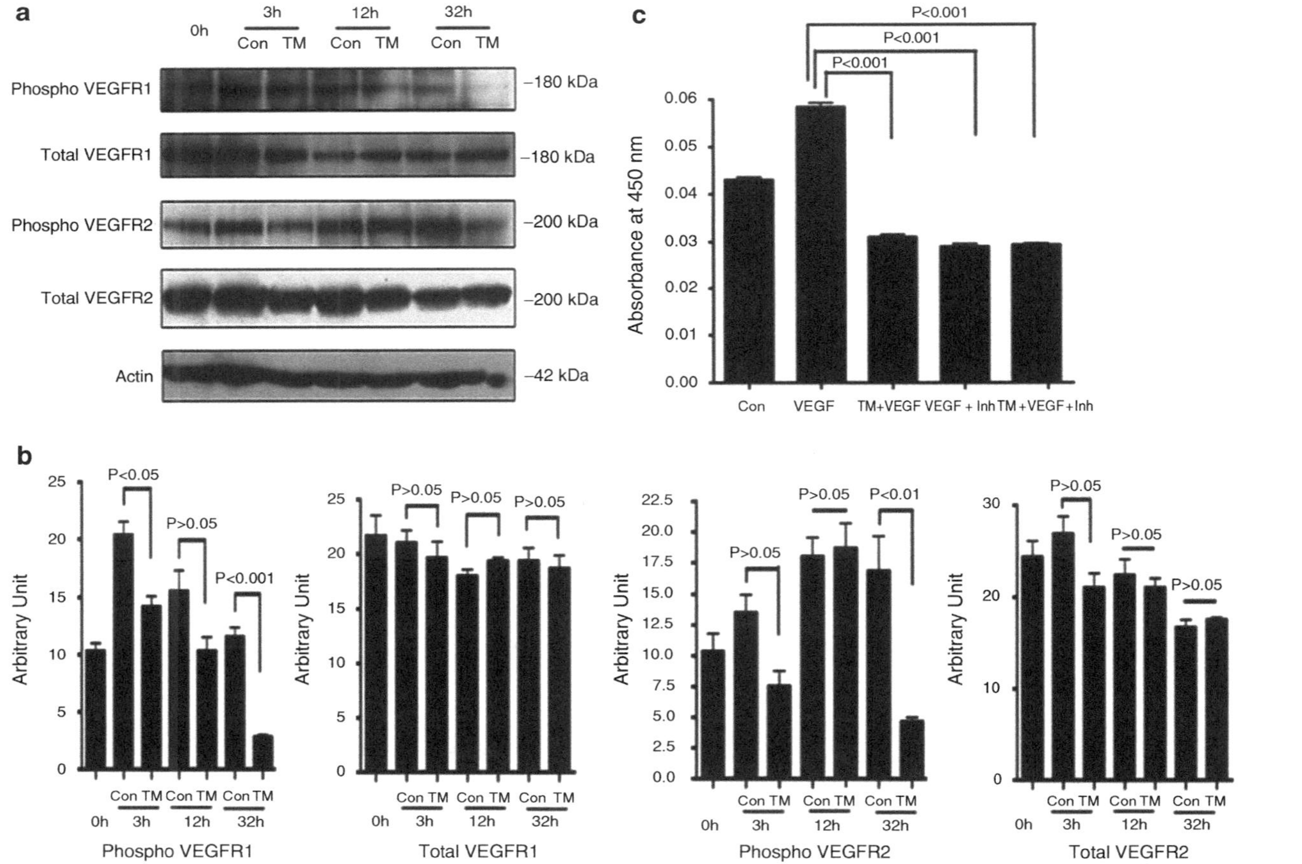

Fig. 22.5 Tunicamycin treatment down regulates VEGF receptors I and II expression (**a**), phosphorylation of VEGFRI and II (**b**) and $VEGF_{165}$-specific protein tyrosine kinase (**c**) (Banerjee et al. 2011)

Table 22.1 Ratio of phosphoVEGFRs to total VEGFRs

	Samples					
Ratio	Con-3 h	TM-3 h	Con-12 h	TM-12 h	Con-32 h	TM-32 h
Phospho-VEGFR1/ total VEGFR1	0.96±0.09	0.72±0.04	0.85±0.09	0.53±0.06	0.59±0.06	0.15±0.01
Phospho-VEGFR2/ total VEGFR2	0.49±0.02	0.36±0.07	0.80±0.06	0.88±0.06	0.99±0.13	0.26±0.01

The ratios are calculated from the values in Fig. 22.5a. The results are expressed as mean ± SE

was observed in cells treated with tunicamycin for 3 h (Fig. 22.5c). To evaluate the specificity of VGF_{165}-stimulated up-regulation of the tyrosine kinase activity, a 17 amino acid peptide, CBO-II (50 μM), which blocks the binding of $VEGF_{165}$ to its receptors, was used. Cells pretreated with CBO-II inhibited the tyrosine kinase activity following $VEGF_{165}$ stimulation ($p<0.001$) to an extent similar to that observed in tunicamycin-treated cells. No further reduction in $VEGF_{165}$-stimulated activity was observed when tunicamycin treated cells were preincubated with CBO-II (Fig. 22.5c). This suggested no changes in the receptor level, but alteration in its signaling ability.

Tunicamycin treatment decreases invasion and chemotaxis of capillary endothelial cells and inhibits angiogenesis in vivo: Capillary proliferation and tissue invasion are important for solid tumor progression. We have studied the Matrigel™ invasion of capillary endothelial cells as well as their chemotactic activity in the presence or absence of tunicamycin. Invasion through the transwell membrane in control plate (Fig. 22.6a(A)) remained almost the same, whereas those migrated through the Matrigel™ reduced significantly in the presence of tunicamycin (Fig. 22.6a(B), far right; $p<0.001$). This supported that tunicamycin makes capillaries less capable of invading the extracellular matrix. In addition to measuring invasion, the cells were also tested for their chemotaxis, i.e. the migratory activity. The rate of migration was ~75 % higher in the presence of $VEGF_{165}$. It was only ~60 % of the control in tunicamycin treated cells and was further reduced (i.e., ~25 %) when both tunicamycin and $VEGF_{165}$ were present (Fig. 22.6a(C), far right; $p<0.001$).

To study the anti-angiogenic effect of tunicamycin *in vivo*, the Matrigel™ plug assay in nude mice was used. Angiogenesis was induced by injecting Matrigel™ containing $VEGF_{165}$. Intense vascularization was observed 10 days post-implantation in Matrigel™ plugs containing $VEGF_{165}$. In contrast, the plugs with Matrigel™ alone appeared pale. Tunicamycin inhibited neo-vascularization, and the plugs appeared pale as well even when $VEGF_{165}$ was present (Fig. 22.6b(A)). Histological analysis of H&E-stained paraffin sections also supported an increased number of vessels in the plugs containing $VEGF_{165}$, as opposed to the plugs with Matrigel™ alone. Tunicamycin treatment caused considerable reduction of neo-vascularization in the plugs containing $VEGF_{165}$ (Fig. 22.6b(B)). Quantitative analysis supported ~75 % reduction in the vessel density in tunicamycin treated Matrigel™ plugs (Fig. 22.6b(B) far right; $p<0.001$).

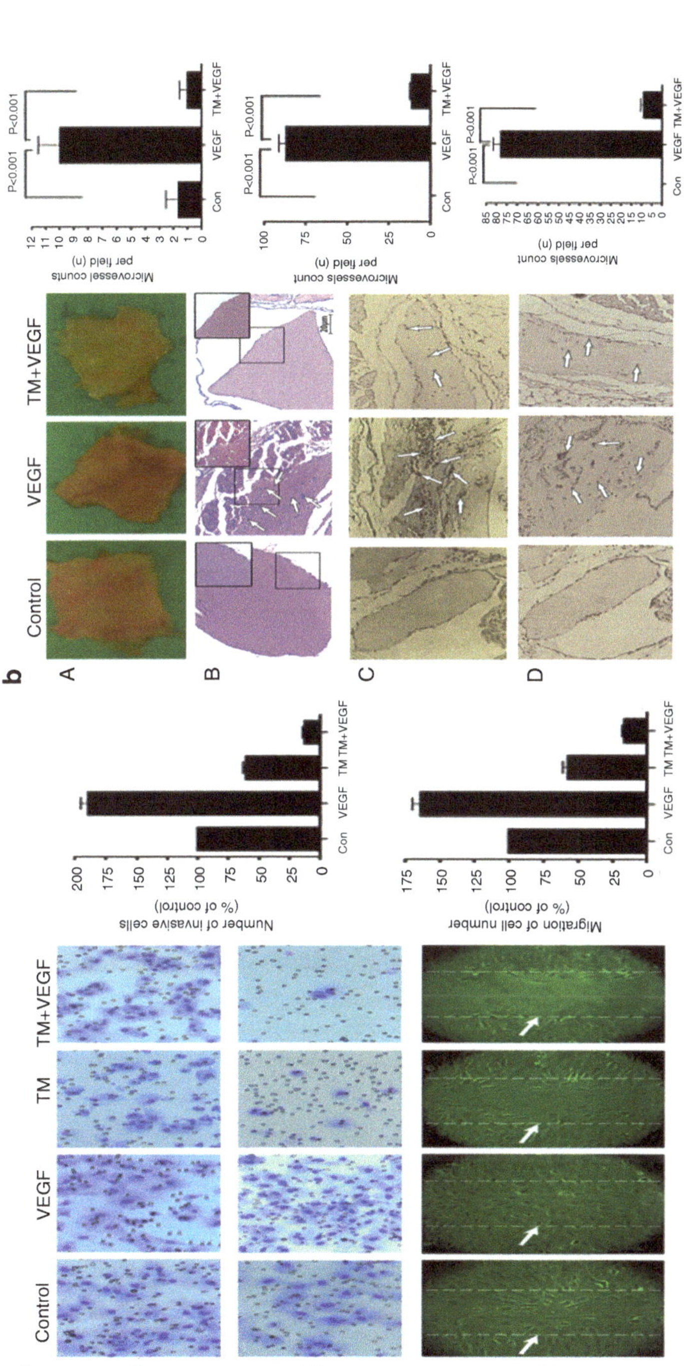

Fig. 22.6 Tunicamycin inhibits (**a**) migration and chemotaxis of capillary endothelial through Matrigel™ matrices in a wound healing assay; (**b**) angiogenesis in Matrigel™ plug assay (Banerjee et al. 2011)

Microvessel density is a vital prognostic biomarker correlating the growth and development of the tumor (Kerbel 2000) and is often used to evaluate the tumor growth (Weidner and Folkman 1996). To assess the tunicamycin-mediated inhibition of vessel formation in Matrigel™ plugs, the microvessel density in the plugs was examined immunohistochemically after staining for CD34 and CD144. Microscopic examination revealed numerous CD34- and CD144-stained vessels in the positive control group ($VEGF_{165}$+heparin) (Fig. 22.6b(C); left to right), but only a few vessels in the group treated with tunicamycin (Fig. 22.6b(C); right). 87±4.04 vessels stained for CD34 and 78±3.4 vessels for CD144 in the control group as opposed to only 12±1.2 vessels in CD34 or 8±1.5 vessels in CD144 in the group treated with tunicamycin, respectively. Quantification (Fig. 22.6b(D)) indicates ~87.5 % decrease ($p<0.001$) in microvessel density because of tunicamycin treatment.

Tunicamycin inhibits breast tumor growth in athymic nude mice: *In vitro* cell culture and *in vivo* Matrigel™ implant models raised the question whether tunicamycin could successfully treat breast tumors. To answer, we took the following approaches: (i) developing orthotopic breast tumor in nude mice with MDA-MB-435 ($ER^-/PR^-/EGFR^+$) cells and treating with tunicamycin by intravenous injection (*group a*; Fig. 22.7); and (ii) developing breast tumor xenografts in nude mice with MDA-MB-231 ($ER^-/PR^-/EGFR^-$) cells and treating with tunicamycin orally (*group b*; Fig. 22.7). In *group a*, tunicamycin treatment was started after the tumor size reached ~3 mm in diameter in 6 days at doses of 0.1, 0.5, and 1.0 mg/kg, respectively, once a week. The results (Fig. 22.7A) indicated that tumor growth progressed in untreated controls and also in the group treated with a low dose of tunicamycin (i.e. 0.1 mg/kg). However, it was markedly reduced (~56 % in 23 days) when treated with 1.0 mg/kg of tunicamycin. Tumor volume regressed ~17 % at 0.5 mg/kg of tunicamycin. It is important to note that 15 times more taxol was needed to match the effect of 1 mg/kg of tunicamycin. The histogram in Fig. 22.7(A) (far right) recorded the tumor weight after each treatment and at the end of the treatment period. H&E staining of paraffin sections of the tumor tissue indicated reduced microvascular density as the tunicamycin concentrations were increased from 0 to 1 mg/kg (Fig. 22.7(B, C)). Figure 22.7(C) (far right) recorded the mitotic index of tumor cells per 10 high power field (HPF) and explained that the mitotic index of the tumor cells declined as a consequence of tunicamycin treatment ($p<0.001$). Ki-67 and VEGF expressions were reduced following tunicamycin treatment (Fig. 22.7(D, E)). In *group b*, tunicamycin treatment was started 7 days after tumor development and was given orally twice a week at a dose of 5 μg/mouse (≈0.25 mg/kg). In the control group (Fig. 22.7(1, 2)), the tumor growth almost doubled in 4 weeks, whereas in the treatment group the growth was reduced by ~65 % in 1 week and maintained (Fig. 22.7(3, 4)). Graphical representation of the observation is in the panel below.

Surface expression of N-glycans on tumor microvasculature and the development of unfolded protein response in tumor microvasculature: Tunicamycin inhibits LLO biosynthesis and consequently the glycosylation of N-linked glycoproteins. To verify the status of N-glycans on breast tumor microvascular endothelial cells, the tissue sections were stained with Texas-Red conjugated WGA and examined

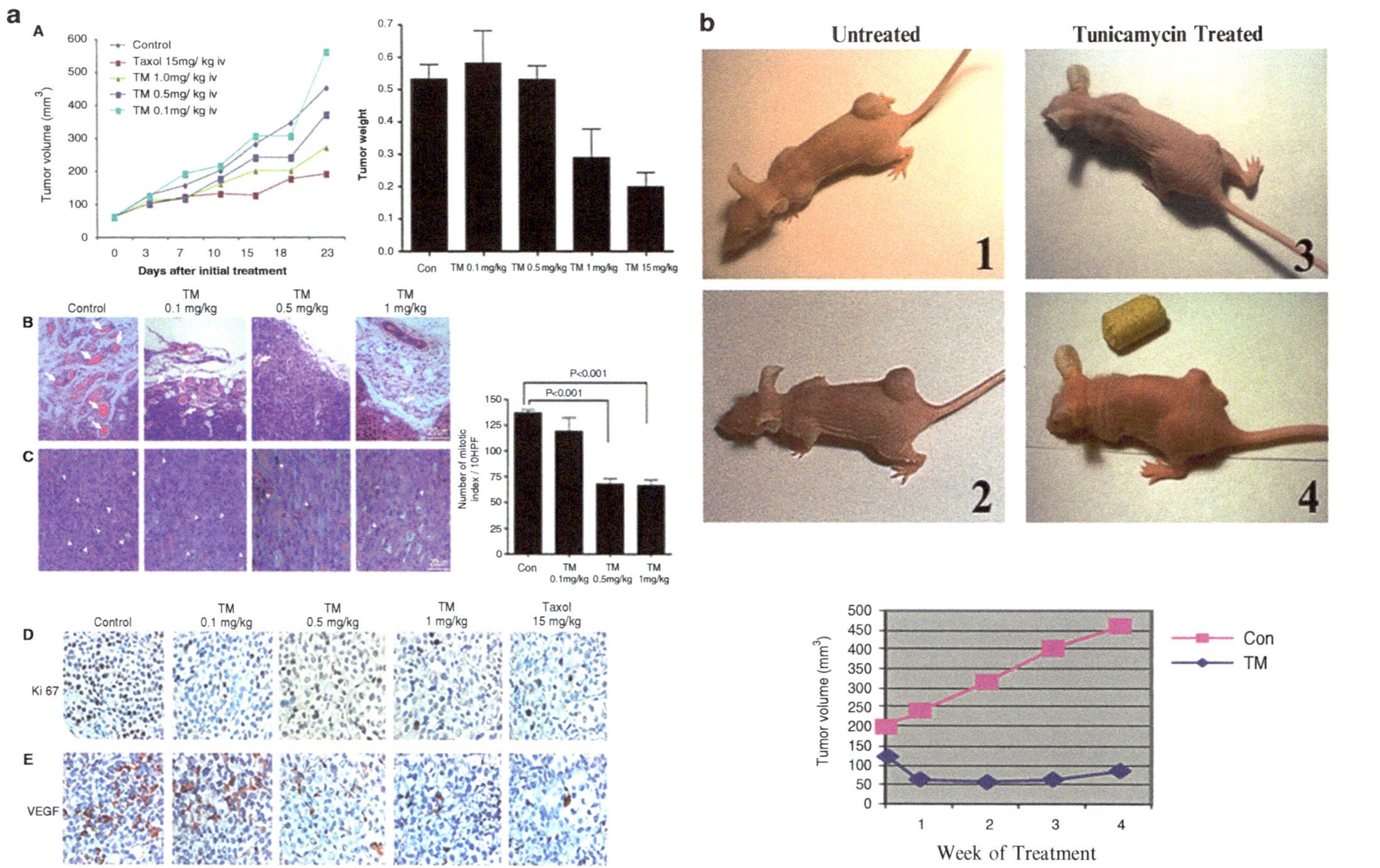

Fig. 22.7 Tunicamycin inhibits double (orthotopic tumor) and triple negative (xenograft) breast tumor progression. Staining of paraffin sections exhibits reduced microvessel density, reduced mitotic index as well as reduced Ki-67 and VEGF expression (a). Caliper measurement of tumor size (b) (Banerjee et al. 2011)

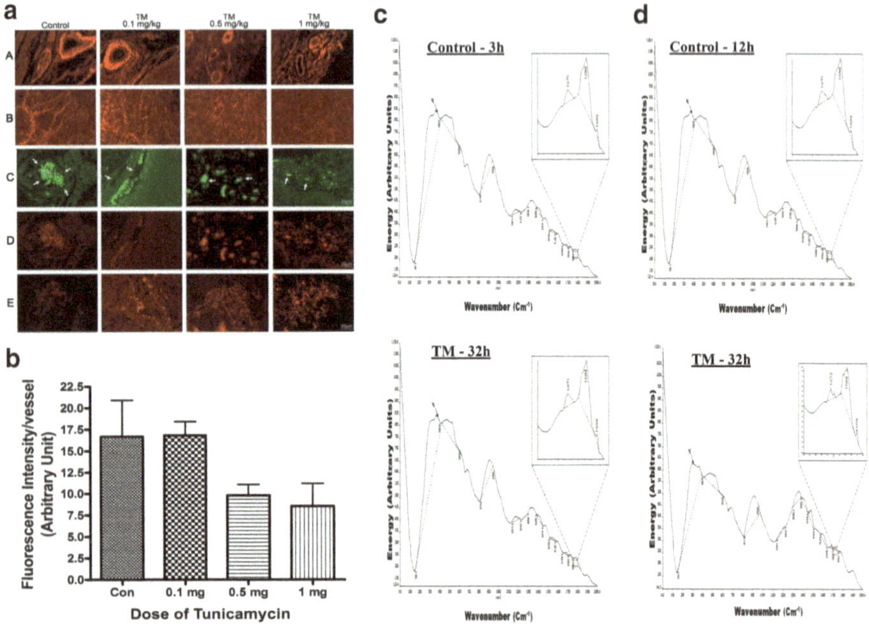

Fig. 22.8 Tunicamycin treatment develops "ER stress" in tumor microvasculature as well as in tumor cells (**a**, **b**) (Banerjee et al. 2011). Analysis of cellular proteome by Raman spectroscopy supports protein denaturation following tunicamycin treatment (**c**, **d**) (Longas et al. 2012)

under a fluorescence microscope. Tumor microvessels in untreated controls were stained markedly but the staining intensity per vessel was reduced almost 50 % upon treating with tunicamycin (Fig. 22.8a(A)). This was further confirmed by quantifying the images with NIH-IMAGE J program (Fig. 22.8b). Tumor cells from untreated control (Fig. 22.8a(B)) exhibited positive WGA staining but the intensity was much less than the endothelial cells. Furthermore, tunicamycin treatment caused a morphological change on tumor cells, and the WGA staining was not only reduced but appeared amorphous.

Inhibition of N-glycan biosynthesis with tunicamycin develops ER stress-mediated *upr* (Zhang and Kaufman 2004). This led us to investigate whether ER stress-mediated *upr* exists in breast tumor microvasculature of the mice receiving tunicamycin. Tumor tissue is heterogeneous. To answer the question, the tumor microvascular endothelial cells were first identified by staining for CD144 (green fluorescence; a marker for endothelial cells) and then stained for the GRP-78 (red fluorescence), an ER chaperone. In untreated control, CD144-stained endothelium appeared as a thin line around the vessel as did the GRP-78 (Fig. 22.8a(C, D)). But, as the tumors were treated with tunicamycin, a high level expression of GRP-78 was observed in microvascular endothelial cells of the tumor (Fig. 22.8a(D)). CD144 also stained the same area (Fig. 22.8a(C)). Both green and red fluorescence of CD144 and GRP-78 co-localized in tumor microvasculature and supported *unfolded protein response* mediated by ER stress. GRP-78 staining in tumor cells following tunicamycin treatment appeared to be increased (Fig. 22.8a(E)). This may be an indirect effect because of nutritional deprivation due to reduced blood flow in the tumor and not necessarily due to an induction of *upr*.

Raman spectroscopy of control and tunicamycin-treated capillary endothelial cells: About one-third of newly synthesized proteins translocate to the lumen of the ER where they are subjected to post-translational modifications and folded into correct 3D structures before being targeted to various cellular organelles or transported to the surface of the cell. This process is highly sensitive to alterations in the ER luminal environment. Among the variety of insults that disrupt protein folding in the ER lumen and activate the *upr*, the most notable chemical inducer is the protein N-glycosylation inhibitor tunicamycin. There is no established analytical tool currently available to detect the ER stress-mediated *upr*. Raman spectroscopy targets protein conformation, we therefore used Raman Spectroscopy to assess the cellular *milieu* following tunicamycin treatment. The spectra were collected from capillary endothelial cells after culturing them in the presence or absence of tunicamycin (1 μg/mL) for 3 and 12 h. Figure 22.8c, d (upper panel) display the spectra in cells cultured for 3 and 12 h in the absence of tunicamycin. The inset is an enlarged 1,700–1,600 cm^{-1} region showing bands at 1,694, 1,684, and 1,672 cm^{-1}, respectively. These bands originate mainly from the C=O stretch of mono-substituted amides as in proteins (Carter and Edwards 2001; Sane et al. 1999). The spectral band near 1,650 cm^{-1} is characteristic of proteins in their secondary and/or tertiary structures (the amide I band). For example, Raman spectra of α-helices peak around 1,655 cm^{-1}, but those of β-sheets peak around 1,670 cm^{-1} (Sane et al. 1999). The band at 1,672 cm^{-1} in the spectra obtained in this study may originate from disordered α-helices (Sane et al. 1999), but the ones at 1,684 and 1,694 cm^{-1} certainly arise from completely disordered protein chains (Sane et al. 1999). The intensities of these bands decreased significantly, when the cells are cultured in the presence of tunicamyin (Fig. 22.8c; panel below), the C=O moieties of their amide groups are not energetically suitable to absorb infra-red light in the 1,700–1,600 cm^{-1} region. The Raman spectra of cells cultured for 12 h without tunicamycin showed typical protein bands; those in the 1,700–1,600 cm^{-1} region are enlarged in the inset (Fig. 22.8d; panel below). The intensities of these bands, however, decreased significantly when the cells were cultured in the presence of tunicamycin (1 μg/mL; Fig. 22.8d, TM). Table 22.2 summarizes the details of these results (Longas et al. 2012).

Effect of tunicamycin nanoparticles on <u>in vitro</u> angiogenesis: Gold nanoparticles (Au NPs) were synthesized by using wet chemical methods (Banerjee et al. 2013). The average diameter of the Au NPs used for functionalization with tunicamycin was 20 nm. The scheme for attachment of tunicamycin to the Au NPs is shown in Fig. 22.9a. To study the cell viability synchronized culture of capillary endothelial cells were incubated either with native tunicamycin (1 μg/mL) or with tunicamycin nanoparticles for 1 h at 37 °C in a CO_2 incubator (5 % CO_2 and 95 % air) in 96-well microtiter plates. At the end, the plates were processed for the MTT assay.

Table 22.2 Intensities of Raman spectral bands in the 1,700–1,600 cm^{-1} region in capillary endothelial cells in the presence or absence of tunicamycin

Sample	Wave number (cm^{-1})	Band Area[a] (absorbance)	S.D.	Change in band area (%)
3 h (no TM)	1,694	0.5400	0.0787	33.33↓
3 h (1 μg/mL TM)		0.3600	0.0759	
12 h (no TM)	"	0.6880	0.1610	32.92 ↓
12 h (1 μg/mL TM)		0.4615	0.0710	
3 h (no TM)	1,684	0.1500	0.0022	20.63 ↓
3 h (1 μg/mL TM)		0.1190	0.0022	
12 h (no TM)	"	0.8104	0.1181	40.08 ↓
12 h (1 μg/mL TM)		0.4856	0.0477	
3 h (no TM)	1,672	0.0408	0.0076	41.85 ↓
3 h (1 μg/mL TM)		0.0237	0.0047	
12 h (no TM)	"	0.0607	0.0010	55.39 ↓
12 h (1 μg/mL TM)		0.0271	0.0084	

TM tunicamycin. S.D. = standard deviation from the mean of three different samples analyzed in duplicate. Spectra were normalized using the Raman Spectroscopy band at 400–155 cm^{-1} which did not change

[a]Expressed in arbitrary units determined by Spectrum 1000 Software

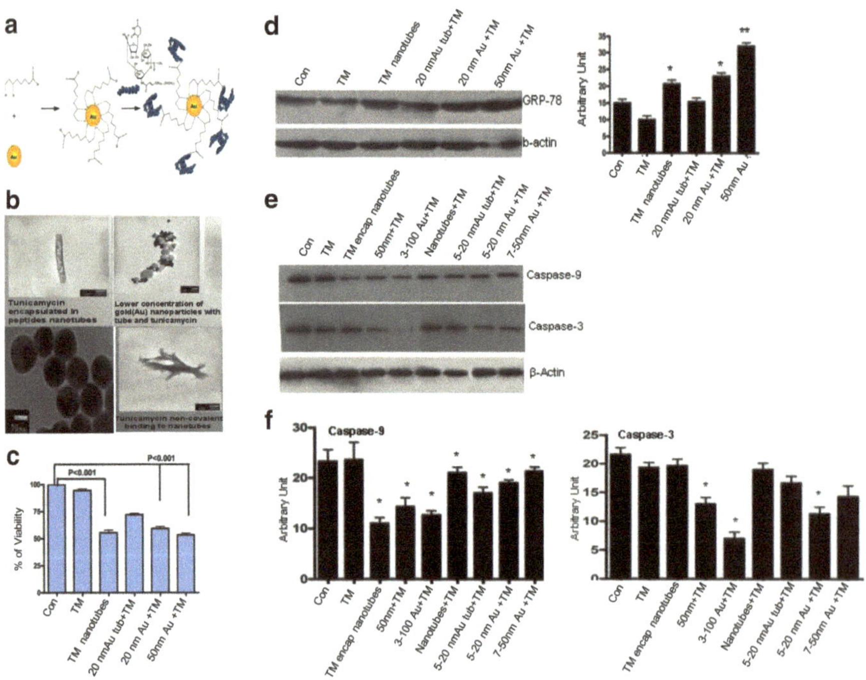

Fig. 22.9 Tunicamycin gold nanoparticles are three times more potent than the native compound. Tunicamycin nanoparticles reduces cellular proliferation by down regulating the expression of cyclin D1 and/or CDK4n (**a**–**c**). The cells experience "ER stress" (**d**) but does not undergo apoptosis (**e**-**f**) (Banerjee et al. 2013)

The results (Fig. 22.9c) indicate that the cell viability was reduced to almost 50 % ($p<0.001$) when treated with tunicamycin conjugated to 20 or 50 nm gold particles. Native tunicamycin under similar condition had no effect. Importantly, both tunicamycin nanotubes and nanoparticles were equally effective in reducing the cell proliferation.

To evaluate, if tunicamycin nanoparticles would induce "ER stress" as the native tunicamycin does, a synchronized population of capillary endothelial cells were treated with various nano-formulated tunicamycin as well as with native tunicamycin just for 1 h. We have analyzed the GRP-78/Bip expression by western blotting as a quantitative measure of "ER stress". Our results indicate that GRP-78/Bip expression was down regulated by ~28 % in cells treated with native tunicamycin, i.e., no detectable "ER stress". The effect of 20 nm gold bound to tunicamycin nanotubes was neutral. On the other hand, the GRP-78/Bip expression was increased by ~126 to 186 % in cells treated with tunicamycin nanotubes, and 20 or 50 nm gold bound to tunicamycin, respectively over the control (Fig. 22.9d).

"ER stress" is associated with apoptotic death of a cell. Since, activation of procaspase 9 to caspase-9 in the aptosome is the initiation of the apoptotic process and the process is concluded by the activation of caspase-3, tunicamycin nanoformulations were tested against caspase-9 and caspase 3 expressions. The expression of caspase-9 was down regulated ($p<0.001$) in cells treated with all tunicamycin nano-formulations but no changes were observed with native tunicamycin (Fig. 22.9e, f). It was selective in the case of caspase-3. The down regulation was markedly enhanced ($p<0.001$) with 50 nm tunicamycin nanoparticles, 100 nm tunicamycin gold nanoparticles, and 20 nm Tunicamycin Au NPs (Fig. 22.9f).

Conclusion and Future Direction

The asparagine-linked protein glycosylation inhibitor tunicamycin targets GlcNAc 1-P transferase and reduces angiogenesis *in vitro* and *in vivo*. It also reduces the progression of double and triple negative breast tumors in nude mice. Importantly, the tunicamycin action survives in tumor microenvironment, and the treated cells do not regain their proliferating activity even after tunicamycin is withdrawn. Tunicamycin gold nanoparticles are three times more potent than its native formulation and the reduction of cellular proliferation in the presence of these nanoparticles is not associated with apoptosis.

Considering the tunicamycin's ability to inhibit both angiogenesis and the tumor cells alike, we are convinced that tunicamycin will be effective against any solid tumor in the clinic serving the concept "one size fits all". But, certainly more work is needed.

Acknowledgement The work was supported in whole or in part, by grants from Susan G. Komen for the Cure BCTR0600582 (to D.K.B.) and NIH/NIMHD 2G12MD007583 (to K.B.)

References

Aksoy S, Dizdar O, Harputluoglu H, Altundag K (2007) Demographic, clinical, and pathological characteristics of Turkish triple-negative breast cancer patients: single center experience. Ann Oncol 18:1904–1906

Baksi K, Tavárez-Pagán JJ, Martínez JA, Banerjee DK (2008) Unique structural motif supports mannosylphospho dolichol synthase: an important angiogenesis regulator. Curr Drug Targets 9:262–271

Banerjee DK (1988) Microenvironment of endothelial cell growth and regulation of protein N-glycosylation. Indian J Biochem Biophys 25:8–13

Banerjee DK, Vendrell-Ramos M (1993) Is asparagine-linked protein glycosylation an obligatory requirement for angiogenesis? Indian J Biochem Biophys 30:389–394

Banerjee DK, Martinez JA, Baksi K (2007) Significance of protein N-glycosylation in breast tumor angiogenesis. In: Maragoudakis ME, Papadimitriou E (eds) Angiogenesis: basic science and clinical applications. Transworld Research Network, Trivandrum, pp 281–301

Banerjee A, Lang JY, Hung MC, Sengupta K, Banerjee SK, Baksi K, Banerjee DK (2011) Unfolded protein response is required in nu/nu mice microvasculature for treating breast tumor with tunicamycin. J Biol Chem. Aug 19;286(33):29127–29138. https://doi.org/10.1074/jbc.M110.169771 for a holistic analysis and conclusion.

Banerjee A, Johnson KT, Banerjee IA, Banerjee DK (2013) Nanoformulation enhances anti-angiogenic efficacy of tunicamycin. Transl Cancer Res. August 21;2(4):240–255. https://doi.org/10.3978/j.issn.2218-676X.2013.08.16098.

Berkowitz GS, Kelsey JL (2006) Epidemiology of breast cancer. In: Marchant DJ (ed) Diagnosis and management of breast cancer. Elsevier, New York

Boring CC, Squires TS, Tong T (1991) Cancer statistics. CA Cancer J Clin 41:19–39

Broomhall SR, Hallissey MT, Whiting J, Scholefield J, Tierney G, Stuart RC, Hawkins RE, McCulloch P, Maughan T, Brown PD, Baillet M, Fielding JW (2002) Marimastat as maintenance therapy for patients with advanced gastric cancer: a randomised trial. Br J Cancer 86:1864–1870

Carey LA, Perou CM, Livasy CA, Dressler LG, Cowan D, Conway K, Karaca G, Troester MA, Tse CK, Edmiston S, Deming SL, Geradts J, Cheang MC, Nielsen TO, Moorman PG, Earp HS, Millikan RC (2006) Racce, breast cancer subtypes, and survivaal in the Carolina Breast Study. JAMA 295:2492–2502

Carter EA, Edwards HGM (2001) Biological applications of Raman spectroscopy. In: Gremlich H-U, Bing Y (eds) Infrared and Raman spectroscopy of biological materials. Marcel Deker, New York, pp 421–475

Cazet A, Julien S, Bobowski M, Burchell J, Delannoy P (2010) Tumour-associated carbohydrate antigens in breast cancer. Breast Cancer Res 12:204–220

Cleator S, Heller W, Coombes RC (2007) Triple-negative breast cancer: therapeutic options. Lancet Oncol 8:235–244

Colleoni M, Rocca A, Sandri MT, Zorzino L, Masci G, Nolè F, Peruzzotti G, Robertson C, Orlando L, Cinieri S, de BF, Viale G, Goldhirsch A (2002) Low-dose oral methotrexate and cyclophosphamide in metastatic breast cancer: antitumor activity and correlation with vascular endothelial growth factor levels. Ann Oncol 13:73–80

Dent RD, Trudeau M, Pritchard KI, Hanna WM, Kahn HK, Sawka CA, Lickley LA, Rawlinson E, Sun P, Narod SA (2007) Triple-negative breast cancer clinical features and patterns of recurrence. Clin Cancer Res 13:4429–4434

Erlichman C, Adjei AA, Alberts SR, Sloan JA, Goldberg RM, Pitot HC, Rubin J, Atherton PJ, Klee GG, Humphrey R (2001) Phase I study of the matrix metalloproteinase inhibitor, BAY 12-9566. Ann Oncol 12:389–395

Folkman J (1990) What is the evidence that tumors are angiogenesis dependent? J Natl Cancer Inst 82:4–6

Folkman J (1992) The role of angiogenesis in tumor growth. Semin Cancer Biol 3:65–71

Forbes JF (1997) The control of breast cancer: the role of tamoxifen. Semin Oncol 24(Suppl 1): S-15–S-19
Fotsis T, Zhang Y, Pepper MS, Adlercreutz H, Montesano R, Nawroth PP, Schweigerer L (1994) The endogenous oestrogen metabolite 2-methoxyoestradiol inhibits angiogenesis and suppresses tumour growth. Nature 368:237–239
Fuster MM, Esko JD (2005) The sweet and sour of cancer: glycans as novel therapeutic targets. Nat Rev 5:526–542
Gastl G, Hermann T, Steurer M, Zmija J, Gunsilius E, Unger C, Kraft A (1997) Angiogenesis as a target for tumor treatment. Oncology 54:177–184
Hakomori S (1996) Tumor malignancy defined by aberrant glycosylation and sphingo(glyco)lipid metabolism. Cancer Res 56:5309–5318
Hortobagyl GN, Buzdar AU (1995) Current status of adjuvant systemic therapy for primary breast cancer: progress and controversy. CA Cancer J Clin 45:199–226
Ingber DE, Folkman J (1989) How does extracellular matrix control capillary morphogenesis? Cell 58:803–805
Jemal A, Siegel R, Ward E, Murray T, Xu J, Smigal C, Thun J (2006) Cancer statistics. CA-Cancer J Clin 56:106–130
Kalluri R, Weinberg RA (2009) The basics of epithelial-mesenchymal transition. J Clin Invest 119:1420–1428
Kelsey JL, Gammon MD (1991) The epidemiology of breast cancer. CA Cancer J Clin 41:146–165
Kerbel RS (2000) Tumor angiogenesis: past, present and the near future. Carcinogenesis 21:505–515
Klauber N, Parangi S, Flynn E, Hamel E, D'Amato RJ (1997) Inhibition of angiogenesis and breast cancer in mice by the microtubule inhibitors 2-methoxyestradiol and taxol. Cancer Res 57:81–86
Livasy CA, Karaca G, Nanda R, Tretiakova MS, Olopade OI, Moore DT, Perou CM (2006) Phenotypic evaluation of the basal-like subtype of invasive breast carcinoma. Mod Pathol 19:264–271
Longas MO, Kotapati A, Prasad KP, Banerjee A, Santiago J, Baksi K, Banerjee DK (2012) Balancing life with glycoconjugates: monitoring unfolded protein response-mediated anti-angiogenic action of tunicamycin by Raman spectroscopy. Pure Appl Chem 84(9):1907–1918. https://doi.org/10.1351/PAC-CON-12-01-06.
Lynn AG, Ries MS (1995) Top 5 cancers for females and males in the US. J Natl Cancer Inst 87:867
Martínez JA, Torres-Negrón I, Amigó LA, Banerjee DK (1999) Expression of Glc3Man9GlcNAc2-PP-Dol is a prerequisite for capillary endothelial cell proliferation. Cell Mol Biol (Noisy-le-Grand) 45:137–152
Martínez JA, Torres-Negrón I, Amigó LA, Roldán RA, Mendéz A, Banerjee DK (2000) Tunicamycin inhibits capillary endothelial cell proliferation by inducing apoptosis. Targeting dolichol-pathway for generation of new anti-angiogenic therapeutics. Adv Exp Med Biol 476:197–208
Muramatsu T (1993) Carbohydrate signals in metastasis and prognosis of human carcinomas. Glycobiology 3:291–296
Nguyen M, Folkman J, Bischoff J (1992) 1-Deoxymannojirimycin inhibits capillary tube formation in vitro. Analysis of N-linked oligosaccharides in bovine capillary endothelial cells. J Biol Chem 267:26157–26165
Nguyen M, Strubel NA, Bischoff J (1993) A role for sialyl Lewis-X/A glycoconjugates in capillary morphogenesis. Nature 365:267–269
Oliveira CM, Banerjee DK (1990) Role of extracellular signaling on endothelial cell proliferation and protein N-glycosylation. J Cell Physiol 144:467–472
Pili R, Chang J, Partis RA, Mueller RA, Chrest FJ, Passaniti A (1995) The alpha-glucosidase I inhibitor castanospermine alters endothelial cell glycosylation, prevents angiogenesis, and inhibits tumor growth. Cancer Res 55:2920–2926

Rhee J, Han SW, Oh DY, Kim JH, Im S-A, Han W, Park IA, Noh D-Y, Bang Y-J, Kim T-Y (2008) The clinicopathologic characteristics and prognostic significance of triple-negativity in node-negative breast cancer. BMC Cancer 8:307

Ries LAG, Kosary CKL, Hankey BF et al (1999) SEER cancer statistics review 1973-2996. National Cancer Institute, Bethesda

Robinson DR, Wu YM, Lin SF (2000) The protein tyrosine kinase family of the human genome. Oncogene 19:5548–5557

Rudek MA, Figg WD, Dyer V, Dahut W, Turner ML, Steinberg SM, Liewehr DJ, Kohler DR, Pluda JM, Reed E (2001) Phase I clinical trial of oral COL-3, a matrix metalloproteinase inhibitor, in patients with refractory metastatic cancer. J Clin Oncol 19:584–592

Sane SU, Cramer SM, Przybycien TM (1999) A holistic approach to protein secondary structure characterization using amide I band Raman spectroscopy. Anal Biochem 269:255–272

Sondik EJ (1994) Breast cancer trends: incidence, mortality and survival. Cancer 74:995–999

Stockmans G, Deraedt K, Wildiers H, Moerman P, Paridaens R (2008) Triple-negative breast cancer. Curr Opin Oncol 20:614–620

Taniguchi N, Honke K, Fukuda M (2002) Handbook of glycosyltransferases and their elated genes. Springer, Tokyo

Tiganis T, Leaver DD, Ham K, Friedhuber A, Stewart P, Dziadek M (1992) Functional and morphological changes induced by tunicamycin in dividing and confluent endothelial cells. Exp Cell Res 198:191–200

Uhr JW, Scheuermann RH, Street NE, Vitetta ES (1997) Cancer dormancy: opportunities for new therapeutic approaches. Nat Med 3:505–509

Van Calster B, Vanden Bempt I, Drijkoningen M, Pochet N, Cheng J, Van Huffel S, Hendrickx W, Decock J, Huang HJ, Leunen K, Amant F, Berteloot P, Paridaens R, Wildiers H, Van Limbergen E, Weltens C, Timmerman D, Van Gorp T, Smeets A, Van den Bogaert W, Vergote I, Christiaens MR, Neven P (2009) Axillary lymph node status of operable breast cancers by combined steroid receptor and HER-2 status: triple positive tumours are more likely lymph node positive. Breast Cancer Res Treat 113:181–187

Wandall HH, Blixt O, Tarp MA, Pedersen JW, Bennett EP, Mandel U, Ragupathi G, Livingston PO, Hollingsworth MA, Taylor-Papadimitriou J, Burchell J, Clausen H (2010) Cancer biomarkers defined by autoantibody signatures to aberrant O-glycopeptide epitopes. Cancer Res 70:1306–1313

Weidner N, Folkman J (1996) Tumoral vascularity as a prognostic factor in cancer. Important Adv Oncol 3:167–190

Zhang K, Kaufman RJ (2004) Signaling the unfolded protein response from the endoplasmic reticulum. J Biol Chem 279:25935–25938

Zwick E, Bange J, Ullrich A (2001) Receptor tyrosine kinase signalling as a target for cancer intervention strategies. Endocr Relat Cancer 8:161–173

Chapter 23
Involvement of Vascular Endothelial Growth Factor in Serotonin 1A Receptor-Mediated Neuroproliferation in Neonatal Mouse Hippocampus

S. Samaddar, B. Ranasinghe, S.J. Tantry, P.R. Debata, and P. Banerjee

Abbreviations

5-HT_{1A}-R	Serotonin 1A receptor
D	8-OH-DPAT (5-HT_{1A}-R agonist)
M	Myr-εV1-2 (N-Myr-EAVSLKPT) (a PKCε translocation inhibitor)

S. Samaddar
The College of Staten Island (CUNY), Staten Island, NY 10314, USA

B. Ranasinghe
CUNY Graduate Center, The College of Staten Island (CUNY),
Staten Island, NY 10314, USA

S. Tantry
Department of Chemistry, The College of Staten Island (CUNY),
Staten Island, NY 10314, USA

P. Debata
Center for Developmental Neuroscience, The College of Staten Island (CUNY),
Staten Island, NY 10314, USA

P. Banerjee, Ph.D. (✉)
The College of Staten Island (CUNY), Staten Island, NY 10314, USA

CUNY Graduate Center, The College of Staten Island (CUNY),
Staten Island, NY 10314, USA

Department of Chemistry, The College of Staten Island (CUNY),
Staten Island, NY 10314, USA

Center for Developmental Neuroscience, The College of Staten Island (CUNY),
Staten Island, NY 10314, USA
e-mail: probal.banerjee@csi.cuny.edu

A. Chakrabarti, A. Surolia (eds.), *Biochemical Roles of Eukaryotic Cell Surface Macromolecules*, Advances in Experimental Medicine and Biology 842,
DOI 10.1007/978-3-319-11280-0_23

SU5416/S8442	(Semaxanib, VEGFR1/2 inhibitor)
SGZ	Subgranular zone
U	U0126 (inhibitor of MEK)
W or WAY	WAY100635 (5-HT$_{1A}$-R antagonist)

Introduction

Serotonin is one of the neurotransmitters that profoundly affect brain development. Serotonin influences neurogenesis, cell migration, dendritic refinement and synaptic plasticity. At a later developmental stage, it is shown to influence the length and branching of dendrites in the hippocampus and cortex (Sodhi and Sanders-Bush 2004). The serotonin 1A receptors (5-HT$_{1A}$-R) are found in both pre-synaptic and post-synaptic neurons of the hippocampus. While studies performed by other research teams have elucidated the signaling pathways linked to the presynaptic 5-HT$_{1A}$-R on the raphé-derived projections (Kushwaha and Albert 2005), our publications have systematically elucidated multiple 5-HT$_{1A}$-R-linked signaling pathways in the post-synaptic neurons located in the hippocampus and prefrontal cortex (Adayev et al. 1999, 2003, 2005; Mehta et al. 2007; Mogha et al. 2012; Purkayastha et al. 2012).

Anxiety is one of the psychiatric disorders that manifest early, even before the age of 11 years (Kessler et al. 2005). Thus, neonatal brain development is likely to play an important role in the precipitation of this disorder. Many studies have linked impaired 5-HT$_{1A}$-R expression or function to the occurrence of anxiety and depression. Mice deficient in 5-HT$_{1A}$-R have been shown to display heightened anxiety-like behavior (Heisler et al. 1998; Ramboz et al. 1998; Parks et al. 1998). A conditional rescue strategy involving 5-HT$_{1A}$-R expression only in the hippocampus and cortex, but not in the raphé nuclei, demonstrated that neonatal 5-HT$_{1A}$-R activity in the forebrain was crucial for the development of normal anxiety-like behavior (Gross et al. 2002). Antidepressants act through the serotonergic pathway to increase neurogenesis in the hippocampus and attenuate anxiety-like behavior (Santarelli et al. 2003). Selective desensitization of the 5-HT$_{1A}$-R autoreceptors in raphé facilitates antidepressant action (Lemonde et al. 2004). Thus serotonergic signaling *via* 5-HT$_{1A}$-R during the neonatal period is important for the developmental programming of anxiety-related behavior (Lo Iacono and Gross 2008).

Our prior studies have suggested that PKCε plays an important role in early hippocampal development (neuroblast proliferation) and that the isoforms α, β, γ, and δ are involved later in synaptic plasticity (Mehta et al. 2007; Purkayastha et al. 2009). We also elucidated the expression profiles and intrahippocampal location of the isoforms ε and α at the two important time points P6 (peak of cell proliferation) and P15 (peak of synaptogenesis) (Purkayastha et al. 2009). Our subsequent studies confirmed the involvement of PKCα in synaptogenesis in the P15 mouse hippocampus (Mogha et al. 2012). Furthermore, our earlier studies also indicated the involvement of PKCε upstream of ERK1/2 in 5-HT_{1A}-R-mediated neuroproliferative signaling in the hippocampal neuron-derived HN2-5 cells and cultured hippocampal slices (Mehta et al. 2007; Samaddar et al. 2013). We then used a selective activator of PKCε, DCP-LA (Shimizu et al. 2011; Hongpaisan et al. 2011), in the 5-HT_{1A}-R(-/-) mice to stimulate PKCε. This treatment corrected neonatal hippocampal development and alleviated the hyper-anxiety phenotype that would otherwise develop in the 5-HT_{1A}-R(-/-) mice in adulthood (Samaddar et al. 2013).

However, a major question still remains: "is there any post-ERK1/2 amplification of the 5-HT_{1A}-R→→PKCε→→ERK1/2 signaling activity that can account for the massive proliferative effect of 5-HT_{1A}-R stimulation observed by us in the P6 hippocampus?" Apart from its well-studied angiogenic role in other regions, vascular endothelial growth factor (VEGF) has been shown to have a neurogenic effect in the hippocampus (Cao Lei et al. 2004). In fact these two VEGF-regulated processes might be inter-dependent, as shown by a close association of the vasculature with the neuroprogenitor cells in the subgranular zone (SGZ) of the hippocampus (Palmer et al. 2000). The major VEGF receptor involved in angiogenesis is VEGFR-2 (Tille et al. 2001), which has been believed to play a multifunctional role in neurodevelopment and neurological disorders. Thus, it can modulate various neurodevelopmental processes like neuronal survival, migration, axonal growth, axon guidance, neuroblast proliferation (Carmeliet and Almodovar 2013; Louissaint et al. 2002), and neurogenesis (Cao Lei et al. 2004).

VEGF has been defined as a potential target of antidepressants (Warner-Schmidt and Duman 2008). Pharmacological inhibition of VEGF receptor signaling is sufficient to block behavioral actions of Fluoxetine (Greene et al. 2009). Multiple classes of antidepressants induce VEGF, which in turn stimulates VEGF-Flk1 (a.k.a. VEGFR-2) (Tille et al. 2001; Warner-Schmidt and Duman 2007). Dentate gyrus (DG)-specific knockdown of VEGF has revealed that expression of VEGF is essential during a specific time window of cAMP-CREB signaling for the action of the antidepressants (Lee Jeong-Sik 2009). By blocking VEGFR with the antagonist SU5416 researchers have demonstrated the involvement of VEGFR in neurogenesis after traumatic brain injury, and that this process involves the Raf/MEK/ERK cascade (Lu et al. 2011). Such studies have illustrated the crucial role of VEGF in neuroproliferation and neurogenesis in adult animals. They have also shown that the Raf/MEK/ERK cascade mediates this outcome. Our study establishes VEGF as a post-ERK neurogenic factor (Lu et al. 2011) that amplifies 5-HT_{1A}-R mediated signaling of neuroblast proliferation in neonatal mouse hippocampus, thereby orchestrating early hippocampal development.

Materials and Method

Animals: The wild type (WT) (C57BL/6) mice were obtained from Taconic and bred in CSI Animal Care Facility. Animals were kept in a 12-h light/dark cycle with *ad libitum* access to food and water. Mice were housed in the College of Staten

Island (CSI) Animal Care Facility and handled following a protocol approved by the CSI Institutional Animal Care Committee.

Materials: Reagents and antibodies used were as follows: 5-bromo-2′-deoxyuridine (BrdU), Xylazine, 8-OH-DPAT, WAY 100635, and SU5416 from Sigma Chemicals (St Louis, MO, USA); U0126 from Calbiochem (La Jolla, CA, USA); Neuronal nuclei (NeuN) (Millipore, Billerica, MA, USA), VEGF antibodies from Millipore (Temecula, CA, USA); Ketamine from Hospira Inc. (Lake Forest, IL); Myr-εV1-2 (N-Myr-EAVSLKPT) prepared by solid-phase synthesis using a peptide synthesizer according to published reports (Chen et al. 2001; Johnson et al. 1996; Teng et al. 2008). The Alexafluor-labeled fluorescent secondary antibodies were obtained from Invitrogen (Carlsbad, CA, USA), and antibodies against P-Ser^{729}-PKCε and Doublecortin (DCX) were procured from Santa Cruz Biotechnology (Santa Cruz, CA).

Organotypic cultures of P6 hippocampal slices: Hippocampi were isolated from P6 pups, sectioned, and cultured on membrane inserts and treated with drugs as described in detail in our earlier publications (Mehta et al. 2007; Mogha et al. 2012).

BrdU injection and intra-hippocampal infusion: Mouse pups at post-natal day 6 (P6) were anaesthetized with a mixture of ketamine (100 mg/kg body weight) and xylazine (10 mg/kg). Two hours before intra-hippocampal injection of drugs for neuroproliferation studies, pups were injected with BrdU (100 mg/kg) intraperitoneally. The pups were then anesthetized and stereotaxically infused with the 5-HT_{1A}-R agonist (±)-8-hydroxy-2-(di-*N*-propylamino) tetralin (8-OH-DPAT) (D) (final concentration in the hippocampus: 100 nM) in the absence and presence of the 5-HT_{1A}-R antagonist *N*-[2-[4-(2-methoxy-phenyl)-1-piperazinyl] ethyl]-*N*-(2-pyridinyl) cyclohexane-carboxamide (WAY 100635) (final concentration: 10 μM), PKC epsilon translocation inhibitor Myristoyl-EAVSLKPT (Myr-εV1-2) (M) (final concentration : 400 nM), and MAPK kinase (MEK) inhibitor U0126 (final concentration: 10 μM), VEGFR1/2 antagonist (a potent and selective inhibitor of Flk1/KDR) Semaxanib, SU5416/S8442 (final conc: 10 μM). The stereotaxic coordinates used were: (V) 2 mm under the external surface of the scalp skin in the frontoparietal area of the right hemisphere, (L) 2 mm from the midline in the lateral-medial plane, and (AP) 3 mm from the junction of the sagittal and lamboid sutures in the rostrocaudal plane.

Drug Concentrations: Hippocampi were isolated, weighed, and then divided by the mean brain tissue density of 1.02 mg/μL to obtain the average hippocampal volume of a P6 C57BL6 mouse as 5 μL. Stock concentrations of drugs were calculated based on the final volume of 5 μL for a P6 hippocampus. The total infusate volume was 0.5 μL for each drug or carrier and the final drug concentrations used were as shown below.

8-OH-DPAT: final concentration of 100 nM; **WAY100635**: final concentration of 10 μM; **Myr-εV1-2** or **M**: final concentration of 400 nM; **U0126**: final concentration of 10 μM; **SU5416**: final concentration of 10 μM (Mologni et al. 2006). The infused carrier (vehicle control) contained PBS plus 0.5 % DMSO.

Cryosectioning of brains: 24 h following drug treatment, the pups were transcardially perfused with 4 % paraformaldehyde, the brains isolated, and placed in 4 % paraformaldehyde. After at least 24 h of post-fixing, the brains were soaked in 30 % sucrose until they submerged completely. The brains were separated into left and right hemispheres using a surgical blade, (World Precision Instruments), mounted with Cryomold Intermediate (Tissue-Tek) using dry ice, and cryosectioned using a cryostat (Vibratome, ULTRApro 5000, SIMS Corporation, South Korea) into 30-μm coronal sections for immunohistochemical staining. Before each experiment, one pup was injected with Coomassie blue to confirm intra-hippocampal injection with the coordinates mentioned above (Mogha et al. 2012).

Immunostaining of brain sections: Carefully selected slices were subjected to treatment with 50 % formamide in 2x SSC (0.3 M NaCl, 0.03 M sodium citrate buffer) for 2 h at 65 °C, washed with 2x SSC for 5 min, treated with 1N HCl for 10 min on ice, followed by 2N HCl for 20 min at 37 °C, neutralized by 0.1 M Boric acid for 10 min, and then rinsed three times with 1x TBS buffer (10 min each). Sections were then blocked with a solution of 3 % normal goat serum/0.25 %Triton X-100/1x TBS for 30 min, followed by treatment with primary (1°) antibody in 2 % normal goat serum and 0.25 % Triton X-100 in 1x TBS overnight at 4 °C. The sections were washed with 1x TBS/1x PBS, and then treated with fluorescent secondary (2°) antibodies. After overnight incubation with the 2° antibodies, sections were washed and then mounted on slides with Prolong Gold antifade reagent (Molecular Probes, Eugene, OR, USA) covered with cover slips for Confocal Imaging. Antibody concentrations used: NeuN (1:500), doublecortin (DCX) (1:500), BrdU (1:250), VEGF (1:300), P-PKCε (1:50), and Ki67 (1:150). Secondary antibodies used: Alexafluor 488 (green) (1:500) or Alexafluor 568 (red) (1: 500) or Alexafluor 633 (blue) (1:750).

Confocal microscopy and statistical analysis: The immunostained sections were imaged using the SP2 confocal microscope (Leica Microsystems, Heidelberg, Germany). 8-OH-DPAT-evoked cell proliferation was analyzed in absence and presence of WAY, M, U0126, and SU5416, by monitoring BrdU (proliferation marker) and doublecortin (DCX) (neuroblast marker) double-labeled cells. Experiments and images were analyzed in triplicates for proper statistical analysis for each treatment. Single and volume-rendered images, created from Z-stacks, were used for analysis. 3D-animation videos were created to exactly view the positions of the cells proliferating and also the mature neurons. BrdU and DCX double-labeled cells were counted with ImageJ point selection tool, and then plotted in GraphPad Prism or Excel for bar graph construction and statistical analysis. Depending on the number of groups compared, student *t*-test and One-way ANOVA (along with Tukey as post-hoc test) were used to calculate the statistical significance.

Results

Serotonin 1A receptor mediated signaling through PKCε causes boosted cell proliferation in the DG of cultured P6 hippocampal slices: The proliferative potential of this 5-HT_{1A}-R-linked, PKCε-mediated ERK1/2 activation was tested in

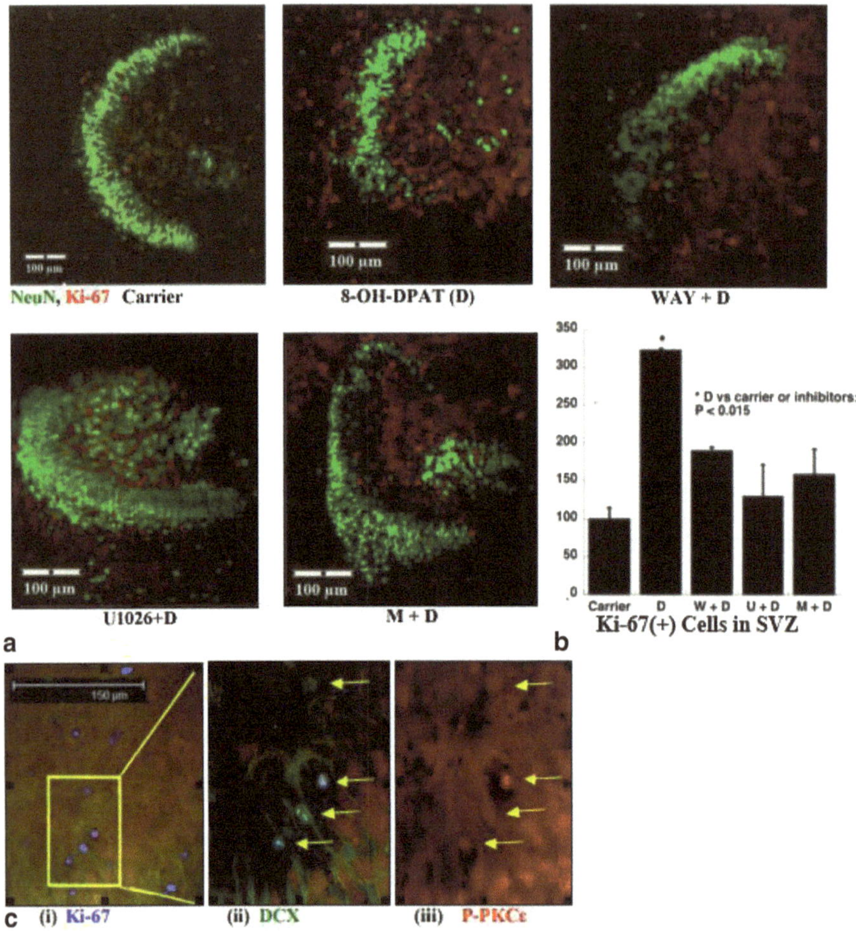

Fig. 23.1 Agonist treatment of organotypic cultures (6DIV) show that 5-HT$_{1A}$-R-mediated and PKCε-relayed stimulation of Erk in the P6 DG causes increased cell proliferation, which is marked by Ki-67 staining. (**a, b**) 8-OH-DPAT (D) treatment (24 h) boosts Ki-67 staining in the subgranular zone (SGZ), which is blocked in the presence of WAY (10 μM), M (400 nM), and U (10 μM) ($n = 3$). (**c**) Ki-67(+) cells near the SGZ (pseudo colored purple from far red) (**i**) were also positive for doublecortin (DCX) (*green*) (**ii**) and P-PKCε (*red*) (**iii**)

organotypic cultures of P6 hippocampal slices (Fig. 23.1) (Mehta et al. 2007; Mogha et al. 2012). Agonist (8-OH-DPAT)-evoked increase in Ki67(+) proliferating cells (red) in the dentate gyrus (DG) was eliminated in the presence of 5-HT$_{1A}$-R antagonist WAY100635 (W), the PKCε inhibitor Myr-εV1-2 (M), and also the MEK inhibitor U1026 (U) (Fig. 23.1a). Statistical analysis revealed a significant increase in Ki67-positive DG cells in 8-OH-DPAT-treated brains with respect to those treated with carrier and inhibitors ($p < 0.015$) (Fig. 23.1b). Since Ki67 is a general proliferation marker, we also tested for the expression of doublecortin (DCX), which is a marker for neuroprogenitors. As shown in Fig. 23.1c, the Ki67(+) proliferating cells (purple) in the DG of the 8-OH-DPAT-treated sections were also DCX(+) (green). Intriguingly, these proliferating neuroblasts also showed nuclear expression of P-^{729}Ser-PKCε (red), which is the activated form of PKCε. This suggests a role of PKCε in neuroproliferation.

Agonist stimulation of 5-HT$_{1A}$-R mediated signaling causes induced neuroblast proliferation in P6 mouse hippocampus: The role of 5-HT$_{1A}$-R mediated signaling in neuroblast proliferation in neonatal dentate gyrus (DG) was next studied *in vivo* using systemic (i.p.) injection of BrdU followed by intrahippocampal infusion of 8-OH-DPAT in the absence and presence of WAY, Myr-εV1-2 (M), and U1026 (U) into the right hippocampus (Fig. 23.2). Immunohistochemical staining of cryo-sections of the ipsilateral lobe revealed a marked increase in the number of cells that were double positive for BrdU (red) and DCX (blue) in the 8-OH-DPAT-treated brain sections ($p<0.05$) (Fig. 23.2a, b, f). This proliferative effect of 8-OH-DPAT was blocked in the presence of WAY, M, and U0126, respectively (Fig. 23.2c, d, e), revealing that this 5-HT$_{1A}$-R signaling activity was mediated by PKCε and ERK1/2. This demonstrated that the identified 5-HT$_{1A}$-R mediated signaling pathway plays a key role in neuroblast proliferation in the SGZ of the DG. We did observe a few cells, which were double positive for NeuN (green) and BrdU (red), but this was indicative of DNA repair occurring in some mature granule neurons of the DG.

In vivo stimulation of the 5-HT$_{1A}$-R in the hippocampus in the P6 mouse causes induction of VEGF and DCX in the dentate gyrus: To study if 5-HT$_{1A}$-R signaling induces VEGF in the Sub-granular Zone (SGZ) of the P6 hippocampus, we stimulated the receptor through intra-hippocampal infusion of 8-OH-DPAT in the absence and presence of 5-HT$_{1A}$-R antagonist WAY and the PKC epsilon inhibitor, Myr-εV1-2 (M). After 24 h, we monitored the expression of VEGF and the neuroblast marker doublecortin (DCX) in the SGZ of the DG (Fig. 23.3). Double staining with DCX (green) (top panel, Fig 23.3a), and VEGF (blue) (bottom panel, Fig. 23.3a) clearly showed the expression of VEGF among the DCX(+) cells in the SGZ. The 8-OH-DPAT-evoked induction of VEGF was blocked in the presence of WAY and Myr-εV1-2 (M) ($p<0.0001$), (Fig. 23.3a, b). Since VEGF induction was also blocked in the presence of M, it was located downstream of PKC epsilon in this 5-HT$_{1A}$-R-mediated signaling cascade.

VEGFR1/2 antagonist SU5416, blocks 5-HT$_{1A}$-R-mediated increase in neuroblast proliferation in P6 mouse hippocampus: Studies in adult mice have demonstrated that VEGF stimulates neurogenesis and VEGFR/Flk1 (VEGFR-2) is the receptor that is involved in the neurogenic effect of VEGF (Tille et al. 2001; Warner-Schmidt and Duman 2007; Lu et al. 2011). We sought to understand if the same receptor was also involved in the 5-HT$_{1A}$-R-mediated neuroproliferation in the SGZ during early brain development at P6. We stimulated the 5-HT$_{1A}$-R with 8-OH-DPAT in the absence and presence of the VEGFR-1/2 antagonist SU5416 and analyzed the number of proliferating neuroblasts (Fig. 23.4). The 8-OH-DPAT-evoked increase in BrdU (red), DCX (blue) double positive (purple-pink) cells (Fig 23.4a) in the SGZ

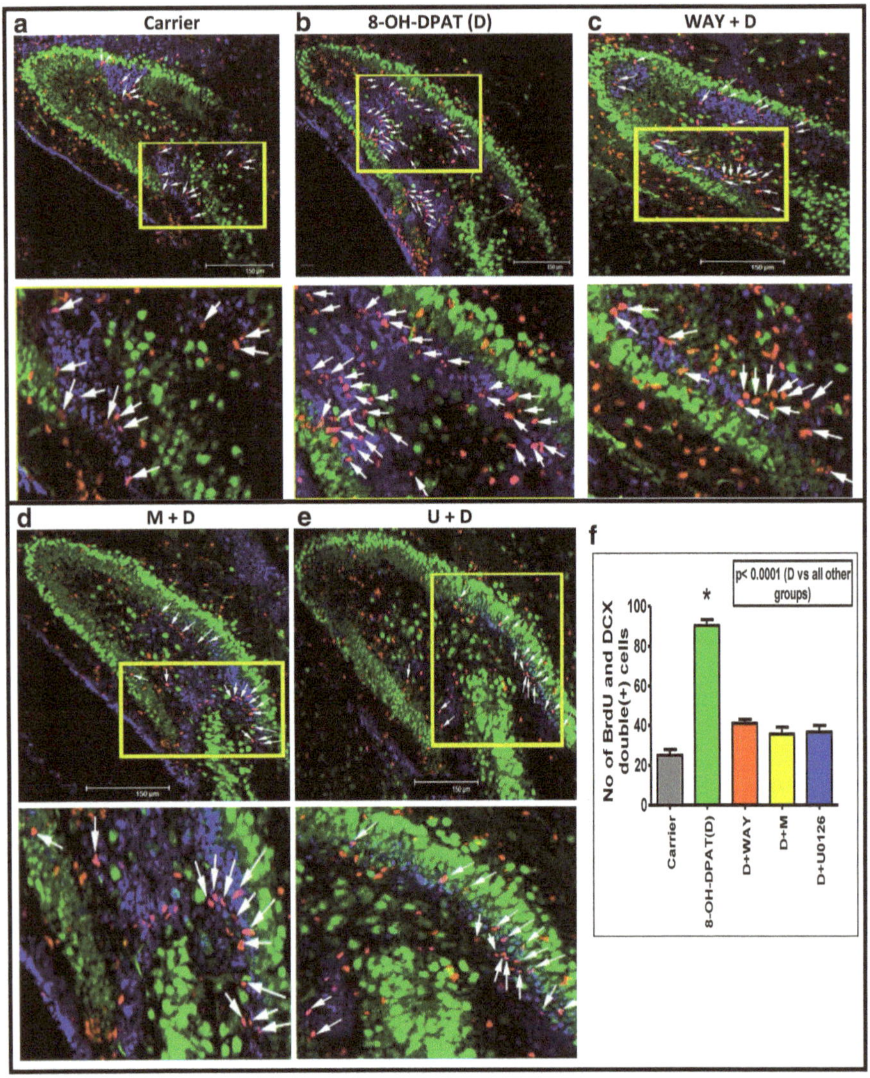

Fig. 23.2 Agonist stimulation of the hippocampal 5-HT$_{1A}$-R causes a dramatic increase in neuroproliferation in the dentate gyrus (DG). Neuroproliferation monitored in the DG, 24 h after intrahippocampal infusion (on P6) of carrier (**a**) and 8-OH-DPAT (D) in the absence (**b**) and presence of WAY (**c**), M (**d**), and U0126 (**e**). Compared to all other treatment groups, a significant increase in the number of BrdU (proliferation marker) and DCX (neuroblast marker) double positive cells was observed in the D-treated group of mice ($p < 0.0001$) (**f**). (**a**–**e**) *Top panels*: Images of the entire dentate gyrus (scale bar: 150 μm). *Bottom panels*: enlarged images of the areas shown within *yellow boxes* in the *top panels*. BrdU and DCX double positive pink cells have been marked with *white arrows*

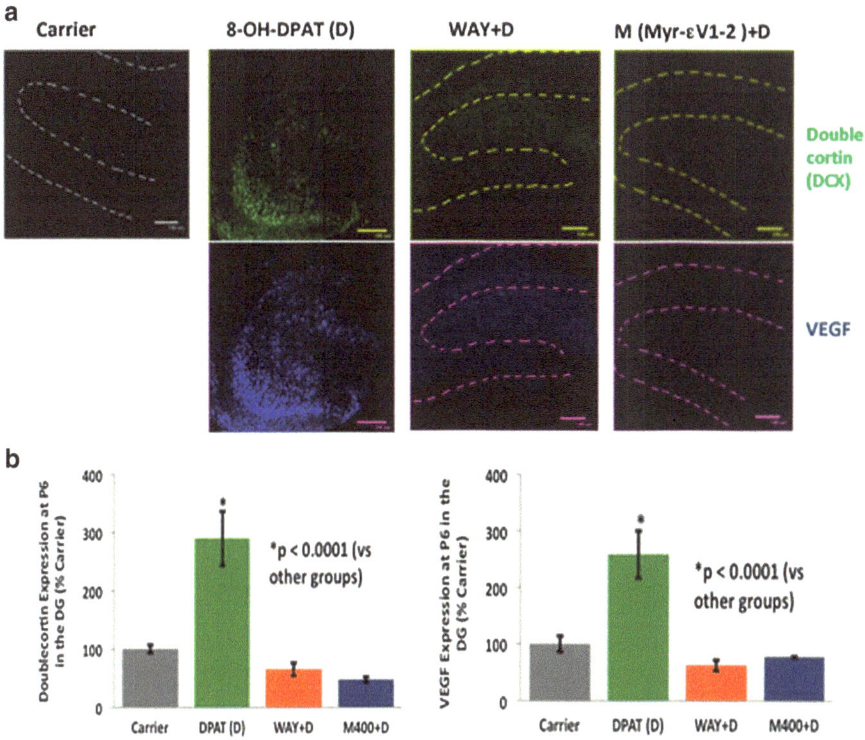

Fig. 23.3 Serotonin 1A receptor mediated induction of VEGF. (**a**) Expression of VEGF in the sub-granular zone of the DG in mouse pups following a 24-h treatment with 8-OH-DPAT from P6 in the absence and presence of antagonist WAY, and PKC epsilon inhibitor Myr-εV1-2 (M400). The *upper panel* shows staining with DCX antibody, and the *lower panel* presents VEGF antibody staining in all the treatment groups except for the carrier for which a merged image has been shown with both the colors. The contours in some images show the location of the DG, especially in case of very low staining in WAY-treated and Myr-εV1-2-treated brains. (**b**) Quantification and statistical analysis for both DCX and VEGF. Scale bar: 108 μm

was blocked in the presence of SU5416 ($p<0.01$) (Fig. 23.4a, b). This result not only shows that VEGFR-1/2 is involved in neuroproliferation in the developmental stages but it also reveals that the 5-HT$_{1A}$-R-mediated neuroproliferative effect at P6 is further amplified by VEGF signaling.

Discussion

With its complex neuronal circuitry, the hippocampus plays an important role in mood disorders. Prior studies have shown the relevance of serotonergic signaling in brain development and illustrated the role of the 5-HT$_{1A}$-R heteroreceptors in the

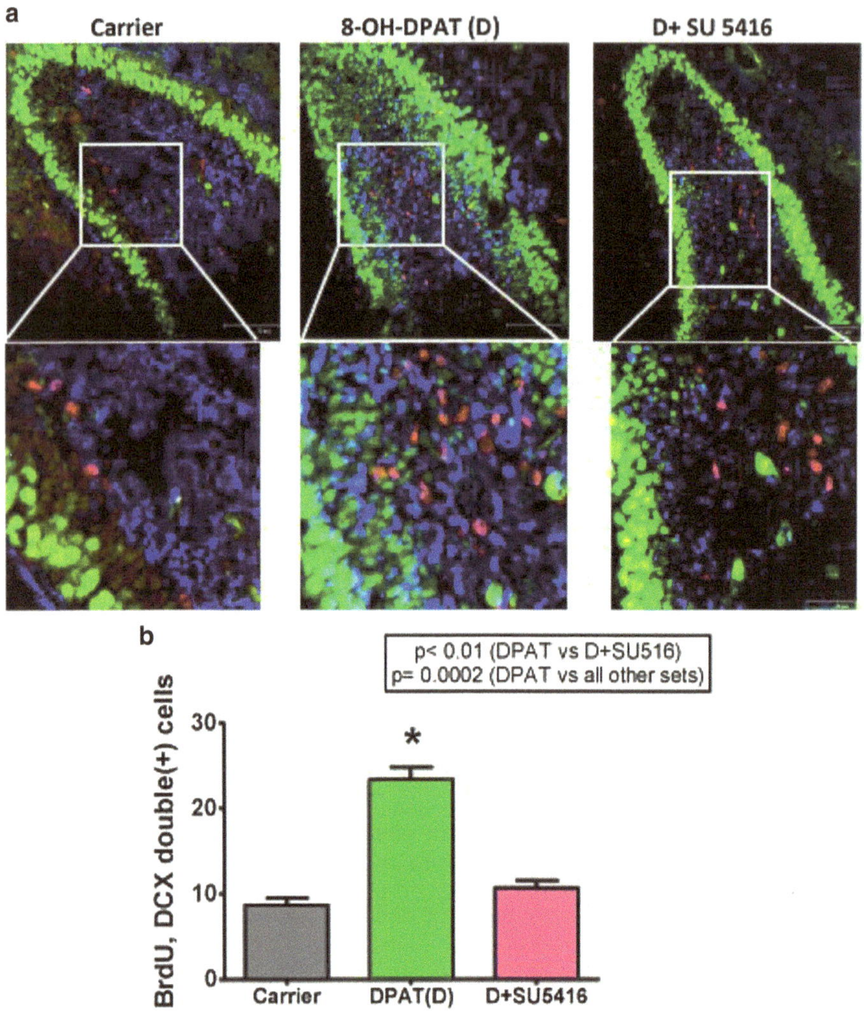

Fig. 23.4 A VEGFR1/2 antagonist blocks 5-HT_{1A}-R-mediated increase in neuroblast proliferation. (**a**) Analysis of the number of neuroprogenitor cells proliferating in a P6 mouse sub-granular zone after agonist stimulation of the 5-HT_{1A}-R, in the absence and presence of VEGFR1/2 antagonist, SU5416. The 8-OH-DPAT (D)-evoked increase in the number of BrdU (*red*) and DCX (*blue*), double positive (*purple*) cells in 24 h was blocked in the presence of SU5416 ($p<0.01$). The *top panels* consist of representative images from the different treatment groups with relevant areas magnified in the *bottom panels* to show the distribution of the BrdU-DCX double positive cells. (**b**) Statistical analysis of the number of BrdU-DCX double positive cells across the different treatment groups is shown using a column graph. Scale bar: 75 μm

development of anxiety in adulthood and have provided evidence for a 'critical time window' when the receptor signaling is most essential (Gross et al. 2002). Antidepressants that are selective serotonin reuptake inhibitors (SSRIs) have been shown to exert their effects *via* elevated neurogenesis (Santarelli et al. 2003).

Our studies have shown that stimulation of the 5-HT_{1A}-R boosts neuroproliferation in the subgranular zone (SGZ), thus revealing a role of the receptor in early postnatal hippocampal development. Furthermore, we have shown that this 5-HT_{1A}-R-mediated increase in neuroproliferation involves the PKCε isoform and ERK1/2 (Samaddar et al. 2013). The current study investigates if a post-ERK1/2 molecule is also involved in 5-HT_{1A}-R-evoked neuroproliferation in the SGZ and identifies VEGF as the post-ERK signaling agent.

Apart from being vasculogenic and angiogenic (Bhattacharya et al. 2009), VEGF is also neurogenic and neuroprotective (Cao Lei et al. 2004; Carmeliet and Almodovar 2013; Louissaint et al. 2002). Most of these experiments have been performed in adult animals and the role of this signaling protein in shaping the central nervous system during the early developmental stages has remained poorly understood. Our results strongly suggest a crucial role of VEGF also in the development of the brain during the early postnatal stages.

Blockade of the 8-OH-DPAT-evoked VEGF induction by the PKCε inhibitor, Myr-εV1-2, establishes that the release of VEGF is controlled by the 5-HT_{1A}-R *via* the epsilon isoform of PKC. Elimination of 8-OH-DPAT-induced neuroproliferation in the P6 DG in the presence of the selective and potent VEGF-1/2 antagonist SU5416 confirms the involvement of VEGF downstream of 5-HT_{1A}-R in neuroblast proliferation, thereby underscoring the importance of the VEGFR-1/2 in the shaping of the neonatal central nervous system.

Previous studies have suggested a close association between the Raf/MEK/ERK cascade and VEGF in hippocampal neurogenesis mediated by VEGFR-2 (Lu et al. 2011), but these studies were conducted in adult animals. Our experiments establish the importance of VEGF in early postnatal hippocampal neuroproliferation that might eventually lead to adult neurogenesis as reported by other groups. Moreover, the current study also sheds light on the fact that activation of the 5-HT_{1A}-R is sufficient to induce VEGF expression in the neonatal hippocampus. Thus, VEGF, which is a crucial neurogenic molecule in adult hippocampal neurogenesis and traumatic brain injury, is equally important in shaping the developing hippocampus and the associated circuitry responsible for various mood and neuropsychiatric disorders.

The kinases ERK1/2 have been known to catalyze phosphorylation of multiple transcription factors that elicit direct induction of cell cycle proteins (Chang and Karin 2001). Our results establish an additional route for 5-HT_{1A}-R- and PKCε-mediated neuroproliferation in the DG through VEGF signaling, which in turn is known to stimulate the MAP kinase pathway leading to the activation of ERK1/2. Such a positive feedback circuit could therefore cause an amplification of the neuroproliferative signaling originally triggered by the 5-HT_{1A}-R pathway. Neuroproliferation in the hippocampus is essential for the maintenance of working memory and also the action of antidepressants (Santarelli et al. 2003; Zhao et al. 2008). Some anticancer agents, such as Avastin (Mukherji 2010), target VEGF signaling in order to control angiogenesis in tumors. Our findings strongly indicate that by suppressing hippocampal neurogenesis, such therapeutic agents could seriously impair working memory in patients.

Acknowledgements The authors express gratitude to Dr. Sudarshana Purkayastha for expert assistance in immunohistochemistry and Dr. Sara Rose Guariglia for help in confocal imaging.

Conflict of Interest None of the authors have any conflict of interest.

References

Adayev T, El-Sherif Y, Barua M, Banerjee P (1999) Agonist stimulation of the serotonin 1A receptor causes suppression of anoxia-induced apoptosis via mitogen-activated protein kinase in neuronal HN2-5 cells. J Neurochem 72:1489–1496

Adayev T, Ray I, Sondhi R, Sobocki T, Banerjee P (2003) The G protein-coupled 5-HT_{1A} receptor causes suppression of caspase-3 through MAPK and protein kinase Ca. Biochim Biophys Acta 1640:85–96

Adayev T, Ranasinghe B, Banerjee P (2005) Transmembrane signaling in the brain by serotonin, a key regulator of physiology and emotion. Biosci Rep 25:363–385

Bhattacharya R, Kwon J, Li X, Wang E, Patra S, Bida JP, Bajzer Z, Claesson-Welsh L, Mukhopadhyay D (2009) Distinct role of PLC 3 in VEGF-mediated directional migration and vascular sprouting. J Cell Sci 122:1025–1034

Cao Lei JX, Zuzga David S, Liu Y, Fong DM, Young D, During MJ (2004) VEGF links hippocampal activity with neurogenesis, learning and memory. Nat Genet 36:827–835

Carmeliet P, Almodovar C (2013) VEGF ligands and receptors: implications in neurodevelopment and neurodegeneration. Cell Mol Life Sci 70(10):1763–1778

Chang L, Karin M (2001) Mammalian MAP kinase signalling cascades. Nature 410:37–40

Chen L, Hahn H, Wu G, Chen C-H, Liron R, Schechtman D, Cavallaro G, Banci L, Guo Y, Bolli R, Dorn GW, Mochly-Rosen D (2001) Opposing cardioprotective actions and parallel hypertrophic effects of delta PKC and epsilon PKC. Proc Natl Acad Sci U S A 98:11114–11119

Greene J, Banasr M, Lee B, Warner-Schmidt J, Duman RS (2009) Vascular endothelial growth factor signaling is required for the behavioral actions of antidepressant treatment: pharmacological and cellular characterization. Neuropsychopharmacology 34:2459–2468

Gross C, Zhuang X, Stark K, Ramboz S, Oosting R, Kirby L, Santarelli L, Beck S, Hen R (2002) $Serotonin_{1A}$ receptor acts during development to establish normal anxiety-like behaviour in the adult. Nature 416:396–400

Heisler LK, Chu H-M, Brennan TJ, Danao JA, Bajwa P, Parsons LH, Tecott LH (1998) Elevated anxiety and antidepressant-like responses in serotonin 5-HT_{1A} receptor mutant mice. Proc Natl Acad Sci U S A 95:15049–15054

Hongpaisan J, Sun M-K, Alkon DL (2011) PKC epsilon activation prevents synaptic loss, A beta elevation, and cognitive deficits in Alzheimer's disease transgenic mice. J Neurosci 31(2): 630–643

Johnson JA, Gray MO, Chen S-H, Mochly-Rosen D (1996) A protein kinase C translocation inhibitor as an isozyme-selective antagonist of cardiac. J Biol Chem 271:24962–24966

Kessler RC, Berglund P, Demier O, Jin R, Merikangas KR, Walters EE (2005) Lifetime prevalence and age-of-onset distributions of DSM-IV disorders in the National Comorbidity Survey Replication. Arch Gen Psychiatry 62:593–602

Kushwaha N, Albert N (2005) Coupling of 5-HT1A autoreceptors to inhibition of mitogen-activated protein kinase activation via G beta gamma subunit signaling. Eur J Neurosci 21:721–732

Lee Jeong-Sik SH (2009) Induction of neuronal vascular endothelial growth factor expression by cAMP in the dentate gyrus of the hippocampus is required for the anti-depressant-like behaviors. J Neurosci 29(26):8493–8505

Lemonde S, Du L, Bakish D, Hrdina P, Albert PR (2004) Association of the C(-1019)G 5-HT1A functional promoter polymorphism with antidepressant response. Int J Neuropsychopharmacol 7(4):501–506

Lo Iacono L, Gross C (2008) Alpha-Ca2+/calmodulin-dependent protein kinase II contributes to the developmental programming of anxiety in serotonin receptor 1A knock-out mice. J Neurosci 28:6250–6257
Louissaint A Jr et al (2002) Coordinated interaction of neurogenesis and angiogenesis in the adult songbird brain. Neuron 34(6):945–960
Lu K-T, Sun C-L, Wo PY, Yen H-H, Yang Y-L (2011) Hippocampal neurogenesis after traumatic brain injury is mediated by vascular endothelial growth factor receptor-2 and the Raf/MEK/ERK cascade. J Neurotrauma 28(3):441–450
Mehta M, Ahmed Z, Fernando SS, Cano-Sanchez P, Adayev T, Ziemnicka D, Wieraszko A, Banerjee P (2007) Plasticity of 5-HT1A receptor-mediated signaling during early postnatal brain development. J Neurochem 101(4):918–928
Mogha A, Guariglia SR, Debata PR, Wen GY, Banerjee P (2012) Serotonin 1A receptor-mediated signaling through ERK and PKCα is essential for normal synaptogenesis in neonatal mouse hippocampus. Transl Psychiatry 2:e66
Mologni L, Sala E, Cazzaniga S, Rostagno R, Kuoni T, Puttini M, Bain J, Cleris L, Redaelli S, Riva B, Formelli F, Scapozza L, Gambacorti-Passerini C (2006) Inhibition of RET tyrosine kinase by SU5416. J Mol Endocrinol 37(2):199–212
Mukherji SK (2010) Bevacizumab (Avastin). AJNR Am J Neuroradiol 31:235–236
Palmer TD, Willhoite AR, Gage FH (2000) Vascular niche for adult hippocampal neurogenesis. J Comp Neurol 425(4):479–494
Parks CL, Robinson PS, Sibille E, Shenk T, Toth M (1998) Increased anxiety of mice lacking the $serotonin_{1A}$ receptor. Proc Natl Acad Sci U S A 95:10734–10739
Purkayastha S, Fernando SS, Diallo S, Cohen L, Levano K, Banerjee P (2009) Regulation of protein kinase C isozymes during early post-natal hippocampal development. Brain Res 1288:29–41
Purkayastha S, Ford J, Kanjilal B, Diallo S, Inigo JDR, Neuwirth L, Elidrissi A, Ahmed Z, Wieraszko A, Banerjee P (2012) Clozapine functions through the prefrontal cortex serotonin 1A receptor to heighten neuronal activity via calmodulin kinase II-NMDA receptor interactions. J Neurochem 120:396–407
Ramboz S, Oosting R, Amara DA, Kung HF, Blier P, Mendelsohn M, Mann JJ, Brunner D, Hen R (1998) Serotonin $receptor_{1A}$ knockout: an animal model of anxiety-related disorder. Proc Natl Acad Sci U S A 95:14476–14481
Samaddar S, Debata PR, Chanthrakumar P, Marsillo A, Tantry SJ, Banerjee P (2013) Serotonin 1A receptor-mediated signaling cascade in neuroblast proliferation and neurogenesis in neonatal hippocampus. 24th Biennial Joint meeting of the International Society for Neurochemistry (ISN) and American Society for Neurochemistry (ASN), Mexico, April 20–24. Abstract Code: PSM08-14
Santarelli L, Saxe M, Gross C, Surget A, Battaglia F, Dulawa S, Weisstaub N, Lee J, Duman R, Aracio O, Belzung C, Hen R (2003) Requirement of hippocampal neurogenesis for the behavioral effects of antidepressants. Science 301:805–809
Shimizu T, Kanno A, Nishizaki T (2011) α, ß-DCP-LA selectively activates PKC-ε and stimulates neurotransmitter release with the highest potency among 4 diastereomers. Cell Physiol Biochem 27:149–158
Sodhi MS, Sanders-Bush E (2004) Serotonin and brain development. Int Rev Neurobiol 59: 111–174
Teng LC-W, Kay H, Chen Q, Adams JS, Grilli C, Guglielmello G, Zambrano C, Krass S, Bell A, Young LH (2008) Mechanisms related to the cardioprotective effects of protein kinase C epsilon (PKCe) peptide activator or inhibitor in rat ischemia/reperfusion injury. Naunyn Schmiedebergs Arch Pharmacol 378:1–15
Tille J-C, Wood J, Mandriota SJ, Schnell C, Ferrani S, Mestan J, Zhu Z, Witte L, Pepper MS (2001) Vascular endothelial growth factor (VEGF) receptor-2 antagonists inhibit VEGF- and basic fibroblast growth factor-induced angiogenesis in vivo and in vitro. J Pharmacol Exp Therap 299:1073–1085
Warner-Schmidt JL, Duman RS (2007) VEGF is an essential mediator of the neurogenic and behavioral actions of antidepressants. Proc Natl Acad Sci 104(11):4647–4652

Warner-Schmidt JL, Duman RS (2008) VEGF as a potential target for therapeutic intervention in depression. Curr Opin Pharmacol 8(1):14–19
Zhao C, Deng W, Gage FH (2008) Mechanisms and functional implications of adult neurogenesis. Cell 132(4):645–660

Chapter 24
Structural Heterogeneity of Glycoform of Alpha-1 Acid Glycoprotein in Alcoholic Cirrhosis Patients

Goutam Mandal, Hirokazu Yagi, Koichi Kato, and Bishnu Pada Chatterjee

List of Abbreviations

2-PA	2-Aminopyridine
AAT	Alpha-1 antitrypsin
ACT	Alpha-1 antichymotrypsin
AGP	Alpha1-acid glycoprotein
ALC	Alcoholic liver cirrhosis
ALT	Alanine aminotransferase
AST	Aspartate aminotransferase
CDT	Carbohydrate-deficient transferrin
DEAE	Diethylaminoethyl
GU	Glucose unit
Hp	Haptoglobin

G. Mandal • B.P. Chatterjee (✉)
Department of Natural Sciences, West Bengal University of Technology,
Salt Lake, Kolkata 700064, India
e-mail: gautammandal2004@gmail.com; cbishnup@gmail.com

H. Yagi
Graduate School of Pharmaceutical Sciences, Nagoya City University,
3-1 Tanabe-diri, Mizuho-ku, Nagoya 4678603, Japan
e-mail: hyagi@phar.nagoya-cu.ac.jp

K. Kato
Graduate School of Pharmaceutical Sciences, Nagoya City University,
3-1 Tanabe-diri, Mizuho-ku, Nagoya 4678603, Japan

Institute for Molecular Science and Okazaki Institute for Integrative Bioscience,
National Institutes of Natural Sciences, Okazaki Institute for Integrative Biosciences,
5-1 Higashiyama, Myodaiji, Okazaki 444-8787, Japan
e-mail: kkato@phar.nagoya-cu.ac.jp

A. Chakrabarti, A. Surolia (eds.), *Biochemical Roles of Eukaryotic Cell Surface Macromolecules*, Advances in Experimental Medicine and Biology 842,
DOI 10.1007/978-3-319-11280-0_24

ODS Octadecyl silica
SNA *Sambucus nigra* agglutinin
Tf Transferrin

Introduction

In recent years, several studies have been reported that excessive alcohol consumption causes liver injury following changes in the concentration of plasma proteins as well as their glycosylation (Turner 1992; Blomme et al. 2009; Comunale et al. 2010). During post translation modification most of the serum proteins are glycosylated which undergo changes during the progression of the disease. Such glycosylation change was observed in patients with various pathologic conditions. Alpha1-acid glycoprotein, transferrin, haptoglobin and α-1 antitrypsin represent the most abundant serum glycoproteins secreted by the liver (Baenziger 1984). Changes in glycan branching, sialylation, and fucosylation of the oligosaccharides in AGP has been studied in patients with acute and chronic inflammation and in different malignant diseases (Brinkman-van der Linden et al. 1996; Ryden et al. 1997, 1999). The glycosylation changes in AGP (Serbource-Goguel Seta et al. 1986), α-2HS glycoprotein (Jezequel et al. 1988), transferrin (Debruyne et al. 1984) and α-1 antichymotrypsin (ACT) (Hachulla et al. 1992) have been observed using crossed affino-immunoelectrophoresis that suggested increased branching of these serum proteins. Desialylation is the most noted alteration observed in alcoholic liver disease (ALD). Carbohydrate-deficient transferrin (CDT) is widely used to detect and monitor chronic alcohol abuse (Arndt 2001). Excessive alcohol intake alters the normal heterogeneity pattern of transferrin. Besides transferrin, many other glycoproteins including AGP, Hp, AAT and ceruloplasmin (Cp) have shown decreased in sialylation in ALD (Patricia et al. 1996; Tsutsumi et al. 1994). The altered glycosylation of serum glycoproteins like fucosylation of AGP and serum cholinesterase]. The altered glycosylation of serum glycoproteins like fucosylation of AGP and serum cholinesterase in liver cirrhosis (LC) and fucosylation of haptoglobin (Hp) in alcoholic liver disease (ALD) has been reported (Thompson et al. 1991; Mann et al. 1994; Kondo et al. 1995; Hada et al. 1999). Previously, an increased fucosylation of AGP and alpha fetoprotein (AFP) in chronic hepatitis B and hepatitis B induced liver cirrhosis was reported by our group (Mondal et al. 2009, 2011). Changes of glycosylation pattern were also monitored in liver cirrhosis and liver cancer patients (Block et al. 2005; Comunale et al. 2006). Decrease in electrophoretic mobility of transferrin (Stibler and Borg 1981), AGP (Serbource-Goguel et al. 1983) and C1-inactivator (Serbource-Goguel Seta et al. 1984) in alcoholic patients has been observed due to loss of sialic acid residues. In alcoholic cirrhosis (ALC), an increase in ConA-non-reactive forms of AGP has been detected using crossed immuno-affinoelectrophoresis (CIAE) that suggested decrease of glycan branching (Biou et al. 1989) and the same type of modification has been observed in α2-HS and transferrin (Maryvonne et al. 1988). On the contrary, in acute inflammatory disorders, a decrease of Con A-reactive components was observed in both AGP and ACT (Nicollet et al. 1981). The increased level of fucosylation was observed in AGP and carbohydrate-deficient transferrin in cirrhosis patients and alcohol abusers (Biou et al. 1987; Stibler and Borg 1986) respectively. It has been reported that Hp binds to *Lotus tetragonolobus* agglutinin in alcoholic liver diseases (Turner 1992). The changes of glycosylation patterns in pathological conditions are mainly studied by lectin-based affinity methods and antibody–antigen interactions. These methods are employed for characterizing particular features of glycans (like degree of branching, sialylation and fucosylation), but provide relatively little information on structural diversity. Mass spectrometry in combination with chromatography provides more detailed information on the glycan structure and heterogeneity (Charlwood et al. 2001). However, detailed information on the structural changes of glycans in AGP in the sera of ALC patients has not been reported yet.

Our aim was to make detailed analysis of different N-glycan structures of AGP in the sera of ALC by HPLC mapping method using three different columns (DEAE, ODS and Amide columns) followed by MALDI-TOF-MS analysis. Change of sialic acid level using sialic acid specific lectins *Sambucus nigra* agglutinin (SNA) by ELISA and lectin blot was also monitored in the present investigation.

Materials and Methods

Patients

Serum samples from 20 ALC patients were kindly supplied by Prof. Dr. Y. K. Chawla, Department of Hepatology, Post Graduate Institute of Medical Education and Research, (PGIMER), Chandigarh, India. The diagnosis of alcoholic cirrhosis is based on the history of habitual alcohol intake (usually more than 80 g/day for >10 years), liver function test (LFT) derangement in the form of elevated bilirubin, AST/ALT (>2), γ-glutamyltransferase (GGT), deranged prothrombin time and USG or CT scan showing features of cirrhosis and endoscopy showing presences of varices. Sera from 20 age and sex-matched healthy individuals were served as controls. All the serum samples were stored at −80 °C till use. Informed consent was obtained from each patient and healthy individual. Ethical committee of PGIMER and West Bengal University of Technology, Kolkata, India approved this study.

Chemicals and HPLC Columns

Proteo-prep blue albumin depletion kit, biotin 3-sulfo-N-hydroxysuccinimide ester, extravidin, proteomic grade trypsin, 2,5-dihydroxybenzoic acid, Sepharose 4B, monoclonal anti-human alpha-1 acid glycoprotein antibody, alpha-1-acid glycoprotein (AGP), *Sambucus nigra* agglutinin (SNA), *O*-phenylenediamine dihydrochloride (OPD) were purchased from Sigma, USA. The TSK-gel diethylaminoethyl (DEAE)-5PW column (7.5×75 mm) and TSK-gel amide-80 were purchased from Tosoh, Tokyo, Japan. Shim-pack HRC-octadecyl silica (ODS) column (6.0×150 mm) from Shimadzu, Kyoto, Japan and pyridylamino derivatives of isomalto-oligosaccharides were purchased from Seikagaku Kogyo Co. and GLYENCE Co., Nagoya, Japan. All other chemicals and reagents used were of high analytical grade and obtained from commercial sources. The reagents used for solvents were purchased from Wako Pure chemical Industries, Japan. Milli Q water was used for preparation of different solvents.

Solvents: solvent A (aqueous ammonia, pH 9.0), solvent B (50 mM ammonium acetate, pH 9.0), solvent C (10 mM phosphate buffer, pH 3.8), solvent D (10 mM phosphate buffer, pH 3.8 containing 0.5 % 1-butanol), solvent E (65 % acetonitrile, 2.9 % triethylamine, pH 7.3, containing 0.6 % acetic acid) and solvent F (50 % acetonitrile, 4.1 % triethylamine, pH 7.3, containing 1.7 % acetic acid).

Purification of Alpha-1-acid Glycoprotein and Preparation of Fluorescent Labeled N-glycans

Alpha-1-acid glycoprotein was purified from albumin depleted pooled sera (n = 20) of ALC patients and healthy controls by affinity chromatography using monoclonal antihuman AGP-Sepharose 4B column. N-glycans were released from purified AGP by heating with 0.2 mL of anhydrous hydrazine at 100 °C for 10 h in an evacuated sealed tube. The excess hydrazine, peptides and other detergents were removed by non-porous graphitized carbon mini-column by using 50 mM triethylamine (pH 7.0) containing 60 % acetonitrile. Released N-glycans were labeled with 2-aminopyridine (2-PA) and purified by the cellulose column using 75 mM ammonium bicarbonate containing ethanol (Yamamoto et al. 1989). The N-glycan mixtures were dried by speed vac and stored at −20 °C for further use.

HPLC Separation of PA-glycans

Three types of columns were used for the separation of N-linked oligosaccharides on the basis of polarity. In the first step, the purified PA-glycan mixture was subjected to anion exchange chromatography on a DEAE column at 30 °C at a flow rate of 1.0 mL/min using solvent A and solvent B. The column was equilibrated with the solvent A and the bound oligosaccharides were eluted as neutral, monosialyl and disialyl fractions using linear gradient (0–20 %) of solvent B for 30 min. In the second step, each fraction separated by the DEAE column was collected and analyzed by ODS column. The column was equilibrated with solvent C and elution was performed at a flow rate of 1.0 mL/min at 55 °C using a linear gradient (0–50 %) of solvent D for 60 min. Each separated oligosaccharide fraction was expressed as glucose units (GU). In the third step, individual fraction from the ODS column was further separated by TSK-gel amide-80 column. In this system two solvents, solvent E and solvent F were used. After injection of sample into the column pre-equilibrated with solvent E, solvent F was applied in a linear gradient 0–50 % over a period of 50 min. The flow rate was 1.0 mL/min at 40 °C. PA-glycans were detected by fluorescence intensity using excitation and emission at 320 and 380 nm, respectively. The elution time of the individual peaks from the ODS and amide silica columns were calibrated with PA-derivatized isomalto-oligosaccharides which was represented in units of glucose (GU) (Yagi et al. 2005, 2010). The molecular weight of each glycan was analysed by MALDI-TOF-MS using AXIMA-CFR plus MALDI-TOF and AXIMA-QIT MALDI quadrupole-ion trap TOF instrument. N-glycan structures were identified by comparing with PA-glycans in the GALAXY database (http://www.glycoanalysis.info/galaxy2/ENG/index.jsp) (Takahashi and Kato 2003).

ELISA and Lectin Blotting

The binding of SNA with AGP was studied using sialic acid-specific *Sambucus nigra* agglutinin (SNA) by ELISA. A lectin immunoassay was performed to determine the change of sialylation level in serum AGP using sialic acid-specific *Sambucus nigra* agglutinin (SNA). Each well of a 96 well microtiter plate (NUNC) was coated with 1 μg monoclonal anti-human AGP antibody (mAb-AGP) in bicarbonate buffer and kept at 4 °C for 24 h. After washing with PBST [10 M PBS, pH 7.4 containing 0.05 % Tween-20], 1 % BSA was added into the wells. On keeping at 37 °C for 1 h, diluted serum (1:10, 100 μL) of individual ALC patients and control group was added into the wells followed by incubation with 100 μL biotinylated SNA (1:1,000 in PBS) at room temperature for 2 h. 100 μL of streptavidin-HRP conjugate (1:5,000 in PBS) was then added to each well and kept at room temperature for 2 h followed by addition of 0.1 % *O*-phenylenediamine dihydrochloride (OPD) (100 μL) and 0.05 % H_2O_2 in 0.05 M citrate phosphate buffer (pH 5.0). The plate was left for 30 min at room temperature. The absorbance of each well was measured at 490 nm in an ELISA Reader. The sialylation level was also monitored in individual ($n = 5$) as well as pooled sample ($n = 5$) of both control and ALC patient groups by lectin blotting (Saroha et al. 2012). The intensity level of band was calculated using Image J software. The statistical analysis was performed by Student's t test and One way ANOVA test ($p < 0.05$).

Results

To identify the structural heterogeneity of N-glycans in serum AGP of ALC patients, multidimensional HPLC mapping protocol was followed which involved sequential HPLC runs (DEAE, ODS and amide silica columns). Three fractions were identified:

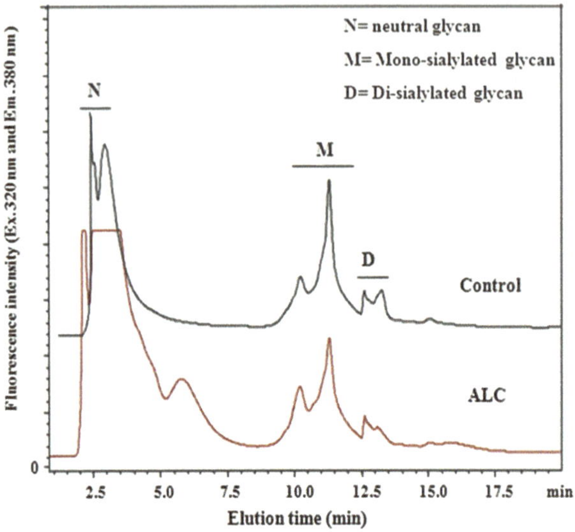

Fig. 24.1 Elution profiles of PA-N-glycans derived from AGP of ALC patients and control on the DEAE column. The PA-glycan mixture was separated according to sialic acid content

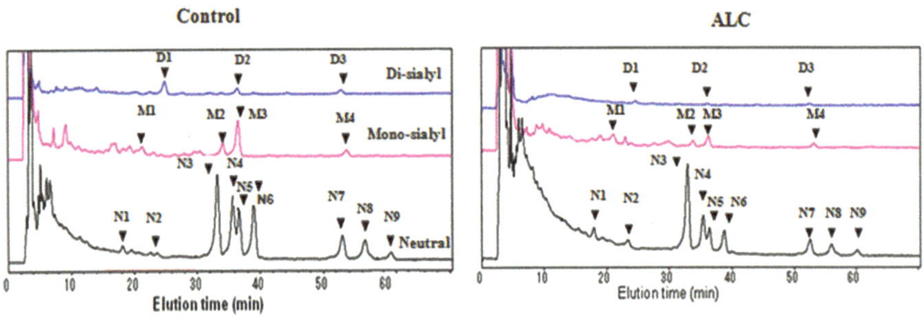

Fig. 24.2 N-glycan profiles of DEAE separated fractions on the ODS column

neutral glycans (N) and two kinds of acidic glycans, monosialylated (M) and disialylated (D) glycans by comparing with the elution profile of standard oligosaccharides (Fig. 24.1). Each fraction obtained from DEAE column was further analyzed by ODS column. Figure 24.2 shows four major neutral glycan peaks (N3–N6) along with other five minor neutral peaks (N1, N2, N7, N8 and N9). Four monosialylated glycans (M1–M4) were found in control as well as in ALC patients group. In control group there appeared three disialylated glycans (D1–D3) whereas in patients group there was very negligible disialylated glycans. Table 24.1 shows the relative percentage of PA derivatized N-linked oligosaccharides of AGP in control and ALC patients. The PA-oligosaccharides were identified by comparing the elution time of isomalto-oligosaccharides normalized in GU followed by the coincidence of elution data with those in the GALAXY data base. For example, the major neutral N-glycan corresponding to peak N6 was eluted at 14.5 GU on the ODS column and 7.34 GU on the amide column. The elution data set was in good

Table 24.1 The structure and relative quantities of PA-derivatized N-linked oligosaccharides of serum AGP from ALC and control groups

				Relative quantity (%)[a]	
	ODS	Measured molecular			
Peaks	Amide (GU)[b]	Mass (m/z)	Structures of PA-oligosaccharides	Control	ALC
Neutral oligosaccharides					
N1	9.0 5.48	1,417	GlcNAcβ1-2Manα1 6 Manβ1-4GlcNAcβ1-4GlcNAc 3 GlcNAcβ1-2Manα1	1.1	3.1
N2	10.2 6.96	1,742	Galβ1-4GlcNAcβ1-2Manα1 6 Manβ1-4GlcNAcβ1-4GlcNAc 3 Galβ1-4GlcNAcβ1-2Manα1	0.7	1.7
N3	12.7 5.45	1,566	GlcNAcβ1-2Manα1 Fucα1,6 6 Manβ1-4GlcNAcβ1-4GlcNAc 3 GlcNAcβ1-2Manα1	20.2	32.6
N4	13.5 6.32	1,729	Galβ1-4GlcNAcβ1-2Manα1 Fucα1,6 6 Manβ1-4GlcNAcβ1-4GlcNAc 3 GlcNAcβ1-2Manα1	15.0	13.3
N5	13.7 6.46	1,710	GlcNAcβ1-2Manα1 Fucα1,6 6 Manβ1-4GlcNAcβ1-4GlcNAc 3 Galβ1-4GlcNAcβ1-2Manα1	8.7	7.2
N6	14.5 7.34	1,890	Galβ1-4GlcNAcβ1-2Manα1 Fucα1,6 6 Manβ1-4GlcNAcβ1-4GlcNAc 3 Galβ1-4GlcNAcβ1-2Manα1	13.3	7.6
N7	19.3 5.78	1,769	GlcNAcβ1-2Manα1 Fucα1,6 6 GlcNAcβ1,4 — Manβ1-4GlcNAcβ1-4GlcNAc 3 GlcNAcβ1-2Manα1	7.1	6.0
N8	20.8 6.54	1,932	GlcNAcβ1-2Manα1 Fucα1,6 6 GlcNAcβ1,4 — Manβ1-4GlcNAcβ1-4GlcNAc 3 Galβ1-4GlcNAcβ1-2Manα1	6.7	4.5
N9	21.6 7.39	2,093	Galβ1-4GlcNAcβ1-2Manα1 Fucα1,6 6 GlcNAcβ1,4 — Manβ1-4GlcNAcβ1-4GlcNAc 3 Galβ1-4GlcNAcβ1-2Manα1	2.1	2.5
Monosialylated oligosaccharides					
M1	9.7 7.49	2,009	Galβ1-4GlcNAcβ1-2Manα1 6 Manβ1-4GlcNAcβ1-4GlcNAc 3 Neu5Acα2-6Galβ1-4GlcNAcβ1-2Manα1	1.4	2.8

(continued)

Table 24.1 (continued)

Peaks	ODS Amide (GU)[b]	Measured molecular Mass (m/z)	Structures of PA-oligosaccharides	Relative quantity (%)[a] Control	ALC
M2	13.0 6.5	1,990	GlcNAcβ1-2Manα1 \ Fucα1,6 6 | GlcNAcβ1,4 — Manβ1-4GlcNAcβ1-4GlcNAc 3 Neu5Ac2-6Galβ1-4GlcNAcβ1-2Manα1 /	3.3	1.8
M3	13.7 7.43	2,155	Galβ1-4GlcNAcβ1-2Manα1 \ Fucα1,6 6 | Manβ1-4GlcNAcβ1-4GlcNAc 3 Neu5Ac2-6Galβ1-4GlcNAcβ1-2Manα1 /	8.9	4.0
M4	19.6 7.55	2,356	Galβ1-4GlcNAcβ1-2Manα1 \ Fucα1,6 6 | GlcNAcβ1,4 — Manβ1-4GlcNAcβ1-4GlcNAc 3 Neu5Ac2-6Galβ1-4GlcNAcβ1-2Manα1 /	1.6	1.7
Disialylated oligosaccharides					
D1	10.5 7.2	2,321	Neu5Ac2-6Galβ1-4GlcNAcβ1-2Manα1 \ 6 Manβ1-4GlcNAcβ1-4GlcNAc 3 Neu5Acα2-6Galβ1-4GlcNAcβ1-2Manα1 /	2.9	0.9
D2	13.6 7.6	2,454	Neu5Ac2-6Galβ1-4GlcNAcβ1-2Manα1 \ Fucα1,6 6 | Manβ1-4GlcNAcβ1-4GlcNAc 3 Neu5Acα2-6Galβ1-4GlcNAcβ1-2Manα1 /	1.3	0.4
D3	19.3 7.72	2,649	Neu5Ac2-6Galβ1-4GlcNAcβ1-2Manα1 \ Fucα1,6 6 | GlcNAcβ1,4 — Manβ1-4GlcNAcβ1-4GlcNAc 3 Neu5Ac2-6Galβ1-4GlcNAcβ1-2Manα1 /	1.2	0.6

[a]Relative quantity (%) was calculated from the peak area in chromatogram of ODS column
[b]Glucose unit (GU) was calculated from the elution time of the peaks obtained from the ODS and amide column

agreement with a known reference Galβ1 → 4GlcNAcβ1 → 2Manα1 → 6(Galβ1 → 4 GlcNAcβ1 → 2-Manα1 → 3) Manβ1 → 4GlcNAcβ1 → 4(Fucα1 → 6)GlcNAc-PA (Code no. 210.4 in GALAXY). By co-chromatography and the MALDI-TOF-MS analysis, we confirmed the structure of the above PA-oligosaccharide. In a similar way, we identified the remaining 15 kinds of the N-glycans derived from AGP, which consists of neutral and sialylated oligosaccharides. It shows that among the neutral glycans percentage of N1, N2, N3 in patients group was higher than those in control group whereas percentage of N4–N9 glycans in patients group was found to be lower than those in control group. The percentage of monosialylated and disialylated glycans was found to be two to three times less in patients group than those in control group except M1 and M4 which was more or less equal in both groups.

This result was substantiated by ELISA and lectin blotting. The binding of AGP with *Sambucus nigra* agglutinin (SNA), which has the specificity to Neu5Acα2,6Gal, in ALC patients group was found to be less than that of control as determined by

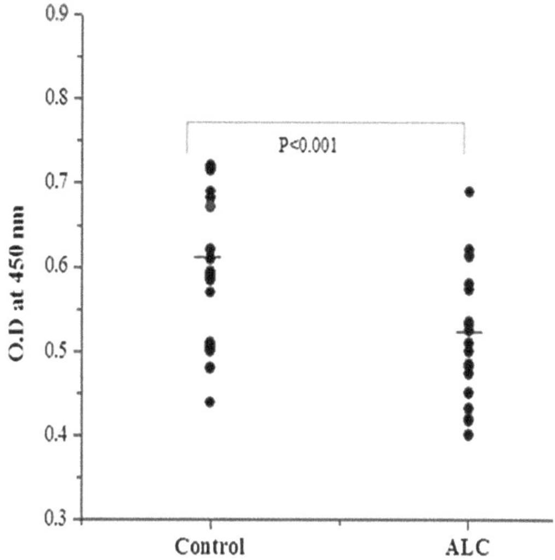

Fig. 24.3 Binding of serum AGP with *Sambucus nigra* agglutinin in control and ALC patients groups by ELISA. The horizontal line indicates the mean value of the binding

ELISA (Fig. 24.3). The band intensity of sialic acid content in AGP of ALC pooled patients was found to be lower than that in pooled control group by lectin blot using SNA. Similarly AGP from individual patients contains less sialic acid compared to that in control as envisaged from band intensity of lectin blot using SNA (Fig. 24.4a, b).

Discussion

The clinical relevance of glycosylation change in various disease processes has been understood by sugar profiling. Alteration of N-linked glycosylation in human serum glycoproteins has been described mostly in inflammation, liver diseases and cancer. Advancement of glycan analysis tools like glycoproteomics and lectinomics has helped in identifying the specific glycan structure which has proved to be essential for understanding the disease mechanism as well as the development of novel diagnostic or prognostic marker. There is a need for accurate biochemical glycan markers to aid in early diagnosis of liver cirrhosis. Currently assessment of structural liver damage by liver cirrhosis involves invasive method like liver biopsy, which is gold standard. Yet it has certain limitations such as lack of sensitivity, risk of complications, and discomfort to patient. The majority of serum proteins are glycosylated and during disease progression subtle changes occur in glycosylation. Analysis of total serum N-glycan is a non-invasive test where blood sample can be obtained without inducing major trauma to the patient. Such an assay is high throughput, sensitive, relatively simple, reproducible, easy to interpret and can be performed in diagnostic laboratory.

The present study has described elaborately the analysis of N-linked glycans to identify any specific change that occurs in cirrhosis developed by patient of high

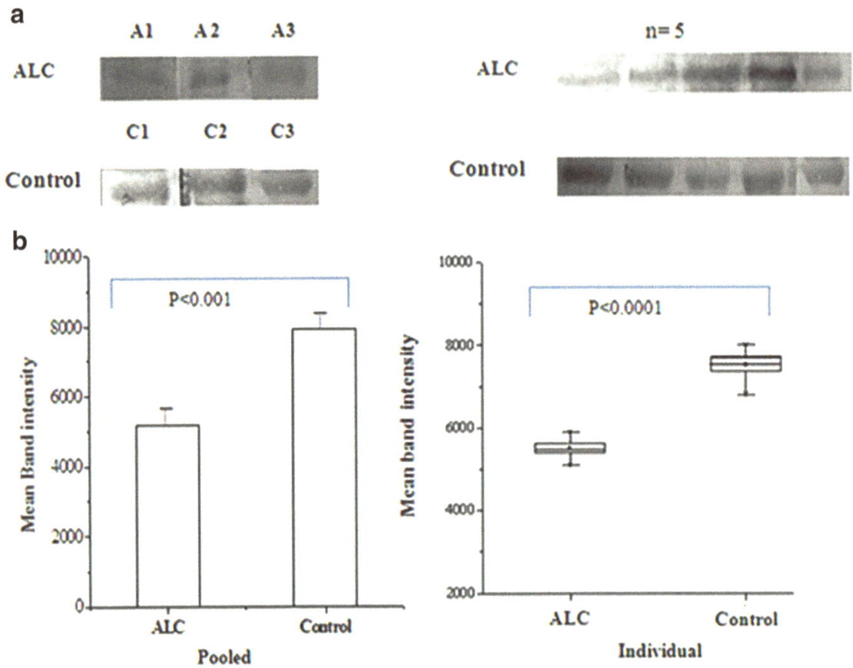

Fig. 24.4 (**a**) Sialic acid content of AGP in control and ALC patients groups using sialic acid specific lectin, *Sambucus nigra* agglutinin by lectin blot. (*Left*) In control ($n=20$), C1, C2 and C3 represent pool consisted of five samples each. Similarly, in ALC patient group ($n=20$) A1, A2 and A3 represent pool consisted of five samples each. (*Right*) Lectin blot image of AGP from individual control ($n=5$) and ALC ($n=5$) samples. (**b**) Mean band intensity of pooled (*left*) and individual (*right*) samples from ALC and control groups

alcohol intake. Our analysis takes the advantage of the widely used reverse phase HPLC by ODS column for identifying, quantifying, and differentiating structural isomers of glycans. Using this highly sensitive method we have analyzed the N-glycans of AGP from ALC patients and control to understand variation of glycan structures in them. Among the neutral fractions, N1 was present threefold higher in ALC than control. The increased amount of galactosylated neutral glycoform (N2) and fucosylated neutral glycoform (N3) was found in ALC patients than in control group. The glycoforms of N3–N9 were found to be fucosylated, of them N7, N8, N9 contain bisecting GlcNAc residue. The percentage of N4–N7 were found to be less in ALC patients than control, of them amount of N6 was found to be half in ALC than control. In monosialylated N-linked glycan fractions, the relative quantity of M1 in patient group was twofold higher than that in control. The relative quantity of the second fraction (M2) and third fraction (M3) was twofold lower in ALC than that in control group. Very less amount of disialylated oligosaccharide in AGP was detected in ALC patients that suggested sialylation level was decreased in ALC patients with respect to control. In both control and patient groups the percentage of

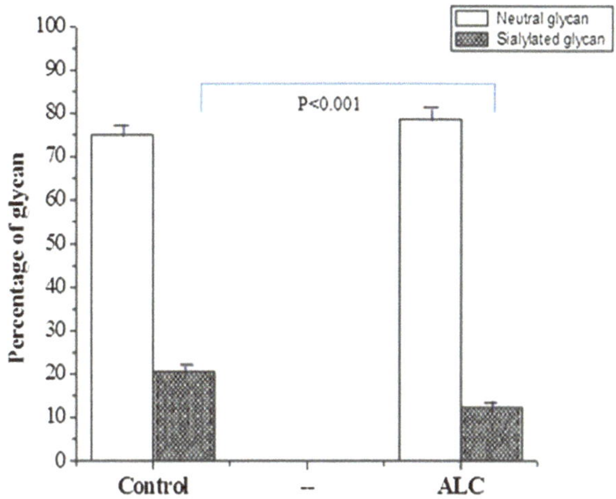

Fig. 24.5 Relative quantity of total neutral and sialylated N-linked oligosaccharides in control and ALC patients

neutral glycans showed nearly the same whereas considering monosialylated and disiasylated oligosaccharides, the overall sialylation level was significantly less in ALC patient groups, almost half than control ($p<0.001$) (Fig. 24.5). The binding of AGP with SNA in ELISA was found to be significantly less ($p<0.001$) in ALC than control groups. Sialylation status of AGP by lectin blotting showed significant reduction in binding of ALC patient groups in both individual ($p<0.0001$) as well as pooled samples ($p<0.001$). This significant reduction in sialylation may be due to enhancement of sialidase enzyme activity or due to release of less amount of sialyltransferase enzyme by the liver during alcoholic liver cirrhosis.

In conclusion, serum N-glycan profiling by multidimensional HPLC is a promising non-invasive method that could be predicted as useful method for diagnosis of alcoholic liver cirrhosis. However, variation in the glycosylation profile of serum glycoproteins in individual patients appeared high in some cases. Therefore, extensive studies of N-glycan profiling on large cohorts of individual patients are needed including longitudinal samplings over long periods. These altered levels of specific glycan variants of AGP could provide a useful method for the early diagnosis of alcoholic cirrhosis eliminating invasive method of biopsy and extending early treatment. HPLC-based glycome profiling in our present study is a highly sensitive and informative method for accurate diagnosis of alcoholic cirrhosis as it provides nearly absolute structural information of glycome in alpha-1-acid glycoprotein.

Acknowledgements The authors sincerely thank Prof. Y. K. Chawla, Department of Hepatology, Post Graduate Institute of Medical Education and Research, Chandigarh, India for providing patients' sera. This study was supported in part by Grants from IMS for IMS visiting Professor from abroad and the Japan Society for the Promotion of Science (JSPS) Invitational Training Program for Advanced Japanese Research Institutes. BPC and GM gratefully acknowledge the research grant (52/22/2008-BMS) provided by the Indian Council of Medical Research, New Delhi.

References

Arndt T (2001) Carbohydrate-deficient transferrin as a marker of chronic alcohol abuse: a critical review of preanalysis, analysis, and interpretation. Clin Chem 47:13–27

Baenziger JU (1984) The oligosaccharides of plasma glycoproteins: synthesis, structure, and function. In: Putnam FW (ed) The plasma proteins, vol 4, 2nd edn. Academic, New York, p 271

Biou D, Konan D, Feger J (1987) Alterations in the carbohydrate moiety of alpha-1-acid glycoprotein purified from human cirrhotic ascitic fluid. Biochim Biophys Acta 913:308–312

Biou D, Chanton P, Konan D et al (1989) Microheterogeneity of the carbohydrate moiety of human alpha 1-acid glycoprotein in two benign liver diseases: alcoholic cirrhosis and acute hepatitis. Clin Chim Acta 186:59–66

Block TM, Comunale MA, Lowman M et al (2005) Use of targeted glycoproteomics to identify serum glycoproteins that correlate with liver cancer in woodchucks and humans. Proc Natl Acad Sci U S A 102:779–785

Blomme B, Van Steenkiste C, Callewaert N et al (2009) Alteration of protein glycosylation in liver diseases. J Hepatol 50:592–603

Brinkman-van der Linden EC, van Ommen EC, van Dijk W (1996) Glycosylation of α-1 glycoprotein in septic shock: changes in degree of branching and in expression of sialyl Lewis(x) groups. Glycoconj J 13:27–31

Charlwood J, Bryant D, Skehel JM et al (2001) Analysis of N-linked oligosaccharides: progress towards the characterisation of glycoprotein-linked carbohydrates. Biomol Eng 18:229–240

Comunale MA, Lowman M, Long RE et al (2006) Proteomic analysis of serum associated fucosylated glycoproteins in the development of primary hepatocellular carcinoma. J Proteome Res 6:308–315

Comunale MA, Rodemich-Betesh L, Hafner J et al (2010) Linkage specific fucosylation of alpha-1-antitrypsin in liver cirrhosis and cancer patients: implications for a biomarker of hepatocellular carcinoma. PLoS One 5:e12419

Debruyne V, Montreuil J, Spik G (1984) Crossed immunoaffinity electrophoresis of human transferring in normal and cirrhotic sera. In: Peeters H (ed) Protides Biol. Fluids Proc. Colloq., vol 31. Pergamon, Oxford, pp 63–68

Hachulla E, Laine A, Hedouin V et al (1992) Variations in the glycoforms of serum alpha 1-antichymotrypsin in liver diseases and after liver transplantation. Clin Sci 82:439–446

Hada T, Kondo M, Yasukawa K et al (1999) Discrimination of liver cirrhosis from chronic hepatitis by measuring the ratio of *Aleuria aurantia* lectin-reactive serum cholinesterase to immunoreactive protein. Clin Chim Acta 281:37–46

Jezequel M, Seta NS, Corbic MM et al (1988) Modifications of concanavalin A patterns of alpha 1-acid glycoprotein and alpha 2-HS glycoprotein in alcoholic liver disease. Clin Chim Acta 176:49–57

Kondo M, Hada T, Fukui K et al (1995) Enzyme-linked immunosorbent assay (ELISA) for Aleuria aurantia lectin-reactive serum cholinesterase to differentiate liver cirrhosis and chronic hepatitis. Clin Chim Acta 243:1–9

Mann AC, Record CO, Self CH et al (1994) Monosaccharide composition of haptoglobin in liver diseases and alcohol abuse: large changes in glycosylation associated with alcoholic liver disease. Clin Chim Acta 227:69–78

Maryvonne J, Nathalie S, Corbic M et al (1988) Modifications of concanavalin A patterns of alpha 1-acid glycoprotein and alpha 2-HS glycoprotein in alcoholic liver disease. Clin Chim Acta 176:49–57

Mondal G, Chatterjee U, Das HR et al (2009) Enhanced expression of alpha1-acid glycoprotein and fucosylation in hepatitis B patients provides an insight into pathogenesis. Glycoconj J 26:1225–1234

Mondal G, Chatterjee U, Chawla YK et al (2011) Alterations of glycan branching and differential expression of sialic acid on alpha fetoprotein among hepatitis patients. Glycoconj J 28:1–9

Nicollet I, Lebreton JP, Fontaine M et al (1981) Evidence for alpha-1-acid glycoprotein populations of different pI values after concanavalin A affinity chromatography. Study of their evolution during inflammation in man. Biochim Biophys Acta 668:235–245

Patricia G, Claude W, Christiane A et al (1996) New alterations of serum glycoproteins in alcoholic and cirrhotic patients revealed by high resolution two-dimensional gel electrophoresis. Biochem Biophy Res Commun 220:78–85

Ryden I, Skude G, Lundblad A et al (1997) Glycosylation of alpha1-acid glycoprotein in inflammatory disease: analysis by high-pH anion-exchange chromatography and concanavalin A crossed affinity immunoelectrophoresis. Glycoconj J 14:481–488

Ryden I, Lundblad A, Pahlsson P (1999) Lectin ELISA for analysis of alpha (1)-acid glycoprotein fucosylation in the acute phase response. Clin Chem 45:2010–2012

Saroha A, Kumar S, Chatterjee BP et al (2012) Jacalin bound plasma O-glycoproteome and reduced sialylation of alpha 2-HS glycoprotein (A2HSG) in rheumatoid arthritis patients. PLoS One 2:e46374

Serbource-Goguel Seta M, Bordas M, Davy J et al (1984) Evaluation of the degree of desialylation of serum C1-inactivator and haemopexin. Clin Chim Acta 143:235–241

Serbource-Goguel Seta N, Durand G, Corbic M et al (1986) Alterations in relative proportions of microheterogenous forms of human alpha 1-acid glycoprotein in liver disease. J Hepatol 2:245–252

Serbource-Goguel N, Corbtc M, Erlinger S et al (1983) Measurement of serum alpha 1-acid glycoprotein and alpha 1-antitrypsin desialylation in liver disease. Hepatology 3:356–390

Stibler H, Borg S (1981) Evidence of reduced sialic acid content in serum transferrin in male alcoholics. Alcohol Clin Exp Res 5:545–549

Stibler H, Borg S (1986) Carbohydrate composition of serum transferrin in alcoholic patients. Alcohol Clin Exp Res 10:61–64

Takahashi N, Kato K (2003) GALAXY (Glycoanalysis by the three axes of MS and chromatography): a web application that assists structural analyses of N-glycans. Trends Glycosci Glycotechnol 15:235–251

Thompson S, Matta KL, Turner G (1991) Changes in fucose metabolism associated with heavy drinking and smoking: a preliminary report. Clin Chim Acta 201:59–64

Tsutsumi M, Wang JS, Takada A (1994) Microheterogeneity of serum glycoproteins in alcoholics: is desialo-transferrin the marker of chronic alcohol drinking or alcoholic liver injury? Alcohol Clin Exp Res 18:392–397

Turner GA (1992) N-glycosylation of serum proteins in disease and its investigation using lectins. Clin Chim Acta 208:149–171

Yagi H, Takahashi N, Yamaguchi Y, Kato K et al (2005) Development of structural analysis of sulfated N-glycans by multidimensional high performance liquid chromatography mapping methods. Glycobiology 15:1051–1060

Yagi H, Yamamoto M, Yu S et al (2010) N-Glycosylation profiling of turtle egg yolk: expression of galabiose structure. Carbohydr Res 345:442–448

Yamamoto S, Hase S, Fukuda S et al (1989) Studies on the sugar chains of interferon-γ from human peripheral-blood lymphocytes. J Biochem 105:547–555

Correction to: *N*-Acetylglucosaminyl 1-Phosphate Transferase: An Excellent Target for Developing New Generation Breast Cancer Therapeutic

Aditi Banerjee, Juan A. Martinez, Maria O. Longas, Zhenbo Zhang, Jesus Santiago, Krishna Baksi, and Dipak K. Banerjee

Correction to: Chapter 22 in: A. Chakrabarti, A. Surolia (eds.), *Biochemical Roles of Eukaryotic Cell Surface Macromolecules*, Advances in Experimental Medicine and Biology 842, https://doi.org/10.1007/978-3-319-11280-0_22

Since the publication of our chapter [1], it has come to our attention that a number of figures were reproduced for a holistic analysis and conclusion without the appropriate attribution:

- The data in Figures 22.4b, 22.5a, b, and c, 22.6a and b, 22.7a and b, and 22.8a and b were initially reported in Figure 1A, B, C, and D, Figure 2A, B, C, and D, Figure 4A, B, C, D, and E, Figure 5A and B Figure 6A, B, and C, Figure 7, and Figure 8 of [2]
- The data in Figure 22.2b, Figure 22.3 upper half (bright field), and Figure 22.8c and d were initially reported in Figure 2A and B1, Figure 3, Figure 4A and B, and Figure 5 of [3]
- The data in Figure 22.9a, b, c, d, e, and f were initially reported in Figure 1, Figure 2D, Figure 5A, and Figure 6B of [4]

The updated online version of this chapter can be found at
https://doi.org/10.1007/978-3-319-11280-0_22

A. Chakrabarti, A. Surolia (eds.), *Biochemical Roles of Eukaryotic Cell Surface Macromolecules*, Advances in Experimental Medicine and Biology 842,
https://doi.org/10.1007/978-3-319-11280-0_25

1. Banerjee A, Martinez JA, Longas MO, Zhang Z, Santiago, J, Baksi K, Banerjee DK (2015) N-Acetylglucosaminyl 1-phosphate transferase: An excellent target for developing new generation breast cancer therapeutic. Advances in Experimental Medicine and Biology 842: 355–374. https://doi.org/10.1007/978-3-319-11280-0_22.
2. Banerjee A, Lang JY, Hung MC, Sengupta K, Banerjee SK, Baksi K, Banerjee DK (2011) Unfolded protein response is required in nu/nu mice microvasculature for treating breast tumor with tunicamycin. J Biol Chem. Aug 19;286(33):29127–29138. https://doi.org/10.1074/jbc.M110.169771 for a holistic analysis and conclusion.
3. Longas MO, Kotapati A, Prasad KP, Banerjee A, Santiago J, Baksi K, Banerjee DK (2012) Balancing life with glycoconjugates: monitoring unfolded protein response-mediated anti-angiogenic action of tunicamycin by Raman Spectroscopy. Pure Appl Chem. 84(9):1907–1918. https://doi.org/10.1351/PAC-CON-12-01-06.
4. Banerjee A, Johnson KT, Banerjee IA, Banerje DK (2013) Nanoformulation enhances anti-angiogenic efficacy of tunicamycin. Transl Cancer Res. August 21;2(4):240–255. https://doi.org/10.3978/j.issn.2218-676X.2013.08.16098.

Index

A
Acinetobacter baumannii, 86, 87
Aeromonas
  A. hydrophila, 85, 92
  A. salmonicida, 85
AGP. *See* Alpha-1 acid glycoprotein (AGP)
Alcoholic liver cirrhosis (ALC)
  AGP
    chemicals, 392
    disialylated oligosaccharides, 394, 396
    ELISA, 393
    elution profile, 394
    fluorescent labeled *N*-glycans preparation, 392
    HPLC columns, 391–392
    HPLC separation, 392–393
    lectin blotting, 393
    monosialylated oligosaccharides, 394–396
    neutral and sialylated glycans, 399
    neutral oligosaccharides, 394, 395
    purification, 392
    SNA, 396–398
  CDT, 390
  diagnosis of, 391
  informed consent, 391
Alpha-1 acid glycoprotein (AGP)
  chemicals, 392
  disialylated oligosaccharides, 394, 396
  ELISA, 393
  elution profile, 394
  fluorescent labeled *N*-glycans preparation, 392
  glycosylation changes, 390
  HPLC
    columns, 391–392
    separation, 392–393
  lectin blotting, 393
  monosialylated oligosaccharides, 394–396
  neutral and sialylated glycans, 399
  neutral oligosaccharides, 394, 395
  purification, 392
  SNA, 396–398
Apoptosis
  Caspase-3/8 cascade pathway, 335
  Caspases activation, 332, 334–335
  CPKAP, 332, 333, 335–337
  DNA fragmentation, 332, 334
  DNA replication complex, 332, 335
  EBRAP, 332, 333, 335–337
  estrogen receptor, 335
  flopping out, 332, 334
  fragmentation, 332, 334
  IMCAP, 332, 333, 335–337
  inactivation of golgi bodies, 335, 340
  membrane blebbing, 332, 334
  NFkBP, 332, 333, 335–337
  phosphatidyl serine detection, 336
  p^{53} protein expression, 335
  progesterone receptor, 335
Atrial natriuretic peptide (ANP), 134, 135
AXIMA-QIT MALDI quadrupole-ion trap TOF instrument, 393

A. Chakrabarti, A. Surolia (eds.), *Biochemical Roles of Eukaryotic Cell Surface Macromolecules*, Advances in Experimental Medicine and Biology 842,
DOI 10.1007/978-3-319-11280-0

B

β_2-adrenergic receptor
  analysis, 250
  cellular signaling, 256
  coarse-grain methods, 248
  energetics, 250, 256
  hydrophobic segment, 252
  lipid occupancy site, 254, 256
  muscle relaxation, 247
  POPC bilayers, 250–251
  POPC occupancy, 255
  protein-lipid interactions, 252–254
  simulation parameters, 249–250
  spatial density function, 250
  spatial distribution function, 253
  system setup, 249
Betulinic acid (BA), 332, 334, 337, 338
Biofilms
  composition, 312
  EPS, 312
  human pathogens, 313
  lipoarabinomannan, 314–315
  molecular structures, 317–318
  passive dispersion, 312
  quantitative assays, 319
  quorum sensing, 318
  reversible and irreversible stages, 312
  sensorgrams, 316–317
  survival mechanism, 318
  synthetic arabino and arabinomannan glycolipids, 314–316
Biological membranes
  feature of, 28
  FRAP (*see* Fluorescence recovery after photobleaching (FRAP))
Biosimilar apoptotic agents
  apoptosis induction
    Caspase-3/8 cascade pathway, 335
    Caspases activation, 332, 334–335
    CPKAP, 332, 333, 335–337
    DNA fragmentation, 332, 334
    DNA replication complex, 332, 335
    EBRAP, 332, 333, 335–337
    estrogen receptor, 335
    flopping out, 332, 334
    fragmentation, 332, 334
    IMCAP, 332, 333, 335–337
    inactivation of golgi bodies, 335, 340
    membrane blebbing, 332, 334
    NFkBP, 332, 333, 335–337
    phosphatidyl serine detection, 336
    p^{53} protein expression, 335
    progesterone receptor, 335
  betulinic acid, 332, 334, 337, 338
  in cancer chemotherapy, 330
  disialogangliosides, 334, 335, 339–341
  drug delivery systems, 348–349
  GD3, GD1b, and melphalan, 332, 334
  GLT, 345–348
  Lewis antigen epitopes, 330–331
  Lewis glycosphingolipid regulation
    biosynthetic pathways, 340, 342, 343
    disialoganglioside GD3, 340, 343
    fluorescent dyes PSS-380, 343, 344
    functions, 343
    GalT-2 inhibition, 340, 343
  membrane damage monitoring
    fluorescence staining, 341
    image capture and processing, 338, 342
    PSS-380/AKS-0 staining, 341–342
    PSS-380/propidium iodide (PropI) staining, 341
  metastatic breast carcinoma cells, 332, 334, 335
  proteins and peptides, 339
  tamoxifen, 332, 334
Bragg's law, 236
Brain natriuretic peptide (BNP), 134, 135
Breast cancer
  angiogenesis switch, 357
  asparagine-linked glycoproteins, 357–358
  incidence, 355
  study design, 358
  therapy, 357
  TNBC, 356–357
  tunicamycin treatment
    capillary endothelial cells, 359
    cellular proteome analysis, 368
    clonogenic assay, 359, 360
    ER stress mediated *upr*, 368
    FGF-2 and $VEGF_{165}$, 360, 362
    gold nanoparticles, 369–371
    Matrigel™ plug assay, 364, 365
    microvessel density, 366
    migration and chemotaxis, 364, 365
    monitoring cell number, 359, 360
    morphological changes, 359
    nuclear fragmentation, 360, 361
    orthotopic and xenograft breast tumor progression, 366, 367
    phospho-VEGFR1 and phospho-VEGFR2 receptors, 361–364
    Raman spectroscopy, 368–370
    time and dose dependent manner, 359
    VEGF-stimulated tyrosine kinase activity, 362–364
Brucella abortus, 81

C
Candida albicans, 30, 147–148
Carbohydrate deficient transferrin (CDT), 390
Carbohydrate recognition domain (CRD), 200
Cascades of protein kinase activation pathways (CPKAP), 332, 333, 335–337
CD33, 6–7
Cholesterol
  atherogenic conditions, 239
  biological membranes, 233–234
  lipid rafts, 232–233
  membrane effects, 232
  membrane function, 234
  ocular lens fiber cell membranes, 239–241
  oxidative modification, 231
  vascular smooth muscle cell membranes, 238–239
  X-ray diffraction
    Bragg's law, 236
    diffraction data, 237
    membrane structural analysis, 234–236
Chondroitin sulfate (CS)
  biological functions, 186
  cellular degradation, 187
  and dermatan sulfate
    2AB-labeled octasaccharide fraction, 176
    biosynthetic pathways, 166
    D-disaccharide units, 177
    disaccharide composition analysis, 166
    embryonic pig brains, 175–177
    hexasaccharde sequences, 179
    immunohistochemical studies, 168, 169
    mAbs oligosaccharide sequences (*see* Oligosaccharide sequences)
    MALDITOF-MS analysis, 176
    neurite outgrowth-promoting activity, 174, 175
    PTN-bound and unbound octasaccharides, 176, 177
  glycosaminoglycans, 194
  HYAL1 activity
    2-aminobenzamide, 191
    GalNAc residue, 194
    HA tetrasaccharide substrate, 192
    in vivo functions, 195
    kinetic analysis, 192
    quantitative method, 191, 192
  HYAL4, cellular localization of, 188–189
  systemic catabolism, 190–191
Chronic obstructive pulmonary disease (COPD), 8
Circular dichroism (CD) spectroscopy, 282, 286–287, 290
Colon cancer, 140
Crossed immuno-affinoelectrophoresis (CIAE), 390
CS. *See* Chondroitin sulfate (CS)

D
Detergent resistant membrane (DRM)
  electron microscopy studies, 63
  lipidomics of, 62
  MPP1, 70
  phosphatidylethanolamine, 62
  phosphatidylserine, 62
  synchrotron X-ray diffraction studies, 63
1, 2-Diarachidoyl-sn-glycero-3-phosphocholine (DAPC), 264, 266, 268
Differential scanning calorimetry (DSC), 282, 287–288
DRM. *See* Detergent resistant membrane (DRM)
Dynamic light scattering (DLS), 299, 300

E
Enzyme-linked immunosorbent assay (ELISA), 393, 396–397
Erythrocytes
  membrane rafts (*see* Membrane rafts)
  membrane skeleton (*see* Membrane skeleton (MS))
Ethanolamine phosphate (EtNP), 19
European Synchrotron Radiation Facility (ESRF), 295
Ewald method, 265
External Bad-receptor activating pathway (EBRAP), 332, 333, 335–337
Extracellular polymeric substance (EPS), 312

F
Fatty acids (FA), 132–133
Fibroblast growth factor (FGF)-induced angiogenesis, 128
Flotillins, 65
Fluorescence polarization anisotropy, 67
Fluorescence recovery after photobleaching (FRAP)
  apparent diffusion coefficient, 28
  cytoskeletal destabilization, 36–37
  GPCR activation
    cholesterol depletion, 33–35
    5-HT_{1A}R-EYFP and *p*-MPPI, 31–33
  lateral diffusion coefficient, 28, 36–37
  lateral dynamics, infection, 37

Fluorescence recovery after photobleaching (FRAP) (*cont.*)
lipid dynamics
$DiIC_{18}(3)$ and *FAST* DiI, 29–30
NBD-PE and 25-NBD-cholesterol, 30, 31
mobile fraction, 28, 36–37
principles of, 28, 29
Fluorescence spectroscopy, 281, 284–286
Francisella tularensis, 98
FRAP. *See* Fluorescence recovery after photobleaching (FRAP)

G

GBS. *See* Group B *Streptococcus* (GBS)
Gel phase (Lβ), 264
GLTs. *See* Glycosyltransferases (GLTs)
Glycolipids and pppGpp analogs
bacterial growth phases, 320
biofilms
composition, 312
EPS, 312
human pathogens, 313
lipoarabinomannan, 314–315
molecular structures, 317–318
passive dispersion, 312
quantitative assays, 319
quorum sensing, 318
reversible and irreversible stages, 312
sensorgrams, 316–317
survival mechanism, 318
synthetic arabino and arabinomannan glycolipids, 314–316
cyclic-di-GMP, 322–324
Michelis Menten kinetics, 321–323
mycobacterial cell wall components, 310–312
sliding motility, 319–320
stringent response and persistence, 320–321
Glycosphingolipids (GSL)
biosynthesis of, 127
cardiac hypertrophy, 134
Glycosylphosphatidylinositol (GPI) anchors, 17
Als5-SS
β-aggregation, 151, 160
BIG-PI predictor, 152
collagen type IV, 156
GPI-SOM, 152
GST-Als5 self-aggregation, 157–159
GST-Als5-SS mutants, 156, 157
L1326R and F1327R, 153
mutagenic primers, 154
protein expression, 154
secondary structure prediction, 155
biosynthesis of, 18–19
C. albicans, 147–148
C-terminal signal sequences, 148–149
GPI-APs (*see* GPI-anchored proteins (GPI-APs))
signal sequences mutation, 149–151
Glycosyltransferases (GLTs)
DNA microarrays, 345
embryonic development, 345, 346
enzymatic assays, 347, 348
gene expression, 345, 346, 348
metastatic processes, 345, 346
post-translational activities, 345, 346
transcriptional regulation, 345, 347, 348
GPI-anchored proteins (GPI-APs)
fatty acid remodeling in Golgi
C84 CHO cells, 20–21
GlcN-(acyl)PI, 20
PGAP2-and *PGAP3*-deficiencies, 21–23
raft association and homodimerization, 21–22
functions, 17
PGAP5, EtNP elimination, 19
PGAP1, inositol-deacylation, 19
GPI-APs. *See* GPI-anchored proteins (GPI-APs)
G protein-coupled receptors (GPCRs)
FRAP measurements
cholesterol depletion, 33–35
5-HT_{1A}R-EYFP and *p*-MPPI, 31–33
spatio-temporal organization, 248
Group B *Streptococcus* (GBS)
Siglec-5, 8
Siglec-9, 4, 10
Group B *Streptococcus* serotype III (GBS-III), 4

H

Hereditary elliptocytosis (HE), 49–50
laboratory features and treatment, 52
molecular defects, 50–51
αβ-spectrin self-association, 50, 51
protein 4.1R deficiency, 51
Hereditary spherocytosis (HS), 46–47
hemolytic anemia, 47
laboratory features and treatment, 49
membrane loss, 49
molecular defects
α-spectrin, 47
amino acid substitution, 48
ankyrin, 47
β-spectrin, 47
pathophysiology, 48–49
HS. *See* Hereditary spherocytosis (HS)

5-HT_{1A} receptor. *See* $Serotonin_{1A}$ receptor
5-Hydroxytryptamine-1A receptor tagged to enhanced yellow fluorescent protein (5-HT_{1A}R-EYFP), 31–33
Hyperphosphatasia with mental retardation syndrome (HPMRS), 22–23

I

IL-1 receptor associated kinase 1 (IRAK1), 87
Immunoreceptor tyrosine-based inhibitory motifs (ITIMs), 3
Intercellular cell adhesion molecule-1 (ICAM-1), 128–130
Internal mitochondrial-caspase activation pathway (IMCAP), 332, 333, 335–337
Isothermal titration calorimetry (ITC)
  equilibrium association constant, 283, 284
  MicroCal VP-ITC instrument, 281
  'one set of sites' model, 283
  profile, 282, 283
  stoichiometry, 283, 284
  thermodynamic parameter, 283, 284

K

Klebsiella pneumoniae, 81, 85
Knob-associated histidine-rich protein (KAHRP), 53

L

LacCer synthases (LCS), 128
Lactosylceramide (LacCer) research
  cardiac hypertrophy, modulation on
    ANP and BNP, 134, 135
    decreased fractional shortening, 133
    FA, oxidation of, 132–133
    fat diet intake, 133
    high blood cholesterol, 133
    increased left ventricular mass, 133
    in vivo and in vitro studies, 134
    PKC, 135
  GalT-V and GalT-VI, 128
  ICAM-1, 128–129
  metabolic pathways, 128
  Neu3 activation, 129
  perspectives, 135
  PLA2 connection, 130–132
  reactive oxygen species, 129, 132
  signaling pathways, 129, 131
  VEGF and FGF-induced angiogenesis, 128
Late-onset alzheimer's disease (LOAD), 6–7
Lectin blotting, 393, 396–397
Legionella pneumophila, 85
Legume lectins
  amino acid
    composition, 204
    sequences, 201, 203
    variability, 209–211
  binding site loops, 206
  Canavalia, 200
  FU lectin group, 209
  full protein sequences, 206
  GA and GAN lectin group, 208–209
  GLN lectin group, 207–208
  MG lectin group, 206–207
  multiple sequence alignment and analysis, 203
  pattern recognition and clustering, 204
  percentage identity matrix, 204, 205
  phylogenetic analysis, 204
  protein secondary structure prediction, 203
  source and monosaccharide, 202–203
Lewis glycosphingolipid regulation
  biosynthetic pathways, 340, 342, 343
  disialoganglioside GD3, 340, 343
  fluorescent dyes PSS-380, 343, 344
  functions, 343
  GalT-2 inhibition, 340, 343
  GLT, 345–348
Lipid bilayer theory, 235
Lipoarabinomannan (LAM), 314–315
Liquid-crystalline phase (Lα), 264
Low-density lipoproteins (LDL), 238

M

Mabry syndrome, 22–23
Maclura pomifera agglutinin (MPA), 115
Major outer membrane protein (MOMP), 85, 87, 98
Malaria, 52–53
Mature parasite-infected erythrocyte surface antigen (MESA), 53
Membrane associated guanylate kinase (MAGUK), 66
Membrane attack complex (MAC), 85
Membrane palmitoylated protein-1 (MPP1/p55)
  HEL cells, 70–71
  palmitoylation activity, lateral membrane organization, 70–71
  physiological roles, 66
Membrane rafts, 63–64
  actin involvement, 72
  BODIPY-GM1, 67
  FCS, 67
  flotillins, 65
  MPP1/p55 (*see* Membrane palmitoylated protein-1 (MPP1/p55))

Membrane rafts (*cont.*)
pathology
DHHC17, 69–70
STOM gene, 69
signaling pathway
lidocaine treatment, 68
mature red cells, 68
ROS, 68
STED, 67
stomatin, 64–65
visualization, 67
Membrane skeleton (MS)
adducin, 43, 44
ankyrin complex, 43
band-3 dimers, 43
characteristic feature of, 45
dematin, 43
hereditary elliptocytosis, 49–50
laboratory features and treatment, 52
molecular defects, 50–51
hereditary spherocytosis, 46–47
laboratory features and treatment, 49
molecular defects, 47–48
pathophysiology, 48–49
lipid bilayer–membrane skeleton interactions, 44
and malaria
intracellular growth, 52
KAHRP, 53
MESA, 53
PfEMP-3, 53
protein phosphorylation, 53
RESA, 53
structural and morphological changes, 52
protein 4.1, 43–45
roles for, 42
spectrin-actin network, 43
structure of, 42
tropomodulin, 43, 44
tropomyosin, 43, 44
Molecular dynamics simulation
area per lipid, 266–267
atom density distributions, 267, 268
DAPC lipid bilayer, 266, 268, 273, 274
energy interaction, 272–273
Ewald method, 265
gel phase, 264
headgroup orientation, 277
hydrocarbon tail and bilayer, 268, 269
Langevin dynamics, 265
lipid hydrophobic thickness, 267–268
lipid tail order parameter, 269–270
lipid-water interaction, 273–274
liquid-crystalline phase, 264
RATTLE and SETTLE algorithm, 265
ripple phase, 264
rotational autocorrelation function, 271–272
structure and dynamical properties, 264
subgel phase, 264
system setup, 265
trajectory analysis, 265
MS. *See* Membrane skeleton (MS)

N
N-acetylglucosaminyl 1-phosphate transferase. *See* Breast cancer
N-acetylneuraminic acid (Neu5Ac)
Siglec-1, 5
Siglec-9, 10
Neisseria
N. gonorrhoeae, 80, 85, 98
N. meningitidis, 81, 85
NFkappaB activation pathway (NFkBP), 332, 333, 335–337
Nuclear magnetic resonance (NMR) spectroscopy
carbohydrate–protein interactions, 218–219
glycolipid clusters, 227–229
GM1 and GM2 oligosaccharides, 226
GM1 pentasaccharide, 223
GM3 trisaccharide, 226
molecular dynamics (MD) simulation, 222, 224
nuclear Overhauser effect, 220
pseudocontact shifts, 220
replica-exchange MD simulations, 225
sample preparation, 219–220
scalar coupling, 220

O
Oligosaccharide sequences
β1,4-GalNAc transferase-I, 170
C4ST-1 and C4ST-2, 171
GD3G7 epitope, 172, 173
in situ hybridization, 171
sulfotransferases, 171
Open reading frame (ORF), 12
Origin 7.0 software, 282
Outer membrane proteins, porins. *See* Porins

P
Pasteurella
P. haemolytica, 81
P. multocida, 81

Pathogen associated molecular patterns
(PAMPs), 80, 87, 88
Pattern recognition receptors (PRRs), 87
PDC-109/*O*-acylcholine interaction
  CD spectra, 282, 286–287, 290
  cholesterol efflux, 279
  DSC, 282, 287–288
  fluorescence spectroscopy, 281, 284–286
  heat capacity and enthalpy-entropy
  compensation, 288, 289
  ITC
    equilibrium association constant,
    283, 284
    MicroCal VP-ITC instrument, 281
    'one set of sites' model, 283
    profile, 282, 283
    stoichiometry, 283, 284
    thermodynamic parameter, 283, 284
  materials, 280–281
  NMR and X-ray studies, 288
  structure of, 279
  thermodynamic data, 284, 289
Peripheral blood mononuclear cells (PBMCs), 81
PFTs. *See* Pore-forming toxins (PFTs)
Phospholipase-A-2 (PLA2) connection, 130–132
PISA webserver, 296
Plasmodium falciparum erythrocyte membrane
protein 3 (PfEMP3), 53
Pore-forming toxins (PFTs), 109–110
  α-PFTs, 110, 111
  bacterial toxins, 110
  β-PFTs, 110, 111
  primary amino acid sequence, 110
  structural and functional
  considerations, 110
  structural classification, 110, 111
  VCC (*see Vibrio cholerae* cytolysin (VCC))
Porins
  adaptive immune responses, 89–91
    antigen presenting cells, 86
    Salmonella porins, 86
    Shigella porins, 86
    Th1 and Th2 cells, 86, 87
  innate immune responses, 82–84
    A. hydrophila, 85
    A. salmonicida, 85
    B. abortus, 81
    complement system, 85
    IL-8, 81
    IL-12, 80
    IL-1β, 80, 81
    K. pneumoniae, 81, 85
    MIP-1α, 81
    MIP-1β, 81
    nitric oxide, 80
    N. meningitidis, 81, 85
    P. haemolytica, 81
    P. multocida, 81
    RANTES, 81
    TNFα, 80, 81
    V. cholerae, 81
  P.aeruginosa, 80
  signaling cascades
    IRAK1, 87
    TLRs, 87, 88
    TRAF6, 87
    UBC13, 87
  S. marscecens, 80
  as vaccine candidates, 92–99
Prostate cancer, 141
Protein acyl transferase (PAT) activity, 69
Protein kinase C (PKC), 135
Protein Secondary Structure PREDiction
server (PSSPRED), 203
Pseudomonas aeruginosa, 80, 98

R

Radial distribution function (RDF), 270
Reactive oxygen species (ROS), 68, 129, 132
Renal cell carcinomas (RCC), 141
Replica-exchange MD (REMD)
simulations, 225
Ring-infected erythrocyte surface antigen
(RESA), 53
Ripple phase (Pβ), 264

S

Saccharomyces cerevisiae strains, 220
Sambucus nigra agglutinin (SNA), 396–397
Serotonin$_{1A}$ receptor
  agonist stimulation, 381, 382
  in vivo stimulation, 381, 383
  Ki67(+) proliferating cells, 378, 379, 381
  8-OH-DPAT (D) treatment, 378, 380
  organotypic cultures, 379–380
  VEGFR1/2 antagonist
  block, 381, 383–384
Serotonin$_{1A}$ (5-HT$_{1A}$) receptor, 31–32
Serratia marscecens, 80
Sialic acid-recognizing immunoglobulin-like
lectins (Siglecs)
  activating Siglecs, 4
  CD33-related Siglecs, 1–2
  DAP12, 3
  family of, 1, 2
  GBS serotypes, 4

Sialic acid-recognizing immunoglobulin-like lectins (Siglecs) (*cont.*)
  genomic localization of, 1–2
  human-specific changes, 4–5
    Siglec-6, 9
    Siglec-9, 9–10
    Siglec-13, 12
    Siglec-17, 12
    Siglec-5 and-14, 7–8
    Siglec-11 and-16, 10–11
    Siglec-3/CD33, 6–7
    Siglec-1/sialoadhesin, 5–6
    Siglec-XII, 11–12
  ITIM motifs, 3
  Neu5Gc, 5
  SLAM-like motifs, 3
Sialidase-3 (NEU3)
  cancer initiation and promotion, 142–144
  cancer progression
    breast carcinoma, 140
    colon, 140
    prostate cancer, 141
    renal cell carcinomas, 141
Sialoadhesin (Siglec-1), 5–6
SIGLEC17 pseudogene *(SIGLEC17P)*, 12
Siglecs. *See* Sialic acid-recognizing immunoglobulin-like lectins (Siglecs)
Signaling lymphocytic activation molecule (SLAM) receptors, 3
Size exclusion chromatography, 299, 300
Spatial distribution function (SDF), 253
Spex model, 281
1-Stearoyl-2-linoleoyl-sn-glycero-3-phosphoethanolamine (SLPE), 62
Stimulated emission depletion (STED), 67
Stomatin, 64–65
Subgel phase (L_c), 264

T

TIP3P water model, 265
TNF receptor associated factor 6 (TRAF6), 87
Transmission electron microscopy (TEM), 117
Triple negative breast cancer (TNBC), 356
Tropomyosin, 43, 44

U

Ubiquitin conjugating enzyme 13 (UBC13), 87

V

Vascular endothelial growth factor (VEGF), 128
  antidepressants, 377, 384
  anxiety, 376
  brain cryosection, 379
  BrdU injection and intra-hippocampal infusion, 378
  confocal microscopy and statistical analysis, 379
  definition, 377
  drug concentrations, 378
  hippocampal neuroproliferation, 385
  immunostaining sections, 379
  kinases ERK1/2, 385
  materials, 378
  organotypic cultures, 378
  pharmacological inhibition, 377
  PKC, 376–377
  serotonin$_{1A}$ receptor
    agonist stimulation, 381, 382
    in vivo stimulation, 381, 383
    Ki67(+) proliferating cells, 378, 379, 381
    8-OH-DPAT (D) treatment, 378, 380
    organotypic cultures, 379–380
    VEGFR1/2 antagonist block, 381, 383–384
  wild type (WT) (C57BL/6) mice, 377–378
VCC. *See Vibrio cholerae* cytolysin (VCC)
Vc-RK. *See Vibrio cholerae* ribokinase (Vc-RK)
VEGF. *See* Vascular endothelial growth factor (VEGF)
Vibrio
  V. alginolyticus, 99
  V. cholerae, 81
  V. harveyi, 99
  V. parahaemolyticus, 99
  V. splendidus, 80
Vibrio cholerae cytolysin (VCC)
  enterotoxic activity, 111
  lipid components, 119
  lytic activity, 111
  pore formation mechanism, 117–119
  proteolytic maturation, 112
  structural features
    β-barrel pores, 117
    β-prism lectin-like domain, 115–116
    β-trefoil lectin-like domain, 115
    cytolysin domain, 113–114
    monomeric and oligomeric pore form, 112, 113
    oligomeric pore formation, structural reorganizations, 116
    Pro-domain, 114–115
Vibrio cholerae ribokinase (Vc-RK)
  active site
    monovalent cation binding, 302–303
    ribose binding, 298, 301–302

ADP and sugar molecule, 303, 304
ATP and monovalent metal binding loop, 304, 305
β-sheet region, 294, 297–299
biphasic life style, 294
catalytic α/β domain, 294, 297–299
cloning, expression and purification, 295
ClustalW, 296
crystallization, 295
data collection, 295–297
dimerization, 299, 300
double-stranded helical pattern, 299, 300
folding patterns and substrate specificity, 293
'half-way closed' substrate dimer, 296, 299–301
lid residues, 304, 305
open apo dimer, 297, 299, 300
Phe residue (F21), 304, 305
PISA webserver, 296
Pymol, 296
structural features, 293–294
structure determination and refinement, 296–298

W
Western blot analysis, 70

X
X-ray crystallography, 218

Kolibri 360 GmbH
Ferdinand-Jühlke-Straße 7
99095 Erfurt, Deutschland
produktsicherheit@kolibri360.de